PHENOTYPES

THEIR EPIGENETICS, ECOLOGY AND EVOLUTION

PHENOTYPES

THEIR EPIGENETICS, ECOLOGY AND EVOLUTION

C. DAVID ROLLO

Department of Biology, McMaster University, Hamilton, Canada

CHAPMAN & HALL

London · Glasgow · Weinheim · New York · Tokyo · Melbourne · Madras

Published by Chapman & Hall, 2–6 Boundary Row, London SE1 8HN, UK

Chapman & Hall, 2–6 Boundary Row, London SE1 8HN, UK

Blackie Academic & Professional, Wester Cleddens Road, Bishopbriggs, Glasgow G64 2NZ, UK

Chapman & Hall GmbH, Pappelallee 3, 69469 Weinheim, Germany

Chapman & Hall USA, One Penn Plaza, 41st Floor, New York NY 10119, USA

Chapman & Hall Japan, ITP-Japan, Kyowa Building, 3F, 2-2-1 Hirakawacho, Chiyoda-ku, Tokyo 102, Japan

Chapman & Hall Australia, Thomas Nelson Australia, 102 Dodds Street, South Melbourne, Victoria 3205, Australia

Chapman & Hall India, R. Seshadri, 32 Second Main Road, CIT East, Madras 600 035, India

First edition 1994

©1995 C. David Rollo

Typeset in Palatino by Florencetype Ltd, Stoodleigh, Devon

Printed in Great Britain by St Edmundsbury Press, Bury St Edmunds, Suffolk

1-800- 842-3636

ISBN 0 412 41030 3

2 $83.00

A catalogue record for this book is available from the British Library

Library of Congress Catalog Card Number: 94-72663

∞ Printed on permanent acid-free text paper, manufactured in accordance with ANSI/NISO Z39.48–1992 and ANSI/NISO Z39.48–1984 (Permanence of Paper).

This book is dedicated to my grade five teacher, Mrs Wright, who got me started making ant colonies. My parents Chris and Marion encouraged me by allowing me to keep the ants in my bedroom, wasps in the windows and cecropia caterpillars on the paeonies. My children Oriana and Tynan put up with a busy dad. Barb Reuter helped with the typing. Finally, my special lady, Helen Radnor helped me with mastering various software, typed seemingly endless drafts and put up with the stacks of papers and books that gradually encroached upon the kitchen, the living room, the bedrooms, the basement . . . Thank you.

Contents

Preface   xi

**1 Hierarchical reductionism and the evolution of biological
    organization**   1
    1.1 Introduction   1
    1.2 Hierarchical structure and reductionism in biology   4
    1.3 Scale and causal mechanisms   10
    1.4 Interaction effects, emergent properties and complexity   14
    1.5 Organizational overlays   16
    1.6 Evolution of biological organization   23
    1.7 Conclusion   57

2 Regulatory interactions, epigenetics and genomic integration   59
    2.1 The basis of heredity: changing concepts of the gene
         and their implications for evolutionary theory   59
    2.2 Genomic organization and subspecific lineage
         selection: a new evolutionary framework   62
    2.3 Operons and gene regulation in prokaryotes   63
    2.4 Eukaryotic regulatory processes   64

3 Epigenetics and developmental hierarchies   82
    3.1 Overview   82
    3.2 Sex determination   83
    3.3 Dorsoventral differentiation in *Drosophila*   85
    3.4 Anterioposterior differentiation in *Drosophila*   86
    3.5 The epigenetic hierarchy: a fundamental genomic
         organization for multicellular eukaryotes   91

4 The subspecific lineage paradigm and the genetic templet **110**
4.1 Overview 110
4.2 Genetic relatedness in sexual species 115
4.3 Genetic inertia and retention of quiescent genetic
programs 125
4.4 Genetic feasibility: cost 129
4.5 Intraspecific genomic variation and stability 130

5 Metalineage selection and sexual reproduction **138**
5.1 Sexual communication and metalineage organization 138
5.2 Hierarchies and lineage structure 144
5.3 Lineage selection and sexual reproduction: an overview
of the selective advantages of sex 153
5.4 The genic framework 157
5.5 Hypotheses of sexual evolution 162
5.6 Müller's ratchet 183
5.7 The Red Queen 188
5.8 Tangled Bank 194
5.9 Conclusion 196

6 Phenotypic plasticity, allelic switches and canalization:
a metalineage perspective **198**
6.1 Introduction 198
6.2 Some examples of adaptive phenotypic flexibility 200
6.3 Maternal and paternal influence 205
6.4 Polymorphic switches 209
6.5 Norms of reaction and plastic developmental trajectories 213
6.6 Canalization 218
6.7 Genetic assimilation 223
6.8 Mesoevolution of plastic organizations 229
6.9 Genetic correlations and guided epigenetic trajectories 245

7 Adaptive suites, coadapted genomes and dynamic genetic
templets **252**
7.1 Introduction: coadapted genomes and phenotypic
adaptive suites 252
7.2 Coadaptation and genetic revolutions 258
7.3 Genetic revolutions and metapopulation dynamics 267

8 Trade offs and synergisms in phenotypic organization:
an evolutionary framework **272**
8.1 Introduction 272
8.2 Sources of synergism and trade offs: specialization 275
8.3 Sources of synergism and trade offs: morphology 276

8.4	Trade offs in physiological and life-history attributes	279
8.5	Sustainable scope	288
8.6	The safe tuning hypothesis	294
8.7	The phenotypic balance hypothesis	297
8.8	The principle of allocation: lessons from the Supermouse	303

9 The evolution of senescence and longevity assurance **310**
9.1	Overview of senescence and longevity	310
9.2	The cost of longevity: factoring longevity into the resource allocation paradigm	321

10 The principle of allocation: limitations, life-times and metabolic relativity **336**
10.1	Introduction: the rate of living theory and life-time metabolic scope	336
10.2	Allometric analyses	342
10.3	Evidence for alterations in longevity associated with metabolic rates	349
10.4	Stress and the dietary restriction paradigm	365
10.5	Conclusion	373

11 Life-times and limitations: speculations in a lineage paradigm 375

12 The ecological triumvirate: phenotypic correlations of stress, disturbance and biotically diverse habitats **387**
12.1	Overview	387
12.2	The stress paradigm	389
12.3	The Red Queen paradigm	392
12.4	The disturbance paradigm	393
12.5	General conclusion	397

References	**399**
Index	**453**

Preface

Why was this book written? As an ecologist mainly interested in whole organisms, I found that current evolutionary theory seemed insufficient to address the design of phenotypes (the bodies elaborated by the genome), as I appreciate them – complex and highly integrated entities in which the integration itself appears to be the dominant aspect of fitness. A focus on individual selfish genes tends to ignore the various levels of interaction among genes where the major differences among species may reside. I wanted to develop a view that seemed more appropriate to the real focus of selection – whole organisms and their regulated development.

As a whole-organism biologist in a department that was strongly biased towards the molecular, I realized that the time was ripe for a unified approach to organismal design that synthesizes knowledge across all levels. In particular, many of my ecological colleagues find seminars on molecular biology unintelligible, and molecular biologists working in the trenches often do not appreciate how their areas of expertise fit into the larger picture, or how questions from higher levels may be relevant to them. This book attempts to forge a linking framework between the new picture emerging at the molecular level with phenomena relevant to ecology and evolution.

All of this generates a model of evolution based on regulatory aspects of the genome. Rather than stressing the contributions of individual genes mainly in terms of the structural proteins they sometimes make, a regulatory framework stresses the way that genes and networks of genes are interconnected into complex circuits. Such circuitry is subject to selection for precise control of metabolism, differentiation and morphogenesis that may be the main target of phenotypic evolution. Although not entirely new, a holistic framework of evolution based on genomic regulatory structure is largely lacking.

When the genome is viewed as a complex of regulatory integration, it becomes apparent that selection may act on hierarchies of regulatory circuitry, various levels of which are relevant to differing levels of phenotypic organization spanning the molecular realm to communities. Such a framework changes our perception of what constitute the units of selection and the relevant time scales for their evolution.

This book really started out more than a decade ago when I worked with Dr I. Vertinsky and Dr W. G. Wellington at the University of British Columbia on developing a comprehensive computer simulation intended to explore the ecological functioning of some simple invertebrates. I came away from that exercise with the profound intuition that these organisms functioned in ways that minimized risks of failure rather than in ways that maximized immediate individual success (although the two are not always mutually exclusive). Specifically, in an unpredictable world where organisms have variously uncertain information (and all environments are ultimately uncertain), strategies that ensure being approximately right most of the time are favoured over those that may be locally superior, but that also run the risk of being precisely wrong.

As it turns out, life-history theory is rapidly shifting to an evolutionary framework that recognizes that variance in fitness is just as crucial as mean success in any given generation or population. This framework of geometric mean fitness, when combined with a paradigm of regulatory evolution, allows for larger selective units than individual genes, and highlights variation among demes (local breeding populations), as key sources of regulatory variation (i.e. evolutionary fuel). The implications are that sub-specific lineages, defined both by intrademic variation in regulatory hierarchies, and by interdemic diversification, are important elements of selection. In other words, evolution may act mainly on lineage persistence rather than the maximal success of individual selfish genes.

A further plank relevant to this framework is a shift in both evolutionary and ecological theory stressing non-equilibrium temporal dynamics and spatial heterogeneity (patchiness). Much of what follows owes some of its development to discussions with Dr G. Harris related to non-equilibrium ecology (Harris, 1986), and with Dr J. Kolasa with respect to hierarchy theory and its applications (Kolasa and Rollo, 1991). My friend Anthony Black provided enthusiastic feedback as to how various ideas were understood by an educated layman.

This book amounts to one biologist's attempt to understand the design and evolution of life. Clearly, such an undertaking cannot address every aspect of form and function. I have chosen four major topics that I believe are the most relevant to the theory developed here, and which I judge to be the most difficult and controversial in evolutionary theory generally. These are: the evolution of biological organization; the evolution of sex; phenotypic plasticity or the deployment of several body forms by the same genetic organization; and aging and senescence.

It was not my intention to write anything controversial. The proposed shift in evolutionary framework, however, favours a number of interpretations that have been variously rejected or debated in the evolutionary literature. These include an argument for progress in organizational evolution, rejection of a selfish gene or genic model of evolution in favour of selection of higher-order units, a view of sex that includes the value of long-term evolvability as a selective advantage, and the idea that lineages may evolve meta-adaptations that transcend individuals. In addition I argue for a return to a 'rate of living' paradigm with respect to senescence, and the idea that longevity may be selected as a programmed feature of life histories. I hope the reader will judge the interpretation outlined here on the basis of overall consistency with respect to these difficult subjects rather than on the basis of alternative explanations for individual details.

Chapter 1 outlines a holistic framework of organization in biological systems and how this organization might evolve. Chapters 2 to 4 explore the structure and dynamics of the modern genome and the way that multicellular entities are developed. The emphasis is on regulatory integration, developmental hierarchies, and how all this new knowledge changes our view of evolution. Chapters 4 and 5 argue that sexual species represent an organizational transition, a perspective requiring an evolutionary framework where selection acts on lineages with multigenerational time scales. Such a view generates some interesting hypotheses that are excluded by more reductionist approaches. In particular, this view suggests that the lineage level of organization may impose features on organismal design that transcend their value to individual organisms (e.g. functions that serve evolvability).

Chapter 6 extends this perspective to consider adaptations that can be considered to be lineage-level attributes (e.g. reaction norms and genetic correlations which allow individual phenotypes to be appropriately modified according to environmental and genetic variation). The picture that emerges is that genomes represent highly coadapted complexes with holistic integration. The reality of such a view and its evolutionary implications are explored in Chapter 7. Forces driving such coadaptation are explored in Chapter 8.

A major feature of organisms, senescence, has not been well integrated into this paradigm and consequently Chapters 9 and 10 attempt to integrate this feature into a global framework of organismal design. Chapter 11 explores some questions concerning senescence that arise from a perspective of sexual lineages as evolutionary entities. Finally, Chapter 12 explores three major ecological forces (stress, disturbance and biotic interactions) shaping recognizable constellations of attributes of organisms. The purpose of the book is to return the organism to its rightful place as a centrepiece of selection and evolution. In doing so it becomes apparent that some organismal properties may be imposed as constraints by features selected at an even higher level – that of sexual lineages and interactions among them. I hope to convince the reader that a framework of hierarchical reductionism generates many different questions and possible biological insights that are lacking in a genic, reductionist model.

There are several glaring omissions. A chapter on non-equilibrium ecology has been deleted due to lack of space. These new perspectives of community dynamics radically alter our traditional views of the framework shaping the evolution of organisms and higher levels of entification, particularly with respect to the power of competitive exclusion (which may be reduced in non-equilibrium systems). The evolution of dispersal and habitat choice emerge as absolutely crucial dimensions in a lineage framework, but again space limitations preclude inclusion here. The evolution of dominance among alleles is also lacking, even though this may well represent a phenomenon equivalent to reaction norms and genetic correlations in terms of features that may be relevant to lineage-level attributes.

The overall focus is an understanding of organismal design. This is embodied in the multicellular phenotype(s) that function in ecological habitats. Understanding of such phenotypes requires consideration of the lower-level molecular systems that generate them. The idea that sexual species may represent a level of organization that constrains the evolution of individual phenotypes is not new, but it has not been strongly explored. I have limited discussion to these levels of biological organization because discussion of even higher-order organization will first require acceptance that sexual lineages represent consolidated entities.

1

Hierarchical reductionism and the evolution of biological organization

1.1 INTRODUCTION

The life sciences are in the throes of a revolution, an upheaval that will radically alter the nature of biology and the future of mankind. This revolution is largely fueled by rapid developments in molecular biology – specifically the discovery of the genetic code and recombinant techniques allowing us to examine, interpret and directly manipulate the fundamental genetic programming of organisms. It is probably no exaggeration to rank these advances somewhere close to the discovery of fire in terms of key milestones for humanity.

The biological revolution involves not one, but three key elements, that of a perspective emphasizing organismal design, that of molecularly fueled power and that of a coalescing and integration of biological knowledge. The perspective of organismal design is a very old one that is finally being allowed out of the closet. In 1828 the theologian William Paley used the analogy of finding a watch to argue for the existence of God; obviously, anything so well designed must have had a maker. Paley observed that: 'every indication of contrivance, every manifestation of design, which existed in the watch, exists in the works of nature; with the difference, on the side of nature, of being greater or more, and that in a degree which exceeds all computation'.

Paley's argument was so powerful, that following the publication of Darwin's theories, it became unacceptable among scientists to refer to the design of organisms, for fear that such reference implied belief in a Creator or some vital force guiding evolution. This sin was termed teleology. Despite this, it has always been the design of life, its beauty and complexity, that has inspired our awe and interest. Thus, Darwin concluded *The Origin of Species*: 'Thus, from the war of nature, from famine and death, the most exalted object which we are capable of conceiving, namely the production of the higher animals, directly follows. There is a grandeur in this view of life, with its several powers, having been originally breathed by the Creator into a few forms or into one; and that, whilst this planet has gone cycling on according to the fixed law of gravity, from so simple a beginning endless forms most beautiful and most wonderful have been, and are being evolved.'

Darwin's reference to a Creator aside, we now think that nature does not design anything, it simply proliferates randomly and leaves those things that work. Design emerges because of the congruence of the survivors' characteristics with the rules of the universe. Thus, the environment acts as a dynamic template that excludes incompetence and leaves well designed phenotypes. Biologists have come far enough to recognize that a scientist who refers to organismal design is referring to adaptations evolved through the joint action of mutation and recombination producing new variations, and by natural selection favouring differential mortality and reproduction among these forms. Richard Dawkins in his book, *The Blind Watchmaker*, provides an eminently readable account of how these processes produce marvels of organic form and function that far outstrip even our most elaborate applied technology, including watches (Dawkins, 1986).

In the last 20 years, addressing organismal design has become not only openly acceptable, it is fast becoming the unifying central project. There are a number of reasons for this, but three major developments are particularly important. Firstly, the power of molecular genetics automatically evokes a design perspective because it is the promise of bold new innovations and the ability to re-design life that has captured the greatest attention. Twenty years ago developmental genetics was still in its infancy, there being relatively little knowledge of the connection between ontogeny and its genetic control (Bonner, 1974; Lovtrup, 1974; Gould, 1977). Since then, molecular epigenetics has opened a floodgate of information and opportunities representing one of the most rapidly burgeoning areas of knowledge. Even though we have barely etched the surface of this treasure trove, linkages spanning vast phylogenies are already apparent, and our ideas of how evolution operates are being fundamentally altered (Raff and Kaufman, 1983; Arthur, 1984; De Pomerai, 1990; Gilbert, 1991; Lawrence, 1992; Kauffman, 1993).

Molecular biology has revealed that the genome is not simply a parking lot for circulating alleles as the genic program envisioned, but a cohesively integrated network of genes and regulatory cascades. Genes are now known to have complex structure, and besides those regions coding for protein products, other regions acting in regulatory handshaking have been revealed. This has important evolutionary implications. Furthermore, some genes may produce products that function in the metabolism or structure of the body, while others derive products that serve exclusively as regulatory signals. Molecular-level advances are rapidly altering our views of how morphological evolution may proceed. Although there will be those who still object to the use of the term design with respect to organisms, that is what modern biology is largely concerned with, either that of interpreting existing structures, or that of diversification and novelty.

A second major development is that knowledge accumulating in the scientific literature has reached a critical mass allowing broad syntheses and integration. For example, Huxley (1932) demonstrated the power of very simple equations to describe relationships among design features either among conspecifics of various ages or across species of various sizes. The application of such allometric equations has culminated in several important books (Peters, 1983; Calder, 1984; Schmidt-Nielsen, 1984; Reiss, 1989). Such important features as respiration, growth rate, longevity, physiological rates, territory size, organ mass, litter size, heat transfer and population size can be related rather precisely to body mass by equations of the form:

$$Y = aX^b$$

where Y is the feature in question, X is the mass of the body, a is a fitted constant and b is an exponent, usually with the value 0.75 (for comparisons among species) or 0.66 (for comparisons within species). Such allometric analyses pave the way for more elaborate comparative syntheses and strongly suggest that organismal design is sufficiently constrained so that general principles may indeed be formulated (see Chapter 10).

A third major factor influencing the rapid advance and increasing integration in biology is the parallel revolution in computer technology. Even at the molecular level computers enhance such tasks as sequence analysis, and for studies involving higher levels of organization such as organisms, populations or communities, computers are essential tools for analysis and prediction. The use of computers to manipulate and display positional information (e.g. the amazing advances in computer graphics and animation) appears to be exactly what developmental genetics is revealing as the necessary framework underlying complex morphological differentiation (e.g. Raff and Kaufman, 1983; De Pomerai, 1990; Lawrence, 1992; Kauffman, 1993). There is clearly a potential here for a powerful marriage.

Engineers solve the problem of designing and building complex machinery by applying a methodology known as systems analysis. This often involves the construction of comprehensive computer models that allow simulation of potential dynamics of a system or the impact of modifications. Engineers proceed by embodying known principles of good design into blueprints or models. The power of their approach is evident in the complex technology that envelopes us. Biologists are rather 'inverse engineers' in that we are presented with a myriad of complex biological machinery that comes without technical manuals (Berryman, 1981). Organisms are probably the most complex machines in the known universe (Dawkins, 1976). Even the most advanced human technology (e.g. a shuttlecraft or Viking lander) is considerably less complex than even a lowly bacteria.

Biologists can learn a great deal about the principles of good design underlying organism form and function by applying systems analysis to create dynamic computer blueprints. Such blueprints can be used as organism impact statements to explore how various modifications might influence other elements or processes. We used this approach, for example, to construct a comprehensive computer model of the ecology of a terrestrial slug (starting slow and simple) (Rollo *et al.*, 1983).

Kingsolver (1985), in an article entitled *Butterfly engineering*, elegantly showed the value of applying engineering principles to organisms. That such principles are indeed incorporated in the design of organisms was also nicely illustrated by Welty (1955) in another article (*Birds as flying machines*). Every day on my way to work I pass through an architectural web of triangular steel supports known as the Warren truss, an invention that engineers discovered provides maximum strength for minimal materials. Welty (1955) reveals that the internal architecture in the wing bones of birds is similarly based on the Warren truss (Figure 1.1).

Given the intrinsic complexity of biological entities, it is difficult to see how progress can be made without the use of computers.

1.2 HIERARCHICAL STRUCTURE AND REDUCTIONISM IN BIOLOGY

1.2.1 Genic and molecular reductionism

Unfortunately, the amazing pace of discoveries in molecular biology has been interpreted by many to mean that extreme reductionism (the division of a system into its smallest functional components) is the correct paradigm for biological science. Population genetics has leaned in this direction since its inception (Fisher, 1930), but the genic perspective was really consolidated by Williams (1966a, 1985) and then hammered home by Dawkin's (1976, 1989a) classic, *The selfish gene*. The essential assumptions of this view are that higher levels of biological organization, such as organisms, can be completely understood from the behaviour of individual genes. Organisms, according to this view, are mosaics of separable,

BIRDS AS FLYING MACHINES

WARREN TRUSS

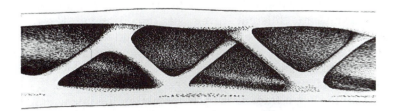

INTERNAL STRUCTURE OF A VULTURE'S WING

Figure 1.1 Birds as flying machines. That organisms may embody principles of good design is nicely illustrated by the wing bones of birds. The internal support structure conforms to the Warren truss, an engineering device that provides maximum strength for minimal materials. The internal structure of a vulture's wing is reproduced with permission of Scientific American Inc.

independent traits that can be teased out and examined as discrete units, or simply added together to achieve the functional whole.

The combined impetus from population genetics and molecular biology has lead to this genic paradigm and extreme molecular reductionism occupying a dominant position in biological theory and research. Thus, Weinberg (1985), in an issue of *Scientific American* heralded as the flagship of the new biology, outlined how molecular biologists: 'work with a certainty that the invisible, submicroscopic agents they study can explain, *at one essential level*, the complexity of life' (my italics).

A similar sentiment applies to the selfish gene paradigm – that is that all higher levels may be reduced, without loss of understanding, to an analysis of the selective pressures acting on individual alleles (Dawkins, 1989a). This book is largely an argument against such a perspective. This stems from an appreciation that organisms are highly complex systems characterized by intricate interactions among parts, that the integration of features and subsystems is highly specific (e.g. precise signals may be emitted by one component that are received and interpreted in specific ways by other components of the coadapted system), and that this integration itself is a primary target of natural selection. As Thompson (1982) remarked, interactions, although less tangible, must be considered to be evolutionary products as concrete as morphologies.

Ironically, what is being uncovered at the molecular level is not some fundamental level of invariant, ultimate control, but complex interactive systems with cascading levels of organization and numerous feedback systems. Genes themselves have complex phenotypes (Chapter 2). Moreover, many of these molecular complexities reflect the need to generate multicellular phenotypes that are flexible or resistant to ecological contingencies. Evolution of integrated cellular communities (i.e. organisms) hinges upon the selection of particular kinds of regulatory interactions relevant to different levels of organization. These include molecular-level processes relevant to genomic and metabolic functioning, cellular-level processes, intercellular differentiation and coordination, organ function, organ-system coordination, organismal-level operations, environmental interfacing and inter-organismal interactions.

Selection in such a framework acts mainly on the interaction systems relevant to the holistic organization. Selection acting on regulatory integration never acts on single genes, and as higher levels of organization are considered, the effective genomic units may conform to progressively larger assemblages of interactive machinery. Thus, hierarchies of biological organization appear to be mirrored by hierarchies of molecular organization. The ultimate integration of biology across various levels of integration requires that a genic perspective be replaced by one of hierarchy. Hierarchy theory will be a central theme in what follows, and consequently, this chapter will provide an overview of the hierarchical approach.

1.2.2 Hierarchical reductionism

The fundamental organization of life is hierarchical (Figure 1.2), and so it is not surprising that hierarchy theory is being successfully applied in nearly every realm of biology (e.g. developmental genetics, evolutionary biology, ecology, behaviour) (Feibleman, 1955; Rowe, 1961; Pattee, 1973; Allen and Starr, 1982; Raff and Kaufman, 1983; Depew and Weber, 1985; Eldredge, 1985; Salthe, 1985; Buss, 1987; Grene, 1987; May, 1989; O'Neill, 1989; Liem, 1990; Barlow, 1991; Kolasa and Rollo, 1991; Yeakley and Cale, 1991). A hierarchy is a system of organization in which different levels can be distinguished (e.g. species, genera, families, orders), or where lower levels are sequentially nested within levels above. Each recognizable level in a hierarchy is called a holon (Allen and Starr, 1982).

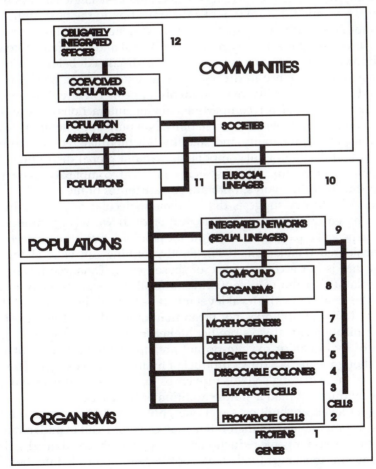

Figure 1.2 A hierarchical presentation of the levels of life, with demarcation of critical transitions in organizational complexity.

There are two kinds of biological hierarchies. Taxonomic or classification hierarchies are essentially groupings imposed by an observer for cognitive expedience (classification). The subcomponents of such hierarchies do not necessarily have any functional relationship to one another. The other kind of hierarchy, and the only kind that will be considered here, are organizational or control hierarchies where there is a functional integration of parts, with each level of organization arising from that of lower-level components (Kolasa and Pickett, 1989).

The widespread existence of hierarchies is not trivial, but probably represents a necessary conformation for organizations of great complexity (Simon, 1962, 1973; Koestler, 1967, 1978). Hierarchical organization allows higher levels to be achieved in smaller consolidated steps, rather than requiring that the entire system be elaborated as a single successful construct. Thus, initial failure to make the transition to the next higher level does not necessarily mean that the entire system dissociates. Instead the lower-level components may remain functional and continue to generate conditions likely for success. As will be discussed below, increasing degrees of connectance in diverse systems tend to reach a critical point beyond which instability may suddenly appear. Hierarchical organization may be crucial for compartmentalizing interactions and allowing large, complex systems to retain stability. Not only does hierarchical organization provide reliability and stability but modularized structure also allows modification of subcomponents without global disruption. Thus, hierarchical structure allows systems of great complexity, which also retain the ability to evolve. Both are key attributes of life.

Hierarchy theory suggests that extreme molecular or genic reductionism is based on a fundamental logical error. If we wish to understand some aspect of a particular level (the focal holon), two other aspects must be considered (Feibleman, 1955). Firstly, nothing can exist at the focal holon that is not a consequence of those below. Thus, the lower levels determine the realm of potentialities a focal level can possibly express. Thus the level below the focal level might be termed the generative holon (Figure 1.3). This still does not mean that knowledge of the lowest level of organization can always predict higher levels. For example, the properties of hydrogen or oxygen do not anticipate the properties of water, even though water is a direct product of hydrogen and oxygen. The interaction of these materials yields properties that are qualitatively and quantitatively different than a simple addition of the parts.

Higher levels are built on lower ones, however, and knowing the properties of components and their interaction can often lead to understanding. This is the basis for genic reductionism. Anything observed in cells, tissues or organisms must eventually have roots extending firmly into the molecular realm. It is significant, however, that the relevant molecular realm on which biological organizations rests is a small subset of all possible mole-

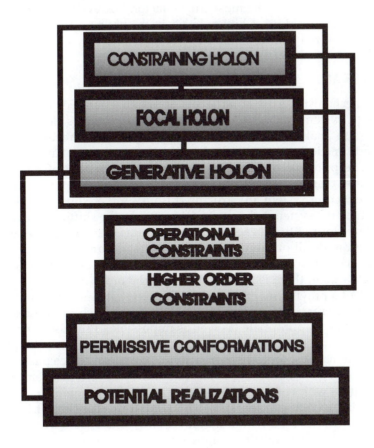

Figure 1.3 The hierarchical triad. Understanding focal-level processes requires knowledge within the focal level (proximal causal mechanisms), at upper adjacent levels (higher organizational constraints) and at lower adjacent levels (historical or compartmental constraints and realization).

cular conformations. In fact it may be argued that reductionism only works at all because the molecular realm that must be examined is a small subset that has been selected by the higher levels of organization resting upon it.

Reductionist research and theory work, and this connectedness across scales is why. Even if the universe were deterministic and we had complete and perfect information, however, the pursuit of ultimate reductionism would be inappropriate. There are numerous reasons for this but some key ones involve emergent properties, interaction effects and the relative independence of causal mechanisms at different temporal/

spatial scales. Probably the most crucial factor, however, is the existence of organization residing at higher levels in the biological hierarchy. This will be considered below.

1.3 SCALE AND CAUSAL MECHANISMS

Causal mechanisms exist at all levels of organization. For example the success of systems analysis – a cookbook of techniques developed by engineers to analyze complex systems – hinges on the recognition that form and function are best studied at their relevant spatial and temporal scales. The principle of hierarchical organization or that of integrative levels states that it is not necessary to understand exactly how particular components of a system are structured from simpler subcomponents to predict how the system behaves. In other words, the shifting sand grains at one level form a firm beach at the next. If this were not so, we would have to examine all phenomena in terms of quarks and gluons.

Just as there is no need to understand the subatomic structure of a key to unlock a door, there is no need to know the order of nucleic acids in the DNA of a praying mantid to study how one catches a fly. Similarly, a molecular approach to a community phenomenon such as forest succession would be inappropriate (as well as impossible). Processes important at this level include plant competition for light and nutrients, predator–prey dynamics, decomposer activity, dispersal mechanisms, fire and meteorology. If not seeing the forest for the trees is a problem among dendrologists with restricted vision, then molecular myopia would compound this problem considerably. Thus, causal mechanisms at the focal level may be addressed within that level or by reference to closely adjacent holons. Levels at greater distances from the focal level usually have strongly diminishing relevance.

Clearly, different spatial and temporal scales have their own unique components, structure and dynamics. It is not only valid, but necessary, to understand the processes of life on all these levels within their relative context. At ecological levels, for example, much of the system's dynamics is among organisms or with environmental factors. Thus, the relevant mechanisms often lie outside the scope of a single genome, or outside the biological organization altogether. Molecular biology has little power to address such processes. For example, fish tails show an evolutionary progression from the primitive state of simply tapering to a point (diphycercal tail), to a lobed tail with a larger upper fin (heterocercal tail), to the equal-lobed tails of advanced fish (homocercal tails) (Figure 1.4). The causal mechanisms determining the design of tails are associated with hydrodynamics and large-scale physical properties of fins. Molecular-level scrutiny might suggest how fish grow tails and move them, but says nothing about why a particular conformation may be favoured. This

EVOLUTION OF FISH TAILS

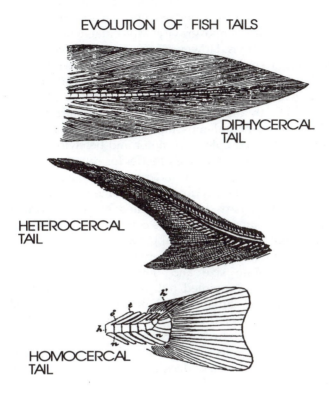

DIPHYCERCAL
TAIL

HETEROCERCAL
TAIL

HOMOCERCAL
TAIL

Figure 1.4 The tails of fish represent a design forged at the level of hydro-dynamics with complex morphology. The attainment of more effective designs has proceeded via the modification of previously elaborated structure. After Schmidt (1909).

example is particularly instructive in that it also illustrates how evolutionary options may be constrained by pre-existing structure.

Similar arguments apply to the causal mechanisms influencing the design of leaves, flowers or seeds. Leaves may have problems with competitors (shade), herbivores or shedding rain (to prevent fungus) that require particular features like large size, protective hairs or drip spouts. Flowers mirror the free-enterprise economics of flower–pollinator interaction. Seeds are a compromise of size and features needed to foster germinating seedlings, enhance dispersal and defend against predators.

Alexander (1982), in an elegant little book entitled *Optima for animals*, explored the theory developed regarding the design of muscular and skeletal systems, their power, speed and reliability. After all, the purpose of muscles and skeletons presumably has to do with movement and support. The causal mechanisms underlying their design have to do with body size, internal stress and gravity. Molecular biology is not directly relevant to such questions. However, if we ever decided to redesign blue whales so that they could walk out of the ocean, we would have to know a lot about how skeletal architecture changes with body mass. We might even have to give them more than four legs.

Gallistel (1980), in an article entitled *From muscles to motivation*, explored how basic neural–muscular organizations are hierarchically integrated to produce complex locomotor and behavioural patterns. The outlined design provides an elegant model for robotics. Ultimately, such compounding complexity may even derive consciousness.

In terms of development, comparative allometric studies suggest that the large brain of humans is not attained by any intrinsically faster growth. Instead, we simply extend the common primate line or prenatal brain development into the postnatal growth period (Gould, 1977). Such knowledge is not molecular, but it is certainly relevant for understanding the evolution of brain size and ontogeny of brain development. Numerous authors have pointed out just how relevant understanding evolution and potential shifts in development (termed heterochrony) are to understanding evolution (Lovtrup, 1974, 1987; Gould, 1977; Reid, 1985; Raff and Kaufmann, 1983; Arthur, 1984; McKinney and McNamara, 1991). Presumably genetic engineering of brain size in primates could profit from such knowledge if society decided that such a venture was acceptable.

The idea that causation ultimately resides at the molecular level implies that the genetic code somehow contains all of the information necessary to understand organismal form and function. Undergraduates are often profoundly surprised when they are told this is not so. The genetic code is just that, a one dimensional code that specifies the necessary steps to produce a functioning organism. It is not feasible to deduce the adaptive function of a structure or process from the code. It is necessary to examine the products of the code directly at the level at which the structure or process actually functions.

Consider a model airplane. If the airplane is analogous to a bird, then the assembly manual is analogous to the bird's genetic code. Wing length is undoubtedly a feature that influences the flight characteristics of both birds or planes. Each generation of birds is born with a variety of wing lengths. Species have wings of a particular length because longer or shorter wings were eliminated by natural selection. Airplanes have wings of a given length because engineers applied known aerodynamic principles to their design. These principles, however, are not stated in the instruction manual. To build a better airplane, you must either know these principles (that come from studying and experimenting with airplanes), or you must build a very large number of airplanes and select those that fly best (expensive).

Similarly, the genetic code of a hummingbird nowhere states the principles of its design. If we wished to modify hummingbirds, we would have to apply known aerodynamic principles (that come from studying birds), or we would have to produce a wide assortment of mutated hummingbirds and see which one was best. This is equivalent to blind mutation followed by artificial selection. The evolution of hummingbirds has involved millennia of ecological interactions between birds and their environment. To argue that the causal mechanisms governing this process are molecular is a case of arguing that the tail wags the dog. Incidently, the fact that tails may serve as flyswatters, heat dissipators, warning signals or balancing organs is not specified in the genetic code of any animals with tails.

If the molecular promise is to be used more effectively than simply as a tool to increase the rate and kinds of possible mutations, higher level studies must be harnessed to provide ecological vision. The truth of this can be seen by considering a few examples. Imagine examining the sequence of nucleic acids in DNA to discover why:

1. Damselflies are brightly coloured. If you paint blue dots on other insects, damselflies with blue dots on their tails will chase them. Male damsel flies are territorial and the release of aggression is colour coded.
2. Viceroy butterflies look like monarchs. Could you explain their appearance by also examining the code of monarchs? Would that tell you that milkweeds make poisons that monarchs store? Would the code of milkweeds tell you that predators do not eat these insects because of the poisons?
3. Crickets sing. Does the code explain why some male crickets do not sing? Would the code hint that parasitic flies are attracted by the song so that some male crickets play mute?
4. Can the code explain the numerous designs of spiders' webs? Can it tell us how effective various webs are at catching particular prey?
5. Can the code of an acacia tree tell us why it grows hollow thorns? Can it tell us that ants that live in these thorns protect the tree from herbivores?
6. Could the genetic code for dandelions tell us much about the best design for parachutes? Could it predict the rate of dispersal of dandelions to distant habitats?

Clearly, the application of recombinant tools to redesign higher-level features will largely be limited by biological knowledge at the relevant level of organization. Similarly, understanding the design of naturally evolved species may require a framework that is not simply decomposable into selfish genes. Instead, the interaction systems relevant to the adaptations in question may form units of selection in their own right.

1.4 INTERACTION EFFECTS, EMERGENT PROPERTIES AND COMPLEXITY

One reason why higher levels of organization have properties not predicted by examining underlying components is that components interact. The importance of interactions can be demonstrated with very elementary mathematics. Probably the simplest relationship between two variables is a linear one (i.e. values of the dependent variable Y plotted against values of the independent variable X lie on a straight line):

$$Y = a + bX$$

where a represents the intercept of the line and b represents how much Y changes with each step in X (i.e. the slope). If Y is influenced by two independent variables, X_1, and X_2, the simplest linear equation describing the relationship must include an interaction term so the equation is not simply:

$$Y = a + b_1X_1 + b_2X_2$$

but:

$$Y = a + b_1X_1 + b_2X_2 + b_3X_1.X_2$$

As more independent factors are added, the number of terms needed to account for interactions rises exponentially as follows: 1, 3, 7, 15, 31 for factors 1, 2, 3, 4 and 5, respectively.

Thus, if there are only five main effects, there are 26 potential kinds of interactions among them. Moreover, interactions can be even more important than main effects. In fact, interactions can be important even if direct effects are relatively neutral. Thus, we are familiar with how wind and temperature combine to produce wind chill, and most know that drinking alcohol while taking tranquillizers can induce coma. These linear models represent the simplest way in which complexity increases with additional factors. If factors have nonlinear dynamics or discontinuities, if order is important, or if feedback and time lags are introduced, very complex or even chaotic results may emerge (Gleik, 1987; Kauffman, 1993).

Higher-level systems are intrinsically more complex than those at lower levels. Part of the success of molecular biology, in fact, is related to its simplicity relative to upper levels of organization. As we ascend the biological hierarchies, the number of basic components increases astronomically:

1. Nucleic acids in DNA, 4
2. Essential amino acids coded by DNA, 20

3. Classes of small molecules in a cell, about 750
4. Classes of large molecules in a cell, about 2000
5. Number of single-celled animal species, about 20 000
6. Number of multicellular animal species, about 2 000 000

Now add in interaction effects and redundancies (i.e. the brain contains millions of neurons and one species may contain millions of unique individuals). It is difficult to see how molecular biology could hope to describe upper levels of organization when so much of these systems is comprised of interaction dynamics among higher-order components.

Higher-order systems are not only intrinsically more complex, they tend to operate on larger and longer scales of space and time. Thus, forest succession may require several hundred years compared with time scales of seconds for molecular processes associated with photosynthesis in chloroplasts. Given the appropriate techniques, progress in molecular biology would be expected to progress faster than studies in development, ecology or evolution. However, all such studies are of value. The value of a science cannot be measured by the rate of scientific progress or the precision with which measurements or predictions can be made. In fact, a greater degree of reductionism applied in higher-level studies would probably impede progress.

There is also good evidence that at each higher level, properties emerge that are not predicted by examining lower levels (emergent properties), even if interaction terms are considered (Feibleman, 1955). Use of the term, emergent properties, has been strongly criticized by ecologists (e.g. Salt, 1979; Edson *et al.*, 1981). It remains, however, that there is no way to predict that the reactive metal sodium and the poisonous gas chlorine interact to produce table salt unless the frame of reference is expanded to include the interaction term and allow for non-additive properties of the system as a whole (Rensch, 1960). This effectively defines the use of this terminology here. At higher levels of organization even more significant properties emerge. Thus, picture a rose in your mind. Now imagine that the rose is coloured like a candy cane. The image of the candy-cane rose exists in a very real sense. However, no examination of the structure of neurons or their interconnections will reveal the rose. Consciousness and candy-cane roses are emergent properties of higher levels of neural organization. The fact that this image was elicited in your mind by the scribbles on this page is an emergent property of the social organization to which we both belong.

The evolution of perception, cognition and memory have allowed the development of a separate information system at a much higher level of organization. The complexity of the human brain is so great that the genes may not always explicitly code the fine detail. Instead, neurons grow in abundance according to some simple rules, and then neurons that are not linked into functional circuits may wither away. The development of an organism from a single cell involves an exponential compounding of complexity and a generation of numerous higher-order holons. Some developmental processes involve interactions among higher-order components that are somewhat independent of the genetic code that initiated them. The intrinsic potential for deviations at numerous decision points, and interactions of development with the environment may mean that we are all mentally as unique as snowflakes.

The cognitive system is not only somewhat independent of its molecular roots, it is this emergent intelligence that is now applying itself to modify the lower molecular codes. It may well be that the complexity of life in the future will be a reflection of this higher-level information system – the mind of man – rather than of any blind watchmaker that preceded biotechnology. It seems exceedingly unlikely that even complete molecular knowledge would allow prediction of such a significant feedback loop. Biotechnology is not simply a convenient boon to human enterprises, it represents the metamorphosis of a blind biosphere to one of intelligent self awareness.

1.5 ORGANIZATIONAL OVERLAYS

Lower levels of organization generate potentialities that yield higher-order realities, although these may not be predictable. The focal level may have properties reflecting all of these potentials, or if only some subset is involved, then the focal level may have properties associated with compartmental or historical constraints. In this case the focal level is based on only a subset of lower-level potentials. For example, populations of sexually reproducing animals constitute only a tiny fraction of potential genotypes. There is a very good chance that the perfect partner that everyone is searching for simply was never born. In fact, Gould (1989a) suggests that if the evolution of life on earth were to be replayed, it is highly unlikely that the sequence of alternatives leading to humans would be re-derived.

There is yet another factor limiting the set of lower-level potentials that a focal level displays. The evolution of life has involved successive overlays of organization, each with important attributes of specific relevance to the existence, structure and processes of that level. Biological entities arise via the progressive integration of lower-level entities (Kolasa and Pickett, 1989). Biological entities are never homogeneous. Organizational structure at higher levels imposes constraints on lower levels of organization, a critical aspect overlooked by molecular reductionism (Feibleman, 1955; Salthe, 1985). What this means is that not all those potentialities of lower levels can be realized at the focal level because they are precluded by higher-order organizational constraints (Figure 1.3).

Consider, for example, that activities advantageous to lower-level components might be detrimental to higher levels of organization. Thus, selfish DNA, the activities of transposable elements, unrestricted proliferation of mitochondria or unconstrained cell division (cancer) would interfere with the success of organisms if such features had unlimited action. Higher-order organizations require that such features are restricted within acceptable boundaries. Although there may be considerable scope for variation within a focal level, higher levels must regulate or circumscribe lower levels to ensure their own integrity.

Thus, in vertebrates, genes that control cell division rate often function as hormone receptors. Such genes have been assimilated by retroviruses that use them to induce cancer in their hosts (Loomis, 1988). This aids the propagation of the virus, and also illustrates the importance of regulatory control over cell division for the survival of the animal.

The constraints imposed by higher-orders of organization may take the form of lower-level organization itself. Thus, the genetic code for amino acid transcription has been conserved across such diverse organisms as bacteria and whales. Most of the biochemistry involved in the basic metabolism of life is shared by both prokaryotes and eukaryotes, and the genes coding for particularly crucial enzymes represent molecular structures that have been strongly conserved by higher-order constraints.

Selection acting only on the speed of DNA replication in a primordial soup would likely select for tiny lengths of largely unorganized nucleic acids. It is selection of organizational attributes at higher levels that imposes the complex genomic structure found in eukaryotes (e.g. genomic concentration in nuclei, the double helix, chromosomes, regulatory elements, transcription apparatus, DNA domains).

Further proof of the importance of higher-level organization is exemplified by evolution itself. Darwin's (1859) theory that evolution occurs via the action of natural selection acting on variation among organisms is the cornerstone of biological science. Fundamentally, this process occurs at the level of organisms and upwards (e.g. kin groups). That Darwin (1859) formulated his theory without any knowledge of molecular biology or even genes is a fact underscoring that selection acts directly at higher levels and only indirectly via constraints on lower levels. The fact that natural selection acts on organisms and that individual organisms provide highly variable genetic backgrounds means that genes cannot be individually selected.

Although natural selection must have scrutinized molecules directly in the primordial soup, the genome is presently insulated from natural selection by numerous layers or organization. Conversely, these same layers impose successive layers of constraint. Natural selection cares nothing about how an organism achieves a particular adaptive feature, only that it is successful at its relevant level of operation. Genes can be considered to be tools of organisms and not vice versa (Haukioja, 1982). Thus, plethodon salamanders have exhibited a remarkable stasis of body morphology over the last million years. Despite their lack of morphological change, there has been remarkable genetic divergence. The genetic change that has occurred, however, has been constrained by the requirement to maintain the successful body form (Wake, Roth and Wake, 1983).

The unit and object of natural selection is the phenotype as a whole. Phenotypes do not serve genes, but rather the opposite is the case. A genic theorist in a library, for example, would undoubtedly conclude that books are vehicles for the transmission of selfish words. However, to argue that the unit of selection is a letter or word (substitute codon and gene) misses the point that all books are written with essentially the same code, but as far as literary success is concerned, the individual words may be relatively unimportant. The same book or a computer program may have essentially the same phenotype and be equally successful if translated into other languages. If books are analyzed as frequencies of words or letters, or if chess games are analyzed as frequencies of squares occupied by the queen, the patterns are random (Salthe, 1975). The meaning of a page or the strategy of a game emerges at higher levels of organization and is lost by a reductionist view. Higher-order features may be selected in their own right, and not simply for the code that they contain. Otherwise, it is very difficult to explain how both marsupial and eutherian lineages independently derived identical best sellers that were large saber-toothed cats. The convergence of insects in distantly related orders of insects onto a mantid-like design is equally striking (Figure 1.5). Other notable convergence has occurred between hummingbirds and hummingbird hawk moths, moles and mole crickets and fish and cetaceans.

Roth (1991) emphasized that changes within one holon may be relatively screened off or insensitive to changes in others. In this case, the maintenance of phenotypic stasis in the face of genetic change argues that the level of selection is the organismal holon and not the genic. A second example is provided by the haemoglobin of horses and man. Firstly, if haemoglobin was not required to carry oxygen in the vascular system of multicellular organisms, or in other oxygen-poor habitats, it is unlikely that the haemoglobin gene would exist as a molecular organization. Haemoglobin is not directly relevant to molecular processes, but it is a crucial factor at the level of multicellular organisms with large bodies and high metabolisms. That selection is on the higher-level respiratory function of the molecule is illustrated by the fact that horse and human haemoglobin are nearly indistinguishable in function, even though these genes vary at 43 out of 287 amino acid positions (A.C. Wilson, 1985).

These arguments suggest that not only is the logic of molecular reductionism seriously flawed, the reality of evolution itself is the inverse of this

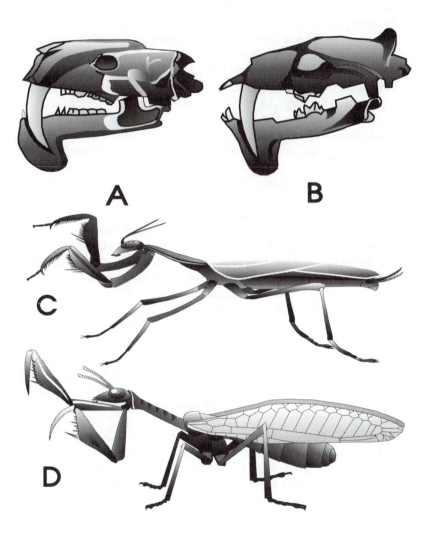

Figure 1.5 Convergent evolution of saber-toothed cats and mantid-like designs. The availability of a particular niche (i.e. spearing prey), has forged similar saber-toothed phenotypes in (A) the marsupial, *Thylacosmilus*, and (B) in the mammal, *Eusmilus*. Redrawn after Rensch (1960) with permission of Ferdinand Enke Verlag, Stuttgart. The design of the true mantids in the order Orthoptera (C), has been remarkably converged upon in members of the Mantispidae (D), that belong to the order Neuroptera.

paradigm. In biology, natural selection acts directly only on higher levels of organization, and molecular organization is largely a product of hierarchically compounding constraints derived from upper levels. Because genomes are only selected as wholes, the conformation of particular parts and their compounding levels of interaction are always selected simultaneously. Arnold and Fistrup (1982) captured this idea when they recognized that all levels below the focal holon for selection are collapsed into a single process of change in components internal to the whole system. Goodnight, Schwartz and Stevens (1992) drew a similar conclusion. They identified the relevant units of selection as individuals, lineages, groups and communities. In each case, the interactions among components constituting the selective unit are usually the largest aspect. The lesson is that complete understanding requires not only knowledge of how molecular attributes yield higher-order potentialities, but also how higher levels impose molecular constraints and organization.

These three perspectives of lower-level potentialities, focal level processes and higher-level constraints can be rationalized into a paradigm for global understanding (Figure 1.3). Mayr (1961) pointed out that considerable controversy in biology arises because there are different perspectives from which a problem can be addressed. Scientists frequently misunderstand one another because they are asking different questions about the same thing. Thus, one researcher might be interested in exactly how a praying mantid captures flies. Questions asked might involve the orientation to prey, the timing of the strike or the effectiveness of the spines on the forelegs with prey of varying size. Such questions address what Mayr (1961) termed proximal causal mechanisms. Specifically what is the structure and function of the feature or process involved?

Alternatively, evolutionary ecologists address questions related to what Mayr (1961) identified as ultimate evolutionary functions. Thus, optimal foraging theory tests the evolutionary hypothesis that animals are selected to maximize their net intake of energy (Pyke, 1984), because ultimately this maximizes reproduction and thus, fitness. Such questions transcend the particular proximal mechanisms used by species (e.g. jaws, mantids' forelegs, frogs' tongues).

The hierarchical perspective suggests that there is yet a third question that can also be asked; how is the feature of interest assembled and maintained? This last question is that most often asked by molecular reductionists. Significantly, these three perspectives are somewhat analogous to the fundamental triadic unit that has been identified as the minimum perspective needed for understanding focal levels in hierarchies (Figure 1.3).

Butterfly wings provide a very nice illustration. Firstly, wings vary remarkably in shape and colour and serve numerous adaptive functions (e.g. flight, mate recognition, territorial displays, camouflage, predator warnings, predator disruption, defence and thermoregulation to name a few). Butterflies in the family Pieridae have two approaches to the problem of thermoregulation. In the sulfur butterflies, the wings are darkly pigmented. These butterflies absorb radiation by orienting their wing surface to intercept sunlight maximally. The heat then diffuses to the body through the wings. Another group of butterflies in this same family, however, are white. Since white butterflies are not as effective at absorbing radiation (they actually reflect it), what is the selective advantage of being white? The answer, surprisingly, is still thermoregulation. These butterflies angle their wings in such a way as to reflect the sunlight directly onto the insect's body (which is darkly coloured). Thus, the heat is reflected and focused directly, rather than absorbed and conducted (Kingsolver, 1985; Kingsolver and Wiernasz, 1987, 1991). Clearly, understanding proximal mechanisms is both fascinating and useful. It is also clear that most questions about proximal mechanisms reside within a single focal level of organization, and that the same function can be achieved by completely different solutions even among closely related organisms.

Another way to approach the question of thermoregulation in the Pieridae would be to ask what are the evolutionary advantages conferred by thermoregulatory abilities and does the insect's behaviour maximize these advantages? Likely candidates include improved flight efficiency, the ability to fly across a greater range of environmental temperatures, better predator avoidance, improved mate acquisition, greater dispersal ability, or faster maturation of eggs. Clearly, the questions concerning the evolutionary function of thermoregulation reside at higher levels, and impose constraints on the focal holon. For example, if thermoregulation is very important, wings will be restricted in how much they can be modified for other purposes. Furthermore, wings are but one structure in a complex anatomy that requires a high degree of integration. Other features will also restrict the extent of wing modification as part of the organismal level of organization. Thus, many butterflies have wings with the upper surface serving functions of sexual display and/or thermoregulation, while the lower surfaces provide crypsis when the wings are folded. Questions of evolutionary function are valuable because they provide global understanding that cannot be achieved even with complete knowledge of the proximal mechanisms.

The third question is how the proximal mechanism is assembled and maintained. This involves the entire developmental process beginning with the genetic code and ending with the functional structure. Molecular reductionists are largely concerned with this question of epigenetic realization, specifically in this case, how are butterflies and their wings built? Clearly, this is a fundamental question of great interest and importance and one necessary for complete understanding. However, getting from the genes to the final structure tells us nothing about why some Pieridae butterflies are yellow (*Colias* spp.), while others are white (*Pieris* spp.).

Full understanding of design at any given focal level requires that all three questions be asked: how is the phenomenon realized and maintained; how does it function; and what purpose is served (i.e. how is it built, how does it work, and what is it for)? All three questions are equally relevant, worthwhile and necessary.

Let us consider one more example in the light of the idea that there are three solutions to every biological puzzle. Molecular knowledge regarding haemoglobin is relatively extensive. Molecular techniques reveal the genetic code and protein structure for the various chains that make up the molecule. They reveal the likelihood that gene duplications were involved in various divergences and allow construction of phylogenetic histories. However, the existence of haemoglobin is largely a constraint imposed by the need for oxygen transport in multicellular organisms.

Take a relatively recent innovation for example. In placental mammals, the partial pressure of oxygen declines about fivefold before reaching the developing embryo. In some species, including humans, the problem has been solved by producing a specialized fetal haemoglobin that has higher oxygen affinity than that of adults. After birth this fetal haemoglobin is rapidly replaced by the adult form. In some mammals such as rodents and horses, the problem of fetal respiration is solved not by making different haemoglobins, but by modulating the affinity of the blood for oxygen using small cytoplasmic molecules (Raff and Kaufman, 1983, p. 305). As in butterflies' wings, proximal solutions to the same problem may vary. The evolution of fetal haemoglobin is related to the evolution of internal embryogenesis, and as such, understanding does not reside entirely at the molecular level.

Sickle cell anaemia further illustrates this point. Molecular biology reveals that disease is 'caused' by substitution of the amino acid valine for glutamic acid at position six of the ß-haemoglobin chain. This causes red blood cells to deform into a sickle shape. Malaria parasites are less effective in attacking these cells so the altered gene is propagated in populations where incidence of malaria is high, even though the gene also causes numerous deleterious effects on health. The proximal mechanism of interest is at the cellular level. The evolutionary question pertains to the survival of organisms against disease in an ecological context.

The application of this fundamental triadic perspective as a paradigm for scientific understanding might be termed hierarchical reductionism. The relative perspective may slide across levels of focal interest. Each level of organization has questions of proximal relevance that are independent of other levels in some sense, but global understanding requires triumvirate vision. The key ingredient lacking from previous discussions is probably the idea that organizations may reside with some independence at higher levels and exert downward constraints. Thus, a discussion of organization and its evolution in biological systems is required.

1.6 EVOLUTION OF BIOLOGICAL ORGANIZATION

1.6.1 Characterization of organization and the depths of overlays

Von Bertalanffy (1952) recognized that the key to understanding life was an understanding of organization and he called for a general system theory to address it. The sequential addition of hierarchical overlays of organization has been a critical aspect of evolutionary progress. Anyone who has studied introductory biology will be familiar with the ubiquitous categorization of life into levels associated with genes, cells, tissues, organs, organisms, etc. In fact, various attempts to classify life hierarchically have derived two separate schemes, one based on ecological units (e.g. populations, communities, biomes), and one based on genealogy (e.g. genera, families, orders) (Eldredge, 1985; Liem, 1990). Species emerge as a common unit in both classifications. Interpretation in this framework considers interactions between the ecological and genealogical hierarchies (Eldredge and Salthe, 1984; Eldredge, 1985).

It is important to differentiate, however, between simple categories or assemblages that have little functional integrity in a coevolutionary context and levels of organization where components display complementary coevolution and functional integration. Thus, classification of the hierarchy of life becomes much more enlightening if the levels are identified by their degree of organization and by critical evolutionary inventions marking significant transitions in complexity (Kolasa and Pickett, 1989).

During the course of evolution there have been at least 12 major transitions associated with additional organizational overlays and increasing complexity (Figure 1.2). Although we tend to think of organisms as the ultimate achievement of evolution, the organizational hierarchy suggests that there are perhaps four additional transitions above that of the organism (five if we consider symbiotic organismal units). Given that higher levels impose constraints on lower levels, genes are clearly insulated and constrained by a relatively immense amount of compounding organization. Moreover, this process of stepwise hierarchical elaboration has certain common elements that have been reiterated with each new higher rung up the ladder.

Before proceeding with a discussion of specific organizational transitions, it is first necessary to define exactly what is meant by an organization as opposed to an unorganized or non-functional assemblage. This is a more difficult proposition than it appears, partly because the degree of organization varies remarkably depending on the forces shaping the system and the length of time that the association has had to evolve. Identification of functional organization is confounded by the existence of self-organizing properties of matter that can generate order in the absence of natural selection (e.g. crystals).

This problem is accentuated even more if natural selection has acted to consolidate systems harnessing naturally ordered states. Thus, Kauffman (1993), based on more than 20 years of considering complex systems, has put forward the bold hypothesis that much of biological organization (spanning genomic integration, epigenetics and community organization) may be built upon tendencies for innate order to emerge in complex inter-active networks.

Kauffman (1993) begins with a simple model in which N components are interconnected to various degrees (K), where K = N corresponds to a state where every component is directly connected to every other. He then assumes that the on–off state of every component is dictated by a logic of Boolean functions (the same universal language used in computers). For example, the status of the component A might be determined by positive inputs from both B and C, or alternatively, by a positive input from either B or C. The functions 'and' and 'either/or' constitute Boolean logic.

Extensive simulations of such NK models show that intrinsic order may emerge, even in ensembles governed by completely random networks of Boolean logic. For example, random feedback circuits will emerge that force all of the components in the circuit into a constant status (forcing structures). Such structure tends to isolate local islands of unordered dynamics. If the system is biased away from neutrality (e.g. a high weighting of on or off status among components), similar ordered patterns may arise within certain ranges of the biasing parameter (P). Across other values of P, islands of frozen order may be derived, embedded in a lattice with unordered dynamics.

Kauffman (1993) then explores the stability characteristics of NK systems with various values of N, K and P. Order and stability are restricted to systems of low N or decreasing K. As N and/or K increase, the system's dynamics shift from highly ordered to states with a mix of frozen and ordered elements and then to chaotic systems whose order can only be characterized by strange attractors (see Gleik (1987) for an introduction to chaos theory).

Kauffman (1993) points out that highly ordered systems may be relatively inflexible, whereas those in the complex realm bordering on chaos have the mixed advantages of both order and intrinsic flexibility. A simple conceptual model is to consider these states as analogous to solid, liquid and gas. Solid systems have high order, but are difficult to change. In a system that is melting, some areas may remain highly ordered while others become fluid. Chaos emerges as the liquid phase becomes turbulent. A perfect gas may correspond to true randomness. Kauffman (1993) applies the sweeping generalization that complex biological systems may be driven by natural selection to the complex realm bordering on the edge of chaos.

Kauffman (1993) points out that the easiest course for natural selection is to harness features deriving innate order. A crucial insight is that the degree of order (or lack thereof) might appear as a necessary consequence of the number of components and their connectance. Such models not only define the easiest routes for achieving stability, they also highlight possible limitations on what sorts of organization are possible.

Kauffman's (1993) thesis may provide the appropriate backdrop outlining the landscapes that natural selection has to explore. Nevertheless, it remains doubtful whether anything resembling organizations selected for adaptive function often emerge in random NK networks. While it is true, for example, that the ensemble space of all possible chess moves necessarily contains every chess game that can ever be played, the likelihood of any randomly generated play winning against a grand master is rather close to zero. Similarly, all possible organisms must derive from a much larger ensemble of all possible forms, but only a very narrow edge of possibilities is realized. The existence of such enormous complexity reflects the ability of nature to generate order of incredible magnitude, but only under the guidance of selection. Thus the direct products of selection (e.g. Dawkin's most complicated machines) and its indirect derivatives (e.g. Paley's watches) leap out as constructs that differ from randomly generated patterns by many orders of magnitude. The emphasis throughout this book will be on adaptive function.

Rather than attempting to define adaptive organization explicitly as a single construct, it is probably better to describe the general attributes associated with biological organizations which can then be used as a multifaceted lens with which to gage systems of interest. In fact, in any assemblage of interacting biological entities, there will usually be some element of organization. For the purposes of comparison, this may be relatively trivial compared with highly evolved systems that have achieved an independent identity (termed entification). Table 1.1 provides a list of important attributes associated with biological organizations, and their comparable state in less organized assemblages.

The evolution of an organization and ultimately entification is associated with a number of key criteria. The application and degree of development for any single factor, however, will vary among specific systems. Because most organizational developments involve melding subcomponents together, transitions usually involve a size increase (e.g. individual cells to multicellular organisms).

Considering organisms, even within a given organizational type or

Hierarchical reductionism

Table 1.1 Attributes associated with organizations as opposed to unorganized assemblages

Attributes	Assemblages	Organizations
Physical size and spatial scale	Smaller	Larger
Relevant temporal scale	Shorter	Longer
Inter-component redundancy	High	Reduced or regulated to specific levels
Independence of components	Autonomous	Interdependent or non-dissociable
Number of different components	Lower	High
Information transfer or connectedness	Low	High
Complementary specialization	Low	High
Degree of order, pattern or structure	Lower, but may be highly structured	High
Spatio/temporal contiguity	Random or non-causally correlated	High, at least intermittently
Unit of natural selection (entification)	Individual components (single phenotypes)	Organization as a whole
Genetic relatedness of components	Random	Either highly or distantly related
Basic interactions	Local rules (competitive, exploitative, parasitic, cooperative)	Regulated homeostasis global rules
Environmental filtering	Low	High
Constraints on lower levels of organization	Low or none	New constraints added. Variable effects on old constraints.
Centralization of functions	None	May be highly advanced

'Bauplan', simple increases in body size also involve increases in organizational attributes. Because larger body size is mainly achieved by increased cell number rather than cell size, larger organisms have more cells. This increases the number of interconnections as well as subcomponents (Bonner, 1988). Consequently, there may be qualitative changes in numerous aspects such as the discriminatory ability of perceptual systems (e.g. better image formation in eyes) and greater capacity for cognition and memory (reviewed by Rensch, 1960). The number of cell types also increases in larger, more advanced forms from three in yeast, 15 in hydras, 60 in worms and insects to 250 in humans (Kauffman, 1993). Thus, the well documented trend for phylogenies to increase in size over evolutionary epochs (Cope's rule), is associated with a more general trend of progressive increases in the number and kinds of cellular components.

The time scale relevant to the new organization is typically longer than for either free-living precursors or their related components in the organization. For example, the cell replication rates (and generation time) of bacteria are more rapid than for cell division rates in a eukaryote organism like an elephant. Thus, organizational transitions are usually associated with transitions in scale. The initial criteria for organization require some sort of interaction or communication. As entification proceeds, the degree of interconnection or communication tends to increase and the signals used or aspects exchanged tend to become more specific and refined (e.g. use of hormonal signals of one organism by its symbiont). Atlan (1974) suggests that such connectedness increases the information content of organizations.

The question of redundancy is rather thorny. Among eusocial insects, redundancy emerging from the reiterated activities of numerous workers is a key factor yielding system reliability (E.O. Wilson, 1985). Such systems still show specialization of parts (e.g. various worker castes), but it remains that organizations may employ redundancy as one dimension of system performance. The degree of redundancy may well be related to the efficiency of the deployed subcomponents. Atlan (1974) suggests that optimal organizations may strike a compromise between maximum information content (diversity of parts) and maximal redundancy. Redundancy is generally regarded as reducing information content, and consequently may be taken as evidence of reduced organization (Pahl-Wostl, 1990). However, organizations may deploy redundancy to minimize system failure, so this interpretation may be mistaken. Pahl-Wostl (1990) also suggests that apparent redundancy may be reduced if the redundant pathways act on different temporal scales. Redundancy may also provide advantages associated with parallel processing and the interaction of such parallel dynamics (Moritz and Southwick, 1992).

In highly evolved organizations the components become specialized in complementary ways, redundant parts tend to be lost and the scope for independent existence is lost. For example, in modern eukaryotes, mitochondria and chloroplasts have lost their independence but still contain their own DNA (Margulis, 1981). However, they are not permitted to replicate independently. Their number is regulated in the best way to serve the cell. Allometry provides some evidence of such regulation. Smaller animals have higher specific metabolisms than larger animals, and the concentration of mitochondria in their cells is appropriately higher to fuel this metabolism (Calder, 1984). Similarly, chloroplasts may be regulated according to the availability of light.

The complementary nature of this control is evident in that some of the genes regulating the number and function of cell organelles are contained in the nuclear genome. Thus, eukaryote cells constitute an integration of several coevolved genomes. There is strong evidence that in organizations with obligately close association, genes may even be transferred among the components. Thus, some genes necessary for mitochondrial or chloroplast functioning have been transferred to the nucleus (Baldauf and Palmer, 1990; Goff, 1991; Covello and Gray, 1992).

Connectedness involves regulatory feedback that controls component dynamics to meet the best needs of the organization itself, not necessarily the best needs of the components (e.g. programmed cell death during embryogenesis or ejection of drones from beehives are detrimental to these components but benefit the organization as a whole). The idea of connectedness relevant to the global fitness of the organization is central to the identification of an entity.

Frequently, the presence of order or predictable structure is interpreted to imply the existence of adaptive organization. However, order may be achieved by several means other than organization (in the sense used here) so the concepts are actually dissociable (Banerjee, Sibbald and Maze, 1990; Kauffman, 1993). For example, order could arise via the interaction of an assemblage according to immediate local rules governing direct interaction of components. Alternatively, order may emerge from imposition of constraints from outside the focal level, or as a simple consequence of elaboration from lower levels of organization. Organization is only present when the components are integrated via global rules relevant to the adaptive function of the holistic system at its focal level.

Although each new level of organization imposes a new constraint structure on lower levels, constraints that already exist in lower hierarchies may be either increased or decreased by a new organizational overlay. For example, consider the discovery that eukaryote genomes contain vast amounts of apparently useless code. The molecular reductionists explain this within a framework of either neutral or selfish DNA (Doolittle and Sapienza, 1980). The fact that prokaryotes do not have this extra DNA is somewhat disturbing for this hypothesis. Possibly the selection pressure on microorganisms for extreme biochemical efficiency precludes the existence of lines that makes extra genetic 'garbage' (e.g. Loomis, 1988).

More recently, regulatory functions have been postulated for considerable amounts of this non-coding DNA. The control of DNA domains needed to regulate cell differentiation and morphogenesis in eukaryotes may require large chunks of non-coding DNA to function in unravelling banks of genes, otherwise maintained in a supercoiled, inactive state (Bodnar, Jones and Ellis, 1989). Similarly, repetitive DNA interspersed throughout the genome may serve as transcription regulators so that banks of genes may be coordinated and integrated during epigenesis (Britten and Davidson, 1969, 1971). Transposable elements may not be simply selfish genes, but may function to improve the evolvability of the species' lineage (Campbell, 1985; Dawkins, 1989b; Wills, 1989). Thus, an alternative interpretation of this 'extra' or apparently selfish DNA in eukaryotes is that it was required for the transition to multicellularity. Clearly, both views have merit and the ultimate explanation probably requires both.

There is little doubt that organized eukaryotic cells in organisms are not under the same selective pressure as bacteria in terms of maximizing their replication rate. This could consequently allow greater genome size and greater redundancy than occurs in prokaryotes (Loomis, 1988). At the same time, the movement, division and differentiation of cells is strongly constrained (regulated) in organisms (Lovtrup, 1974; Raff and Kaufman, 1983; Buss, 1987). Thus, organismal organization appears to release some constraints on cells while imposing others. One general trend is that as an organization increases in complexity, subcomponents become simplified (Moritz and Southwick, 1992). Thus, single celled organisms such as protozoans are generally more complex than cells in any single tissue in a Metazoan.

A global view of organizational transitions reveals that there is no need for genetic relatedness of components as a prerequisite (Figure 1.6). Although there is a strong tendency for organizations to evolve as elaborations of a single genetic lineage, the coevolution of very distantly related lineages to form single phenotypes or polyphenotypic entities is ubiquitous. The only real question is whether organizations may also evolve based on associations of intermediately related lineages. The occurrence of cooperative species of ants and mixed flocks of birds suggests that this may be possible, but generally, most entified organizations appear to be based on either very closely related lineages or very distantly related ones.

In most lower levels of the biological hierarchy, organizational transitions have involved components that associate to produce a single entity. The maintenance of close association across numerous generations has been termed parasexuality by Margulis (1981). Some systems may have components that are independent for various periods, but then fuse into single entities during specific periods (e.g. slime molds). Many authors have recognized a tendency towards centralization as an organizational trend (Rensch, 1960). This is probably true for all lower levels of

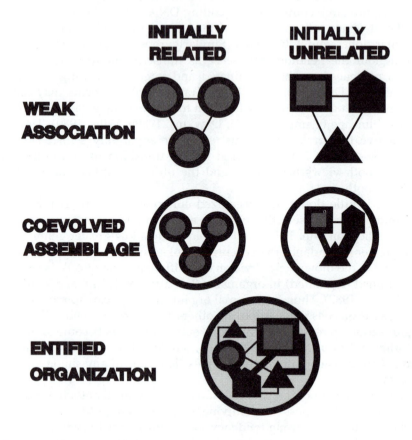

Figure 1.6 Convergence of systems composed of close relatives or distant species on a system with organizational attributes.

organization where single phenotypes are the rule, but this rule does not necessarily hold at higher levels. At the highest levels of organization that have yet evolved, entification may involve distributed sub-entities (e.g. ants in an ant colony), that never coalesce into a single unified phenotype. This is an important point because it means that an organizational entity does not necessarily have to reside in a single body, but may have a distributed existence. This will become an important point in Chapter 5.

In unorganized assemblages, local interactions may be competitive, exploitative (predation, parasitism), or selfishly cooperative. In organized systems, the interaction of parts requires global homeostatic regulation, otherwise imbalances would eventually become disruptive (Margulis, 1981). In such integrated systems, natural selection acts on the organization as an entity. If the organization is polyphenotypically distributed, selection can also act directly on lower levels to a certain extent but all levels below that of the discrete phenotype will be selected only indirectly by constraint. A key concept is that the interactions among components are themselves important objects of selection and constitute evolutionary units. Models reducing the system to individual components may capture the dynamics of unorganized assemblages (Huston, De Angelis and Post, 1988), but such approaches are unlikely to address evolved organizations adequately. Specific regulatory integration does not magically pop out of selection on individual components.

Reductionist views of evolution tend to emphasize the priority of selfishness of lower-level components as determining whether a particular state is evolutionarily stable. Thus, the evolution of positive benefits to a higher-order system is usually seen as unlikely if subcomponents can cheat and usurp benefits without paying for them (e.g. Bull and Rice, 1991). The history of life, however, illustrates the attainment of complex organizations like organisms, despite the fact that subcomponents retain some ability to convert into cheaters. The key that is missing in this interpretation is that when selection acts on the organization as a whole, it not only selects for positive interactions among the parts, it also selects for features that prevent cheaters from arising or defends against their invasion. Wilson (1977) articulated this in the context of group selection, where a selfish element lowers the potential fitness that could be achieved in the absence of the cheat. In this case, selection on other elements for improving fitness would also favour those able to resist or exclude the cheat. Progressively more complex overlays of organization have been consolidated despite continual threats from dissolution by cheats, and attributes of inter-holonic defence may be a ubiquitous hallmark of consolidated biological organization. The initial or continued existence of cheaters is not excluded (e.g. selfish DNA, cancer, meiotic drive, asexuality), but this does not necessarily preclude organizational advance and consolidation. Rather than accepting the real or potential existence of cheats as evidence against possible higher-order organization, the key criteria may be instead the existence of systems erected to minimize their impact.

Figure 1.6 outlines the stages in the evolution of a new level of organization as defined by attributes in Table 1.1. Despite considerable theory arguing for various prerequisites for organizational transitions, the discussion suggests that organizations are selected mainly for their global properties and these may be independent of the properties of the individual parts or their specific dynamics. Thus, organizational transitions may converge on viable solutions independently of the genetic relatedness of their components, or the kind of local interactions governing the initial association.

1.6.2 Selective value of organizations: fitness and filters

What are the factors that favour the evolution and selective success of an organization? What benefits do organizations provide that fuel the progressively tighter association of components towards entification? There are three key pathways favouring organizational evolution. Firstly, an association could improve the immediate fitness of participants, either by increasing the level and/or efficiency of available resources (which is eventually translated into improved reproduction), or by reducing mortality risks to improve the likelihood of reproduction or offspring survival. The defence of desert toads by tarantulas is probably such a case. The toads eat ants that prey on baby tarantulas, but avoid eating tarantulas because of their urticating body hairs. Adult tarantulas will actually confront snakes interested in a toad tidbit (Hunt, 1980). Wilson (1980) argues that interactions that indirectly elevate fitness in such a fashion can lead to organizational evolution, even among unrelated components.

Cooperation may also be favoured by increased transmission of genes contained in close relatives (inclusive fitness benefits), again either through modification of resources or mortality rates (Hamilton, 1964a,b). Kin selection can offset loss in individual fitness that might be involved in organizational participation, and this may be a crucial reason why higher-order entification often involves closely related organisms.

Finally, organizations may reduce variation in success over longer time scales, reducing the risk of failure and improving persistence of the participating lineages. The relevant dimensions are stabilization of functions and resources or avoiding conditions of high risk. Of these, most discussions have emphasized immediate or inclusive fitness. The importance of geometric mean fitness was introduced by Gillespie (1974, 1977) and Slatkin (1974) immediately recognized its relevance to life-history evolution. The concept of geometric mean fitness is similar to risk averaging in utility theory, where the costs of failure are greater than the rewards of success. For biological lineages, the ultimate proof of this is that zero fitness in any one generation represents permanent extinction, regardless of the degree of successful performance at other times.

The geometric mean (calculated as the nth root of the product of fitness in n successive generations) is the appropriate measure of fitness whenever there is temporal variation in the fitness of a genotype (Seger and Brockmann, 1987; Philippi and Seger, 1989). These authors point out, for example, that a genotype that produces nine offspring in good years, but one in bad years, has an arithmetic mean fitness of five. Another genotype that produces only five offspring in good years, but three in bad years has an average fitness of only four. However, the latter genotype has a geometric mean fitness of 3.87 compared with 3.0 for the former.

The genotype with the greatest geometric mean will win in the long term, even though it has lower arithmetic mean fitness.

Since all life consists of lineages with temporal variation in fitness, the appropriate paradigm is one where selection acts to maximize geometric mean fitness. In other words, selection may minimize the risk of extinction in lineages rather than maximize the performance of individuals. This can be achieved either via conservative genotypes that trade off a higher mean for reduced variance across generations (a bird in the hand is worth two in the bush), or by diversification of components which spreads the risk of poor performance within generations (do not put all your eggs in one basket) (Philippi and Seger, 1989).

Frank and Slatkin (1990) recognized two relevant phenomena here. Bet hedging (Slatkin, 1974; Stearns, 1976) reduces the variance of success, whereas risk aversion (Real, 1980a,b; Real and Caraco, 1986) increases average success. Thus, both the mean and variance contribute to long-term fitness (Real and Ellner, 1992). Bet hedging and risk aversion appear to be major factors in many organizational transitions, but remain relatively unexplored in this context. Addicott (1986) provides a discussion suggesting that mutualistic associations may stabilize the population dynamics of the species involved and also increase their ecological resilience.

The value of organizational progress might be best appreciated by considering highly advanced species. We might consider humans, one of the most successful mammals, to be a special case because of our greater mental abilities. However, it is undoubtedly our social organization that is the crucial ingredient. Perhaps as much as 40% of all terrestrial plant productivity is now harnessed by humans. Among insects, the most successful forms are eusocial species (ants, bees, wasps and termites). E. O. Wilson (1985) estimates that about one third of the total animal biomass in the Amazon rain forest consists of ants and termites. Moritz and Southwick (1992, p. 173) estimate that honey-bees consume as much as 11% of net primary production in tropical forests.

It is important to emphasize that a genotype that employs bet hedging or risk aversion may do so via more variable or diversified phenotypes. Thus, organizations may sometimes achieve greater fitness by increased phenotypic variance. Redundancy also fits in here because it is a mechanism that may spread risks in variable environments, even though deployed features or phenotypes may be similar. Such attributes may occur at any and all levels of hierarchical organization deployed by genomes (e.g. cells, organisms or multispecies associations) (Lacey et al., 1983). Thus single genotypes may be selected for diversification (Chapters 5 and 6) or similar yields may be harvested by separate genomes that deploy a mutually integrated organization (or multispecies phenotype).

A second important implication is the time scale of selection. Traditional genic perspectives consider that the only logical temporal unit of selection is a single generation (Salthe, 1975; Van Noordwijk, 1990). Models of geometric mean fitness have been applied to changes in gene frequencies and evoke a multigenerational time scale. If we consider that the genome may be constrained by organization emerging at numerous levels of organization, the time scale of a holon yielding the bet hedging or risk aversive benefits may also have a relevant time scale that spans generations. For example, Chapter 6 argues that species that deploy different phenotypes in successive generations cannot have a selective time scale of one generation. If so, the relevant unit of evolution becomes lineages, the time scale becomes multigenerational, and the relevant criteria for success is some function of both abundance and persistence.

A genic evolutionary model set in a uniform, constant environment leads to a simple maximization interpretation of organismal design (i.e. global optimization of adaptation). Bookstaber and Langsam (1985) pointed out that what is optimal in the long term may not appear optimal on short time scales.

Adaptations relevant to lineage persistence encompass forces for maximal short term success, but also require four additional categories. These include: (1) resilient features that allow return to particular states or trajectories following disturbance (e.g. repair, homeostasis); (2) plastic responses allowing deployment of different features or trajectories in different environments; (3) robust features that can accommodate a wide range of eventualities with little need for change (e.g. flight responses based on any sudden movement or puff of wind); and (4) adaptability (e.g. sexual reproduction, molecular mechanisms for evolvability, cognitive systems). These last two categories lie outside even existing models that take into account variance in fitness. Such models assume that variance conforms to a particular known distribution. In a non-equilibrium world, the nature of the variance itself will also change so that there is uncertainty of uncertainty. Successful lineages have evolved in a framework of compound uncertainty, which may be reflected in features that are not amenable to simple mathematical models.

A simple conceptual model for organizational evolution is provided in Figure 1.7 based on the concept of filters. In information theory, filters are devices that transform data to eliminate noise, amplify weak signals or remove those that are not of interest. If the environment is considered as some variable attribute of biological relevance, then organizations may be valuable in ways that can be conceptualized as filtering.

Within the hierarchical framework being applied here, the environment of a focal level includes everything external to it, including higher levels of organization. Thus, the immediate environment of eukaryotic cells is a

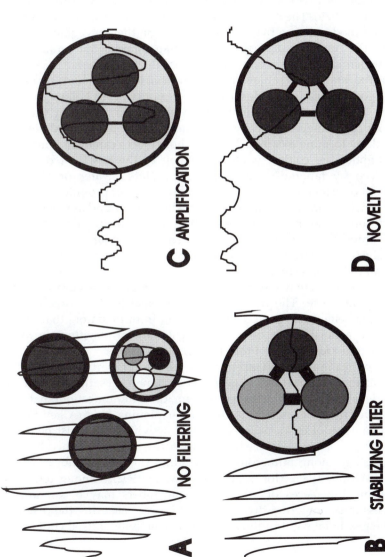

Figure 1.7 Filtering properties of organizations. Unorganized components have little ability to modify environmental dynamics in their favour (A). Organizational systems may be selected if they stabilize environmental variations, thus improving the long-term operating environment of the components. Organizations that reduce the impact of an unfavourable environmental aspect are likely to be selected for (B). Organizations that markedly improve utilization of advantageous environmental aspects (amplification), are likely to be favoured (C). Organizations that open the door to novel opportunities are likely to evolve (D). The concept of filters was developed jointly with J. Kolasa and N. Waltho.

C AMPLIFICATION

D NOVELTY

A NO FILTERING

B STABILIZING FILTER

tissue inside an organism. There is considerable evidence that organizations often function to increase the stability of the internal milieu. For example, consider thermoregulation in bees (reviewed by Moritz and Southwick, 1992). Free-living individual cells are far too small to generate sufficient heat to offset rapid losses because of their small surface to volume ratio. Note, however, that large bacterial populations can generate considerable heat in situations favouring local abundance and environmental insulation (e.g. compost heaps).

Individual bees can produce sufficient heat to elevate their flight muscles to 30°C and higher. Bumble-bees are particularly good at thermoregulation as a consequence of their larger size and furry insulation (Heinrich, 1979). The heat can even be shunted to the abdomen which has a bare brood patch so that the bees can sit on their eggs just as birds do. Thus, the organismal level allows for an internal temperature that is somewhat independent of environmental fluctuations.

Further stability is achieved at the eusocial level. By choosing nests that have good insulative properties (e.g. old mouse nests, hollow trees) or by using insulative building materials (wax in bees, paper in wasps), the colony can achieve relatively good thermal stability for the developing brood. By forming a compact swarm, honey-bees can even remain warm throughout the coldest winters. Cooling of the nest if it overheats is achieved by fanning, evaporating regurgitated water or altering the amount of insulation on the nest. Von Frisch (1974) provides an overview of animal architecture, the crowning achievement perhaps being the air-conditioned high rises of some termites that rival skyscrapers in relative magnitude and constructive design. Thus, one adaptive value of progressive organization in such animals is a stabilization of temperature and all of its associated benefits.

Organizations may also act to reduce the impact of a negative environmental influence. The transition to a colonial level of organization in *Volvox* is believed to protect the organism from predators that would be effective against single cells. Young colonies form within the parental colony and are thus also protected from predation.

Amplification or enhancement is also a property associated with organizational transitions (Figure 1.7). In honey-bees, the eusocial colony has developed a sophisticated system of communication that greatly improves the efficiency of finding and exploiting food. A key feature in the evolution of eusocial insects is that activities that must be achieved in a single correct sequence by solitary forms can be obtained by parallel processing among numerous workers and this is less subject to failure (E.O. Wilson, 1985; Moritz and Southwick, 1992). Among termites, re-ingesting the faeces of other colony members may be an important process improving digestive efficiency. Besides ensuring the distribution of their symbiotic gut protozoans, such an activity of itself could amplify the amount of nutrients gleaned from their food, particularly if fungi and symbionts continue to act on the faeces after they are deposited.

Finally, organizational transitions can free a system from some limiting factor or allow exploitation of a completely new resource or niche (Figure 1.7). Thus, transitions to larger multicellular organization may open the door for opportunities to prey on smaller organisms. Alternatively, the eusocial transition in ants allows them to prey on animals many times larger than any individual could subdue. McKinney, Broadhead and Gibson (1990) documented that a mutualistic association between a coral and bryozoan allows them to form upright structures and thereby escape limitations imposed by available substrates. Organizational transitions often open new niches because they provide a mechanism for changing scales. Partitioning of temporal and spatial scales is a key mechanism for niche specialization or diversification (Harris, 1986; O'Neill, 1989). Ruse (1993) emphasizes that innovations are often associated with transitions to new niches, and such innovations are often associated with organizational progress. A nice example reviewed by Moritz and Southwick (1992) is the cooption of thermoregulatory functions for defence in *Apis cerena*. Workers of this species kill large wasps by clumping around the victim and generating a heat load that is lethal to the wasp.

Many of the advantages provided by organizational transitions are interactive. Thus, greater internal stability can improve system efficiency. This in turn can allow the organization to be more competitive, invade previously inaccessible habitats, improve resource acquisition or help avoid risks. Thus, the value of an organization lies in its ability to filter the environment in a beneficial way compared with the function of a relatively unorganized system. The cost of the organization is associated with constraints arising from connectedness, specialization and interdependence and impacts mainly on subcomponents. If the organizational advantages outweigh the costs, then its evolution should be favoured.

1.6.3 Pathways to organizational entification

A global view of biological organization suggests that initial associations may be of four basic types that are not necessarily mutually exclusive: kin-based altruism; reciprocal cooperation; parasitism; or exploitation/ enslavement. The literature on the evolution of higher-level organizations via reciprocal cooperation and/or kin selection is relatively enormous (e.g. Hamilton, 1964a,b; Trivers, 1971, 1985; E.O. Wilson, 1975; Holldobler and Wilson, 1990). The other mechanisms appear to be the initial basis for numerous organizations, but they have been generally neglected in the theoretical literature. There are two key problems to consider. One is related to why entities might associate in an organization, particularly if free-living forms have higher immediate fitness. The second is related to the evolution of cheaters who may reap the organizational benefits without paying for them.

The reductionist program has stressed that only associations of mutual benefit to the selfish components are likely to provide a basis for further progress. Individual selection will dominate over the good of the assemblage unless there is something sufficient to offset it. In this context, the evolution of altruistic behaviour where a component sacrifices part of its fitness for others is problematic (e.g. asexual workers in social insects). The favoured explanation in such cases is that by helping others the component can offset its loss of individual fitness by enhancing the success of its genes that reside in relatives (inclusive fitness and kin theory) (e.g. Hamilton, 1964a,b).

Trivers (1971) first proposed a model of reciprocal altruism that does not require that participants are related. That cooperative behaviour may evolve among selfish components is wonderfully illustrated by a game theory model known as the iterated 'prisoner's dilemma' (Axelrod and Hamilton, 1981; Axelrod, 1984; Axelrod and Dion, 1988). The model is based on the situation where two prisoners have the option of implicating one another in a crime or refusing to cooperate with the authorities. A payoff table can be constructed where both prisoners will go free if they support one another, both will receive long prison terms if they both defect and one will receive a reduced sentence for ratting on his uncooperative accomplice. Although the greatest reward is obtained by both parties not defecting, the best individual strategy is to defect in the hopes of obtaining a reduced sentence.

This situation can be abstracted into a generalized reward table, and the game can then be played iteratively. In this case there is feedback between the players and each has the ability to reward or punish the other. The best strategy in this game is one known as 'tit for tat'. A player employing this strategy always cooperates unless the partner defects. In this case, tit for tat retaliates. As soon as the other party resumes cooperation, so does tit for tat (i.e. there is no spiteful memory). The key to such cooperation is partner fidelity. Bull, Molineux and Rice (1991) addressed this by examining the virulence of phage in bacterial populations where numerous hosts were available for horizontal infection, or in which the phage was limited to vertical transmission within host lineages. Phage associated with particular bacterial lineages were selected to reduce their impact on the host.

Although such insights are noteworthy, a hierarchical perspective suggests a larger framework might be necessary for complete understanding. For example, if bacteria contribute to the digestive efficiency of inhabited animals, it seems likely that phage would also be selected by their impact on the larger vessel, and by the resulting interactions of these inhabited animals within and among species. Such dynamics might be quite complex. For, example, Washburn, Mercer and Anderson (1991) found that parasitic cilia had a negative impact on mosquito populations with abundant food. Under food limitation, however, parasitized populations produced more and fitter adults because larval density, and consequently competition, was reduced. Similar unexpectedly positive impacts often occur in plants for which herbivory may amount to adaptive pruning (Vail, 1992).

A global view of biological organizations suggests that these paradigms of kinship or mutual reciprocity do not address the full range of organizational frameworks. For example, kin theory has no power to address organizations consisting of distantly related parts and reciprocity has limited application to interactions that are initially of a negative nature. Individually oriented models ignore the fundamental nature of organizations. Such models hold that cheaters will invariably replace altruists unless they can recognize their own type and/or discriminate against cheaters. Thus, individual selfishness appears inescapable. However, organizational evolution appears to involve the acquisition of mechanisms specifically deployed to exclude cheats or prevent their genesis. Such features are especially likely to evolve where there is an asymmetry of power among components, an aspect that is largely absent from reductionist treatments.

A second aspect relates to why entities might associate. If the association improves the performance of all parties, then there is little problem. Could organizations evolve, however, where the immediate fitness of a component is lower than its free-living relatives? Several lines of reasoning suggest that the answer is yes. Firstly, organisms engaging in an association may no longer be competing directly with their free-living relatives. Such a distinction is evident in the symbiotic association between the protozoan, *Paramecium burseria*, and the photosynthetic algae, *Chlorella*. The algae engaging in the association are not harmed, but free-living forms, even of the same species, may be eaten and digested (Margulis, 1971).

Proto-mitochondria that invaded ancient host cells entered a different niche than their free-living contemporaries. Unless there was very high exchange of these incorporated elements with free-living populations, the appropriate criterion for success is relative only to the new niche. For example, the replication rate of these entities inside host cells could be vastly slower than free-living relatives, but this in no way would imply that they would not persist (unless they disrupt the cell). In lichens for example, a rapid-growing fungus and fast-growing alga forge a symbiotic organization with a remarkably slow growth rate but high environmental tenacity (Margulis, 1981).

Thus, if an organization has sufficient survival value in its own right, it might evolve even if all of the components had lower immediate fitness (i.e. their rate of immediate replication), than their free-living relatives.

The loss of immediate fitness relative to free-living populations may be offset by enhanced long-term persistence (higher geometric mean fitness), or the ability of the partnership to exploit entirely different niches.

This leads to the second point. Entified organizations, and to some extent more primitive associations, are selected as higher-order units. Gause's hypothesis or the competitive exclusion principle holds that no two species that utilize the same niche in exactly the same way can coexist. Within a species, if all members are identical and utilize the environment in the same way, then only a single fittest type should evolve. This is a very simplistic view, however. Whenever components form an association, selection is not just on these components individually, but on the association as well. If the association uses the environment in even a slightly different way than do non-associated entities then competitive exclusion may be bypassed. The organization may not be displaced by free-living components if there is a niche shift.

Simple models of altruism predict selfish individuals win via competition. This occurs mainly when populations approach the carrying capacity of the environment. Recent theories of non-equilibrium ecology suggest that populations may be prevented from reaching carrying capacity due to the activities of predators, disease, stress or environmental disturbance. Under these conditions, organizations could persist even if they were somewhat less fit (slower growing) relative to components not participating in the collaboration. In such conditions, selection could greatly favour altruists if the organization acted as a filter allowing attainment of higher average carrying capacity over the longer term. A slower-growing organization that is highly stable and which is more efficient or competitive with regards to resources would also be expected to displace free-living subcomponents, even if these otherwise have the potential for realizing higher short-term rates of increase.

A crucial aspect of organizational evolution is selection of mechanisms that prevent the selfish activities of components from disrupting the system (i.e. substitution of global rules for local rules). If an organization has very high survival value in its own right, there will be very strong selection pressure for such mechanisms. Crossing the bridge from local to global rules is a crucial period and it is likely that numerous transitions fail. However, once consolidated, this new entified organizational base is available for elaboration in new directions (e.g. the morphological diversification of multicellular organisms).

The transition from cooperative assemblage to a new functional entity would occur as components lost their ability for independent existence. At this point, the functioning of the organization imposes global design constraints on the interdependent components, including adaptations to prevent any selfish activities or proliferation that reduce system efficiency or integrity.

A second avenue for organizational evolution is the possibility that one component is capable of enslaving others for its own needs. In this instance, the stronger or enforcing component is able to impose constraints on other components to derive some benefit at their expense. In the classic association of fungi and algae to form lichens, for example, the fungal partner appears to impose a controlled exploitation on the algal partner (Kendrick, 1991). Similarly, parental manipulation is an alternative hypothesis to altruism for the evolution of eusociality (Andersson, 1984). Human agriculture is an excellent example of enslavement, that in some cases has led to extreme specialization of subcomponents and loss of their dissociability from the imposed system. The incorporation of chloroplasts in eukaryotic cells may have occurred via the host engulfing proto-chloroplasts and enslaving them (Margulis, 1981).

A third route towards organization is via parasitic or predatory interactions (e.g. Thompson, 1982; Guerrero, 1991; Price, 1991). For example, some cell organelles may have originated as parasites that utilized the larger cellular environment as a filter to improve their own proliferation, while operating without constraint. Mitochondria probably followed this path (Margulis, 1981). Interestingly, I am unaware of any documented cases of mitochondrial cancer in modern organisms. In general, parasites tend to evolve reduced harmful impacts on their hosts (Price, 1991). Thus, most epidemic diseases arise via the introduction of a relatively novel pathogen into a host population. The same pathogen in its coevolved host is often relatively harmless or only mildly symptomatic. Examples include yellow fever, bubonic plague and the myxomatosis virus of rabbits. If such parasites also have the potential to improve their own success via some contribution to the host's function, it would be favoured by evolution and at some point the original selfish parasitism of one component could be transformed to a beneficial integration for the whole organization.

The pathway to organization is often associated with benefits accruing from stabilization (Addicott, 1986). Although relationships like predation and parasitism have local impacts that appear highly negative, there is good evidence that such interactions can lead to associations of increasing stability. Thus, Pimentel (1968) examined the population dynamics of houseflies and the wasp, *Nasonia vitripennis*, that parasitizes fly pupae. The dynamics of populations that had little coevolutionary experience with one another showed strong oscillations and little stability of numbers. Remarkable stabilization of numbers in both the predator and prey species emerged when coevolution was allowed.

Gilbert (1966) documented a case of a rotifer (*Branchionus calciflorus*) that produces a spined phenotype in response to a chemical produced by a predacious rotifer, *Asplanchna* spp. He argued that the adaptation would benefit the predator via stabilization of predator–prey dynamics, another example of Pimentel's genetic feedback hypothesis relevant to regulation of population size. In this example several features of higher-order organization are present (complementary specialization of parts, specific chemical information transfer and increased stabilization of the association). A complementary example where a predator uses specific information from its host is exemplified by the rabbit flea. The flea uses the host's own hormones to coordinate reproduction to coincide with the appearance of a new litter of rabbits (Rothschild and Ford, 1964).

Significantly, particular components may not have a choice about participating in associations where there are strong asymmetries and initial interactions do not need to be reciprocally beneficial. If the organisms are strongly associated with one another, organizational progress can proceed to fulfil most of the criteria of entification (e.g. yucca moths and yucca plants, fig wasps and fig trees). Thus, it appears to matter little how the original association leading to organization arises. If the association has potential advantages, or the interactive system performs like an entity, organization is likely to be consolidated, effectively harnessing the components and constraining their future evolution to avenues promoting the persistence and success of the new functional system. Once consolidated, the higher level of organization can evolve even in directions that are not necessarily in the best interests of particular components.

1.6.4 Organizational hierarchy: climbing the helical ladder

The evolution of biological organization began perhaps four billion years ago. The first stages involved the gradual accumulation of complex, stable organic molecules and was initiated when some of these molecules attained replicating properties. Various elaborations were selected that improved the accuracy, rate and efficiency of replication. The replicating molecules extended their reach to other molecules, notably proteins, that of themselves had no replicatory power, but improved survival and evolvability. Cycles of such biochemical evolution added successive levels of complexity (Eigen and Schuster, 1979). Numerous key transitions followed (cell membranes, cytoplasm, cell walls).

There are numerous textbooks devoted to this primordial evolution (Margulis, 1981; Loomis, 1988; Kauffman, 1993), but this lies outside the scope of the current discussion. Consequently, numerous, crucial early transitions are treated here as two simple steps, harnessing of proteins to DNA via RNA intermediates and the subsequent elaboration of primitive cells (Figure 1.2). In fact, most of the history of life was associated with these early stages.

The earliest cells, the prokaryotes, lack nucleic and have a simpler more diffuse organization than the more sophisticated eukaryote cells.

Eukaryotes have their DNA centralized within a nucleus, and they contain complex organelles such as mitochondria and chloroplasts. These organelles are believed to have originated as independent prokaryotes that gradually became obligatory components of the eukaryote organization (Margulis, 1981). Chloroplasts may have derived from cyanobacteria-like ancestors whereas the cytochrome c of mitochondria is similar to that of the purple eubacteria (Woese, 1983; Gray, 1989). Coevolution among the various components proceeded to increase their complementary integration until a point was reached where these units were forged into a new, non-dissociable organizational entity – the eukaryote cell.

The requirement for a system of communication between the nucleus and cytoplasm may have been a critical preadaptation for the evolution of the next stage of organization, multicellularity (Raff and Kaufman, 1983). A simple extension of such capabilities would allow cells to communicate with one another. It is quite possible that the nucleus also serves to sequester the cell's DNA away from mitochondria which generate potentially harmful free oxygen radicals (Chapter 9). Buss (1987) explored the evolutionary resolution of the conflict between the self-interests of cell lineages and the stable existence and efficiency of multicellular organisms. Buss (1987) suggests that during the early stages of organismal evolution, cell lineages could act selfishly (e.g. cancer) or compete for access to the germ line. Such activities would destabilize organismal integrity. He suggests that adaptations such as very early sequestering of the germ line are organizational adaptations that close the evolutionary gates on possible renegade cell lineages and consolidate the organismal level of integration.

Once consolidated, much more elaborate adaptations could be built on this new stable foundation. Although the details may be debatable, this idea that early stages in the evolution of new organizations is a period of instability is undoubtedly true. The new organization is open to invasion by components that can obtain short-term gain at the expense of the long-term stability and yield of the higher-level organization. Examples of such cheaters can still be found in primitive systems that form dissociable or weakly integrated colonies or temporary cell aggregates such as in the slime molds (Bonner, 1974; Buss, 1987).

If a new level of organization has high selective value, the bridge from the next lower level or organization may be crossed numerous times in evolution, but this bridge may be relatively flimsy. Depending on the value of the higher level for long-term survival, however, adaptations to improve the integrity of the organization and protect it from dissolution will be strongly favoured. At some critical point, the level may be consolidated and subsequent evolution will act directly on that level, and on underlying levels only via constraint. With increasing specialization, integration and interdependence of components, the option for autonomous action is revoked. Just as various components of the eukaryote cell differentiated into more specialized components subject to interactive control, the evolution of organismic complexity has involved an elaboration of an increasing number of specific cell types.

Gould (1989a) rejected the idea that evolution has proceeded by steady improvement and diversification of a few simpler ancestors. Instead, he documents that numerous fundamental body plans were generated in a Cambrian explosion of diversity, with only a small subset of these Bauplans eventually elaborating existing organisms. He suggests that there may have been an unpredictable element determining which lineages persisted, in particular, survival during extinctions possibly being related to fortuitous features that were not a key aspect of adaptation to previous conditions. A similar phenomenon has shaped the evolution of microcomputer and VCR technology. However, the pattern of initial proliferation followed by success of a few does not dismiss the exponential progress in the technology of these devices. Nor does the continued market for calculators (i.e. technological prokaryotes), despite the advent of more powerful machines, mean that progress has not been taking place.

In this respect, Gould (1989a) asks why, if multicellularity is so advantageous, the transition lagged behind the evolution of single-celled eukaryotes by perhaps 700 million years. The answer may be that the transition to multicellularity is fraught with problems associated with the control of competition for the germ line (Buss, 1987), and magnification of the impact of Muller's ratchet (Muller's ratchet refers to the accumulation of deleterious mutations which is magnified in lineages with larger genomes and smaller population sizes). The latter may have required the consolidation of effective sexual reproduction and improved DNA repair/protection (see Chapters 5 and 12). A simple explanation could also be that sufficient oxygen was required to support mutlticellular entities. The rapid proliferation of numerous basic body plans during the Cambrian (even more diverse than at present) is consistent with the expansion of a newly consolidated level of organization into unoccupied niches at larger environmental scales. Gould (1989a) provides a complete discussion of alternatives.

The process of differentiation largely involves the interpretation of positional information in the organism, switches controlled by cellular level clocks and induction of one cell lineage by others. These processes required changes in genetic structure to allow cells that all contain identical genetic code to activate and repress various programs selectively in particular cell lineages. The evolution of morphogenesis, although represented separately in Figure 1.2, probably evolved in parallel with processes of cell differentiation. The two are represented separately because the differentiation of cells into particular tissues and the shaping of these tissues into a diversity of physical structures are potentially distinct and dissociable processes (Lovtrup, 1974; Raff and Kaufman, 1983; Arthur, 1984; Lawrence, 1992).

The genetic organization responsible for the integrated control of differentiation and morphogenesis involves structural genes, integrated cell batteries or gene nets (Britten and Davidson, 1969, 1971; Davidson and Britten, 1973, 1979; Bonner, 1988) or blocks of genes contained in regulated domains (Bodnar, Jones and Ellis, 1989). Development proceeds via a cascade of binary switches associated with regulatory genes (Raff and Kaufman, 1983; De Pomerai, 1990; Kauffman, 1993). The structure of the genome constitutes a hierarchical organization that reflects the increasing organizational complexity of the organismal phenotype. Relatively small changes in regulatory genes can lead to radically altered morphologies. Consequently, the diversification of multitudes of species of organisms can be obtained via relatively minor genetic shifts. Moreover, much of this diversification is independent of the basic bio-chemical capabilities of the species. Thus, phylogenesis among lineages of organisms largely represents changes in the packaging of a relatively conserved biochemical machinery.

The evolution of sexual reproduction has been an enigma because it appears to involve very high costs (Chapter 5). Specifically, the repro-ductive contribution of an organism to the next generation is halved by sex (because each offspring only contains half of each parent's genome). Moreover, the organism's identity is rapidly diffused in subsequent generations, and where males are involved, half of the population may make little contribution to production of offspring if parental care is low. Sexual reproduction makes perfect sense, however, as a feature that inte-grates organisms into a higher-order organization. Sex provides a genetic conduit for communication, and organismic components simultaneously lose their option for independent action.

As in other organizational transitions, specialization (via sexual selec-tion in this case) ensures complementary integration of non-dissociable, specialized components (males and females). The organization that is formed is essentially a lineage, and the ultimate integration of subspecific lineages constitutes a species. The selective value of the lineage is associ-ated with long-term persistence as opposed to the immediate short-term maximization of organismal fitness assumed by the traditional adapta-tionist program. It is interesting that genic reductionism owes its mandate to the fact that the sexual transition supposedly negates organisms as evolutionary entities. This view holds that organisms have no long-term identity but are temporary holding vessels for genes. Thus, genes must be the appropriate units of evolution.

The hierarchical view suggests that the truth is in the opposite direction. Sexual reproduction adds yet another overlay of biological organization forging organisms into a cohesive, communicating system (Figure 1.2). The idea that sex gives the genic view a mandate is incorrect. Sex adds yet another higher level of organizational constraints on genes. The search for some sort of individual-level selective advantage to explain why organisms engage in sex may be fruitless. Sex may be advantageous at the lineage level, and consolidation of this organizational transition could impose features that are not in the best interests or of immediate relevance to organisms. Thus, sexually reproducing species may be considered as a new organizational transition that might be better regarded as a meta-organism.

Even further levels of organization have been added in some lineages. Two major pathways are available. One involves elaborations involving genomes derived from a single species (eusociality) and the other involves forging together unrelated genomes. Both avenues have been explored. In numerous species, symbiotic organismic associations have evolved. In these cases, what is apparently an organism is actually an integrated community. In fact, many 'species' represent integrated communities of species that inhabit a single body. For example, cockroaches contain symbiotic bacteroids that live within cells of the fat body. These bacteroids allow roaches to metabolize uric acid stored in the body for amino acid synthesis. This allows cockroaches to store and re-utilize nitrogen, an ability lacking in most other organisms. The bacteroids are transmitted directly with the eggs. The more sophisticated descendants of the cockroach, the termites, similarly use protozoans in their guts to enable them to digest wood.

The literature on symbiosis is replete with examples of such associations, including lichens and the cellulose-digesting gut flora of ungulates. A striking case is the deployment of viruses by parasitic wasps which neutralizes the immune responses of victims (Price, 1991). The degree of coevolution in such associations is variable, but in highly developed systems the relationship is obligatory, there has been strong complementary specialization and the free reign of one or either component is constrained by regulatory interfeedback (e.g. Schwemmler, 1991; Tiivel, 1991). As in other transitions, the higher-level system could be approached along avenues of parasitism, cooperation or enslavement.

In eusocial lineages many offspring do not become reproductively functional, but serve in specialized roles that improve the success of their mother. Moritz and Southwick (1992) suggest that the term superorganism should be reserved for organizations where individuals specialize in reproductive *versus* non-reproductive functions. Examples include termites, some aphids, ants, bees, wasps and naked mole rats. The genome in most of such lineages is organized to differentiate specialized adaptive phenotypes via inter-organismal regulation (pheromones), although in primitive systems dominance hierarchies also serve to communicate organizational information. The main hypotheses for this evolutionary transition are based on parental manipulation (enslavement) or cooperation (kin theory and altruism) (Andersson, 1984; Holldobler and Wilson, 1990; Moritz and Southwick, 1992). Although such hypothesis are areas of intense debate within the areas of interest concerned, from an organizational perspective, it is clear that the end result could arise via either pathway.

Although some eusocial organizations retain some degree of dissociability (e.g. multiple foundresses in polistine wasps), others have progressed to a level where functional independence of individuals is entirely precluded. For example, there are no known species of solitary ants. Although there is evidence for conflict between individual and colony levels of selection even among the ants, some species have progressed to a point where workers lack ovaries (E.O. Wilson, 1985). Reproductive revolts are no longer an option. As in other transitions, the organization is characterized by specialization of parts and regulatory communication among components. In some eusocial systems, transitions 8, 9, 10 and 12 are all evident (Figure 1.2). For example, individual termites are functionally integrated entities consisting of insects and protozoans, and sexual lineages of termites are also eusocial. Such compound organisms may become very complicated indeed. The termite protozoan, *Myxotricha paradoxa*, is itself actually a conglomeration of four symbiotic organisms (Margulis, 1981). Such eusocial systems are often further associated with separate populations of other species (e.g. fungi). Ants that tend acacia trees may also be mutualistically associated with homopteran insects on these trees (Thompson, 1982). In fact, eusocial insects appear to have an unusually high degree of symbiotic and mutualistic associations, a trend which requires some theoretical assessment.

Populations may be composed of a single, sexually integrated lineage or groups of variously discrete subspecific lineages. Such lineages may entail simple organisms, compound organisms or eusocial colonies. Populations may represent simple assemblages, but there may also be organizational elements and specialization of components. Thus, fungus associations with various insects, such as the parasol ants, represent an obligatory association of separate populations (Kendrick, 1991).

Within sexual lineages communication among components may be developed to support social organizations that improve defence, food acquisition, reproduction, etc. It is at this level that cognitive processes and communication coalesce to provide a higher-level information system that may have considerable independence from that of the genome. Specialization of parts may also occur, but this is most often overlaid on the primary specialization already evoked by sexual reproduction. Thus, larger size and aggressiveness of males frequently serves multiple functions (e.g. mate acquisition, family defence, food acquisition in predators and social defence as in baboons). Alternatively, specialization may be largely behavioural with little morphological divergence (e.g. humans, honey-bees). The transition at the population level (and in eusocial species at the lineage level) to societies is a crucial one from a human point of view.

Cognitive information systems allow organizational developments that are no longer strictly genetic, but the kinds of phenomena associated with organizational transitions remain (e.g. behavioural or morphological specialization of parts, communication among components, loss of dissociability and defence against cheats). The basis of this type of organization on a new higher-order information system also frees the system from requiring any high degree of genetic relatedness among components. The point that organizations are somewhat independent of the best interests of their components is nicely illustrated in human societies in that historically governments have been more often based on exploitation (e.g. slavery) than on cooperation.

The current convergent evolution of democracies and communist systems also suggests organizations have selective forces that override the selfish interests of individual components. Those living in 'free' societies need only withhold their income taxes, refuse to serve in the armed forces, or decline jury duty to experience the strength of organizational constraints on personal freedom. That humans are not the only species to evolve social systems that are somewhat independent of relatedness is illustrated by the system of nestmate recognition and food sharing in vampire bats (Wilkinson, 1988).

There is little problem in identifying the existence of organization at the level of organisms downward. Organizational transitions above the level of symbiotic composite organisms, however, involve distributed individuals. Such systems can be regarded as entities without walls (corrupted from Holt, in Wilson, 1990). Because the organization is dispersed over space and time it is more difficult to conceptualize and quantify. Eusocial species and slime molds illustrate that such systems can still represent integrated entified organizations. As we move to even higher levels, the question of organization becomes hotly debated.

Probably the central question in community ecology recently has been whether there are coevolved functional assemblages that limit community membership, or whether communities represent assemblages governed only by local interspecific rules (Underwood, 1986, Roughgarden, 1989). A related question is whether communities reflect long-term evolutionary stasis and stable species composition, or whether species constantly drive one another's evolution and species composition frequently shifts (Stenseth and Maynard Smith, 1984). The assembly/armsrace perspective appears to dominate, although there are many arguments suggesting that organizational evolution should be expected at the community level (e.g. Ulanowicz, 1980; Moore and Hunt, 1988). Goodnight (1990a,b) demonstrated the effectiveness of community-level selection in the laboratory. Key insights from these studies include: that the main features selected involve interactions among species (competing *Tribolium* spp. in this case); that the integrated response derives from genes distributed among genomes; that the integration is not detectable from examining isolated populations; and that the relevant selective forces were ecological pathways that do not contribute to individual selection.

Kauffman (1993) expanded his NK models to consider the linkage among species (S) where some proportion of components (C) within each genome affect one another. Thus the internal dynamics of one species is determined partially by interactions with others. Simulations of such NKSC networks suggest that each species might rapidly attain and be held at some local stable conformation (i.e. stasis). Large perturbations, however, could lead to avalanches of reorganization within and among species. Thus, the models predict periods of stasis punctuated by periods of rapid change. Alternatively, some parts of the global network may be frozen in stasis while others change freely. A crucial conclusion of such analysis was that stability characteristics of communities might emerge from metadynamics of coevolution that push the system into the complex realm poised between order and chaos (Kauffman, 1993). In other words, the stability relations of the system might be selected on system-level scales.

The controversy regarding the reality of community-level organization may partly stem from differences in perspective. In nearly all documented cases, the reality of ecological circuits has what might be called vertical rather than horizontal structure. Thus, goldenrod plants have several highly specific gall insects, each of which has its own specific parasites and hyperparasites. The elaboration of producer organisms with guilds of fungus, animal herbivores, pollinators, parasites or perhaps dispersing agents constitutes a vertical organization or at least a proto-organization.

Growing next to the goldenrod might be a burdock plant with its assemblage of stem-borers, leaf miners, seed predators and pollinators. Alternatively the neighbour might be a milkweed with its well known insect associates that utilize the plant's defensive chemistry as one aspect of their extended community connectance. The picture might be somewhat garbled by generalists like honey-bees that utilize all three species of plants, but the vertical associations are well documented and relatively stable.

What is less obvious is whether plants such as goldenrod, milkweed and burdock might be subject to organizational integration among themselves. This would constitute a horizontal aspect of community organization. Community ecologists, perhaps because of their interest in succession, have focused almost exclusively on horizontal organization (i.e. among competing plants at a single level of trophic structure). It may well be that community superstructure constitutes the horizontal competition of vertical organizations. However, horizontal organization might still be expected where vertical components have inverse patterns of resource utilization, waste products from one system are beneficial to another, or the organisms can benefit from the overlap or sharing of deployed filters. Thus, one might ask whether understorey plants like trilliums or jack-in-the-pulpits are simply competitively adapted to low light levels in woodlands, or whether they benefit from the ameliorating influence of the forest canopy on levels of light and temperature. Organization is possible if such organisms were configured so as to benefit the trees as well.

It is probably rare that a species occurs in communities where it has little coevolutionary history. However, the colonization of islands or the introduction of alien species to ecosystems are situations where such associations do occur. Organizational transitions may occur, however, where different species coevolve to become mutually interdependent and ultimately, obligatorily linked. Thus, the parasol ants and their fungus gardens represent an obligatory association where parts are specialized via complementary coevolution and independent existence is not possible. Acacia trees and their ant associates are another example with perhaps some remaining dissociability. The symbiotic relationship between the caterpillars of some butterflies and ants is stamped with the clear trademark of organization: the caterpillars emit specific signals that call ants (DeVries, 1990). Other examples include the yucca moth and the yucca plant, and fig wasps and fig trees.

A critical aspect of organizational transitions is the presence of regulatory communication among components (Bonner, 1984). Although not investigated with this perspective in mind, it is apparent that many species respond to information from other species to engage coevolved adjustments in behaviour, morphology, physiology and life history. Thus, numerous species of prey produce phenotypes defensively adapted to the presence of specific predators, utilizing chemical cues directly associated with the presence of their predators (Chapter 6).

Piper ant plants are remarkable in that they produce lipid and protein rich cells that provide food for symbiotic ants. Moreover, these cells are produced only in response to a specific species, *Pheidole bicornis*, or to a beetle that has broken the organizational code (Letourneau, 1990). Such examples illustrate a pervasive trend for genomes to acquire capabilities that complement associated species and to regulate deployment of these capabilities based on specific signals. It should be recognized, however, that information exchange need not involve a chemical specifically designed for communication (i.e. pheromones) but could simply involve the complementary exchange of fluxes involved with resource acquisition and excretion. In many systems expressing specific information exchange, the components are both sexual species with polyphenotypically distributed organization and individuals need not be constantly associated.

Predator–prey systems are perhaps not the best examples to use to argue for organizational evolution. Undoubtedly similar communication and adjustment occurs with respect to mutually beneficial relationships, but ecological studies have tended to focus on the negative interactions, specifically competition and predation. Thus, much more information is available on such interactions. There is no doubt that at least some coevolved populations have reached a level of organizational entification. Thus, the question is not whether community-level organization exists but to what extent various species engage in ecological circuits that can be regarded as superorganisms (Wilson and Sober, 1989). In many such systems, recognition of evolved organization may be overlooked because the relevant features may only be expressed under specific conditions and the various species may also function in other associations where such attributes are not expressed.

In this regard, Janzen (1985) argued that an alternative to specialized coevolution among species is a flexible 'species fitting' achieved via robust or plastic attributes. Thus, some species are found in diverse communities without displaying significant changes. There may be a continuum of strategies spanning narrowly specialized to robustly generalized types.

Another factor is the length of time available for organizational evolution. Communities are likely to evolve on very slow time scales. Higher-order organizations may still be in the process of evolving. Alternatively, in a non-equilibrium world, disturbance may continually disrupt organizational evolution at higher levels. In that case, greater organization and more entified population circuits should be expected in older, more stable communities (i.e. tropical rain forests, some marine communities). This appears to be true. Organizational evolution at higher levels could still progress in the face of disturbance, but lower-level holons must become more sophisticated and provide a foundation less amenable to disruption. My own feeling is that sexual species reflect such an organization.

In this context it might be imagined that non-equilibrium environments may even drive organizational evolution because of the ameliorating role of derived filters. Graham-Smith (1978) makes the point, for example, that organisms persist by maintaining a complex flux with their ever-changing environments. If so, very stable habitats might actually select for simplification. Parasites represent one possible example. On a global level, however, such simplification is only possible because parasites hide behind the filters of their hosts (Kolasa, personal communication).

Connell (1978) and Moore (1983) argue that maximal species diversity occurs at intermediate levels of community disturbance. At high levels of resources and low disturbance, a few highly competitive forms may dominate, whereas other species may persist in non-equilibrium environments that reduce the effectiveness of competitive displacement (Wiens, 1977). Severe stress, low resources or disturbance may also eliminate all but a few robust species. Thompson (1982) concluded that intermediate levels of environmental disturbance might promote mutualisms, which might be envisioned as selection for damping filters. The ultimate truth concerning what circumstances may favour organizational advance remains obscure.

A fundamental ecological question is whether greater community complexity might yield greater stability and improved persistence of component populations (Rahel, 1990). Such a feature would argue for the existence of higher-order filters amenable to organizational evolution. The belief that greater complexity is associated with greater stability at one time enjoyed the status of a folk law (Cohen and Newman, 1985). The simplicity of this view was radically challenged by May's (1974) mathematical analysis suggesting that larger systems are inherently less stable.

May (1974) used systems of models based on the logistic equation to simulate the dynamics of interacting predators and their prey or competition among species in systems with increasing diversity. The results were resoundingly clear: more complex systems were less stable than simpler ones. The crucial elements of complexity identified by May (1974) were the number of components, the number of non-zero connections among them, and the relative strength of the interaction coefficients (the latter is missing from Kauffman's (1993) models). The general conclusion was that systems of greater diversity must have decreasing degrees of connectance to maintain stability. As connectance increases beyond a critical threshold, instability may suddenly appear. In fact, the idea that interactive systems become critically unstable as they grow has been formulated into a general theory of 'criticality' that has applications to processes such as earthquakes, economics, ecology, genome organization and turbulence (Bak and Chen, 1991; Kauffman, 1993). Kauffman's (1993) NK models derive the same predictions as those of May (1974): stability declines as the number of components and/or their degree of connectance increases.

The real contribution of May's (1974) modelling was to demark clearly that increasing complexity, although a property of organizational advance, is not equivalent to organization. May (1974) recognized that real communities do appear to have an association between increasing complexity and stability, but he showed that such properties do not mystically materialize with increasing diversity. The opposite is expected. Consequently, natural communities must represent a rather restricted kind of complexity that is amenable to maintaining stability (May, 1974,

p. 173; Cohen and Newman, 1985; Kauffman, 1993). If so, then such attributes must have been derived by natural selection acting at the level of such integration. May (1974) even suggested one possible mechanism: the transition from a stable equilibrium to limit cycles of increasing amplitude would amplify the risk of extinction for components. Associations with unstable parameters have lower geometric mean fitness and could be selected against by natural selection. Gilpin (1975) has extended such an interpretation.

Organizations entail specific connectedness that involves regulatory feedbacks. If one were to throw together random assemblages of pond organisms with no coevolutionary experience, it is likely that local extinctions and chaotic population dynamics would increase with greater initial diversities. Such a test (comparable to May's analysis) does not address whether real pond communities with extensive coevolutionary experience might represent systems with some degree of organizational development. Analysis of real communities suggests that connectance declines as diversity increases, a finding consistent with the idea that selection has acted on stability criteria (Yodzis, 1980).

Significantly, Yodzis (1980) suggested that the crucial feature yielding reduced connectance was the local integration of feeding guilds (i.e. vertical organizations) with weaker connections horizontally. Thompson (1982) also recognized that most strong connections among species are associated with feeding guilds and their hosts. Such results are consistent with the hypothesis that hierarchical organization is a crucial feature allowing highly complex systems to maintain stability. Locally strong integration may be restricted to particular compartments, reducing the overall connectance of the system and among higher-order levels of interaction (Moore and Hunt, 1988; Pahl-Wostl, 1990). Pahl-Wostl (1990) suggested that maximization of temporal organization may involve minimization of niche overlap among competitors. Recent evidence suggests that even connectance within food webs may be relatively feeble. To persist, herbivores must ultimately exert weak or neutral impacts on their hosts. This appears to hold (Lawton, 1992; Paine, 1992). By also impacting on the competitors of a host, herbivores may even exert a net positive impact.

Wilson (1980) pointed out that May's (1974) models provide no avenue for interactions among species that might raise the local carrying capacity and indirectly select for positive integration. In fact, logistically based models predict that symbiosis and mutualism are inherently unstable because such interactions drive exponential growth in both parties (May, 1974, p. 224). An ecological circuit providing indirect benefits as envisioned by Wilson (1980) might well be viewed by natural selection in its own right. In the extreme, such a framework would suggest the reality of superorganisms, particularly in settings where components are closely associated with particular ecological niches such as carrion, or communities of bark insects (Wilson and Sober, 1989). Thompson (1982) envisioned that mutualisms yielding improved access to resources would be particularly favoured in stress environments.

In biotically rich habitats, Thompson (1982) observed that a high degree of mutualism is associated with antagonistic interactions among species. In such cases the mutualist is involved in an alliance with one party that ameliorates the antagonistic impact of another species. Such a configuration could be viewed as reducing the interaction strength of the antagonistic linkage, a factor that May (1974) identified as one avenue for promoting system stability. Dodds (1988) developed a model that suggests that communities should tend towards connectance structures where negative impacts are diffuse and weak, whereas positive interactions might be stronger and more direct.

None of the models tested so far embody all of the features that appear crucial for understanding community integration and stability. What is needed appears to be a model that considers the number of components, the degree of connectance, the strength of such interconnections, and perhaps most importantly, the structure of interconnections. A crucial component is incorporation of how the strength and degree of connectance changes with density and diversity, a feature that cannot be captured by models of components that have simple on–off status where connectance is treated as a constant. Biological systems are characterized by density-dependent, frequency-dependent and stress-dependent selection such that the strength of interactions may not only change, it may reverse sign. Kauffman's (1993) NKSC models underscore a further complication. Complexity and stability across levels is probably interrelated. Thus, genome structure, organismal complexity, intraspecific interactions and interspecific integration are undoubtedly interrelated, and any successful model might have to assume such a hierarchical framework.

The picture that emerges is that species within communities may well be engaging in ecological circuits that reflect organizational evolution. Other peripheral species may have weaker associations such that they may contribute to the core organization, but may dissociate and be replaced by other species in different environments. Thus, flowers may be fertilized by different insect pollinators at different times or places, and particular pollinators may not be limited to single species of plants. Similarly, ant–plant associations may vary from those that are highly specific to relatively loose interactions (Schemske, 1982; Thompson, 1982). Does this mean that no organizational evolution has occurred? A particularly telling example is that of the mutualistic association of mycorrhizas that facilitate nutrient uptake by the roots of plants. Much of the success of terrestrial plants hinges on such associations (Lewis,

1991). Endophytes have an obligatory requirement for a host, and their densities may be regulated by their host. However, such endophytes commonly display a lack of host specificity (Law, 1988; Kendrick, 1991). Such circuits, where more than one combination derives the same result, are similar to what Kauffman (1993) termed canalysing functions with respect to genomic interactions. Kauffman (1993) concluded that the prevalence of such circuitry may reflect selection for system stability.

In some sense, a system that has evolved general purpose plugs and sockets might still be considered in an organizational framework. It seems very likely that higher-level population circuits would interact with other species or other circuits so that the organization might grow, or form even higher orders of integration. High connectance tends to be unstable because perturbations are rapidly transmitted throughout the system. In a non-equilibrium world, it may well be that organizational selection would take the form of compartmentalized or weak linkages spread among components (i.e. general purpose plugs and sockets). Such a situation, however, does not necessarily reject the existence of real organization underlying such a system.

Such discussion points out that detecting possible organizational evolution at higher levels may be obscured by the sheer complexity of interactions. In many well established cases, the relative units of integration represent three or more species (Thompson, 1982). At each trophic level the sign of interaction between species associations changes (i.e. herbivores have negative impacts on plants, but predators of herbivores represent positive interactors as far as the plants are concerned). Specific signalling between plants and the predators of their herbivores would be indicative of organizational evolution. Corn plants, for example, produce a chemical that attracts parasitic wasps that prey on herbivorous caterpillars. Significantly, the plant does not produce this material in response to just any injury, but only when the caterpillar's saliva is detected in the wound (Turlings, Tumlinson and Lewis, 1990).

The information content of specific signals is highly relevant to whether communities are randomly assembled or not. A suggestive study was performed by Dixon and Payne (1980) who baited traps with six volatile extracts from the southern pine beetle and its host in various combinations. Fifteen insect species known as predators or parasites of the beetle were caught. An additional 13 species that were food competitors, scavengers or mycetophytes were also collected. Such experiments argue for a relatively high degree of specific connectance in vertical community circuits (Schoonhoven, 1990). Norris (1990) points out that each message acts in multiple ways. Thus, a chemical produced by species A may act as an allomone for species B (invokes a response in B that favours A), but as a kairomone for species C (evokes a response by C that is unfavourable for A). A single chemical may be indicative of favourable habitats to one species, but may be interpreted as aversive for others (Norris, 1990).

The most radical extension of organizational ideas (and the most controversial) is Lovelock's hypothesis that life on earth is structured to alter the environment for its own support on global scales (Lovelock, 1979, 1988; Barlow, 1991). That is, the earth's biosphere, soils, oceans and atmosphere are modified to support the common interests of a single organizational entity that Lovelock called Gaia after the Greek earth goddess. Most mainstream biologists reject Gaia as a level of entification because natural selection only works when there are alternatives to select from, and because individual selection would be expected to act more strongly than any forces favouring global organization. The current deleterious impacts of humans on the biosphere, and our measurable alteration of key environmental features such as carbon dioxide, ozone and pH are probably proof of this. It remains, however, that most recognize the value of a perspective emphasizing global-scale interactions (Baerlocher, 1990).

A major criticism of Gaia is that it requires particular species to assume conformations that mainly serve to benefit others. Natural selection might only favour such attributes if there are sufficient indirect returns to the participating components, or if those generating imbalances select their own demise (e.g. possibly ourselves). Gaia is rejected because it is viewed as requiring some sort of cooperative planning. In this respect it seems rather profound that one of the first proposed applications of genetic engineering was intended to alter the impact of frost on crops.

Although Gaia may not currently exist as a functional entity, she may have been recently born by the fusion of naturally evolved DNA with the seed of human thought. It is difficult to imagine what several millennia of genetic engineering might add to her stature. Even conventional agriculture has already reshaped the surface of the earth and its biota. The success of such a process will require that individual components do not behave with complete selfishness. In particular, the brain of the system must come to terms with regulating its own size and activities. Regardless of the value of intelligence, cancer of the brain must ultimately lead to dissolution.

The idea that there are inherent evolutionary forces driving progressive evolution of life has been almost universally rejected (Williams, 1966a; Benton, 1987; Ayala, 1988; Hull, 1988; Ruse, 1988, 1993; Gould, 1989a; Maynard Smith, 1991). Such a formulation smacks of vitalism, and it has also proved extremely difficult to derive a working definition of progress. While it is unlikely that there might be significant pressure for bacteria necessarily to evolve into a higher or more complex form (bacteria may be nearly perfect at what they do after 3.5 billion years of evolution), it remains that hierarchically structured systems tend to convert upper holons into entified organization under appropriate conditions. Because the consolidation of organization at one level then allows organizational evolution to proceed at the next (and not before), there is an element of progressive organization driven by an intrinsic organizational ratchet. There is no need of lower-level subcomponents to increase in complexity; rather, they may be selected for simplification or specialization instead. One of the most exciting questions in biology is not whether there has been organizational progress in life, but to what extent entification has proceeded.

1.7 CONCLUSION

The emphasis of this chapter has been to show that there are interesting and essential questions to ask at every level of biological resolution. Moreover, real understanding requires a perspective that spans the entire range of temporal and spatial scales, from the molecular to the community. A unifying vector cutting through all of these levels is organizational evolution. The reality of organization is undeniable at lower levels of entification, but remains vaguely understood or even vehemently denied at levels of species, their interactions or community integration.

A hierarchical framework suggests, however, that entification can be expected across numerous levels and the evidence shows that in at least some cases transitions have been consolidated. A crucial conceptual jump is required in considering higher-order entification because even sub-components (e.g. species) may have phenotypically distributed organization and individual phenotypes may or may not be interacting in a particular coevolved circuit in any given circumstance. Thus, the inducible defences of some invertebrate species are deployed in response to a diversity of particular predators and the form of the response may vary according to contingencies (Chapter 6). Such responses demonstrate that species-specific coevolved features have been achieved even where associations vary and/or occur intermittently.

To forge a unified science of biology, a broad base of understanding that transcends scales is required. Biologists can learn some strong lessons from engineers. One of the Viking Landers bore a microdot with the signatures of 10 000 people contributing to the project. If you think about it, even mundane machines like automobiles are emergent properties of a socially distributed intelligence and accumulated culture. It may seem paradoxical, given the abundance of cars, that nobody on earth knows how to build one (consider metals, glass, plastics, rubber, fabrics, silicon chips, radios, lights, batteries, transmissions, paint, etc.).

If systems analysis is needed to construct complex machines, redesigning organisms or understanding extant species will require an even stronger cooperative focus. Perhaps through a marriage of our two most powerful new technologies, computers and recombinant DNA, we can make rapid progress in both the design and development of new life forms, linking ecological vision to the molecular miracle. The impact of biotechnology will ultimately be at the level of organismal function. The Aswan Dam gave considerable impetus to ecology as an instance of engineering proceeding blindly. It will be especially necessary that genetic engineering does not proceed with similar myopia, because unlike the Aswan Dam, the products of biotechnology may reproduce themselves without our further cooperation or permission.

2

Regulatory interactions, epigenetics and genomic integration

2.1 THE BASIS OF HEREDITY: CHANGING CONCEPTS OF THE GENE AND THEIR IMPLICATIONS FOR EVOLUTIONARY THEORY

Charles Darwin and Arthur Wallace formulated their theory of evolution via natural selection without any knowledge of the genetic basis of heredity. Mendel's demonstration that the units of heredity were particulate and relatively independent was not intuitively obvious because most phenotypic traits are quantitative, meaning that they show continuous variation in populations rather than discontinuities or discrete features (e.g. wing lengths, height, body weight). When numerous genes contribute to a given character, the distribution of the trait in a population tends to a normal distribution. This allows the heredity of such attributes to be analyzed statistically as if their underlying determination actually was continuous. This is the basis of quantitative genetics (Falconer, 1981), which is currently in the midst of a vigorous renaissance and extension to life-history theory (Stearns, 1992; Roff, 1993).

The quandary that faced Darwin (1859) was how to explain why genetic variation persisted when the continual blending of parental contributions would lead inevitably to an average phenotype with little variability. The Mendelian breakthrough solved this problem: genes do not mix, they maintain their individual character across generations. Quantitative genetics works when applied at the level of the phenotype, even though it does not directly reflect the causal mechanisms of heredity (i.e. it involves statistical description rather than functional representation).

The Mendelian paradigm gave birth to population genetics, which is largely concerned with how the frequencies of particular alleles change in populations. The point of all this is that our concept of the basis for heredity is crucial to our understanding of evolution. Even if quantitative genetics can provide statistical descriptions or predictions of phenotypic evolution, the concept of particulate genes, and circulating variations of particular genes (alleles), changes our interpretation, and alerts us to potential paradigm failures. Similarly, new molecular evidence related to the structure and integration of the genome is altering our perspective even further (Shapiro, 1992).

Modern population genetics crystallized with the classic publications of Wright (1931, 1932), Fisher (1930) and Haldane (1932). As will become apparent, the models developed by Sewal Wright differ radically in that they have always incorporated elements of both non-equilibrium ecology and group selection (e.g. fragmented populations with spatially and temporally variable selection pressures). The neo-Darwinian view of the genome has been that genes represent discrete regions that lie along chromosomes like beads on a string. In its extreme form, this paradigm has been disparagingly refered to as 'bean bag genetics' by Mayr (1963), a slight that was immediately addressed by Haldane (1964). Different versions of each gene may exist, and in sexual species these alleles can be exchanged independently at each locus to create new combinations of loci. New alleles are created by mutations. The idea that new loci can also be created by mutation or duplication was always recognized, but for the most part, evolution has been effectively treated as population-level changes in allele frequencies at existing loci (Roughgarden, 1979). In keeping with the observations of quantitative genetics, such a process leads to gradual evolution by incremental small steps.

Neo-Darwinism or 'the modern synthesis' was subsequently consolidated during the 1930s to the early 1950s (classic contributions include Fisher, 1930; Wright, 1931; Haldane, 1932; Dobzhansky, 1937; Huxley, 1942; Mayr, 1942; Simpson, 1944, 1953a; Schmalhausen, 1949; Stebbins, 1950; Waddington, 1940, 1957). Even during this period of synthesis, however, there was a major grumble of dissent. Goldschmidt (1940) was unable to explain numerous observations of development and heredity within the existing paradigm and proposed that some form of genetic structure at a level higher than alleles was responsible for much of phenotypic evolution. Furthermore, Goldschmidt (1940) proposed that empirical observations and the existence of such heredity supported the idea of large evolutionary steps, mainly associated with speciation events. This was extremely unpopular because it implied that allele frequencies (i.e. what everybody had spent most of their time studying), might be largely irrelevant to macroevolution. Moreover, there was no known mechanism of heredity that conformed to Goldschmidt's (1940) chromosome repatterning, whereas the existence and behaviour of alleles was easily recognized and amenable to mathematical analysis. Consequently, Goldschmidt's ideas were demoted to a level somewhere close to those of Lamarck.

Significantly, this entire early phase of evolutionary theory was built with no knowledge of the genetic code. Beginning in the 1950s the discovery of the structure and composition of DNA led to an apparent confirmation of the validity of the gene/allele model, and a refinement of the concept of mutation to that of changes in base sequences that could alter protein structure. Probably the biggest surprise came during the 1960s when electrophoretic techniques detected considerably more genetic variation than predicted by the classic genetic model. This led to some interesting theoretical battles between those supporting a modified classical model (i.e. considerable polymorphism was actually neutral) (Kimura, 1983) *versus* those supporting a model of balanced polymorphism to explain how this variation was maintained (Dobzhansky, 1970; Lewontin, 1974).

These rapid developments also ushered in a new wave of evolutionary theory contributed by molecularly oriented biologists (e.g. Ohno, 1970; Kimura, 1983). Particularly important ideas were that selection (or lack thereof) at the molecular level could propagate neutral alleles or allow large quantities of 'junk' DNA to accumulate. In addition it was recognized that 'selfish' code that might have no higher-order adaptive purpose could proliferate in a genome (Kimura, 1983). The basic conceptual framework of relatively independent circulating alleles remained viable throughout this period and in fact may have peaked with the classic books of Williams (1966a) and Dawkins (1976, 1982).

The first complications for the paradigm probably occurred with the demonstration of regulated gene activity and the concept of the operon developed by Jacob and Monad (1961a,b) (see Miller and Reznikoff, 1980). Since then the molecular scrutiny of the structure and function of the genome has revealed a complexity, our appreciation of which is still growing exponentially. There has been a major lag in incorporating these molecular revelations into the synthesis (Shapiro, 1992). Another major dissatisfaction with the synthesis has been the relative omission of development as an integral component. Epigenetics (the genetic control of development) suggested that the intervening organization spanning the gap between the genotype and the adult phenotype represented a secondary structure of high regulatory relevance. Environmental insults were found that produced phenocopies – deviant phenotypes mimicking genetic mutations. The phenomena of phenotypic plasticity and canalization were identified as key evolutionary subjects with developmental underpinnings.

Some of Goldschmidt's (1940) ideas were born in these contexts, but the real champions of development and its evolutionary implications were Schmalhausen (1949) and Waddington (1940, 1957, 1975). Others added their voices, notably Bonner (1974), Rendel (1967), Gould (1977), Lovtrup (1974, 1987), Raff and Kaufman (1983), Reid (1985), John and Miklos

(1988), Arthur (1984, 1988), Hall (1992) and Stearns (1992). However, the real crystallization of this area has exploded over the last 15 years as the molecular control of development is being rapidly unravelled (see books by Raff and Kaufman, 1983; Arthur, 1984; Davidson, 1986; John and Miklos, 1988; Loomis, 1988; De Pomerai, 1990; Gilbert, 1991; Slack, 1991; Hall, 1992; Lawrence, 1992).

This new knowledge lays one cornerstone for a new evolutionary synthesis. The structure and function of the new gene, its dynamics in hierarchical regulatory gene nets, and the further regulatory levels of control associated with RNA and protein interactions requires that the genic view of independent circulating alleles be extended to a new conceptual model (Hunkapiller *et al.*, 1982; Shapiro, 1992). Just as the Mendelian demonstration of particulate genes did not preclude the application of quantitative genetics, the new gene does not preclude the application of earlier approaches, but it does greatly alter our interpretation. In particular, the realm of mechanisms and phenomena of evolutionary relevance is greatly increased and this will fundamentally alter the questions that seem most interesting. In particular, the older view of genes mapping to the phenotype in a one-way one-gene-one-enzyme framework is being replaced by a view of complex integration and feedback at multiple levels of control (Alberch, 1991; Shapiro, 1991, 1992).

2.2 GENOMIC ORGANIZATION AND SUBSPECIFIC LINEAGE SELECTION: A NEW EVOLUTIONARY FRAMEWORK

The controversy regarding what are the units of natural selection rages on (e.g. Brandon and Burian, 1984; Barlow, 1991; Ereshefsky, 1992), but the dominant paradigm has been the highly reductionist genic view and its kin extensions championed by Fisher (1930), Hamilton (1964a,b), Williams (1966a, 1985) and Dawkins (1976, 1982, 1989a). A powerful argument supporting this view has been that sexually reproducing organisms are temporary entities whose identities are rapidly shuffled back into the population gene pool. Because sexual reproduction apparently negates organisms and their individual genotypes as evolutionary entities (Salthe, 1975; Sibly, 1989), the focus has shifted downwards to genes. Despite the success of the genic paradigm as a source of explanation and prediction (e.g. Haldane, 1964; Dawkins, 1982), recent molecular advances reveal that the perspective of the genome as a relatively independent assembly of loci determined by competition among selfish alleles is far too simplistic. Instead, eukaryote genes have complex structures associated with regulation of their rates, timing of transcription and integration of their interactions (Rendel, 1967; Watson, Tooze and Kurtz, 1983; Raff and Kaufman, 1983; Arthur, 1984, 1988; Ohta, 1988; John and Miklos, 1988; De Pomerai, 1990; Latchman, 1990; Dickinson, 1991;

Gilbert, 1991; Slack, 1991; Hall, 1992; Lawrence, 1992). Complexes of genes are organized in interacting nets or batteries (Britten and Davidson, 1969, 1971; Davidson and Britten, 1973, 1979; Davidson, Jacobs and Britten, 1983; Kauffman, 1985, 1993; Davidson, 1986; Arthur, 1988; Bonner, 1988; Dickinson, 1991). Important epigenetic decisions may not be made at the genic level, but may reside at higher levels of regulatory organization (Alberch, 1991; Hall, 1992). Knowledge of genes and genome organization has progressed to the point where the genic paradigm is now not only inadequate, but it is also inconsistent with respect to these revelations. For example, even in relatively simple eukaryotes, such as yeast, the determination of fundamental features such as mating types involves regulated transposition as well as genomic imprinting (discussed below) (Klar, 1990). Both of these are new phenomena that lie outside the classic genic paradigm.

To explore this and elaborate an extended framework, I will review some major features of the new gene and the modern genome. This will then be extended to argue that eukaryote evolution mainly involves regulatory hierarchies of epigenetic genes. Adopting this framework then requires a shift in what we consider evolutionary units and the temporal framework relevant to natural selection.

2.3 OPERONS AND GENE REGULATION IN PROKARYOTES

Understanding of regulatory gene interactions began with the classic models of Jacob and Monad (1961a,b) concerned with control of the enzymes involved in utilizing lactose as an energy substrate in the bacteria *Escherichia coli*. They postulated that much of the coordination in metabolic pathways in bacteria is obtained by linkage of relevant structural genes into operons that act as large transcription domains (Jacob and Monad, 1961a,b). In eukaryotes (multicellular organisms), regulation is more complex with interacting genes being widely dispersed, or linked perhaps by a history of common ancestry via gene duplications. Significantly, fundamental developmental units such as the Antennapedia and Bithorax complexes in *Drosophila* do consist of linked regulatory genes that are transcribed initially as very large transcription units. These are subsequently processed into smaller units. This linkage may be relevant to regulatory aspects (John and Miklos, 1988).

Another feature of the operon model is that most control is negative, 'inhibiting rather than provoking, specific protein synthesis' (Jacob and Monad, 1961a). The operon model has proved to have widespread application in prokaryotes (e.g. Miller and Reznikoff, 1980). Evolution of metabolic capabilities in bacteria can serve as simple models for understanding higher organisms. In particular, the ability to utilize a substrate involves the integrated evolution of several key features (which must all change if the substrate changes). These include the structural genes coding for particular enzymes, receptors able to detect the potential resource, cell membrane transport or permeability features and regulatory systems to obtain on–off or modulated control in place of inefficient constitutive (always on) function. Notably, the necessity for a number of integrated changes involves transgenerational evolutionary steps and some responses (such as gene duplications) are non-genic (e.g. Lerner, Wu and Lin, 1964; Clarke, 1983; Hall, 1992).

Significantly, different bacterial lineages from the same initial population may show distinct genetic solutions when faced with the problem of utilizing a novel substrate (Campbell, 1985). The separate solutions are compartmentalized among sublineages within the selected cultures. As will be seen, eukaryotic evolution shows similar trends. Although instructive, prokaryote evolution is far too simple to apply to multicellular eukaryotes. Prokaryote evolution mainly pertains to biochemical capabilities associated with metabolic/catabolic pathways. Eukaryotes have many more levels of regulatory control, more complex genetic organization, and their evolution is largely concerned with the epigenetics of multicellular phenotypic structure rather than simply biochemical aspects.

2.4 EUKARYOTIC REGULATORY PROCESSES

2.4.1 Chromatin structure, genomic imprinting and the regulatory domain model

A simplistic way to view prokaryotes is that their regulatory capabilities are largely inhibitory. Turning something on usually involves turning off another inhibitory gene. Eukaryote genes are much more complex. Such genes have large regulatory regions for handshaking with other distantly located genes and coding regions that contain introns which are later deleted from mRNA (Figures 2.1 and 2.2). Furthermore, eukaryotic chromosomes consist of a matrix of protein intermeshed with DNA (chromatin) and organized into repeating units called nucleosomes (Felsenfeld, 1978; Nicolini, 1988; Gilbert, 1991; Grunstein, 1990, 1992). The nature of chromatin can be altered so that the DNA is highly condensed and unavailable for transcription (=heterochromatin) or it may exist in less condensed states that may be available for transcription (=euchromatin) (see Nicolini (1988) for a more sophisticated discussion). Some areas of the genome are permanently heterochromatic such as near centromeres or at the tips of chromosomal arms (Ohno, 1970), but transitions between heterochromatin and euchromatin may be achieved under regulatory control.

Only in the last 5 years has the potential regulatory significance of

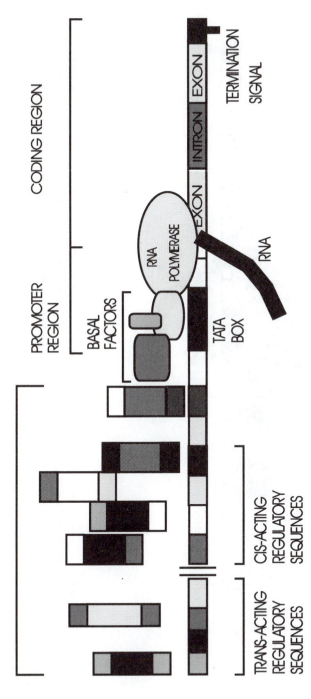

Figure 2.1 Key features of a typical eukaryotic gene. Transcription of DNA coding regions (both introns and exons) into RNA by the enzyme, RNA polymerase, requires the binding of basal factors to the promoter region. The binding of these factors into a functional initiation complex allowing transcription, may be controlled by regulatory elements in nearby upstream sequences (cis-acting regulatory sequences), and/or by more distantly located control sites (trans-acting regulatory regions). Regulatory regions bind the products of other genes, and so such regions allow integrative handshaking among numerous genes. Such complexity may not simply control initiation, but may allow regulated modulation of transcription rates or differential utilization of different promoters within complexly organized genes.

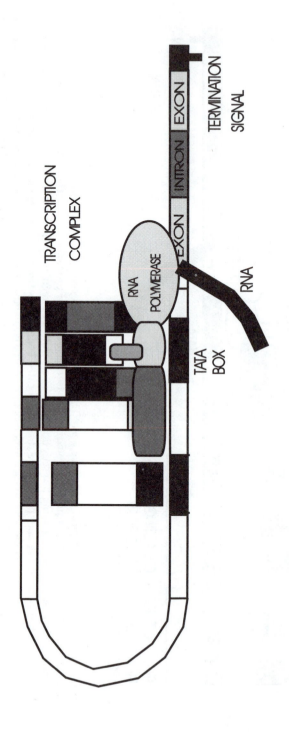

Figure 2.2 Formation of transcription complexes. The most widely accepted model of how regulatory sequences interact in transcription complexes envisions that DNA looping brings transcription factors that bind to upstream control elements into ordered juxtaposition with the promoter region. More distant regulatory sequences might also act via looping, or alternatively, by modification of local chromatin structure.

chromatin structure been even remotely appreciated (Grunstein, 1992; Kennison, 1993). Four histones (H2A, H2B, H3 and H4) bind as octamers into discrete protein cores. A stretch of DNA, some 146 nucleotides long, wraps almost twice around this spool, forming the fundamental unit called a nucleosome (Figure 2.3). These are linked together by intervening DNA segments 54 bases long. Thus, Gilbert (1991) points out that our view of DNA organization has changed from one of DNA beads on a string, to strings of DNA wrapped around beads of protein.

Another histone protein (H1) appears to be involved in linking nucleosomes together and consequently changes the degree of compaction of chromatin. It may also be involved in anchoring DNA tightly to the protein core of each nucleosome. Highly condensed chromatin is transcriptionally inactive. For genes to be transcribed, condensed DNA must be unfolded and then unwound from nucleosome cores (Grunstein, 1992) (Figures 2.3 and 2.4).

A model of how such events are mediated by regulatory control suggests that particular DNA sites exist (enhancers) that are accessible to specific DNA-binding proteins. Such sites may control the local compaction of specific areas of chromatin. One possible mechanism might involve modification of how chromatin is bound to the nuclear scaffold (Bodnar, Jones and Ellis, 1989). Histone proteins also have tails that extend outside of the nucleosome structure. An interaction between these tails and protein complexes formed at upstream control sites may modify the binding of DNA to nucleosomes (Figure 2.3A). This process then exposes previously inaccessible promoter regions of genes to other regulatory proteins and RNA polymerase which are required to effect transcription (Gilbert, 1991; Grunstein, 1992) (Figure 2.3B).

The best known example of regulation of chromatin structure is the conversion of one copy of the X chromosome into heterochromatin which is required to maintain balance in the dosage of X-linked genes (Ohno, 1973; Holliday, 1990). In some insects the entire set of paternal chromosomes may also be heterochromatinized or even deleted, and in ascarids and cyclops, all regions of heterochromatic DNA may be selectively deleted in somatic cells (John and Miklos, 1988; Nur, 1990). There has been a surge of interest in the regulation of chromatin structure, largely associated with its probable role in the phenomenon of genomic imprinting (Solter, 1988; Holliday, 1989a; Sapienza, 1990; Peterson and Sapienza, 1993). In this case, the maternal and paternal genomes pass copies of homologous genes that differ in their chromatin structure. Differential patterns of imprinting on the genome may have important developmental consequences (Surani *et al.*, 1990) or may change the activities of different enzymes (Holliday, 1989a; Monk, 1990). Imprinting may result in certain genes being expressed in only one sex (Peterson and Sapienza, 1993). Imprinting may be involved in differentiating male and female

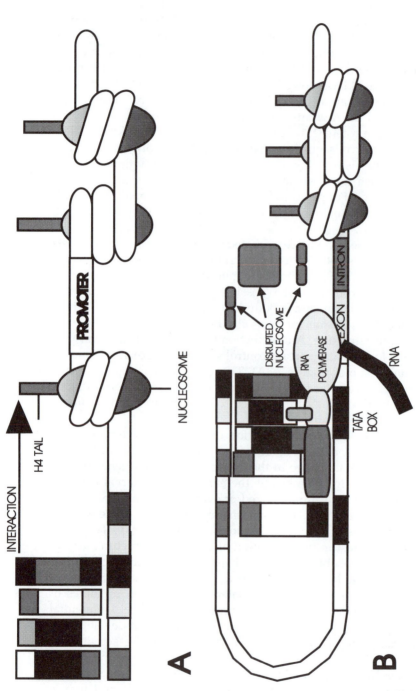

Figure 2.3 Transcriptional regulation and nucleosomes. (A) Local activation of a gene may be controlled by interactions between proteins bound to regulatory sequences and local nucleosomes, possibly via tails of H4 protein that extend outside the nucleosome core. (B) Such interactions may promote uncoiling of DNA from the nucleosome or disruption of the nucleosome. In this case, naked DNA may then form transcription complexes similar to the model presented in Figure 2.2. Highly modified after Grunstein (1992) with permission of Scientific American Inc.

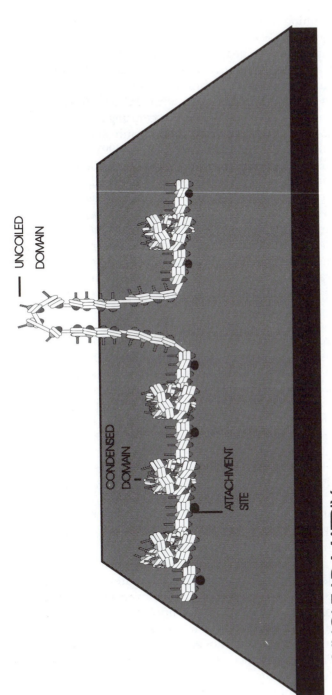

UNCOILED DOMAIN

CONDENSED DOMAIN

ATTACHMENT SITE

NUCLEAR MATRIX

Figure 2.4 The domain model of eukaryotic gene regulation. This model proposes that chromatin may have several states associated with the degree of coiling of DNA onto nucleosomes, and the degree of compaction of nucleosomes. Condensed clusters of nucleosomes are unavailable for transcription and may be uncoiled under regulatory control. Activation of particular genes in such uncoiled domains may further require the unwrapping of DNA from nucleosomes (Figure 2.3). Modified from Bodnar, Jones and Ellis (1989) with permission of the *Journal of Theoretical Biology*, Academic Press.

phenotypes, determining sex ratios or even mediating dominance (Chandra and Nanjundiah, 1990).

The potential importance of genomic imprinting for developmental expression is illustrated by the occurrence of non-reciprocal phenotypes obtained by inter-and intra-specific hybridizations, such as the production of either mules or hinnies in donkey × horse crosses (Reik, Howlett and Surani, 1990). In vertebrates, methylation of cytosine residues may be somehow involved in shifting between euchromatin and heterochromatin (John and Miklos, 1988; Holliday, 1990; Gilbert, 1991; Peterson and Sapienza, 1993). It is perhaps significant then, that methylation changes in specific loci have been noted during embryogenesis (Shemer *et al.*, 1991). This cannot be the whole story, however, since *Drosophila* obtains comparable changes in chromatin structure without methylation, so methylation may represent a different level of control. Genomic imprinting represents a mode of epigenetic information transfer that can act as an overlay on the basic genetic state and thus falls outside the genic paradigm (Sapienza, 1990; Hoffman, 1991a). Thus, imprinting may be reversible, but may also exhibit transgenerational stability (Peterson and Sapienza, 1993).

The extent to which genomic imprinting may serve as a general feature of eukaryote gene regulation remains obscure, but the available evidence is startling. At least 20 to 30 genes regulating the degree of heterochromatinization have been inferred from work on *Drosophila* (Tartof and Bremer, 1990). Results from insertions of transgenes suggest that relatively localized alterations of chromatin structure are possible (Reik, Howlett and Surani, 1990). Such mechanisms may be important for regulating the degree of expression in blocks of repetitive DNA or inhibiting otherwise mobile elements.

For eukaryotes, the greatest potential significance of the phenomenon lies in epigenetics. Bodnar, Jones and Ellis (1989) have put forward an elegant model for the regulation of cell differentiation and morphogenesis (the Domain model) which proposes that DNA domains 25 000–125 000 base pairs in length may be activated or repressed by regulatory genes controlling the degree of coiling in chromatin (Nicolini, 1988; Grunstein, 1990, 1992). This model is highly consistent with available evidence (Holliday, 1990; Slack, 1991; Grunstein, 1992) (Figure 2.4). The major conflicting evidence is that genes can often be moved without affecting their activity (see John and Miklos, 1988). Significantly, however, this may require inclusion of flanking regions (which contain regulatory regions). Alternatively, if a reporter gene is moved around via attaching it to a transposable element, the existence of numerous cell and tissue-specific enhancers is revealed (Lawrence, 1992).

Position effect variegation is also well known in *Drosophila*. In this phenomenon, the normal expression of a gene is inhibited when a chromosomal rearrangement brings the gene close to a heterochromatic region (Tartof and Bremer, 1990). The key significance of chromatin in eukaryotes is that most eukaryote genes may have a base state of inactivity rather than constitutive function (Ohno, 1970; Renkawitz, 1990; Grunstein, 1992). Thus, regulatory control can be more complex than in prokaryotes, both activation and repression being flexible features (Davidson, 1986). Evidence for processes such as regulated chromatin structure reflects the existence of higher-level control and coordination, to which particular loci are subservient. In fact, whether genes are expressed in any particular generation may be regulated by such mechanisms.

2.4.2 Promoters

Once chromatin has been converted to an activated state, transcription of messenger RNA requires the binding of RNA polymerase II to a promoter region (Figure 2.1). We now know that regulatory genes may have more than one promoter allowing different transcripts to be obtained from a single gene, or for these to be utilized at different times or in different kinds of differentiated cells (e.g. Slack, 1991; Lawrence, 1992). Although many promoter regions for protein-coding genes have conserved elements (such as the TATA box), there is also considerable variability among promoter regions. In prokaryotes, the core RNA polymerase may be directed to different classes of promoters by protein initiation factors (Travers, 1985).

2.4.3 Transcription factors and regulatory recognition sites

Prior to 1980, no proteins regulating transcription or their DNA binding sites had been defined (Latchman, 1990). In eukaryotes, recognition of the promoter region by RNA polymerase II is now known to require the cooperation of five basic transcription factors (basal factors) to form a functional initiation complex (TFIIA to TFIIE) (Gilbert, 1991). The process begins with the binding of TFIID to the TATA box. Histone proteins may compete with these transcription factors in binding to the promoter region, and enhancers may modulate the outcome of this competition (Grunstein, 1992).

The effectiveness of the promoter may be further synergized or inhibited by interactions with other transcription factors (Figure 2.2). Very early on it was proposed that integrated regulatory circuits could exist in eukaryotes, despite the lack of physical linkage among these elements. This could be achieved if recognition sequences, acting as identifying codes, were incorporated into various genes in a network (e.g. Britten and Davidson, 1969, 1971; Davidson, 1982; Davidson and Britten, 1973; Davidson, Jacobs and Britten, 1983). This has proved to be a fundamental truth, although the sizes of regulatory sequences are smaller than those originally suggested by Britten and Davidson (John and Miklos, 1988; Latchman, 1990). Most genes contain various regulatory sites (i.e. specific sequences of DNA), that are recognized by proteins with domains that preferentially bind to these sites. These regulatory sequences are often found upstream and surrounding the promoter region, but may also exist within introns or in downstream locations. Thus, the initiation and regulation of transcription involves protein transcription factors with multiple domains that either bind DNA control regions, or interact with domains of other transcription factors (reviewed by Mitchell and Tjian, 1989; Latchman, 1990; Beardsley, 1991; Harrison, 1991).

Broad classes of DNA-binding domains include helix-turn-helix motifs, two subclasses of zinc-binding domains (zinc fingers), leucine zippers, B-ribbon proteins, helix-loop-helix domains, and POW domains. This list is still expanding. Within each class there may be subclasses and RNA-binding domains (that may function in post-transcriptional control) are also known. Transcription factors composed of RNA have also been discovered (Hoffman, 1991b; Young *et al.*, 1991). The diversity of structure is probably related to requirements for differing flexibility needed to achieve particular functional utility. Thus, leucine zippers are not individually stable, but require complementary partners for effective DNA-binding. This makes them amenable to the construction of complex regulatory interfaces (Harrison, 1991). Tandem zinc fingers can yield highly accurate binding to long recognition sites (Nardelli *et al.*, 1991).

Regulatory genes may also produce proteins that do not contain any known DNA-binding domains. Such products could possibly interact indirectly by binding to transcription factors or by linking them together. Very complex interactions are possible. For example, active protein domains may be only unfolded into an active configuration when particular transcription factors interact at enhancer sites (Schaffner, 1989). Such complexes may then act directly on promoter regions by looping out the intervening DNA (Grunstein, 1992; Kennison, 1993) (Figures 2.2 and 2.3B). Thus, transcription factors may operate via interactions with one another, directly on the promoter, or via alteration of chromatin structure at more distant sites (Schaffner, 1989; Grunstein, 1990, 1992). Combined with further controls at the loci coding for the transcription factors themselves, target loci may be either activated or repressed (Renkawitz, 1990).

The ubiquity of transcription factors, and the necessity for complementary recognition sites to forge interactive circuits, is difficult to interface to a genic framework that has traditionally focused on only the structural regions of genes. Firstly, it appears that many gene products and considerable genetic structure that codes nothing are solely concerned with communication and regulation. Secondly, the independent evolution of complementary recognition mechanisms throughout appropriate circuits is difficult to explain other than via sequence duplications and transposition, both of which are largely ignored in the traditional paradigm.

2.4.4 Enhancers and silencers

Enhancers are DNA sequences that markedly increase the transcription of a gene. Analogous regions that suppress genes (silencers) are also known. Unlike promoter elements, enhancers act relatively independently of their position or orientation relative to the gene. They may show tissue-specific activity (John and Miklos, 1988; Gilbert, 1991), strongly suggesting a role in epigenetics. Regulation via modification of chromatin structure or via transcriptional regulation are complementary kinds of control (Davidson, 1982; Kennison, 1993). Together these two tiers of regulation provide an immense capacity for combinatorial definition of numerous states or pathways using a relatively small number of regulatory genes (Bodnar, Jones and Ellis, 1989) (Figure 2.4). Thus, tissue-specific expression may involve nets of enhancers (Gilbert, 1991, p. 423) which would then expose further nets of promoter control elements (North, 1984). Genomic rearrangements may bring various promoters under the control of new enhancers or generate new combinations of enhancers (Dickinson, 1991).

2.4.5 Processing of messenger RNA

The immediate RNA products of eukaryote transcription do not represent functional messenger RNA (mRNA) but require further processing. In fact, much RNA transcribed in the nucleus is never processed into functional mRNA, indicating that considerable regulation of development occurs at this level (Davidson and Britten, 1979; Gilbert, 1991). There are two important features associated with this. Firstly, eukaryotic genes are segmented. Regions that eventually become translated into protein (exons) are separated by regions (introns) that do not contribute to the final protein product (Gilbert, 1978). Introns are spliced out of the original RNA transcript in small structures known as splicesomes (Wassarman and Steitz, 1991).

Significantly, exons often represent an active domain of a protein product such as a DNA-binding sequence or an enzymatically effective region (Phillips, Sternberg and Sutton, 1983; Wills, 1989; Slack, 1991). A single protein may have multiple active domains that allow it to catalyse a chemical reaction, bind to the domains of other proteins or bind to particular DNA sequences. The segmental structure of genes could be important for DNA repair processes (see section 5.6), but such compartmentalization also allows effective domains to be individually excised and transposed into other genes. Such a process, termed exon shuffling

(Gilbert, 1978, 1985), could produce novel proteins with remarkable new potentials. Tittiger, Whyard and Walker (1993) provide empirical support for the reality of exon shuffling, having found an intron at a specific site predicted by the hypothesis.

A deletion or addition of a single exon could change the metabolic function of a protein, change its interaction with other proteins or change the status of DNA recognition capabilities (i.e. a new regulatory circuit could be established). Thus, exon shuffling may synergize regulatory evolution, and, because exons represent the business end of functional gene actions, their shifting into other roles constitutes a mutation process highly biased towards potentially useful impacts. This has been delightfully articulated by Wills (1989) who views such evolution as analogous to exchanging working heads on existing tools in the genomic toolbox. Doolittle (1987) provides some examples.

Post-transcriptional RNA processing also provides an additional level of regulatory control (Davidson and Britten, 1979; Brody *et al.*, 1988; Smith, Patton and Nadal-Ginard, 1989; De Pomerai, 1990; Holliday, 1990; Gilbert, 1991). Although not as well studied as the action of transcription factors, regulatory genes producing protein or RNA products that regulate or modify RNA molecules are being rapidly identified (reviewed by Bandziulis, Swanson and Dreyfus, 1989). Such RNA products might never themselves be translated into proteins (Davidson and Britten, 1979).

Although of potential evolutionary significance, the segmental structure of eukaryote genes appears to be mainly an adaptation providing yet another level of regulatory control (e.g. Smith, Patton and Nadal-Ginard, 1989; Holliday, 1990). By differentially splicing RNA transcripts from a single gene, several different proteins can be produced (Gilbert, 1978). This conflicts with the genic view that one gene makes one product. In many cases, it is at this level that the regulation of gene on/off expression is determined (Smith, Patton and Nadal-Ginard, 1989). Differential splicing is not only important for producing diversity in gene products; major developmental decisions may be made at this level of organization. In particular, developmental switches involved in elaborating either male or female phenotypes in flies involve sex-specific alternative splicing (Baker, 1989; Smith, Patton and Nadal-Ginard, 1989; De Pomerai, 1990; Slack, 1991). Alternative splicing adds yet another layer of regulatory complexity which is only just beginning to be appreciated.

The genic paradigm argues that genes can be considered the ultimate evolutionary entities because they show long-term conservation. If this is the criterion, then perhaps exons are even more appropriate fundamental units. This illustrates nicely that there may be important evolutionary units that reside at different levels of functional organization. If both genes and exons are viable functional units, then a paradigm based on only one level of genetic organization must be abandoned. The evolution of tRNA is a telling example. The complementary domains that link a particular amino acid to a specific nucleotide triplet constitute the two working heads of a tRNA molecule. The domains themselves can be viewed as conserved features, but the appropriate coupling of domains that recognize a specific amino acid and a domain that recognizes a particular mRNA triplet is the basis of the genetic code itself. As such, specific combinations have been conserved for billions of years because they are needed by other genes to make proteins (Crick, 1968; Ohno, 1970, 1973).

Because of their nuclear/cytoplasmic compartmentalization, eukaryotes may exert further regulatory control by varying the rate of release of mRNA to the cytoplasm. The modern view suggests that transcription of mRNA, and subsequent processing and transport, takes place on a solid substrate provided by a nuclear cytoskeleton or matrix. Such a mechanism could allow direct control over processing rates, transport of mRNA to the nuclear envelope and selective release through nuclear pores (Gilbert, 1991). Holtzman (1992) provides a review highlighting the importance of transport and delivery of various macromolecules in the regulatory control of cell function. In some cases, relatively complex systems have evolved that achieve rather precise targeting and delivery. The control of chromatin structure may also involve regulation of DNA sequences that control the anchoring of domains to the nuclear matrix (Bodnar, Jones and Ellis, 1989).

A complexity of mechanisms may also regulate translational processes. The availability of ribosomes and appropriate classes of tRNA may provide one level of control. The stability of particular mRNA may also be an important feature that allows flexibility in removal or storage of such products. Finally, the translated proteins may have varying activity associated with the cleavage of specific components, phosphorylation, addition of monosaccharide residues or binding with other proteins. The latter may change the structural conformation of a protein, exposing functional domains that were previously inactive or inaccessible (John and Miklos, 1988; Latchman, 1990; Gilbert, 1991). Thus, the binding of steroid hormones to specific cell receptors is a crucial aspect of intercellular communication.

2.4.6 Higher-order regulation

Regulatory interactions extend beyond the realm of single cells in eukaryotes, because of the necessity for developmental integration. Differential concentrations of cytoplasmic factors can lead to cells with different information contents during cell proliferation. Cell–cell interaction is a fundamental level of regulation in development that involves numerous features such as membrane receptors, secreted proteins, cell adhesion molecules and contact induction (John and Miklos, 1988; Gilbert, 1991; Slack, 1991; Hall, 1992).

Given that each level of regulation elaborates a further order of magnitude in combinatorial interactions, the potential for complex protein–protein interactions to contribute further to epigenetic regulation is enormous. Gilbert (1991, p. 503) refers to this level of integration as a regulatory ecosystem. Of particular importance for epigenetics is integration via growth factors and hormones. Some hormones may act via binding to cell-surface receptors which in turn may alter cell activities via secondary messengers. Others, such as the steroid hormones, may pass through cell membranes and bind with specific cytoplasmic receptors to produce an activated hormone-receptor complex. Such complexes may then enter the nucleus and bind to DNA recognition sites, thus altering the information status of the cell (De Pomerai, 1990, p. 116). Lawrence (1992) reviews evidence that the ultimate deployment of terminal structures like bristles and ommatidia in a fly may involve complex cell–cell communication.

The point of this discussion is to convey that the morphology, physiology and behaviour of phenotypes arise via complex overlays of regulatory control extending from the genome upwards even to social and ecological interactions (Figure 2.5). All of this evidence attests to a complex overlay of regulatory controls which may be the key target of phenotypic evolution and most of which does not seem congruent with the behaviour of an assembly of selfish fundamental units (Alberch, 1991). There are also several other features of the genome that have recently moved to the forefront, which also represent challenges for genic theory.

2.4.7 Gene duplication: redundancy and diversification

At the same time that Britten and Davidson (1969, 1971) were proposing that the functional organization of the eukaryotic genome involved dispersed repetitive recognition sequences, Ohno (1970) published a classic monograph on the role of gene duplication in evolution. Although neo-Darwinism has necessarily recognized that loci may be added or deleted from genomes, this has never been a significant component of theoretical applications. In general, the genic framework considers only the changes in allele frequencies at one particular locus, largely because this has proved a mathematically tractable problem. Evolution is measured as a change in allele frequencies and speciation occurs when enough microevolutionary divergence has accumulated. Thus, the genic framework has largely ignored the evolutionary problems associated with gene duplications and their subsequent diversification.

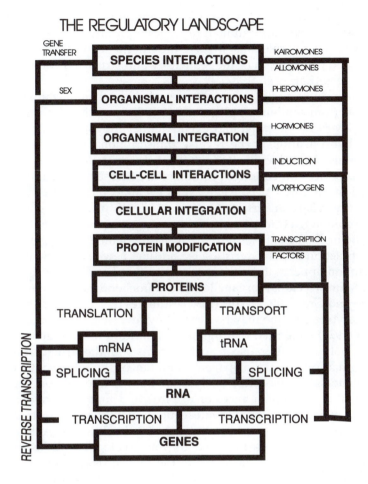

Figure 2.5 The regulatory landscape governing eukaryotic genome evolution. Possible feedback loops between genome structure and function occur at numerous levels of integration, all of which may be visible to natural selection. The specificity of the signals is in fact evidence that organizational evolution has proceeded at these various levels. This suggests that the genomic level is constrained by functional organization realized in higher-order holons (Chapter 1).

Ohno's (1970) main conclusions and insights remain largely applicable. In particular, Ohno (1970, 1973) stressed that loci are highly constrained by selection for conservation of functional domains. Duplications allow one copy of a gene to maintain function while the other(s) diversifies. In most cases this divergence will lead to a functionless hulk or pseudogene, but occasionally a slightly different function may be obtained allowing extended capabilities. Alternatively, an entirely new function may eventually be assumed. Thus, progressive evolution is largely dependent on the process of gene duplication. It must be emphasized that the evolution of a new functional role must involve the simultaneous or stepwise modification and coevolution of the regulatory and structural components of the new gene (Raff *et al.*, 1987).

The main sources of duplicated genes needed for organizational extension arise from tandem duplications (usually associated with unequal crossing over) or via polyploid events (either intraspecific or via interspecific hybridizations). The reverse transcription of mRNA back into DNA has also been documented (Hollis *et al.*, 1982). Reverse transcription usually produces functionless pseudogenes because their promoter is not included in the mRNA. However, such events could be of evolutionary significance where an inserted construct comes under the control of a nearby promoter, a transposable element moves a promoter to the appropriate site or the gene fuses into the structure of an existing gene, changing its functional characteristics by adding new exons. Such reverse transcription appears to be the mechanism that has allowed the transfer of genes from cellular organelles like mitochondria, into the nucleus (Covello and Gray, 1992). Wills (1991) estimates that 10% of the sequences in a eukaryotic genome may be derived by reverse transcription.

Ohno (1970) suggested that the most significant form of gene duplication was polyploidy because in such systems both the structural loci and their regulatory associations were simultaneously released to explore new potentials. A masterful analysis of known information suggested that a fish or amphibian ancestor of birds and mammals underwent a polyploid event that provided subsequent evolutionary flexibility. This has been recently supported by evidence that the Bithorax/Antemapedia complex of *Drosophila* is mirrored by four analogous blocks of genes located on different chromosomes in the mouse (Hall, 1992, p. 135). A further interesting conclusion of Ohno's (1970) analysis was that current organizational complexity largely precludes further reorganization via polyploidy due to disruption of sex determination and dosages of regulatory genes. That is, evolutionary flexibility may be slowing down in more advanced taxa.

It is now recognized that many loci can be grouped into gene families with varying degrees of sequence homology (e.g. Raff *et al.*, 1987; Ohta, 1988). The evolution of such gene families from a common ancestor has been extensively explored, probably the best example being that of the haemoglobins from myoglobin ancestors (Jeffreys, 1982; Jeffreys *et al.*, 1983). Given the very complex structure of existing genes, it seems unlikely that new eukaryotic genes are often obtained by the shaping of a stretch of random junk DNA. Instead, the creation of new loci involves the initial duplication of the complex structure of an existing gene, either entirely or in terms of significant parts (e.g. promoter regions). Subsequent evolution involves modification of this structure, which is already predisposed to have functional consequences. Thus, the evolution of new loci has a strong historical component that must bias or constrain available evolutionary pathways and suggests that various lineages may have structures that differ in evolutionarily significant ways. In a regulatory context, duplication is an essential process. The independent evolution of receptor and recognition sequences at numerous appropriate locations in the genome is probably a statistical impossibility. A process of duplication and diversification (coupled with transposition) is very likely essential to address the regulatory evolution of higher organisms. If so, the genic framework, in its classical form, is largely incapable of addressing the evolution of eukaryote epigenetics.

2.4.8 Gene conversion and molecular drive

Various repetitive sequences or gene families might be expected to diverge randomly, but instead, many sequences or genes show unexpected conservation of structure within species, and sometimes remarkable divergence among species (Hickey *et al.*, 1991). Such unexpected internal homogeneity (or concerted evolution) reflects the operation of several processes, most notably unequal exchange and gene conversion. These can act both within and among chromosomes (Dover, 1982, 1986; John and Miklos, 1988). Unequal exchange is in fact accentuated by the very existence of tandemly repeated genes (Ohno, 1970). Gene conversion involves the modification of one form of repetitive sequence to that of another. Because such conversion can be biased in one direction, fixation of one form of repetitive sequence may be driven in the genome via molecular mechanisms unrelated to selection pressure. Given the potential regulatory implications of repetitive sequences or the degree of gene redundancy, it is not surprising that molecular drive has been postulated sometimes to play a role in speciation and the production of biological novelty (Dover, 1982, 1986). The extent of such contributions remains in doubt, but John and Miklos (1988) consider that this process may rank on an equal footing with natural selection and neutral drift in contributing to evolutionary dynamics.

2.4.9 Transposable elements

Probably the greatest new development in our view of the genome is the diversity and action of transposable elements whose activities dice and splice the genome (Wills, 1989; Fontdevila, 1992; Shapiro, 1992). This aspect will be elaborated in section 4.5. Mobile elements have the capacity to rearrange cytogenetic organization (e.g. to produce linkage groups or changes in chromosome complement), as well as to modify the structure of genes themselves. Thus, enhancers, promoter regions, transcription factor recognition sequences, exons, introns, virtually any functionally significant genetic component, can be rearranged into new potentially useful configurations.

Such dynamics suggests that the genic beanbag is an impossibly simplistic model for considering phenotypic evolution. Although genes may be shuffled as discrete units (i.e. the genome is fluid but genes may still be highly conserved (John and Miklos, 1988)), functional units at both higher and lower levels of organization also exist and their integration may be crucial. Moreover, how does a selfish gene paradigm easily address post-transcriptional RNA processing? As we scrutinize the molecular landscape with increasing sophistication, rather than discovering a lower level of fundamental control, regulation appears instead to recede rapidly into higher levels of organization. Furthermore, these higher units may be selected in their own right. The complex regulatory regions of many genes shows that the beans in the bag are all engaged in vigorous invisible handshaking. Because handshakes are often emitted signals, the bag may even be shaken and stirred to some extent without destroying the reality of coordinated integration.

A really crucial conceptual leap provided by a regulatory perspective relates to the constraints on the organization of the genome imposed by higher-order interactions (Figure 2.5). As Oster *et al.* (1988) point out, in eukaryotes, genes do not make phenotypic structures directly. However, these structures still have a genetic basis. Perhaps the most extreme example of indirect selection on the genome is in social insects where a single genetic program directs morphogenesis into a diversity of specialized forms and these castes often have strong symbiotic or mutualistic integration with other species. All of this organization evolves via selection on queens who reside in protected recesses and have virtually no direct ecological interactions. Another good example of indirect control is the determination of the male sex by the *Syr* gene in mammals (Koopman *et al.*, 1991). Once the gonads have been determined as testes, these then direct further divergence of the male phenotype from the default female form via production of steroid hormones that act among cells and tissues (Gilbert, 1991). Only the lower level of control involves direct genetic determination, following which higher levels of control are released or inhibited.

The significance of this is that the entire epigenetic landscape represents a higher-order field of regulatory potentials, the capture of which requires downward constraints on genome organization (e.g. Katz, 1982; Sander, 1983; Sachs, 1988; Wake and Roth, 1989). What this means is that a considerable amount of pleiotropy and epistasis relevant to the form, function and evolution of the phenotype will reside at these higher levels of organization (i.e. it is epigenetic rather than directly genetic) (Alberch, 1991). Such integration, although not always visible at the genetic level, may nevertheless be a major factor determining genomic coadaptation (Figure 2.5). The potential kinds of higher-order configurations (which involve both internal and environmental interactions) effectively determine the likely pathways available for phenotypic evolution, and as such may constrain the kinds of solutions that are possible. Thus, reaction diffusion models predict that it is possible to obtain a spotted animal with a striped tail, but such a system cannot generate a striped animal with a spotted tail (Murray, 1988). Any appreciation of phenotypic evolution must be firmly based on an understanding of epigenetics, because ultimately, development is both the constraint and the target for natural selection of phenotypes (Levinton, 1988).

3

Epigenetics and developmental hierarchies

3.1 OVERVIEW

The new view of genes and the genome reveals a remarkably sophisticated coordination of metabolism and development via numerous overlays of regulatory integration (Raff and Kaufman, 1983; Arthur, 1984; Davidson, 1986; John and Miklos, 1988; De Pomerai, 1990; Gilbert, 1991; Slack, 1991; Lawrence, 1992; Hall, 1992; Russo *et al.*, 1992). The complex structure of individual genes, the widespread elaboration of gene families and regulatory interconnections via duplications of genetic elements, and processes such as gene conversion or transposable elements are relatively new wrinkles of great evolutionary significance. These developments alone would require extensive modification of the classical evolutionary paradigm.

Development imposes a functional structure on the eukaryote genome that incorporates these multiple levels of control as part of a hierarchical developmental framework. The selfish gene perspective is not only inadequate in its current state to address this framework, it is also inconsistent with a system of hierarchically distributed regulation. We have progressed from knowing a tantalizing smidgen about the genetic control of development to a relatively insightful understanding in only about 15 years. The breakthrough remains confined to a small number of key organisms, notably the nematode *Caenorhabditis elegans*, sea urchins, amphibians (particulary *Xenopus*), the mouse, the chick and more recently, zebra fish. Russo *et al.* (1992) provide an overview across species. The human genome project will also expand this framework (Wills, 1991). By far the most outstanding achievement to date, however, is the work on the fruitfly, *Drosophila melanogaster*, that has progressed to the point that Lawrence (1992) has marshalled a comprehensive account of how to make a fly.

It was my original intention to include several chapters on the epigenetics of *Drosophila*. Not only do several recent and excellent reviews make this redundant (e.g. De Pomerai, 1990; Gilbert, 1991; Slack, 1991; Lawrence, 1992; Kauffman, 1993), but the literature is now far too voluminous. Slack (1991, p. 266) concluded, 'the number of components is large, but not astronomical, and it is even possible for one individual to remember for a few days the names of all the genes and their principle characteristics'. This optimism quickly erodes, however, as one expands the horizon to include mice, fish, amphibians and nematodes. It is already impossible to keep pace with all the literature in this field, and the range of material continues to expand, both phylogenetically and down the developmental hierarchy to incorporate more and more of the developmental program. A concerted effort to put this field into a systems analysis framework would be extremely useful.

The discussion provided here is an executive summary of epigenetics, that amounts to a caricature of the detail available. The purpose is to document the reality of epigenetic regulatory hierarchies, conservation of higher-order features, the ubiquity of complex targeted control and the implications of such phenomena for evolution. The rich regulatory mechanisms outlined earlier need to be placed in an epigenetic context. The best understood system is that of *Drosophila*, and I will provide an overview of epigenetics in that species in support of the idea that regulatory hierarchies impose a functional organization on eukaryote genomes. For greater detail see John and Miklos (1988), De Pomerai (1990), Gilbert (1991), Slack (1991) and Lawrence (1992).

The uninitiated reader may find the following fairly hard to assimilate because of the complexity and number of elements involved. However, it is here that various regulatory tools are deployed in integrated constellations, and these reflect the basic targets of phenotypic evolution. An appreciation of this aspect will be required of anyone interested in the evolution of multicellular organisms so it is well worth the effort to absorb the framework.

The elaboration of an adult fly from an egg initially involves three systems of determination that can be conceptually disengaged and which are somewhat independent in the phenotype. These are: (1) sex determination; (2) dorsoventral differentiation; and (3) anterioposterior differentiation.

3.2 SEX DETERMINATION

For a comprehensive account of sex determination in animals generally, refer to Gilbert (1991). In *Drosophila* the primordial germ cells are derived from 3–12 nuclei that migrate into a region in the posterior pole of the egg and become the pole cells. The cytoplasmic determinants inducing pole cells are maternal gene products, *Tudor* being one candidate. Null mutations of *Tudor* lead to sterility as well as defects in abdominal segmentation. Another maternal gene, *vasa*, is required for proper female development. Thus, the paramountcy of maternally derived information is evident. The pole cells eventually become sequestered in the developing gonads (John and Miklos, 1988; De Pomerai, 1990).

Unlike sex determination in mammals, in *Drosophila* each cell recognizes its sexual status independently (reviewed by John and Miklos, 1988; De Pomerai, 1990; Gilbert, 1991). Female *Drosophila* have an XX constitution and males are XY. In normal flies, cells distinguish their sexual status by interpreting the dosage relationship between the X chromosome and autosome (X/A) that differs between XX females (=dosage of 1) and XY males (=dosage of 0.5). Thus, flies or cell lineages lacking an autosome (i.e. XO) develop male characteristics. The Y chromosome is still needed in later development, however, or spermatogenesis aborts.

The cytogenetic X/A signal is mediated by the X-linked *sisterless a* and *b* genes, the role of *sisterless b* (*sis-b*) being best understood (Erickson and Cline, 1991). *Sis-b* maps to a region within the achaete-scute complex of genes that are involved in later neurogenesis. It comprises one transcription unit in the *scute-a* gene (protein T4) as well as flanking regulatory regions. This protein is a DNA-binding transcription factor that targets the master switch gene, *sex-lethal* (Erickson and Cline, 1991). Single dosages of *sisterless* genes (i.e. one X chromosome) are insufficient to switch on female pathways, but double dosages trigger female determination. The maternal product of the *daughterless* gene (another DNA-binding transcription factor) cooperates with the *sisterless* genes in regulating *sex-lethal*, as do the maternal and zygotic products of the gene *liz*.

The case of *sis-b* is revealing because it illustrates how a single protein (T4) can be used in early sex determination under one set of regulatory controls, and later in neurogenesis under a different set of regulators. Which is the gene: *sisterless-b*, *scute-a* or both? The system also illustrates how different dosages in a single gene may be used as a binary switch mechanism. This kind of organization reveals a critical weakness in the genic paradigm of independent selfish genes because multiple levels of functional integration sometimes may be dissociated for independent purposes. In this case the units of selection are nested within one another.

The switch gene *sex-lethal* stands at the apex of several regulatory pathways, being involved in sex determination of both germ line and somatic cells, and in regulating appropriate dosage compensation in X-linked genes. *Sex-lethal* has two separate promoters. The integrated action of *sisterless a*, *b* and maternal *daughterless* product turn on the early acting promoter in females, but do not initiate such establishment transcription in males (eggs lacking *daughterless* cannot turn on *sex-lethal*). This establishment is associated with a brief pulse of *sis-b* activity very early in development. Transcription from the early promoter of *sex-lethal* derives a product that acts in post-transcriptional RNA splicing. As the early promoter is permanently turned off, the later promoter engages in constitutive transcription. If the earlier *sex-lethal* product is present, transcripts from the promoter are correctly spliced into functional proteins, creating an autoregulatory loop. This loop ensures that female sexual status is permanently 'remembered'. If the early product is not present, RNA from *sex-lethal* is not spliced into a functional product, thus determining permanent male status (Erickson and Cline, 1991). Optimality theorists will not be impressed with this mechanism.

Autoregulation is a common feature of many regulatory genes which means that, once determined, they can maintain their status without constant resetting from other genes. In somatic cells, *sex-lethal*, whose protein product shows evidence of two RNA-binding domains, modulates the switch gene *transformer* via sex-specific RNA splicing.

Differential splicing of the *transformer* transcript under the control of *sex-lethal* leads to a functional mRNA being derived only in females. The presence or absence of this product then modulates the expression of the gene *doublesex*, again via post-transcriptional RNA splicing. Differential splicing of *doublesex* RNA produces two mRNAs, one that represses female expression in male lineages and another that represses male expression in females. A second gene, *transformer 2*, is also involved in regulation of *doublesex* transcription, and both *transformer* genes produce proteins with RNA-binding domains (Hoshijima *et al.*, 1991). *Doublesex* may also be under transcriptional-level control. Downstream of *doublesex*, the gene *intersex* provides a female-specific product that finalizes the sexual differentiation of cell lineages. Numerous downstream genes then act differentially to deploy appropriate male or female anatomy and metabolic functions. Even this simplified summary conveys the sophisticated regulation used in differentiating sexual phenotypes in this species. There is a clear regulatory hierarchy, with levels of control emerging from complex interactions and feedback among genes and among their protein and RNA products at higher levels.

3.3 DORSOVENTRAL DIFFERENTIATION IN *DROSOPHILA*

The first lesson is that the initial patterning of the egg is established by the mother, not by the oocyte genome. Thus, there are at least 15 maternal genes essential for deriving the dorsoventral pattern. The genes *cornichon*, *gurken* and *K10* are required in the oocyte for proper development of the follicle cells. These cells surround the oocyte and interact with particular regions in determining the dorsoventral pattern. The interaction of the above genes activates the receptor gene, *torpedo*, in the follicle cells.

This signal then results in release of a factor that requires the genes *windbeutel*, *nudel* and *pipe* from the follicle cells. This product locally activates the oocyte receptor protein produced by *Toll*. This signal is synergized by local interactions with products of the genes *easter* and *snake*. Within the fertilized eggs the *dorsal* gene product is uniformly distributed but is probably inactivated via binding with that of the gene *cactus*. In those regions where the *Toll* receptor is locally activated, the *dorsal* protein is released from its *cactus* repressor. Interactions with other late-acting maternal genes (*spätzle*, *tube*, *pelle*) are also involved at this stage, possibly modifying the structure of the *dorsal* protein which then allows it to enter nuclei.

In *Drosophila*, the early blastoderm is composed of nuclei that have not yet formed cell membranes (the syncytial blastoderm stage) which facilitates direct inter-nuclear communication. The end result of the above regulatory cascade is that the *dorsal* protein product preferentially enters the blastoderm nuclei on the ventral side. Thus, this complex sequence of maternal gene interactions ultimately serves to provide a smooth ventral–dorsal gradient of *dorsal* product. This ventral–dorsal gradient regulates the expression of zygotic genes in a hierarchy leading to differentiation of: (1) mesoderm; (2) the ventral epidermis and neurogenic region; (3) the dorsal epidermis; and (4) the amnioserosa (an extra-embryonic structure). Thus, the *dorsal* gene product appears to act as a true morphogen.

3.4 ANTERIOPOSTERIOR DIFFERENTIATION IN *DROSOPHILA*

3.4.1 Overview

Anterioposterior differentiation is considerably more complex than that of dorsoventral regulation, but a reasonably clear picture has crystallized (see reviews by Akam, 1987; Scott and Carroll, 1987; Ingham, 1988; John and Miklos, 1988; De Pomerai, 1990; Gilbert, 1991; Slack, 1991; Lawrence, 1992). The initial spatial information is specified by maternal genes that can be classified into the anterior, posterior and terminal systems. These maternal gene products are then interpreted by zygote 'gap' genes, so called because their mutants leave obvious gaps in the body pattern. An immediate lesson is that some maternal effect genes activated in the mother during oogenesis produce the same products that are later deployed by the zygote when it becomes transcriptionally active. This must require different regulatory circuitry to derive the same product at different times in ontogeny.

The gap genes developmentally regulate expression of the pair-rule genes, mutations of which lead to deletions in every other body segment. There appear to be two tiers of pair-rule genes in the hierarchy (Slack, 1991). These lead to clear segmentation in the embryo. The pair-rule genes regulate the segment polarity genes, which finalize the definition of segmental borders and segment polarity. All of this regulatory hierarchy acts to elaborate sequentially and refine spatial information for the segmental anterioposterior pattern. Ultimately, the body is defined as numerous distinct compartments. This information is then interpreted by the homeotic 'selector' genes that switch on downstream pathways leading to features specific to particular body compartments (e.g. antennae, wings, appropriate legs). Thus, a cascade of elaborating decisions leads to coordinated differentiation of segments and segment-specific morphology.

3.4.2 Maternal genes

Most maternal genes act during and previous to the syncitial blastoderm stage, when nuclei divide but cell membranes are absent. The anterior end of the elliptical egg expresses a gradient of the *bicoid* protein. Null mutations of *bicoid* result in loss of the embryonic head and thorax. The genes *exuperantia staufen* and *swallow* may contribute to anterior localization of *bicoid* mRNA perhaps via protein binding. This results in an anterior–posterior gradient of translated protein that acts as a true morphogen. This gradient regulates the zygotic gap genes *Krüppel* and *hunchback* as well as other genes needed to produce the head.

In the posterior pole, mRNA of the maternal gene *nanos* is localized, perhaps via contributions from *pumilio* and *oskar*. *Nanos* product directs development of the abdomen. It acts to ensure the transcription of the zygotic gap gene *knirps* in the prospective abdominal region by inhibiting translation of maternal *hunchback* mRNA in that region (otherwise *hunchback* inhibits *knirps*). If *knirps* is not inhibited in the anterior, mutant 'bicaudal' embryos form that have two posteriors in mirror symmetry. *Caudal* mRNA is initially distributed throughout the egg, but a posterior–anterior gradient forms during the syncytial blastoderm stage, probably under the guidance of *bicoid*. This gene product is also needed for proper segmentation of the abdomen. Lawrence (1992) points out an interesting evolutionary lesson embodied in this complexity. If maternal *hunchback* were simply shut off, then *nanos* and its regulatory associates would not be necessary. Mutant flies deficient in both maternal *hunchback* and *nanos* are completely normal. This says something about parsimony in regulatory evolution.

Maternal genes of the terminal system control development of the acron and telson at the poles of the embryo. The gene *torso* codes for a cell-surface receptor that is selectively activated at the poles by a ligand produced by follicle cells. This system ultimately activates the zygotic gap genes *tailless* and *huckebein*.

The significance of all these maternal genes is that the egg does not generate its own initial information, but a significant imprint must be stamped into the oocyte cytoplasm by the mother. Thus, the genetic program generating an individual phenotype resides partially in the organism itself and partially in the parent (i.e. the information for coding a eukaryote organism is transgenerational, a fact that must ultimately support a lineage paradigm). With respect to that age old question, an egg is in fact partially the last chicken, and the two are not really separable as distinct entities. The chicken and egg represent a continuum as far as epigenetics is concerned, even though they eventually dissociate.

3.4.3 Gap genes

Information bestowed by the maternal genes regulates expression of the zygotic gap genes during the syncytial and early cellular blastoderm stages. Key genes include *hunchback, Krüppel, knirps, giant, tailless* and *huckebein*. Gap genes are all transcription factors (usually with zinc fingers) that regulate both one another and the downstream pair-rule genes. Transcription of *hunchback* is activated by *bicoid* in the anterior embryo, and later, a posterior stripe is activated, probably via *torso*. These products are necessary for development of the labium, thorax and posterior abdomen.

Three other gap genes (*orthodenticle, buttonhead* and *empty spiracles*) function anteriorly of *hunchback* in determining head structure (Cohen and Jurgens, 1991). *Krüppel* is mainly expressed as a central band that arises via a balance between *bicoid* activation and *hunchback* repression. Lawrence (1992) argues that the band margins may be determined by different thresholds of response to *hunchback* and perhaps by *giant* as well. *Krüppel* contributes to the thorax and abdominal seqments 1–5. *Knirps* shows an expression pattern mainly in the ventral head region with two distinct bands located at distances approximately one quarter and two-thirds down the embryo. Expression appears to be via a complex of constitutive transcription, repression by *hunchback*, activation by *Krüppel* and repression by *tailless*. Null mutations of *knirps* have one large segment in place of abdominal segments 1–7.

Tailless is activated by the maternal terminal system (i.e. *Torso*). Null mutations lose parts of the head, the Malpighian tubules and abdominal segments 8–10. This gene may then help position the posterior margins of the *knirps* and *giant* stripes. The expression pattern of *giant* is an anterior and posterior stripe, the latter being repressed by *hunchback*. The anterior stripe appears to be separately activated by *bicoid*.

Finally, *huckebein* is activated by the terminal system, specifying the anterior and posterior midgut. Given that the gap genes are largely DNA-binding transcription factors, their main function is to elaborate and compartmentalize the information initially laid down by the maternal genes.

Thus, these genes represent a holon in the epigenetic hierarchy with strictly regulatory function. The elaborate complexity of integration hinted at here may partially function to provide robust stability to the developmental system (Lawrence, 1992, p. 61). If so, this kind of regulatory integration as well as redundancy (F. M. Hoffmann, 1991) may be one mechanism for achieving phenotypic resistant to environmental or genetic changes (see section 6.6).

3.4.4 Pair-rule genes

There are at least eight pair-rule genes that function as yet another layer of interpretation, in this case linking the gap genes and polarity genes. Mutations usually cause deletions in every other segment of the body. These genes appear to have two tiers, the primary holon involving *hairy, eve* and *runt* which are required for correct expression of the other secondary genes (Slack, 1991). Most of the pair-rule genes ultimately generate patterns of seven stripes that show various degrees of overlap (e.g. *hairy, runt, even skipped,* and *fushi tarazu*). In the case of *paired*, 14 bands are formed. The result is that any anterioposterior location is characterized by different but repeating combinations of these genes.

There has been considerable interest in how such striped patterns could be generated. Turing (1952) demonstrated how relatively simple models of reaction/diffusion involving autocatalysis and inhibition could generate discontinuities. For example, such models can explain patterns of spots or stripes on animals like leopards and zebras (Murray, 1988). Hassell, Comins and May (1991) even showed that such complex patterns could be generated via simple models of population growth and local dispersal in ecological systems. Models of *Drosophila* development based on these principles have remarkable explanatory power, even predicting the transient stages of striped patterns seen in some genes (e.g. Meinhardt, 1986; Hunding, Kaufman and Goodwin, 1990; Kauffman and Goodwin, 1990).

Recently, however, it has become apparent that the expression of individual stripes is specified by particular regulatory regions (e.g. Howard, Ingham and Rushlow, 1988; Dearolf, Topal and Parker, 1989; Pankratz and Jäckle, 1990; Stanojevic, Hoey and Levine, 1989; Stanojevic, Small and Levine, 1991; Lawrence, 1992). All of these genes have relatively immense regulatory regions (Slack, 1991), and individual stripes may be defined by complex combinations of maternal genes and gap genes (Lawrence, 1992). It still seems possible that Turing-type processes could contribute to the overall dynamics, however. In that case, regulated features could act as an underpinning framework to obtain a specific result via further reaction/diffusion interactions. Kauffman (1993) reviews considerable evidence supporting the idea that Turing-like processes provide a scaffold that evolution has utilized for segmental control.

The maintenance of the pair-rule stripes is mediated by regulatory regions conferring autoregulation (Stanojevic, Hoey and Levine, 1989) for *even-skipped* and (Hiromi and Gehring, 1987) for *fushi tarazu*. In the case of *fushi tarazu* there is a regulatory region (zebra) which allows modulation of transcription by at least four other regulatory genes (i.e. *hairy, runt, eve* and *caudal*). *Fushi tarazu* also has a separate regulatory region controlling its expression later in the central nervous system reminiscent of the situation described for *sisterless b*.

3.4.5 Segment polarity genes

There are at least nine segment polarity genes that represent the final overlay that achieves the segmental pattern by the cellular blastoderm stage. A complication arises in that the fundamental subdivision of the embryo corresponds to segments that are out of phase with the final segmentation pattern. These parasegments have their anterior border in the middle of one prospective final segment and their posterior border in the centre of the next. The parasegmental boundaries conform to the borders between the expression patterns of the genes *engrailed* and *wingless*. These segment polarity genes appear to be regulated initially by combinatorial interactions with pair-rule genes. Once established, interactions among the segment polarity genes maintain their expression pattern. Segment polarity genes show expression patterns with 14 stripes, the odd and even stripes being independently regulated. The expression of the genes is out of register with that of others so that each parasegment may be compartmentalized into four regions demarked by unique combinations of regulatory genes (Ingham, Baker and Martinez-Arias, 1988; Martinez-Arias, Baker and Ingham, 1988; De Pomerai, 1990).

3.4.6 Homeotic selector genes

The major function of all of these regulatory holons is to elaborate an increasingly complex pattern of spatial information. The homeotic genes interpret this information locally and probably select the appropriate downstream pathways to derive structures such as antennae, mouthparts and legs in their correct locations and alignment. There are at least nine homeotic genes grouped in two major gene clusters, the *Antennapedia complex* (ANT-C) and the *Bithorax complex* (BX-C). These complexes also contain extensive regulatory sequences in their intergenic regions and introns. Most homeotic genes (e.g. *labial, deformed, sex combs reduced, antennapedia, ultrabithorax,* and *abdominal A* and *abdominal B*) contain homeoboxes, indicating that they are transcription factors. The genes are expressed in a temporal sequence matching their spatial position along the third chromosome.

In keeping with their function as interpreters of information supplied by the combinatorial status of higher genes, the homeotic genes are complexly regulated by activation and repression by gap, pair-rule and segment polarity genes, as well as by interactions with other homeotic genes and autofeedback. Some genes contain multiple promoters.

Mutations in homeotic genes can be striking. For example, mutations of *Antennapedia* may transform antennae to legs (Struhl, 1981), and loss of function in regions of *Ultrabithorax* can produce flies with four wings instead of the usual two. A major lesson emerging from these genes is that numerous mutations previously hypothesized to reflect separate loci are now recognized as mapping to different regulatory regions in single genes. Thus, at least four hypothesized homeotic genes are now known to be regulatory sites in *Ultrabithorax*. Similarly, rather than the expected series of loci, each controlling development of specific abdominal segments, it now appears that abdominal differentiation is largely associated with elaboration of regulatory machinery in only two genes, *abdominal-A* and *abdominal-B*. The full picture of the regulatory epigenetic hierarchy is still incomplete. Downstream of the homeotic selector genes lie other regulatory genes, some of which appear to be growth factors, gradient determinants, and some involved in cell–cell interactions and adhesion (Edelman, 1989; Gould *et al.*, 1990; Reid, 1990; Budd and Jackson, 1991; Lawrence, 1992). Only a broad outline is visible at this time, but the rapidly unfolding panorama is a thing of real beauty. The key linkage between the morphological patterning of the phenotype and underlying genomic organization appears to reside with the mechanisms involved in determining positional coordinates on finer and finer scales (Wolpert, 1983).

3.5 THE EPIGENETIC HIERARCHY: A FUNDAMENTAL GENOMIC ORGANIZATION FOR MULTICELLULAR EUKARYOTES

The above discussion suggests several important aspects that are largely absent from the genic paradigm. Firstly, the complex overlays of regulatory machinery represent a degree of interdependence and cooperation that belies the possibility of selfish genes. Secondly, the hierarchical organization of these regulatory mechanisms imposed by development has considerable evolutionary implications that have not been adequately assimilated into the modern synthesis. The other aspect, that of parental contributions to the phenotype, will be considered in Chapter 6.

Even single-celled prokaryotes exhibit relatively sophisticated regulation of metabolism and cellular functions. The conceptual differentiation of structural genes (that make functional end products) from regulatory genes (that control other genes) is often not possible, however, because genes that contribute structural products contain regulatory recognition sequences, and particular domains of protein products (or even regions on the products of enzyme catabolism or synthesis) may also be recognized and used in feedback control. Thus, a typical metabolic pathway may involve the stepwise interplay of regulatory genes, structural genes and other biochemical factors. This organization allows flexible control of the initiation, rate and switching of metabolic pathways which constitutes the main evolutionary framework for prokaryotes.

The eukaryote cell is a more complex organization that allows even further elaboration of regulatory levels. Despite this, most of the fundamental biochemistry of eukaryotes is remarkably conserved (even compared to that of prokaryotes). Shapiro (1992) reviews some aspects. For example, insects have relatively little need of calcium (because their skeletons are chitin) and they have a few other special requirements. Otherwise, insects largely have nutritional requirements very similar to mammals (Gordon, 1959, 1972). It is only a slight exaggeration to say that most animals reflect differential packaging of the same basic biochemistry and regulatory elements (Alberch, 1991; Shapiro, 1992). The increased genomic organization of eukaryotes is apparently not focused on evolution of novel biochemical capacities, but rather on the elaboration of multicellular phenotypes via differentiation, regional specification and morphogenesis. Rather than simply a loosely coadapted pool of genes, regulatory dynamics at the molecular level involves sequence-specific functional interactions.

Beginning in the 1960s, various authors began to appreciate that evolution of multicellular organizations may be largely associated with regulatory genes rather than the structural genes which were the vogue of population genetics (Britten and Davidson, 1969, 1971; Ohno, 1970; Wilson, Sarich and Maxson, 1974a,b; King and Wilson, 1975; Wilson, 1976; Hedrik and McDonald, 1980; Raff and Kaufman, 1983; Kauffman, 1985, 1993; Davidson, 1986; Thomson, 1987; Dickinson, 1989; McDonald, 1990). As outlined above, rapid progress in developmental genetics has resoundingly confirmed that the development of organisms is controlled by hierarchies of regulatory genes and higher epigenetic structures resting upon them. Eukaryotic evolution must proceed largely by modification of this organization (Akam, 1987; Arthur, 1988; Ambros, 1989; De Pomerai, 1990; McDonald, 1990; Alberch, 1991; Gilbert, 1991; Slack, 1991; Hall, 1992; Lawrence, 1992). In fact, in most important genes involved in development, their regulatory regions are enormous compared with their protein-coding regions. In the bithorax complex, 95% of DNA is regulatory and only 5% is transcribed (Lawrence, 1992, p. 213). Many of these transcribed proteins are themselves regulatory signals that function only briefly in the nucleus. Thus, most genetic information (and evolution) is physically associated with regulatory regions rather than structural regions.

In the traditional paradigm, where genes were viewed as discrete, autonomous units, mutations or alternative alleles were seen as hinging on changes in the transcribed proteins. We now know that whole families of mutations previously thought to reflect changes in separate genes in fact map to changes in regulatory elements of single genes (e.g. the BX-C complex) (Slack, 1991; Lawrence, 1992). Not only is there no need for a change in the transcribed protein, but such regulatory mutations effectively represent a change in the integration of the gene with others. In other words, regulatory mutations reflect changes in genomic organization, and are important because genes are not independent selfish elements. As Oster *et al.* (1988) and Shapiro (1992) point out, the evolution of phenotypes may not emphasize the products of new genes, but rather the timing and location of expression of existing genes (i.e. regulatory reorganizations).

The genetic organization determining segmental identity in *Drosophila* entails unique interactions of genes specifying positional information within specific compartments (Akam, Dawson and Tear, 1988). During this process continuous gradients are converted to an essentially digital encoding of information (Baumgartner and Noll, 1991). Bodnar, Jones and Ellis (1989) claim that such interactions constitute a higher-order genetic language, and in this sense, such genes may be analogous to letters that are only relevant in the context of the words that they comprise.

The genic paradigm holds that speciation involves the accumulation of many different allelic variants in diverging lineages. These are always considered to be individually of small effect. Thus, evolution proceeds as a gradual process over long periods. It now appears that there may be very little relationship between the degree of genic changes and the phenotypic evolution of species. The first tremor came with the revelation that humans and chimpanzees, despite their remarkable behavioural and morphological differences, are genetically 99% identical with respect to their structural genes (King and Wilson, 1975).

Comparative studies of phenotypically diverse phylogenies (e.g. mammals) *versus* phenotypically conservative groups (e.g. frogs), revealed no correlation between genetic and phenotypic diversity (Wilson, Sarich and Maxson, 1974b). More recently, molecular information has clarified some questions related to one of the world's most interesting evolutionary stories. Lake Victoria and other African rift lakes are inhabited by diverse species of cichlid fish that are of recent evolutionary origin (Dorit, 1990). Lake Victoria contains at least 200 species that have evolved in only the last one million years. There has been remarkable trophic diversification in various rift lakes (e.g. herbivorous grazers, snail crushers, insect and fish predators). One species even plucks the eyes from other fish (Avise, 1990). It is rather stunning then, that Meyer *et al.* (1990) found less genetic variation among genera of lake Victoria cichlids than exists within a typical vertebrate species (e.g. humans) or even within a highly conserved species like the horseshoe crab (Avise, 1990). Meyer *et al.* (1990) estimate that all of this explosive speciation arose from a single common ancestor in less than 200 000 years. Moreover, the occupation of parallel niches by cichlid fish in separate lakes appears to be via convergent evolution, rather than shared descent. Shapiro (1992) goes so far as to suggest that the real difference between an elephant and a mouse may not be related to their structural genes (which may be functionally similar) but to the control of their timing and transcription.

The Hawaiian *Drosophila* provide another insightful example. *Drosophila heteroneura* sports a fabulous head shape reminiscent of the hammerhead shark (Figure 3.1). Despite this, it is perfectly interfertile with *D. sylvestris*, which has a normal-shaped head, the large phenotypic difference being attributed to perhaps as few as 10 different genes (Val, 1977; Lande, 1981). Dickinson (1989, 1991) reviews strong evidence that regulatory evolution has contributed to the diversification of Hawaiian *Drosophila*.

Another classic evolutionary system concerns those gems of the Everglades, the tree snails, *Liguus fasciatus*. These snails are renowned for their remarkable diversity in shell colour and they have been variously

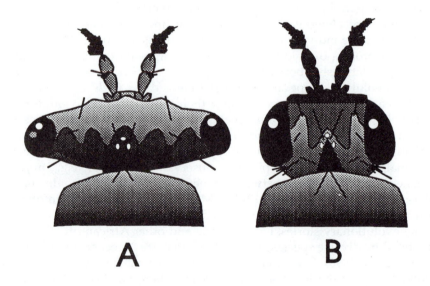

Figure 3.1 Divergent head shapes in (A) *Drosophila heteroneura* and (B) *D. sylvestris*. Despite remarkably different phenotypes, these flies are interfertile. Redrawn after Val (1977) with permission of *Evolution*, Allen Press.

classified into separate species. The Everglades is essentially a very slow-flowing river, covered in sawgrass and peppered with tree islands called hammocks. The tree snails are distributed as isolated demes among these hammocks. Hillis, Dixon and Jones (1991) recently documented a striking paucity of genetic diversity in these snails. Evidently this represents a single species with diversity obtained by isolation and local inbreeding. Colour variants could arise mainly by single allelic variants in structural genes, but the results in *L. fasciatus* suggest that regulatory evolution may also be involved here.

On the other side of the coin, various species regarded as living fossils show no evidence of restricted genetic variation or evolutionary potential. Thus, organisms with remarkably conserved phenotypes, including horseshoe crabs (Selander *et al.*, 1970), and plethodon salamanders (Wake, Roth and Wake, 1983; Wake and Larson, 1987), may show normal levels of genetic variation. Kimura (1983) points out that an ancestor essentially homologous to the highly conserved opossum gave rise to the huge marsupial diversity that developed in South America before the Panama land bridge introduced mammalian competitors. Thus, there is no *a priori* reason to believe that such living fossils lack evolutionary potential.

Species stasis in the presence of high genetic variation is surprising but can be reconciled with the genic view if long-term stabilizing selection exists (Maynard Smith, 1983; Lande, 1986). Some authors have pointed out, however, that stabilizing selection that acts over periods of millions of years seems highly unlikely and that some further factor such as developmental constraints must also be acting (e.g. Arthur, 1988; Maynard Smith *et al.*, 1985). Even if the genic view can stretch to explain stasis, evidence of rapid evolution and high phenotypic variation without associated high genetic variation or change is contradictory. Such results strongly support the paramountcy of regulatory evolution in eukaryotes. The fact that most species differ at many loci has been argued to support the notion of gradual genic divergence, but this is confounded by history. Species separated for long periods will naturally diverge. The crucial question involves the diversity associated with recent speciation and whether detectable changes actually entail changes in epigenetic function. The African cichlids reveal that rapid phenotypic evolution is decoupled somewhat from genic divergence. Although this does not reject the traditional view of speciation via accumulation of allelic variants, it does reject genic claims that all evolution must conform to this model. Alberch (1982) argued that epigenetic systems may tend to show discontinuities, even if genetic changes are gradual.

My original interest in insect epigenetics derived from curiosity as to whether morphological redundancy (as seen in the myriapod-like ancestor of the insects) might be reflected by genetic redundancy, and if so, whether evolution had proceeded by specialization of initially duplicated genes for each body segment. This seemed to me to be very relevant to questions pertaining to the possible role of redundancy in morphological evolution, and whether lineages with such redundancy had a greater potential for evolution. Alberch (1991) suggests that both multicellularity and segmentation were inventions that contribute to evolvability.

The real story is obviously far more complex. There is no one-to-one mapping of a chromosomal region to particular segments. Creatures with identical repeating segments could be generated from elaboration of a common developmental cascade in each segment (although there must still be some way of controlling segment number). For segments to obtain specialized features, however (e.g. as antennae, jaws, mouthparts, specialized legs, wings, genitalia), there must be a diversification of the regulatory hierarchy and imposition of numerous switches so that different pathways are followed in different body regions. Akam (1989) suggests that insects have added genes to a fundamental program diversifying head to tail, so as to highlight segmentation, whereas mammals elaborated the same initial program differently.

An oversimplified conceptual model suggests that early-acting genes of large effect constrain later-acting genes of progressively smaller effect in an epigenetic cascade of bifurcating decisions leading to the adult phenotype (e.g. Rendel, 1967; Gould, 1977; Oster and Alberch, 1982; Raff and Kaufman, 1983; Arthur, 1984, 1988; John and Miklos, 1988; De Pomerai, 1990; Atchley and Hall, 1991; Slack, 1991; Hall, 1992; Lawrence, 1992). This is clear in the epigenetics of *Drosophila* which involves progressive delineation of finer detail which differs among segments. Early-acting genes of large effect determine anterior–posterior and dorsal–ventral axes. Later-acting genes determine particular compartments for development of segments (in insects), or regions and other genes subdivide these compartments (e.g. anterior and posterior of the wing). As a limb develops, other batteries of genes sequentially determine the finer details along a proximal to distal orientation. Thus, Yokouchi, Sasaki and Kuroiwa (1991) observed that one battery of regulatory genes was associated with major limb segments in developing chicks, while another battery was associated with bifurcations resulting in the forelimb, metacarpals and digits (Figure 3.2).

All of the evidence underscores the fact that the epigenesis of eukaryote phenotypes involves a hierarchical cascade of bifurcating regulatory decisions. This concept of hierarchical organization is an old one, dating back to the epigenetic landscape model of Waddington (1940, 1957) (reviewed by Arthur, 1988). Bonner (1973) can be credited with one of the clearest early articulations. With the revolution in developmental genetics in the 1980s, the genetic underpinnings for such a foundation were progressively revealed. The Gestalt was simultaneously snatched from the ether by numerous authors (e.g. Gould, 1980; Hedrick and

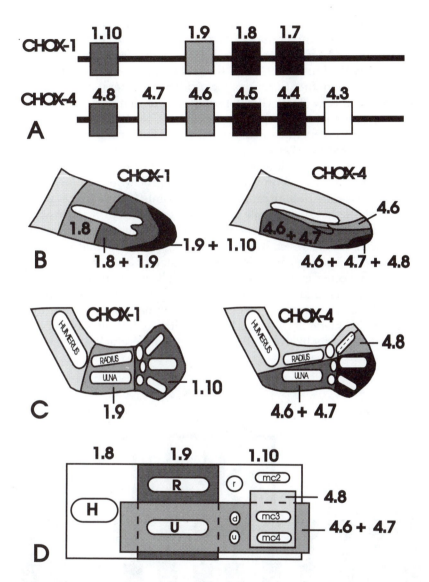

Figure 3.2 Regulatory genes determining vertebrate limb development. Such genes represent a developmental hierarchy specifying progressively finer degrees of diversification. Redrawn after Yokouchi, Sasaki and Kuroiwa (1991) with permission of *Nature*, Macmillan Magazines Ltd. and the author.

McDonald, 1980; Hunkapiller *et al.*, 1982; Raff and Kaufman, 1983; Arthur, 1984, 1988; Wake and Larson, 1987; Blau, 1988; De Pomerai, 1990; Gould *et al.*, 1990; Immergluck, Lawrence and Bienz, 1990; McDonald, 1990; Morata and Struhl, 1990; Slack, 1991; Hall, 1992; Kuo *et al.*, 1992; Lawrence, 1992; Tautz, 1992). Of these, Hunkapiller *et al.* (1982) and Arthur (1988) were first to seriously consider the general implications to evolutionary theory. Hall (1992) provides extensions. Arthur's (1988) conclusion was that the genic framework must be abandoned.

The regulatory structure of eukaryotes could conform to several potential models. Firstly, earlier-acting genes in a given holon might regulate those genes in the next lower holon directly, but genes at even lower levels would be affected only indirectly. Control would follow a strict hierarchical cascade of bifurcating decision points (Figure 3.3). Kauffman (1993) suggests that such simple hierarchies would be difficult to achieve and maintain in a genome where mutations may continuously forge new interconnections. However, Kauffman's (1993, p. 501) models still predict that hierarchical organization is likely to emerge, even in complex networks. At the other extreme the genome might be interconnected into a complex network that obscures hierarchical underpinnings (Kauffman, 1983, 1993). Developmental genetics has discerned that a mixed model dominated by hierarchical cascades but with some trans-holonic interconnections appears likely (Gould *et al.*, 1990; Immergluck, Lawrence and

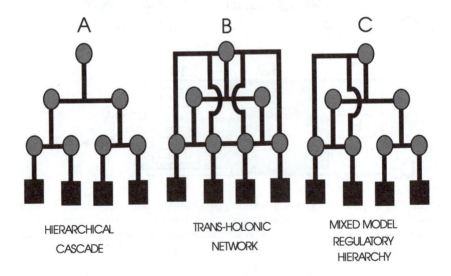

A   B   C

HIERARCHICAL   TRANS-HOLONIC   MIXED MODEL
CASCADE   NETWORK   REGULATORY
                                     HIERARCHY

Figure 3.3 Alternative models of genomic organization. (A) Simple hierarchies, (B) gene networks or (C) mixed models.

Bienz, 1990; McDonald, 1990; Morata and Struhl, 1990; Baumgartner and Noll, 1991). We are only beginning to assemble information on the sequential integration of such hierarchies so the final picture remains obscure (Morata and Struhl, 1990).

Given the tendency of evolution to exploit any potential adaptive avenue, some trans-holonic integration in regulatory hierarchies would be expected. A review by Matsuda (1987, p. 5) also supports a hierarchical cascade model based on endocrine control of sequential gene switching during insect metamorphosis. There may be mixed responses to endocrine signals suggesting that some features are controlled by local gene switches while others (such as pigmentation) may respond to more global signals.

In *Drosophila*, there are cases where genes at one level (e.g. the homeotic selector genes) are regulated by several other levels of control (e.g. gap genes, pair-rule genes and segment polarity genes) (Figure 3.4). There are also interactions among genes within a given holon as well as auto-regulatory feedbacks. Thus, the regulatory structure of the genome is a complex hierarchy with trans-holonic and intra-holonic integration. The

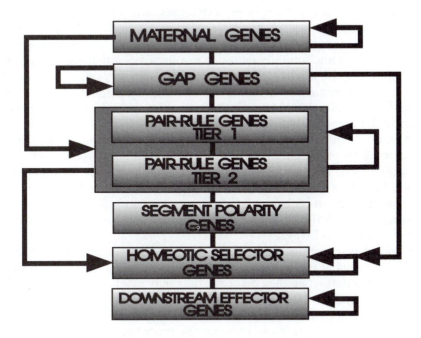

Figure 3.4 The hierarchical organization of epigenetics in *Drosophila*. The system conforms to a mixed model with a discernible hierarchical structure.

degree of integration can be appreciated by the estimate that perhaps 30% of all genes in *Drosophila* contribute to the development of the eye (Lawrence, 1992, p. 195). Kauffman (1993, p. 505) provides evidence that ecdysone (the hormone that controls moulting) may impact directly on at least 155 genes in *Drosophila*.

There are further complications. For example, developmental decision points are generally congruent with regulatory gene activity, but they can also reside at higher levels such as RNA processing or in cell–cell interactions. In some cases it is not simply a switching event that must be considered but the timing, rate and duration of gene activity, and the distribution, duration and binding rates of gene products. Thus, Sachs (1988) points out that major developmental shifts with large phenotypic impact do not necessarily require large genetic changes. For the purposes of discussion, referring to such complex dynamics simply as a regulatory hierarchy captures the key concepts.

An immediate question is, to what extent does evolution proceed via modification of higher-order regulatory features? A key factor that has allowed rapid progress in developmental genetics is that there is considerable conservation of higher-order features even across very broad phylogenies. This pertains not only to various DNA-binding domains such as the homeobox (now known from vertebrates, arthropods, echinoderms and molluscs), but also in actual homology among regulatory genes. Thus, many key regulatory genes found in *Drosophila* have homologous relatives in mice and amphibians. Such results suggest that higher tiers in the regulatory hierarchy are more ancient and represent relatively consolidated organization (Akam, Dawson and Tear, 1988; Akam, 1989; Duboule and Dolle, 1989; Alberch, 1991; Shapiro, 1992). A lot of progress in particular areas is obtained by raiding genetic libraries across species. Even entire regulatory gene complexes (i.e. ANT-C and BX-C) have homologous representation in insects and vertebrates (e.g. Duboule and Dolle, 1989; Graham, Papalopulu and Krumlauf, 1989; Rossant and Joyner, 1989; Kenyon and Wang, 1990), suggesting that there is considerable conservation of higher-order regulatory structure. This would make sense since any radical divergence of a higher-level gene might disrupt all downstream pathways.

This kind of constraining structure suggests that successful contributions of changes in higher-order regulators will decline as we climb that hierarchy, but that when they do occur, they may be of large effect. A note of caution is in order here. For example, the genetic information required to elaborate complex nervous systems may equal all of that needed to specify metabolic functions and phenotypic morphogenesis (reviewed by John and Miklos, 1988). It is significant that rather than creating new genes, nearly every class of gene used in the early epigenetic hierarchy has been re-used for specific functions later in neurogenesis (see examples in De Pomerai, 1990; Slack, 1991; Lawrence, 1992). In many cases this involves completely separate promoters and regulatory regions. Thus, changes in the associated regulatory feature of a gene may completely alter its use, or allow it to be re-utilized without the necessity of initial gene duplications as envisioned by Ohno (1970). This ability to use genes in different ways based on regulatory divergence will be discussed further when considering phenotypic plasticity.

De Pomerai (1990) concluded that strong conservation of elements could either reflect a basic unity of higher-order developmental organization across diverse phylogenies, or it could conceal a bewildering complexity in deployment of key conserved components for novel purposes. The truth probably lies somewhere in between. Multipurpose utilization of a structural element or the entire transcript from a particular gene would add further constraints on possible changes in exons. In this case, regulatory evolution can not only bypass this constraint, but becomes increasingly the only way to proceed.

It seems likely that mutations of large phenotypic impact are less likely to be successfully assimilated into the hierarchical organization than lower-level tertiary deviants. Not only do genes in lower holons have progressively less phenotypic effect, they also impact on fewer downstream genes. Thus, the most likely route for new organizational evolution is via adding on tertiary genes of small effect (Rasmussen, 1987). Conversely, evolutionary simplification may delete such gene functions first. For example, the literature on limb loss in various vertebrate lineages suggests that although selection may access higher levels of control directly, it more often acts progressively up the regulatory hierarchy. Thus, loss of limbs in tetrapods is associated with a common sequence of degeneration proceeding from the outer digits to the main trunk attachments.

Although there may be some variation in the specific sequence of digits lost, there is a generally common pattern (reviewed by Lande, 1978; Vermeij, 1987; Oster *et al.*, 1988). This is inversely correlated with the hierarchical organization of genes controlling development of such structures (Yokouchi, Sasaki and Kuroiwa, 1991) and their appearance in development (Lande, 1978) (Figure 3.2). It is tempting to suggest that the hierarchical model never decomposes into a genic paradigm because higher levels of organization may be embedded in overlays of constraint. Selection may act like peeling an onion. Undoubtedly, the situation is more complicated than this. Darwin (1859) observed increased variability in structures no longer of apparent function, consistent with selective release of lower-level holons. Simplification of the limb is also accompanied by progressive miniaturization (Brace, 1963), and control of such proportionality remains obscure (Slack, 1991; Lawrence, 1992). Organisms at intermediate stages may retain some adaptive use of their reduced appendages which will also modify the course of selection.

Although this example suggests that lower-level holons are generally affected first, it remains that the genic and hierarchical views are fundamentally incompatible. The hierarchical model also does not rule out a macroevolutionary loss of limbs by a higher-order mutation. The fact that whales occasionally express external hind limbs (although reduced in size) suggests that in this lineage, limb loss involved a general reduction in appendage size (Wyss, 1990) followed by a higher-order mutation that bypassed downstream subprograms (Gingerich, Smith and Simons, 1990). A single dominant mutation will also restore clawed digits to the wings of chickens (Cole, 1967). Such evidence suggests that limb reduction proceeds by accumulation of suppressor genes (Lande, 1978).

A hierarchical regulatory organization creates constraints on allowable variation. For example, all genes acting in a given regulatory circuit must have complementary signal and recognition elements or at least systems enabling indirect linkages. Higher-order organization, once consolidated, may persist almost indefinitely by virtue of all of the subsequent organization built on that foundation. Ohno (1970, 1973) refers to such phenomena as frozen accidents. It is not too difficult to extend this reasoning to envision how frozen accidents in higher regulatory organization could promote species stasis or impose developmental constraints on future evolution (especially where genes are re-used).

Further evidence that upper regulatory tiers are subject to modification is provided by the fact that juvenile stages may undergo evolutionary modification (Rice, 1990). It is thought, for example, that the spiral body plan of adult gastropods was derived from selection pressure on juveniles that allowed defensive withdrawal of the head region. The required torsion during development is consequently transmitted to adult phenotypes which then developed for the modifications to displace the anus from the head region and improve the distribution of weight for efficiently carrying the shell. If true, such an event constitutes a macroevolutionary jump of large effect. Among many organisms the problem of modifying juvenile stages without incurring such drastic impacts on the adult are achieved via developmental metamorphosis. In such cases, the regulatory hierarchy is compartmentalized to various degrees. Thus, in *Drosophila*, tissues destined to become adult structures are derived in parallel with larval development, but the primordia are sequestered in quiescent pockets of cells called imaginal discs that are only activated during the pupal stage. In such systems, both the larval and adult stages represent diversified end points in a bifurcating developmental hierarchy, rather than representing a serial continuum.

Such genomic organization allows some freedom for specialized developmental stages, impacts on adult phenotypes depending on the degree that regulatory features are shared. Williams (1992) has suggested that developmental systems may be extremely compartmentalized so as to allow independent evolution of self-constrained parts. In fact, the widespread existence of metamorphosis among diverse phylogenies attests to the hierarchical structure of the genome and solutions to circumvent the constraints such organization imposes (Matsuda, 1987; Arthur, 1988). Kauffman (1993) proposes a very different model that also addresses the dual need for stability and evolvability. His model suggests that highly interconnected genetic networks may generate stable subsystems, but such systems may be capable of rapid reorganization if disturbed.

A particularly illustrative example of the importance of regulatory organization as an evolutionary feature in its own right comes from experiments inducing expression of moth egg chorion genes in *Drosophila*. Remarkably, the moth genes, when integrated into the *Drosophila* genome, are expressed in the correct sex, at the appropriate time and in the proper tissues. Evidently, the relevant regulatory structure has been conserved across flies and moths, even though particular structural genes have markedly diverged (Kafatos *et al.*, 1985; Mitsialis and Kafatos, 1985). Although eukaryotes tend to have genomes with widely distributed structural genes, there is a strong tendency for regulatory genes to be localized in tight batteries. This is undoubtedly related partly to the derivation of new genes or regulatory regions via tandem duplications and subsequent divergence.

Given the apparent antiquity of these genes, and the relative fluidity of the genome, however, it seems highly unlikely that such tight physical linkage can simply reflect their mode of origin. Slack (1991) suggests that the high density of regulatory sequences intermeshed in the introns and intervening sequences between such genes may have contributed to conservation of regulatory complexes as single units. Regardless, the evolutionary significance is that there may be little recombination in these complexes, and subsequently, even if polymorphisms exist, there is a good chance that coadapted complexes could be assembled that physically resist disruption during sexual reproduction. If coadaptation is advantageous, local rates of recombination could also be suppressed, achieving relatively stable fundamental organizations.

It is well established that most sexual species contain a considerable store of genetic variation, both via heterozygosity of alleles within particular loci, and via balanced complexes of coadapted loci (e.g. Hedrick, 1986; Carson, 1990a; Mitton, 1993). There has been considerable support for the idea that in sexual diploids the optimal phenotype is in fact partially dependent on such heterozygosity (see Lerner (1954) for the original synthesis). If so, the regulatory hierarchy might incorporate alleles in a heterozygous state as one dimension of its functioning. There appears to be little information from developmental genetics impinging directly on this aspect, although there is ample evidence of the phenotypic impacts of mutations in these genes. It seems that even a subtle change in a regulatory receptor could change the action of a transcription factor sufficiently to obtain phenotypic consequences. Thus, allelic variants could be easily overlooked at the present state of the art.

Clearly, regulatory hierarchies pose some interesting questions. How far up the hierarchy do heterozygous loci exist? How does the evolution of dominance fit into this framework? Do higher-order variants sequester downstream coadapted complexes? At present we can only ask the questions. In summary, considerable empirical evidence and logic suggest that there will be considerable conservation of higher-order circuitry. Batteries of regulatory genes also show tight linkage that might reduce recombination, and recombination could also be locally suppressed.

So, is there any evidence that evolution has proceeded at higher levels despite all this? The era of comparative regulatory organizations has not yet arrived, but the available evidence suggests a very fruitful field. Seeger and Kaufman (1987) used DNA hybridization to examine the conservation of the ANT-C across a range of species spanning *Drosophila*, houseflies, mosquitoes and a moth. They found strong conservation among closely related species, and diminishing homology for more distantly related organisms. Even then, some homology was detected across phylogenies separated by at least 100 to 250 million years.

Tribolium beetles also show good homology to the *Drosophila* homeobox genes, but they have a single regulatory cluster rather than separate ANT-C and BX-C complexes (Beeman *et al.*, 1989; Stuart *et al.*, 1991). Thus, the separate clusters in *Drosophila* probably were originally derived from a single large complex (=HOM-C). Some of the mutant phenotypes in the beetle differ somewhat from *Drosophila* suggesting that the ultimate developmental functions have diverged (Slack, 1991). The honey-bee also shows very high conservation of some genes found in *Drosophila*, despite a separation of 250 million years (Walldorf, Fleig and Gehring, 1989). Significantly, however, the key *Drosophila* gene, *fushi tarazu*, was not found in the honey-bee, while a honey-bee homeotic gene shows no homologous relative in *Drosophila*. *Tribolium* beetles also appear to lack *fushi tarazu* (Stuart *et al.*, 1991).

In the mouse, there are four main clusters of homeobox genes (HOX clusters) that show close homology to one another and to the ANT-C/BX-C (HOM-C) clusters of *Drosophila* (Akam, 1989; Duboule and Dolle, 1989; Graham, Papalopulu and Krumlauf, 1989; Rossant and Joyner, 1989; De Pomerai, 1990; Holland, 1990). This strongly supports Ohno's (1970) contention that the regulatory evolution of mammals was facilitated by a polyploid event in an ancestral fish. Such duplication would free each gene to pursue divergent roles. Significantly, each HOX cluster contains subfamilies of genes with recognizable members occupying comparable positions in each cluster (Duboule and Dolle, 1989).

Crucial insight into regulatory evolution is obtained from the study of

Patel, Ball and Goodman (1992) comparing the grasshopper, *Schistocerca americana*, to *Drosophila*. In both lineages *even-skipped* functions similarly in neurogenesis in later development, and a neurogenic function is also retained in vertebrates. In *Drosophila*, the segmental pattern is finalized at the end of the blastoderm stage, largely via pair-rule control of *engrailed* by *even-skipped*. In grasshoppers, segments are added on sequentially following the blastoderm stage, and in these insects, *even-skipped* does not serve a pair-rule function (i.e. it does not regulate *engrailed*). Holland (1990) reviewed evidence suggesting that both *fushi tarazu* and *engrailed* were co-opted to serve roles in segmentation in flies, but this function is not universal in other lineages. Such results suggest that genes primarily dedicated to one function may be commandeered for other purposes with no loss of their original role. In the case of the fly, genes with neurogenic functions may have been modified to obtain a more advanced form of segmental pattern definition or vice versa.

Although the evidence remains sparse, it is apparent that even the upper developmental holons are not frozen fast, but undergo significant evolution despite remarkable conservation. There is evidence for diversified function via elaboration of regulatory genes, and deletions and additions of particular kinds of genes. Changes in such fundamental control elements suggest that macromutations of evolutionary significance do indeed occur. A genic, microevolutionary model may have been salvaged if only tertiary, lower-level genes were evolving. The fact that the entire hierarchical structure is open to modification as an integrated organization rejects the genic model as a workable explanatory framework.

Selection can theoretically act on all levels of the regulatory hierarchy simultaneously. Although various levels of organization may have different degrees of variability, mutations will continue to be available for periodic selection even in highly conserved genes. Higher-level mutations are likely to cause radical phenotypic shift, and further evolution would be required to restore coadaptation with downstream components. Previous models based on equilibrium ecology precluded such individuals surviving the rigours of intense competition. In a non-equilibrium paradigm, lower fitness phenotypes of functional competence could persist in otherwise vacant patches or during periods of low competition long enough to consolidate a viable organization (Wiens, 1977). Consider, for example, what I consider to be a 'Goldschmidt toad', that in spite of a monstrous developmental anomaly (i.e. having your eyes in your mouth) was found doing very well in a local garden (Figure 3.5).

Selection on downstream modifiers could follow a pathway returning to the wild type. There are many examples of the impact of a mutation being ameliorated over time as modifiers adjust to damp its effect (Schmalhausen, 1949; Lerner, 1954). However, if another selective equilibrium were close by, the genotype could theoretically shift to a new

Figure 3.5 A 'Goldschmidt toad'. Despite its remarkable developmental alteration, this toad with its eyes in its mouth was found doing very well in a local garden. With permission of *The Hamilton Spectator*, Southam News.

adaptive suite representing a new race, ecotype or species. Such a paradigm is essentially congruent with many of the ideas proposed by Goldschmidt (1940).

Goldschmidt's (1940) ideas were rejected by the majority of evolutionary biologists, largely because there was no obvious mechanism that could support his ideas of 'chromosome repatterning' as a source of higher-order constraint on genic selection. Regulatory hierarchies explicitly provide such a framework, and also suggest that macromutations are likely events. Goldschmidt's (1940) theories remain unpopular, but vindication has proceeded to the point where a recent book on developmental biology was dedicated to him (Raff and Kaufman, 1983).

Another route for macromutation is via heterochrony (Gould, 1977; Matsuda, 1987; McKinney and McNamara, 1991; Hall, 1992). For example, neotony involves maturation at earlier stages of development. In the regulatory hierarchy, lower-level control elements may be deleted, or some intermediate level of control may be bypassed. This is equivalent to reducing the depth of hierarchical development. Its consequences are that the system is reset and can make new, more fundamental choices in direction (since downstream constraints are relaxed). There is good evidence that heterochrony has the potential to reveal numerous pathways that are otherwise buried and unavailable to the organization (West-Eberhard, 1989). Thus, some organizational advances may follow from initial simplifications (Arthur, 1988). Alberch (1991) points out that if there is strong conservation of regulatory elements, then heterochrony may well be one of the major routes for achieving novelty.

As will be discussed later, selection on a given trait will often lead to correlated responses in other traits (genetic correlations) and such linkages are believed to be important constraints on short-term evolution. As the developmental hierarchy proceeds, larger numbers of genes are progressively activated (Hall, 1992), and there may be changes in the quality and quantity of genetic correlations during ontogeny (e.g. Atchley *et al.*, 1984). The nature of such changes may be important in determining avenues for heterochronic shifts.

Given rapid advances in DNA sequencing, we may actually be able to assess the relative contribution of macromutation to various lineages. Although some higher-order mutations will be lethal dead ends or perhaps lead to radical selection on lower-level organization, other such mutations may shut off entire pathways or bypass them by shifting to alternative routes. This would leave some lower-level branches as unexpressed remnants that should be detectable in the genome. The tentative evidence (reviewed in section 4.3) strongly supports the contribution of macromutations to evolution, but more explicit tests are required.

Although macromutations may well be important in lineage evolution, the most prevalent form of selection would likely involve lower-level modifiers of small effect. Higher levels of organization may have their penetrance to the phenotype altered by an overlay of later-acting minor modifiers. Given that rates of recombination and mutation may be selectable or locally regulated (e.g. Brooks, 1988; Shapiro, 1991) it is also possible that the higher-order organization could have reduced levels of recombination and mutation in fundamental control elements, while lower-level elements are varied to a greater extent to explore potential pathways. Although completely conjectural, such a distribution of recombination and mutation might be expected in a hierarchically evolving lineage, and testing these ideas may soon be possible. Certainly we might expect there to be greater uniformity on the upper tiers of the developmental hierarchy if only due to strong stabilizing selection.

All of these ideas fit nicely with the observation by West-Eberhard (1989), that complex phenotypic features are often controlled by conditional switches that are usually environmentally cued (i.e. no genetic variation required), or less often, associated with allelic switches. Most circulating alleles appear relegated to minor tertiary variations such as colour. This would lead to higher-order polygenic features being recognizable as selectable genetic units. I will return to this idea in Chapter 6.

The picture that emerges is that eukaryote evolution involves consolidating and then extending levels of hierarchical organization by exploring modifiers of small effect (i.e. most recent alleles). Sometimes there may be a retraction or simplification that allows escape or diversification into new conformations. Thus, Lerner, Wu and Lin (1964) observed that most known cases of microorganisms evolving novel metabolic functions involve the release of an induced enzyme from regulatory control, so that the enzyme can work constitutively on a new substrate. Such mutants may occur with a frequency of 4×10^5 cells. Because enzymes are highly adapted to catabolize their inducer substrates, their action on new substrates is usually very poor. Subsequent modifications may either obtain increased effectiveness by increases in gene copies (that increase enzyme concentrations) or via changes in enzyme structure to utilize the new substrate better.

If gene duplication occurs first, then evolution to use the new substrate via enzyme modification may not necessitate loss of function for the original substrate. Otherwise, alternative selection for use of either substrate would be disruptive. Such simple systems illustrate how regulatory systems may first back up (loss of induction in this case), and then either abandon an old conformation for a new one, or evolve bifurcating machinery where both the new and old potentials can be accommodated in the same genome. Higher-level mutations might occasionally lead to reorganization as well. Given that all of the downstream control elements are coadapted with particular high-order switches, however, the likelihood of significant macromutations being successful must decrease as we ascend to higher levels of control. Kauffman (1993, p. 82) makes the interesting suggestion that in a hierarchically structured epigenetic landscape, the holon that responds to selection may differ according to the degree of environmental change.

Both Dawkins (1982) and Williams (1992) provide forceful expositions of evolution that partition the evolutionary process into two components: that of genetic information *versus* all the higher-order organization that imposes the framework for the success of selfish genes (i.e. replicators and interactors where selfish genes are the only replicators). In this context, whether one takes a perspective of genic selfishness or that of higher-order organization may simply be semantic, or as Dawkins (1982) suggests, different perceptions of a Nekker cube. The problem with this view is that it underemphasizes the importance of genomic organization as a crucial component of phenotypic evolution.

In a hierarchical model, each level of interactor organization is reflected by relevant replicator organization, most of which cannot be addressed at the level of a selfish gene. Higher-order interactor features may select higher-order genomic organizations that act as replicators with some independence of the lower-level components that compromise them. Thus, the Nekker cube analogy suggests a fundamental dichotomy between replicators and interactors that ignores the mapping of interactor hierarchies onto replicator hierarchies. If the replicator–interactor dichotomy is set aside in favour of parallel hierarchies of genomic and phenotypic organization, then evolution is seen to occur in a continuous field of hierarchical levels of selection spanning the genic to the phyloge-netic levels of organization. In particular, the dichotomy between organ-ismal-level selection and species selection is seen to be linked via a field of subspecific lineages highly relevant to the evolution of regulatory inte-gration. What modern molecular biology teaches us is that even genes are not only replicators, they are themselves interactors, with selectable functions and structure. Thus, the much vaunted conceptual advance of separating replicators and interactors may be a false dichotomy. When the genome replicates (emphasis on genome-level replication), genes are replicated, but so is the regulatory network that determines how the gene functions. Although not as obvious, such genomic replication also dupli-cates all of the higher-order organization that emerges as a consequence of compounding regulatory integration. Such genomic complexes can be expected to act as selective units in their own right. If conservation is the criterion, then epigenetic hierarchies appear to have recognizable structure transcending species and even phylogenies. Chapters 4 and 5 outline a paradigm for evolution incorporating selection at the level of subspecific lineages.

4

The subspecific lineage paradigm and the genetic templet

4.1 OVERVIEW

The epigenetic regulatory hierarchy is the bridge required to span the gap between the genome and elaborated phenotypes. This is a vast improvement because phenotypes and their associated coadapted genomes are the actual targets of natural selection, not individual genes. Internal and external environmental factors may also be important in epigenetics and can be incorporated into this framework (Schmalhausen, 1949; Waddington, 1957; Alberch, 1991; Hall, 1992). Such aspects are deficient in evolutionary models based solely on alleles. The proper linkage between genetics and phenotypes then provides a conduit linking genomic structure to ecological processes through developmental filters.

Traditional evolutionary models emphasized changes in code that modified the structure of functional proteins (i.e. exons). A regulatory model focuses on changes in the regulatory regions that front end genes and modify regions of heterochromatin. Such changes are relevant to how distant genes shake hands, and the elaboration of higher orders of control reflected in gene nets, domains, regulatory cascades and diversifying hierarchies. An evolutionary framework highlighting phenotypes and regulatory hierarchies requires that the units of selection also be cast in a new light encompassing relevant genomic components, spatial scales and temporal integration. Here I will argue that subspecific lineages constitute important units for natural selection. Such lineages may arise in single populations (where even a single individual may found a lineage with a new regulatory jackpot), associated with subspecific variation in genomic connectance or structure. In this context, genomic components of various size may be regarded as selective units that can be traced as lineages within single sexual populations.

The effective compartmentalization of populations into demes of varying size and degrees of sexual communication also allows divergence in regulatory organization (e.g. Wake and Larsen, 1987; Wade, 1992). Thus, subspecific populations may also be regarded as lineages with recognizable identity. Subspecific lineage selection (where lineages differ in genomic structure or degree of isolation) acting on transgenerational time scales may contribute strongly to both the coadaptive aspects of phenotypes and the high degree of regulatory integration in the genome. Chapter 5 argues that sex functions to forge subspecific lineages into a single organization that can at once yield phenotypic stasis while allowing a high degree of regulatory exploration to facilitate evolution. The important impacts of spatial and temporal heterogeneity will be explored in Chapters 5 and 6 in the context of sexual communication.

Here I will mainly explore some important perspectives and assumptions that are relevant to recognizing lineage identity and continuity. To help conceptualize the subspecific lineage paradigm, consider the case of selection for tail length in mice. Rutledge, Eisen and Legates (1974) successfully performed such selection in different lines of mice, an experiment that could be modelled in a quantitative genetics framework. However examination of the proximal mechanisms obtaining longer tails showed that one line had fewer, longer vertebrae and another, more and shorter vertebrae. The changes in vertebral number and conformation clearly reflect the underpinnings of divergent regulatory mechanisms and suggests that, as in bacteria, different regulatory solutions may be discovered in response to the same selective regime. Such variation in genomic response is widespread and has long been recognized as a problem in quantitative genetics because any given experiment may not yield exactly repeatable results. In mice selected for large size, for example, some populations show greater emphasis on skeletal dimensions, and others on fat deposition (Falconer, 1953). Such divergent responses may be obtained even among lines derived from the same base stock.

Lauder and Liem (1989) discuss the potential importance of such 'historical' factors for subsequent evolutionary diversification. Clearly, if such lines represented different isolated demes, there could be a divergence in regulatory organization as the differing solutions were consolidated into various coadapted genomes, followed by downward constraints as further changes were built on these foundations. For example, at the other end of the body, mice, humans and giraffes all have seven vertebrae in their necks, changes in vertebral number perhaps being constrained by their association with important neural and circulatory pathways (Williams, 1992).

It is possible to imagine both solutions for longer tails arising simultaneously in a single population or in two intermittently communicating demes, although the first solution to provide the appropriate response would likely be consolidated. In either instance the different strategies might be either mutually antagonistic or synergistic. In the case of synergism, a greater regulatory response might follow consolidation. In the case of antagonism, the winning strategy might have a higher net fitness. Alternatively, both strategies might be retained for some period until one achieved further regulatory innovations assuring its predominance, the two lines evolved isolating mechanisms, or dominance relations were evolved that reduced disruptive interactions.

Avise (1989) captured this idea of subspecific lineage evolution when he extended the phylogenetic analysis of gene trees downwards to intraspecific twigs. He found evidence for rapid dynamics at this level, and pointed the way for application of such molecular phylogeny to the realm of microevolution. Harvey *et al.* (1992) provide a similar insight. The existence of regulatory hierarchies and the possibility of regulatory divergence within single populations will be controversial to the extent that sympatric speciation is acceptable. Such possibilities are indeed beginning to be more widely recognized (e.g. Felsenstein, 1981; West-Eberhard, 1989; Wilson, 1989).

Diversification of populations into a fragmented demic structure would also allow the divergence, parallel evolution and periodic fusion of independent regulatory elements (see Chapter 5). Recurrent or intermittent selection pressures within single populations may also impart a genetic imprint upon regulatory hierarchies. Together the interaction of environmental heterogeneity and epigenetic hierarchical variation forge a selective framework capable of explaining regulatory evolution relevant to phenotypes. However, the assumptions of what constitutes the physical units of selection, their genomic reflection and the relevant time scale of the selection process need to be reevaluated.

The currently accepted functional unit for selection is individuals in a single generation (e.g. Salthe, 1975; Stearns, 1986; Ghiselin, 1987, 1988; Kirkpatrick and Lande, 1989). The logic is that since natural selection acts on organisms, their longevity or generation time defines the temporal scale and their individual reproductive success or intrinsic rate of increase necessarily constitutes fitness.

There are at least four good reasons why the temporal unit of selection may be considered instead to be multigenerational. Firstly, regulatory circuits, even relatively simple ones such as those that occur in prokaryotes, are usually forged stepwise across generations (e.g. Lerner, Wu and Lin, 1964; Hall, 1982; Clarke, 1983). Thus, Ohta (1988) points out that hormones and their receptor molecules have coevolved. The real stuff of phenotypic evolution does not entail simply the differential assembly of alleles at different loci, but the forging of complex integration that entails variants and actual structural modifications at numerous levels of functional organization (i.e. regulatory receptor sequences, signal sequences, promoters, exons, domains, regulatory complexes and regulatory hierarchies). If so, higher-order regulatory organization largely evolves on multigenerational scales via the stepwise modification of genetic structure. In this context, evolution may act like an organizational ratchet – various lineages randomly exploring different regulatory configurations or sub-components, which are immediately 'frozen' if a fitness 'jackpot' is achieved. Such jackpots might entail genomic units constituting an integration of several or numerous loci. They would emerge as selectable units via their contribution to fitness at particular focal levels and via constraints from higher-order organization elaborated on these units (Chapter 1).

A second consideration is that some regulatory mechanisms may themselves span generations or incorporate an integration adapted to a greater range of environmental or genetic variation than that encountered in single individuals or even single populations (e.g. genetic correlations, plasticity and reaction norms, canalization and dominance relations). It seems likely that directional selection or opportunistic evolution may involve the assembly of genetic units of various sizes across generations, although this would usually entail small add-on components. Once in place, however, stabilizing selection may freeze the higher-order organization, shoring up the unified edifice against failure in any subcomponents. In this respect, different forms of selection may tend to act on selective units of different size and on various time scales.

Thirdly, some features appear to be adaptations for evolvability or long-term persistence that have no (or even negative) consequences for the immediate adaptation or success of individuals (i.e. regulated transposition, sexual reproduction, risky dispersal or even programmed death). Finally, the upper end of the developmental hierarchy leading to an individual phenotype resides in, and is supplied by, parental genomes. Mechanisms exist for transmission of transgenerational information that are not limited to DNA sequences (e.g. heterochromatic states) and these can have wide ranging morphological, ecological or evolutionary impacts.

In a genic framework, although the proportions of existing alleles may change during selection over multigenerational time scales, the effective units are alleles and organization consists only of their combinatorial consequences. In a lineage framework, effective organizational units may vary in size and may not even exist as a complete functional entity within any current organism. Instead they are most often constructed step-wise across generations. Once erected, the organization itself (e.g. a regulatory circuit) may be retained, even if genetic variants or component parts are circulated during sex (e.g. variations of a particular transcription factor).

This represents a fundamentally different paradigm than the orthodox genic tradition. The genomic units in a regulatory paradigm constitute levels of organization spanning from regulatory receptors to entire organismal genomes to organizational elements distributed among separate individuals (e.g. sex chromosomes). From a genic viewpoint it might be argued that each step in achieving a regulatory circuit must entail improvements in fitness, otherwise the variants would be eliminated from the wild type population (Williams, 1985). In a non-equilibrium model, however, local variants could escape competitive exclusion long enough to make the next step, even if they are slightly deleterious. Moreover, some steps in assembly of circuits may be of low cost and phenotypic impact, so assembly can proceed somewhat cryptically. Such a situation does not diminish the importance of such variation for future regulatory organization that may emerge in a phenotype (e.g. Saunders, 1990).

If we relax our biases on what may be considered as an evolutionary entity, it is possible to identify a hierarchy of organization in the genome (both in structural and in functional connectedness), various levels or features of which may act as evolutionary units with varying degrees of discreteness and evolutionary persistence (Hunkapiller *et al.*, 1982). Dawkins' (1976) 'immortal genes' constitute highly consolidated structures that have reached some optimal conformation. Most 'adaptive' alleles circulating in extant species are relatively ancient and have had ample time for their adaptive coevolution into the genome (e.g. Ohno, 1970; Hamilton, Axelrod and Tanese, 1990). However, such genes evolved initially to achieve their highly tailored state, and as such, passed through a transient phase of unstable identity (see Maynard Smith (1989) for a nice example of such molecular evolution). Moreover, even genes may be only strongly constrained within domains of high functional importance, other regions being free to vary (Ohno, 1970; Kimura, 1983).

Thus, genes have no better claim to the title of fundamental units than the ubiquitous Homeobox domain, the supergenes of *Cepaea* (Jones, Leith and Rawlings, 1977), the Bithorax or Antennapedia complexes of the Insecta, the inversion polymorphisms of *Drosophila*, various bacterial operons or nets of genes functionally integrated by reciprocal receptor–signal sequences (see Watson, Tooze and Kurtz, 1983; Kauffman, 1985; McDonald, 1990). In fact, the concept of the gene as an extractable bead that can be indiscriminately juggled among genomes is strained by the reality of developmental genetics. Functions attributed to genes may be localized to small regulatory regions, the same transcription units may be accessed by different promoters and regulatory networks and some 'genes' appear to be nested components of others. Thus, Seidel, Pompliano and Knowles (1992) have referred to exons as micro genes. It may well be that more fundamental aspects may represent more consolidated structures with longer-term conservation of recognizable identities. Recently evolved features or end points of developmental programs may represent more variable aspects. Even then, the species' Bauplan or basic wild type may represent simply a higher level of regulatory consolidation (Ohno, 1973).

There appears to be no conceptual barrier to impede replacement of selfish genes by hierarchically organized genomes. Even Williams (1966a, p. 24) and Dawkins (1976, 1989a) actually advocated a form of sliding reductionism that could consider genes, groups of linked genes, chromosomes or even populations of asexual clones as units of selection. This sophistication and its implications have been overlooked by many in their enthusiasm for the selfish gene, a meme which Dawkins (1989a) suggests was somewhat unfortunate for its emphasis on the word gene. Although it might appear trivial at first glance, the difference between a reductionist view based on a fixed and fundamental focal level (i.e. truly genic) and a view that allows for shifts or multiple focal levels is decidedly different. Even sliding reductionism usually considers selection to be mainly acting on a single selective unit or level. In a hierarchically organized regulatory system selection would act on numerous levels simultaneously, sequentially or intermittently. This thesis, although strongly supported by recent molecular evidence, has not yet been incorporated into a cohesive evolutionary paradigm.

The genic view is pervasive and firmly entrenched in evolutionary theory and is emphatically not simply a convenient straw man for addressing this issue (e.g. Sibly, 1989). Rather it represents a fundamental philosophy of biological explanation based on extreme reductionism. The emphasis here is on explanation. Nearly all successful science utilizes a reductionist approach of divide and conquer. Thus, hierarchical or genic explanations will both largely be derived from reductionist research. The difference in explanation arises when we consider whether all higher-order phenomena can be explained by a single lower level or whether explanation requires consideration of multiple levels (Chapter 1).

The question of whether genome structure itself constitutes a hierarchical organization arising from selection operating at various levels of phenotypic organization, or whether it simply represents the fitting of assembled genes selected for their cooperative coexistence in the gene pool is a crucial one. If selection operates at higher levels, then global rules may operate that constrain lower-level elements like genes (Wilson, 1980, 1988). If the genome constitutes an integrated organization, then its structure and function may transcend selfish interests among genes to achieve more global optima. Alternatively, the prevailing genic view suggests that the genome is the product of an unrestricted civil war among selfish genes (Bell, 1982, p. 439), or at best, cooperation restricted to that predicted by kin theory or the prisoner's dilemma (Axelrod and Hamilton, 1981; Axelrod, 1984; Axelrod and Dion, 1988; Dawkins, 1989a).

4.2 GENETIC RELATEDNESS IN SEXUAL SPECIES

Degree of relatedness is a cornerstone of evolution, the process itself hinging on genetic similarity between parents and progeny. The units of selection problem has, as one major watershed, generated dichotomous approaches to how relatedness should be calculated (reviewed by Shields, 1982). Taking a gene's-eye view, the probability of an allele being transmitted to an offspring from one parent would be 0.5 (because the other parent contributes the other half of the genome). Grandchildren then obtain the gene with a probability of 0.25. Such logic can be used to calculate the likely kinship between various distant relatives and extensions can be made to various aspects of kin theory and the evolution of altruism.

The classic example of applying degrees of relationship to evolutionary problems was Hamilton's (1964a,b) insight that female workers in the Hymenoptera are 75% related to their sisters but only 50% related to their potential direct offspring. This occurs because the fathers have haploid genomes and daughters all obtain this full set of identical chromosomes, plus a contribution from their diploid mothers. The consequence is that (theoretically) workers can obtain better genetic transmission by helping mother to raise fertile sisters (relatedness = 0.75) than in raising their own brood. These papers have fuelled 30 years of intense interest in kin theory.

The basic tenet of kin theory is that the fitness of an allele is not simply related to the immediate fitness of its bearer, but has an extended fitness related to the success of all other carriers as well. The summation of both forms of fitness is termed inclusive fitness. Much of the theory of altruism and cooperation has subsequently been elaborated on the idea that the costs of altruistic behaviour to a selfish allele that resides in a donating organism may be offset by the gains obtained by the improved fitness of closely related carriers (that are also likely to carry the same gene). Kin theory has proved exciting to genic theorists because classical theory had no way to explain the common evolution of social cooperative behaviours or eusocial systems that contain sterile classes (Wynne-Edwards, 1986). A gene's-eye view is consequently a dominant aspect of the entire paradigm.

It is easy to see why such a probabilistic view of relatedness has sidetracked consideration of lineages as selective units. In an asexual clone, the number of ancestors contributing to the current genotype over the last 10 generations is only 10 and variation is confined to mutations. For a sexual species, the number of ancestors contributing to an individual over 10 generations is 2048. Thus, sexual individuals potentially represent a focusing of considerably more genetic information, even though recent contributions count exponentially more. Bell (1982, p. 45) notes, however, that because each sexually produced individual stands on numerous lines of descent, and acts as the founder of many more, the population is knit into a single entity (Chapter 5).

A second way to assess degree of relationship is on the basis of the entire genome. Shields (1982) reviews the statistical approaches to the problem. With improvements in DNA techniques we may be able to compare the totality of shared sequences in the genome empirically. This would be improved by corrections to subtract junk DNA sequences that may lie outside the functional organization. Dawkins (1979, 1982) argues that as far as genic evolution is concerned, it is only the degree of relatedness at the locus in question that is of any relevance. Thus, a gene for altruism should act only on the basis of degree of relationship at this particular locus. In the absence of linkage and with complete recombination, the rest of the genome can (and must) be regarded as background that neither influences nor is affected by selection at the altruistic locus.

There are several tracts that suggest that this genic view and its logical extensions are incorrect, and that degree of relationship should be derived at the genomic level. In a thorough review, Shields (1982) concluded that most organisms have systems of dispersal and habitat choice that promote inbreeding and local lineage identity. A note of caution is required here because the term inbreeding has become almost synonymous with concepts of loss of heterozygosity and inviability. Shields (1982) uses the term in the context of reinforcement of local genetic identity, and that context will be assumed here as well. He argues that species avoid both strong inbreeding and outcrossing depression and that low dispersal and specific habitat choice are adaptations that promote fidelity in the transmission of parental characteristics to their offspring (i.e. lineage identity).

In the examples above, there is little likelihood that the 2048 ancestors contributing to an organism over the last 10 generations are very independent at all (Shields, 1982). In fact, for an offspring to achieve a relationship of 0.5 requires that its parents are completely unrelated. In sexual species this is virtually impossible (Shields, 1982, p. 191). Populations grow, stabilize, contract and undergo bottlenecks. None are infinite in size as classic models assume and few organisms have dispersal characteristics that realize panmixia (Ehrlich and Raven, 1969; Endler, 1977; Shields, 1982; Wynne-Edwards, 1986). In conditions of stability, decline, or low dispersal, inbreeding becomes an important element, strongly reinforcing the local identity of sublineages. The smaller the population, the more important this lineage aspect becomes.

Dawkins (1989a, p. 288) suggests that the probabilistic approach to relatedness is consistent with a high degree of genomic identity because such probability refers to that portion of the genome that is shared, over and above the mean background. This still does not work. Probabilistic calculations predict a rapid diminishment in relatedness of individuals in sexual species that just does not happen. For example, the probabilistic approach predicts that the degree of relationship of a parent to a descendant after 50 generations will be $\frac{1}{2}^{50}$. In fact, they are likely to retain genetic identity at the level of the genome on the order of 90% or more.

We can expect some, and perhaps many, higher-level regulatory features to become fixed. Strong conservation of genes associated with fundamental aspects of metabolism or development appears to be the rule. The actual degree of heterozygosity of populations is usually not greater than 30%, and in fact, many species (especially large vertebrates) have relatively low variability (Kimura, 1983). Even then, much of this measured variation is associated with aspects of structural genes under such neutral or weak selection pressure that they may be effectively invisible to selection (Kimura, 1983). To capture the historical aspect of random drift in such genes, a lineage model would in fact be appropriate.

Hartung (1981) noted that most considerations of genetic relatedness have ignored the existence of fixed portions of the genome. He argued that such genes could enjoy the benefits of sexual recombination without risk, and that a genome parliament dominated by such a majority tyranny might well resist changes in the fixed component (Dawkins (1982, p. 138) reviews earlier related ideas). Thus, whole blocks of genes could form a non-variant component in an otherwise varying lineage. The existence of regulatory hierarchies suggests that the degree of relatedness relevant to ultimate phenotypic organization might also count higher-order genes more strongly. Thus, two individuals with identical upper hierarchies but divergent lower holons might be considered to be more closely related (in an organizational sense) than another individual with common lower-level elements, but divergent higher organization.

Shared lineage identity might be perceived more clearly in these higher-level features. We are in no position to adopt such an approach at present, but this could be the most relevant aspect as far as organizational evolution is concerned. Extrapolating the idea of Hartung (1981) to this framework, the existence of higher-order fixation might mean that a relatively small number of higher-level dictators could disproportionately dominate the organization as a sort of holonic junta. The regulatory hierarchy also allows for compartmentalization of the genome so that different sets of coadapted genes can differentiate in a single genome. One mechanism related to this is the fact that some alleles might constitute higher-level signals that may shunt development into radically different alternative phenotypes (see West-Eberhard, 1989).

To clarify, consider sexual dimorphism. The sisterless genes, for example, act as genetic signals that control subsequent sexual determination. In this case, major genetic structure may be involved (i.e. the XX or XY cytogenetic dichotomy). In some species there are more than two sexes. Thus, three phenotypically distinct male forms are found in one species of marine isopod. The differentiation of these alternative ways of being male has occurred in single populations (Shuster and Wade, 1991). To some extent the complex life-cycle of many insects and invertebrate parasites also demonstrates that single genomes can deploy numerous locally adapted phenotypes of radically different design. Similar differentiation may occur among phenotypes in response to other ecological aspects such as food (e.g. Greene, 1989; West-Eberhard, 1989; Wilson, 1989). Thus, one of the key ideas of genic models, that gene flow must necessarily disrupt phenotypic differentiation, appears to be absolutely wrong (see Chapter 6). Of critical importance is the fact that differentiation of the genome may also involve actual differences in structure, not simply differences in allelic variants. Although some common elements might be shared, such structural alternatives would represent distinct lineages within or among demes. Differentiation of phenotypes may not be impeded by disruption because structurally different forms may not necessarily have competing alternatives (unless their absence is considered as a null alternative).

Other recent evidence has confirmed major failures of the genic paradigm in at least two key areas. Firstly, differentiation of subpopulations in spite of large local gene flow (i.e. 40% exchanges of individuals) has been experimentally and theoretically demonstrated (Endler, 1973, 1977; see also Ehrlich and Raven, 1969; Wade, 1977). In addition, the prediction that inbreeding or founder events necessarily reduce selectable phenotypic variance has also been rejected by empirical and theoretical studies (reviewed by Carson, 1990a). In both cases, it appears that the missing conceptual key was the existence of coadapted gene complexes. In Endler's studies, local modifiers promoted differentiation when a single feature was selected, and in the case of inbreeding, breakdown or simplification of coadapted gene complexes appears to release selectable genetic variation that was previously locked into these complexes. Such results do not reject the traditional emphasis of isolation as a major mechanism promoting local identity. Thus, Hillis, Dixon and Jones (1991) point out that isolated populations of tree snails are themselves like single individuals within a sexually outcrossing species. This interpretation captures the idea of subspecific lineage selection very well – each deme approaching the status of selective entities while maintaining membership in a larger genetic organization. With some gene flow there may be some loss of individual identity, but not all.

The key concept that divides the genic from a genomic view is whether the genome represents a coadapted complex of interdependent units or whether single-locus arguments can be applied to most genetic and phenotypic features. The idea that any complex trait, such as that for altruism, can be realistically modelled as a single locus does not seem feasible. If we consider the minimal requirements for altruism, it must include some sort of sensory system to input information on existing situations, a signal-receptor system to transmit information on relatedness (however that is determined), some sort of decision-making apparatus to weight the pros and cons of response and an effector system to carry out the altruistic program.

When we consider the degree of complex regulatory integration that is involved in the epigenetics of even simple morphology and realize that this becomes even more elaborate as we proceed into neurogenesis, the idea of alleles for altruism at a single locus becomes incomprehensible. Even if we could dissociate the entire regulatory system for this character, the idea that it could be independent of the rest of the genome seems impossible. In terms of whom to donate altruistic acts towards, close kinship is an obvious criterion. However, could we also argue that in an outcrossing sexual species, less related individuals represent potential and necessary genetic resources for breeding and future fitness? In populations limited by factors other than strong competition, altruistic behaviour towards unrelated individuals (e.g a prince from a distant kingdom rescuing the princess from a dragon) could yield large increases in inclusive fitness within a very few generations.

In an evolutionary framework with transgenerational units of selection, it seems feasible that the prince and dragon phenomenon could even be extended to same-sex interactions as long as immediate competition for mates is not an overriding factor. Even if another individual does not represent an immediate genetic resource due to sexual incompatibility, the children of same-sex individuals could benefit from the availability of genetically diverse mates and the integration of their parental genomes. Same-sex individuals cannot mate in the current generation but may still represent genetic resources because they can fuse vicariously via their children.

Given that mechanisms have evolved that are apparently adapted to prevent severe inbreeding and promote limited outbreeding (Thornhill, 1993), such evolutionary pathways seem like veritable turnpikes for kinship theory. Their paucity in the literature on altruism appears to be due to genic myopia rather than the likelihood that evolution has ignored such aspects. The criterion for relatedness is immediately highlighted. In a sexually communicating organization, genetic reinforcement of global and local identity may occur, even if all individuals have programs to select genetically different mates (Shields, 1993). Thus, female mice select males that differ in their immunological constitution (Beauchamp, Yamazaki and Boyse, 1985). This may lead to polymorphism in particular alleles, but such alleles will often be associated with other genetic differences. Although a genetically divergent mate is selected, the successful male is resoundingly part of the organization and the mate-selection system has evolved with such *unrelated* males as part or target of the mechanism.

From a genic view, the princess and foreign prince may be indeed unrelated. From a lineage view, the prospective couple are clearly members of the same organization, share most of their fundamental genomic configuration and carry many alleles that are ancient variants selected to make positive contributions to the overall genomic organization. Thus, an unrelated prince could conceivably improve the fitness of descendants by donating altruism to other unrelated male or female humans, but should have little compunction to do the same for other species that are not members of the shared organization. Over 30 years has been invested in kin theory, mainly in a genic paradigm. The thesis presented here argues that the measure of relatedness of relevance to phenotypic evolution is the regulatory organization in a more holistic sense. The wondrous degree of integration revealed in the developmental genetics of *Drosophila* resoundingly validates the intuition of numerous evolutionary biologists, that the genome represents a highly coadapted complex (Chapter 7).

The genic view has been partially based on the fact that there appears to be no way for coadapted genes to remain together in sexually pan-mictic populations. Based on the arguments that linkage is rare, and differentiation cannot occur in the face of gene flow, the individual locus can be viewed as an essentially independent unit. How can we reconcile the empirical reality of a highly integrated genomic organization with a system of apparently independent components? A first step necessitates a more careful definition of coadaptation. In a genetic sense, coadaptation has been defined largely on the basis of relative fitness. If two genes in combination show greater fitness than they do independently, they are said to be coadapted. If they show no such direct epistasis and linkage, they are considered to be independent. If, however, most circulating alleles are consolidated components of an organization, they may show little direct epistasis and yet they may be coadapted to the organization as a whole on larger temporal and spatial scales.

For example, Dawkins (1989a) used the analogy of a sculling boat crew to argue that genes may behave selfishly as long as they cooperate with the other genes in the boat. Crew members are envisioned as being freely exchangeable competitors for seats. If the boat and crew are viewed as an organization, however, it is clear that the crew are constrained by the structure of the boat to have anatomies that can grasp and pull an oar while sitting in the appropriate position. Thus, substituting crew members with a mouse, a rhino and an ostrich just would not work. Higher-order organization does not preclude some variation, but it may strongly constrain what is allowed. Genes may have independence only within the boundaries of this higher-order constraint, in this case, the design of the boat and its function in maximizing speed.

Rather than being free-ranging selfish outlaws, most consolidated genes probably reside in rather cramped organizational prisons. Selfish DNA (Doolittle and Sapienza, 1980; Orgel and Crick, 1980) and viruses, if they have not coevolved with their hosts, might be viewed analogously as rats scurrying from cell to cell. The existence of free-ranging rats, however, in no way obviates the reality of incarceration for the inmates.

Higher-order organization is not precluded by the existence of alternative alleles at some loci. The existence of alternative alleles for transcription factors that yield different transcription rates, for example, is not inconsistent with higher-order organization. Integrated complexes of structural and regulatory genes may constitute evolutionary units in their own right. An allele that no longer recognized the appropriate control signature for its gene net would probably be rapidly eliminated or cast adrift as a pseudogene.

The linkage of the actual organization is based largely on signal–receptor systems that do not require physical linkage. Kauffman (1985) points out that in such a regulatory organization even relatively low degrees of connectance (i.e. three-way interactions) would allow each gene directly or indirectly to impact 87% of all the genes in the system. Lawrence (1992, p. 195) estimates that at least 30% of all genes in the *Drosophila* genome impact on eye development. Wright (1980) points out that in mice there are 50 loci that affect coat colour, 60% of which also have gross morphological impacts. At least 6500 copies of the proposed target sequence for the *fushi tarazu* DNA binding protein have been detected in the genome (Budd and Jackson, 1991). Hormones have particularly far-reaching impacts. The thyroid hormone, thyroxine, alters the expression of over 100 enzymes as well as other hormones (Weindruch and Walford, 1988). Thus, the genome represents an organization with relatively high functional connectance. I say relatively, because Kauffman's (1993) analysis clearly suggests that very high connectance may lead to system destabilization.

The nearly universal pleiotropy of genes probably arises as a direct consequence of the genome's hierarchical organization (Atchley and Hall, 1991), but such universal pleiotropy is itself one of the main targets of selection (Wright, 1982a,b). In addition, the key regulatory regions do comprise physical linkage groups as well (e.g. HOX and HOM). Thus, the assumption that sexual species show no significant linkages among loci is false for many of the most important epigenetic regulatory genes.

Another important conceptual shift that is necessary in considering sexual species is that the organization is designed to incorporate variability. Thus, Lerner (1954) developed the thesis that both highly canalized phenotypes and genetic variability for evolutionary fuel were obtained by maintenance of heterozygosity at numerous loci. The fact that allelic polymorphism may represent coadapted parts of a dynamic organization is supported by the fact that such systems may transcend speciation events (reviewed by Ohno, 1970; Harrison, 1991). For example, humans and chimpanzees have retained the same polymorphic alleles for a class of heavy-chain immunoglobulin from a common ancestor existing

7 or 8 million years ago. Similarly, humans and gibbons retain the same polymorphism for antigen on the surface of erythrocytes, which dates back to a common ancestor at least 26 million years ago. Thus, polymorphism itself can be considered as a recognizable adaptive feature of these lineages, albeit a dynamic one.

Lerner's (1954) thesis postulated that maintenance of heterozygosity in coadapted complexes might be accommodated by advantages that emerge due to a combination of heterozygote advantage combined with epistatic integration among numerous loci. Hartung (1981) also pointed out that, although an association between two alleles may suggest some coadapted advantage, lack of such an obvious association proves absolutely nothing. If the advantages of coadaptation reside in larger genetic units composed of numerous loci and their interactions, then an examination of direct linkage relations between individual loci may not detect the relevant higher-order organization at all (see also Falconer, 1977). In a system where organizational components are perhaps distributed among individuals and in which variation in components is an integral part of the system's nature, degree of relatedness may have to be extended upwards to incorporate this aspect. Thus, we are faced with the paradox that individuals that vary at particular loci may still be related in a systems sense, because both variants are coadapted to the same fundamental organization.

Even Dawkin's sliding reductionism demands that units of selfish DNA constitute discrete chunks that have long-term recognizable identities. The question as to how much of the genome may constitute a selective unit in sexual species is unknown. Conventionally, selection of entire genomes has been considered to be only relevant for asexual clones or in group-selectionist contexts for selection among sexual populations in genetically isolated demes. Rather than seeking the extent of higher-order identity, the genic view has stressed the high genetic diversity of species, or at best, the importance of close kinship in evolutionary dynamics.

The subspecific lineage paradigm differs from the genic view mainly in considering the hierarchical organization of eukaryote epigenesis. This view dictates that when evolution occurs for particular alleles, what is selected is not that single allele, but the entire regulatory organization for which that allele constitutes one component. Wright (1982b) made a similar point when he argued that phenotypes are not mosaics of unit characters so the value of a particular gene depends on the array of its associates. The point is that any given allele is probably never selected independent of its organizational context. A single gene could give rise to variation among such organizations (Dawkins, 1979), but in regulatory networks such single gene effects are likely the exception rather than the rule.

If organizational features relevant to phenotypic selection are considered, the degree of functional variation within lineages may sometimes be considerably reduced or may be constrained by coadaptation. Within this framework, it is not inconceivable that selection approaching the level of the entire genome (i.e. of lineages), or at least the upper levels of organization, may operate, even within sexual species. Such an idea represents the core of Sewall Wright's life-time work on structured demes (Chapter 5), which was largely driven by a belief that 'there must somehow be selection of coadaptive interaction systems as wholes' (Wright, 1982b).

Cheetahs are an exceptional example of a sexual species with little genetic variation at the genomic level; however, genetic uniformity, particularly of higher-order genomic features of fundamental importance, may not be particularly exceptional at higher levels of genetic organization. Selection on various levels of phenotypic organization will be complemented by various levels of recognizable structural or dynamic organization in the genome. Thus, various organizational features of sexual lineages may be selected over different relevant time frames. If so, sexual species may be profitably considered to be comprised of subspecific lineages with varying degrees of local identity in various organizational attributes (Chapter 5).

Given the reality of coadapted genomes and their phenotypic reflection (adaptive suites), selection on large components of coadapted organization seems like a more realistic model for phenotypic evolution. This requires a genomic measure of relatedness. Given the ancient origin of most genes, their amplification by selection and reinforcement by inbreeding, the degree of relationship among members of a single species is likely to be much higher than the genic perspectives allow. The reality of this situation has in fact required the cumbersome contrivance in population genetics of defining genic relatedness in terms of recent direct descent.

Phenotypes care nothing at all about whether a particular allele is obtained by recent or ancient descent. The necessity for such an awkward prop is symptomatic of a serious conceptual difficulty despite its utility in predicting the short-term dynamics of allele frequencies.

Localized demes may obtain very high degrees of relationship, with local lineage identity reinforced by strong philopatry (the tendency to remain or return to the homesite for breeding) (Endler, 1977; Shields, 1982, 1993; Wynne-Edwards, 1986). Such a reality allows a lineage paradigm to be applied even to sexual forms. There are actually two related aspects of this thesis. One refers to consideration of polygenic units of organization representing different levels or compartments of the regulatory hierarchy. Such genomic lineages can exist in single populations and may even have overlap in genes contributing to the dichotomous forms.

The other aspect involves the more traditional spatial/temporal localization of populations and their differentiation of recognizable phenotypic and genomic identities. The fundamental idea is that the entity selected is the epigenetic organization. This organization is conceptually novel in that variation may be an integral component and it may have mechanisms for self-alteration (evolvability features). I will refer to this organization as the 'dynamic genetic templet' to distinguish it from the traditional concept of the gene pool which lacks the idea of regulated integration. The genetic templet is housed among distributed units and variation may be selected or forged across generations. The templet may also span spatially or ecologically separated demes. In these cases, interdemic communication can forge relevant organizational features (e.g. norms of reaction or adaptive polymorphisms), transcending individual adaptation. The idea is that the genetic templet has the capacity to integrate selection across various scales of space/time, even if there are discontinuities. A framework that allows for responses to intermittent or discontinuous selection and which involves organizational restructuring requires several assumptions that are not normally considered in genic models. These include a memory system that retains imprints of previous selective regimes (=genetic inertia), a low cost for such memory and extraneous genetic features and genetic structural variants that can provide the fuel for regulatory reorganizations needed for epigenetic evolution. These aspects are considered below.

4.3 GENETIC INERTIA AND RETENTION OF QUIESCENT GENETIC PROGRAMS

For natural selection to forge a multigenerational templet requires that programs be cheaply maintained even if they are only intermittently expressed or are only deployed in specific environments or genetic backgrounds. Mesoevolution of regulatory organization also requires that successive steps necessary to form higher-order features can be retained across generations, even if they express nothing or even if phenotypes are mildly maladaptive.

There has been little attention regarding over how many generations inactive genes can be maintained. The most relevant molecular literature is that pertaining to pseudogenes. Such genes arise via gene duplication (Ohno, 1970) or sometimes by reverse transcription of mRNA back into DNA (Hollis *et al.*, 1982). Where these genes are permanently abandoned, they are free to mutate and evolve randomly. Rates of mutation in such genes may amount to 1×10^{-8} per nucleotide per generation (Kondrashov, 1988) or 5×10^{-9} per year (Kimura, 1991). The fact that numerous such genes remain recognizable in the genomes of most species attests to the long-term retention of genetic identity, even under neutral selection (Kimura, 1983). Some of these relics are still recognizable, presumably after millions of years. The potential significance of pseudogenes can be appreciated by considering that the rat contains at least 25 pseudogenes derived from cytochrome *c*, and over 200 for glyceraldehyde 3-phosphate dehydrogenase (John and Miklos, 1988).

The most optimistic evidence for genetic retention is Kollar and Fisher's (1980) discovery that chick epithelium grafted to mouse molar mesenchyme synthesizes tooth enamel, even though the last toothed birds existed 70 million years ago. Grant and Wiseman (1982) questioned this result using theoretical estimates of random mutation rates, but were effectively answered by Kollar and Fisher at the end of their article. Besides, the prevalence of atavisms and the expression of ancestral features during ontogeny or via hormonal manipulation (including ancient muscles and bones in birds, ancient teeth in cats, ancestral toes in horses, fingers in chickens, canine teeth in sheep and external limbs in whales) empirically support Kollar and Fisher's (1980) results (Rensch, 1960; Kurten, 1963; Ohno, 1970; Curio, 1973; Gould, 1977, 1982; Lande, 1978; Alberch *et al.*, 1979; MacBeth, 1980; Raff and Kaufman, 1983; Blackburn, 1984; Hall, 1984; Reid, 1985; Gingerich, Smith and Simons, 1990). Species that are derived from polymorphic ancestors via consolidation of one phenotype may also atavistically express characteristics of the abandoned alternative morph (West-Eberhard, 1989).

Even intermittent selection is sufficient to maintain otherwise quiescent genetic programs in a functional state (Li, 1984). The genetic control of gene expression in response to environmental cues is well known in prokaryotes (Jacob and Monad, 1961a,b) and so not all genes are expressed in every generation. Similarly cryptic or latent genes may remain silent unless activated by mutation, recombination or the action of transposable elements (Hall, Yokoyama and Calhoun, 1983; Moody and Basten, 1990). Such cryptic genes are believed to be a vital element of the genetic repertoire maintained by powerful mechanisms (Hall, Yokoyama and Calhoun, 1983) (e.g. long-term adaptive potentialities). The expression of latent mutations may be under direct genetic control in *Drosophila* (Kubli, 1986). In trypanosomes, quiescent genes are an integral reservoir for scrambling their antigenic identity to foil recognition by their hosts (Pays, 1989).

Among molluscs, there is evidence that behavioural programs used in low-oxygen environments may be deployed following hundreds of generations of disuse (Russell-Hunter, 1978). *Ranunculus flammula* is a plant expressing lanceolate leaves terrestrially and linear leaves aquatically. Plants with long histories of exposure to either terrestrial or aquatic habitats were still capable of expressing the appropriate leaf forms in the alternative environments (Cook and Johnson, 1968), although they did not persist in these habitats. The ecology of inducible defences is also replete with examples of organisms that retain the code for features such as spines or chemical products that are not necessarily deployed in every generation (Chapter 6). Treatment of parthenogenetic wasps with anti-biotics released the expression of male phenotypes that had presumably been suppressed for vast periods of time by cytoplasmic parasites (Stouthamer, Luck and Hamilton, 1990). Lande (1978) has estimated that in mammals, atavistic structures may be retained for roughly 10^6–10^7 generations, a rather substantial period of time.

When I first considered this aspect, I thought of two experiments that could test the strength of such genetic inertia. My first intention was to hormonally manipulate wingless forms of insects to induce super-numerary moults. The prediction was that if the loss of wings was neotenic, such species might still express wings if forced into additional moults. A positive result would strongly support the inertial aspect of the lineage paradigm. It was somewhat startling, consequently, to find that Pantel (in Goldschmidt, 1940, p. 257) documented a species of wingless earwig, *Anisolabis annulipes*, from which an individual spontaneously expressed fully developed wings resembling those of another genus. More recently, Coyne and Prout (1984) showed that in *Drosophila* in which wing expression was suppressed by the recessive mutation, *vestigal*, completely normal wings could be restored after 1000 generations of winglessness. Matsuda (1987) documents that events such as wingless species spontaneously re-acquiring wings are in fact fairly common. The literature on thyroxine-induced metamorphosis of normally neotenic amphibians into adult forms is also of relevance here (Goldschmidt, 1940; Gould, 1977; Matsuda, 1987).

A second possible experiment was to rear *Daphnia* for various numbers of generations with no exposure to cues from predators and then test to see how well the various lineages expressed relevant phenotypes in response to predators. A strong positive result would underscore the importance of previous selective regimes and lineage history in terms of current phenotypic responses. Weider and Pijanowska (1993) recently demonstrated that *Daphnia* from ponds with no recent history of fish pre-dation indeed showed phenotypic and life-historical shifts in response to fish similar to those of *Daphnia* collected from lakes inhabited by fish.

Conversion of heterochromatic states is one mechanism that can differ-entially inactivate or activate genes or gene complexes. Hybridization events often involve changes in heterochromatic patterning that may release hidden genetic variation or result in novel complex expression patterns (Heslop-Harrison, 1990). Treatment of organisms with the demethylating drug 5-azacytidine can result in restoration of wild type in what were initially considered mutant vertebrates (Holliday, 1990), or increased expression of variability in plants (Heslop-Harrison, 1990). Thus heterochromatinization is one mechanism that might be involved in inac-tivation of ancestral programs, with the potential to allow re-expression later. Another unknown factor is whether such inactivated regions might have lower mutation rates by virtue of strong binding with histone proteins. This might yield unexpectedly long persistence in the genome.

Mutation rates in DNA have been variously estimated at 10^{-7}–10^{-10} per base pair per generation (Ohno, 1970; Chao, 1990; Chao, Tran and Matthews, 1992). Ohno (1970) estimated that the rate of mutations affecting gene function is on the order of 10^{-5} per locus per generation, or 10^{-9} for tolerable mutations. Dawkins (1986) suggested that a conservative estimate of DNA fidelity in the absence of natural selection is an error rate of about 1% in 5 million generations. Trivers (1985) estimates that in a section of DNA long enough to code a protein, a base pair mutates about once in 100 000 replications. Kondrashov (1988) estimated the upper limit for deleterious mutations in support of his idea that sex is necessary to eliminate them. He proposed that vertebrates may sustain as many as 100 mutations per diploid genome per generation and invertebrates may obtain 10. These estimates include non-coding regions and neutral mutations as well, and so they are not directly applicable to phenotypes. Even then, such estimates of DNA fidelity show that retention of unselected genetic information is not an obstacle for a lineage perspective. An organizational ratchet would be greatly facilitated if consolidated steps did not rapidly deteriorate in intermittent selective regimes.

A possible lineage-level function of sex related to the genetic repair hypothesis might even improve on this. Dormant programs would inevitably become denatured by random mutation, but different lineages would accumulate different damage. Occasional sexual reproduction could create lineages with restored functional programs, even if the parental lineages all had partially denatured codes. This would be particularly advantageous during times of stress so it may be significant that the timing of sexual reproduction often coincides with such events. Bremermann (1987) has priority for this idea since he proposed that such a mechanism might be important in maintaining quiescent genes in prokaryotes.

These results have some interesting implications for genetic engineering. If genomes retain numerous ancestral genes, then there may be a much larger system available for manipulation than wild type phenotypes suggest. It might matter, for example, whether organisms have already traversed certain epigenetic landscapes during evolution. Thus, if mice are secondarily small, then enhancing their growth with extra genes might produce larger mice while retaining cohesive integration among features. If rats represent the upper size extreme in their particular history, enhancing their growth might disrupt the integration of phenotypic design elements to a greater extent than a species that has been there before.

Cairns, Overbaugh and Millar (1988) recently shocked the neo-

Darwinian camp with data indicating that bacteria in novel situations had mutations that were significantly biased towards functional utility. Despite considerable reluctance by mainstream theorists to accept such a phenomenon, the results of Cairns, Overbaugh and Millar (1988) have been substantiated by others (Gillis, 1991; Hall, 1991). If genetic organizations contain some mechanism to access suppressed genetic information (e.g. via release of heterochromatic suppression or via targeted activities of transposable elements), then a historical landscape highly biased towards adaptive functionality could be uncovered during periods of stress. Cairns, Overbaugh and Millar (1988) recognized that bacteria can access an armoury of cryptic genes to utilize novel substrates, but hold that some additional mechanism allows bacteria to achieve functionally biased mutations.

4.4 GENETIC FEASIBILITY: COST

In terms of the cost of maintaining extra genetic material, there is conflicting information. Reid (1985) points out that the conventional wisdom is that useless code is lost (i.e. the genome is parsimonious). A common trend in symbiosis is reduction of redundancy (Margulis, 1981). Many successful species, however, maintain complete information for multiple ecological designs (e.g. dragonflies and their aquatic nymphs, butterflies and caterpillars), without apparent detriment. The ubiquity of large quantities of apparently non-functional code and the fact that the amount of DNA in a genome is largely unrelated to the organism's complexity suggests that selection may not act strongly at this level.

The amount of ATP required to replicate apparently functionless DNA is trivial relative to general cell economy (Loomis, 1988, p. 214). Although cavernicolous and parasitic animals appear to lose useless features, in some cavernicolous salamanders, at least, functional eyes may sometimes be restored by environmental or surgical modifications (e.g. Goldschmidt, 1940, p. 257; Raff and Kaufman, 1983, p. 178), a result nicely predicted by the domain model of DNA organization. The genome may act much like a modern computer when it comes to deleting programs. Rather than physically removing the code from memory, the file name is simply deleted from the directory and the old code can then be written over or ignored. Bodnar, Jones and Ellis (1989) suggest that unused domains of condensed DNA could be carried almost indefinitely.

Since such features constitute pseudogenes, however, eventual deterioration will be inevitable. In the blind mole rat, *Spalax ehrenbergi*, the αA-crystallin gene is still expressed in the degenerate eyes despite an estimated 25 million years of subterranean life. This has been accompanied by an increased rate of neutral molecular evolution compared with strong conservation in other rodents (Hendriks *et al.*, 1987).

Continued expression of useless gene products must constitute a much larger cost than simply carrying quiescent code. The section on genetic inertia reinforces the generalization that old code is usually repressed prior to functional loss, a situation of immense evolutionary potential.

In bacteria, selection to utilize novel or diluted substrates is often met by gene duplication which provides larger amounts of enzyme. When restored to normal media, such redundant code is usually lost (Clarke, 1983). There is also evidence that bacteria that have lost the ability to synthesize particular products may be superior to wild type if these products are provided (Diamond, 1986). Bacteria, however, may be under strong selection pressure for very rapid replication and high biochemical efficiency that might place a relatively higher premium on parsimony than is required in eukaryotes (Doolittle and Sapienza, 1980; Loomis, 1988, p. 174). Even then, attempts to quantify the energetic savings obtained by null mutants provided with necessary substrates have not detected the expected savings because wild type strains may simply suppress protein production if the necessary substrate is freely available (Diamond, 1986).

Thus, there is evidence either way, but overall, there does not seem to be a prohibitively high cost of extra DNA, particularly in eukaryotes. Consequently, genomes can (and do) support features of little or no immediate value but which pre-adapt the lineage for mesoevolutionary steps. Such steps may not be available to other lineages that are otherwise identical in phenotype and genotype (e.g. differences in the number of gene duplications). Several aspects of an epigenetic circuit or a biochemical capability could evolve cryptically and emerge in a lineage after numerous generations. In fact, persistent lineages may be those that tolerate or even encourage extra genetic code because of its long-term evolutionary usefulness.

Factors associated with persistence of intermittently selected code include recombination rates, the length of the functional sequence, redundancy, mutation rates, how much substitution can be sustained before coded proteins lose function, effectiveness of DNA repair mechanisms, and whether parts of the code are active in other functional gene nets or intermediate epigenetic stages. Even where the cost of inactive code eventually exceeds acceptable limits, its removal via selection of lineages that have deleted blocks of redundant or useless code does not conform to a genic model.

4.5 INTRASPECIFIC GENOMIC VARIATION AND STABILITY

Even within the genic paradigm, environmental heterogeneity or inefficient sexual recombination can lead to recognizable lineages (demes, races, ecotypes, subspecies). Theoretically, sexual reproduction can homogenize populations in remarkably few generations (Fisher, 1930). Theory aside, the reality of subspecific lineages in sexual species is nicely demonstrated by the high correlation of blood types with family surnames in humans (Dobzhansky, 1970, p. 290). The correlation of mitochondrial phylogenies and eukaryotic racial variants also attests to the reality of subspecific lineages in sexual species. In most species, limited dispersal and habitat choice may enhance local lineage identity and differentiation (Shields, 1982). Major differences in genome structure can constitute distinct lineages even where sexual communication exists. A theoretical example was provided by Perrot, Richerd and Valero (1991). These authors modelled selection for diploidy or haploidy where the two genomic variants exchanged genes including deleterious mutations. Their model predicted conditions under which diploids would displace haploids depending on the degree of recessiveness in mutants. Their model is at once an example of individual and genomic-level selection.

Critical to lineage selection is the presence of higher-order genomic variation (i.e. polymorphism in regulatory alleles or variation in larger genome structure). Mesoevolution of higher-order genetic architecture such as regulatory circuits also requires intraspecific variation in genome structure and switching elements. Some may arise via mutation, but there may already be enough variation retained in sexual lineages to reconfigure phenotypes to remarkable degrees.

The literature on phenocopies is particularly relevant to whether higher-order variation is available for selection. Phenocopies are aberrant phenotypes induced by environmental insults such as temperature shocks or ether exposure during critical developmental stages. Such phenocopies are often identical to known mutations and pointed the way to investigations of higher-order regulatory organization in *Drosophila* (Goldschmidt, 1940; Waddington, 1957; Scharloo, 1991). Of themselves, phenocopies do not prove that higher-order variation exists, but the fact that such traits can be genetically assimilated (Chapter 6) is conclusive evidence that sufficient variability exists to achieve truly major reorganizations (Waddington, 1957; Scharloo, 1991).

Homeotic mutants (those where one structure such as an antenna is converted into another such as a leg) demonstrate that potential variation might occasionally derive from changes in higher-order genes. The key question is how much variation may be retained in natural populations. The strong conservation of higher-order switching genes across wide phylogenies (i.e. nematodes to *Drosophila* to mice) suggests that these may be largely consolidated parts of rather fundamental organizations. Alternatively, we have not looked closely enough to be sure that higher-level functional variants do not exist. It seems most likely that the degree of variation decreases as we ascend the regulatory hierarchy, but we have no idea where, or if, it stops. The fact that the mammalian regulatory system appears to be derived from a quadra duplication of the analogous complex in *Drosophila* (Hall, 1992), and that homeotic genes may have been added to derive segmental specialization, suggests that considerable evolutionary activity may occur even at higher levels.

In sexual species, every individual may be unique, so speaking of lineages at first glance appears paradoxical. However, in a hierarchical system with low variation in the upper tiers, identifiable lineages could be differentiated despite lower-level variation in detail. Considerable variation in low-level modifiers may exist, without detracting from the reality of distinct organizational variants.

The concept of regulatory hierarchies is a valuable framework for considering lineage evolution, but this view is also a major simplification. Epigenetics is a physical as well as a genetic and chemical process. Thus, Lovtrup (1974), finding no evidence for the reality of morphogens, derived a credible model of epigenetics based strictly on cell–cell interactions and physical forces. Waddington (1957), although recognizing that genes differed in their degree of epigenetic effects, also argued that numerous small modifiers could achieve similar ends. Thus, if an inducing tissue does not contact another tissue competent to react to it, a particular organ may not appear (Alberch, 1991).

Either a single major gene or a number of modifiers of small effect might bring tissues together resulting in a major epigenetic event (Waddington, 1957, p. 53–54). A similar process could separate an inducer from its normal target, leading to major events such as loss of digits or limbs. This does not dismiss the reality of the regulatory hierarchy (the induction itself must be a regulatory event) but it does point out that some epigenetic information is concealed in the physical structure of the phenotypic organization itself (Alberch, 1991). What might be interpreted as a genetic switch, perhaps associated with an allele at a particular locus, could instead arise from the presence or absence of a layer of tissue inserted between a potential inducer and a competent target. Waddington (1957) applied this argument to explain how the *bithorax* phenotype could be achieved either by a known major mutation, or via selection of appropriate modifiers.

Differences as gross as variation in chromosome number and structure are well documented within single species (Stebbins, 1950; Mayr, 1963; Dobzhansky, 1970; Lewontin, 1974; Bush *et al.*, 1977; White, 1978). In humans, chromosome 21 may vary in size by as much as 25% among individuals (Wills, 1991). Only polymorphisms for chromosome inversions in *Drosophila* have received much attention (reviewed by Dobzhansky, 1970; Lewontin, 1974; Anderson *et al.*, 1991). The consensus is that these inversions protect constellations of co-adapted genes and that various chromosomes are advantageous in different environments.

The relevance to lineage selection is clear. It is important to know the genetic organization of various lineages, not just their genic statistics (Lewontin, 1974, p. 140). Operationally, the genic view requires a genome with fixed harbours for circulating alleles. Early expectations were that the genome would display the same degree of refined parsimony that natural selection has imparted to morphology. The discovery of large quantities of apparently functionless and/or redundant code consequently was surprising, but the idea that the operational genome is constant and parsimonious within a species persists.

Advances in molecular biology have provided a more refined view of genome structure. The panorama is that of a remarkably diverse and complex intraspecific landscape (see review by Britten, 1986a). This information not only documents that there is plenty of fuel to allow meso-evolution via subspecific lineage selection, but that this variation probably reflects the action of this very process. The only limitation is the ability of chromosomes to pair successfully and recombine if the species reproduces sexually. Alleles are really micro-variations in chromosome structure themselves. There is no need, for example, for alternative alleles to have equal numbers of bases for chromosomes to pair successfully and recombine (although this may create instability and lead to problems).

Even high variability (e.g. the widespread occurrence of inversions) does not necessarily preclude successful interbreeding. *Drosophila melanogaster* contains numerous chromosomal inversions, four of which are ubiquitous (Carson, 1990b). Thus, differences in structure exist within a single species. In *D. silvestris*, heterokaryotypic flies may be of high fitness (Carson, 1987b), which extends Lerner's (1954) ideas that heterozygosity may be an integral part of the overall organization. Studies of hybridization show that even species that have radically diverged genetically and/or phenotypically are often capable of successful interbreeding (Stebbins, 1950; Mayr, 1963; Val, 1977; John and Miklos, 1988). The limits of mesoevolution and the transition to speciation and macro-evolution will be demarked ultimately by the ability to maintain sexual communication. There is good evidence that karyotypic reorganizations can induce speciation (White, 1978).

Recently, the question of intraspecific genomic variation has taken a giant step forward. Evidence is rapidly accumulating that genomes are adapted to modify their own structure in very specific ways; they are not only variable, but fluid. Some of this capacity is deployed in response to environmental stress or genetic disruption from damage, virus infections or hybridization events (McClintock, 1984; Cullis, 1986; McDonald *et al.*, 1987; McDonald, 1990). Reorganizations may also be employed for generating diversity in antibodies for immune responses (Wills, 1989). In large part these abilities appear to have evolutionary functions; they are adaptations to facilitate the evolution of lineages and contribute little or nothing to the immediate fitness of organisms (Campbell, 1985, 1987). Conrad (1979, 1983), Campbell (1985, 1987), Wills (1989) and Shapiro (1992) persuasively argue that genomes are designed to facilitate their own evolution. Dawkins (1989b) similarly concluded that higher-order selection may act to favour embryologies with high evolutionary potential, a lineage-level adaptation for evolvability. Such a view is rapidly becoming mainstream (Arnold *et al.*, 1989).

Campbell (1987) documented 27 classes of enzymes which direct every conceivable manipulation of DNA. These enzymes may be harnessed by the genome to genetically engineer itself in ways that facilitate evolution in a much more efficient manner than envisioned by the traditional view of random mutations (Shapiro, 1991, 1992). Campbell (1985) and Shapiro (1991) argued that most mutations are not accidental as was previously believed, but are generated by what can be considered as reprogramming functions. Wills (1989) explores how these abilities may be used for genetic reorganizations such as the evolution of supergenes (shuffling genes around into linkage groups or creating inversions) or for diversifying gene families via exon shuffling. McClintock (1984) pointed out that formation of new species via hybridization may often involve such reorganizations. Significantly, Cullis (1986) found that even the interaction of subspecific lineages could trigger such events.

In *Drosophila*, suppressor genes exist whose products may inhibit the expression of retrovirus-like elements so that otherwise mutant phenotypes are not produced (Kubli, 1986; Corces and Geyer, 1991). There is mounting evidence that such inhibition is released during inbreeding or stress, expressing dormant or novel variation at times when this might be evolutionarily advantageous (e.g. Ratner and Vasilyeva, 1989; McDonald, 1990; Travis, 1992). Such processes extend the breadth of genome differentiation that could be encompassed within one genotype. Adaptations for evofacilitation are lineage-level attributes that are potentially selected for their evolutionary consequences rather than their adaptive function in the organism.

Examination of the phylogeny of chromosome structure in *Drosophila* confirms that such genetic reorganization (i.e. inversions, translocations, duplications, transpositions, conversions and changes in chromosome number) has been a dominant feature of speciation (Dobzhansky, 1970; White, 1978). Lineage selection is also relevant where current genetic structure is biased by environmental, developmental or historical constraints. Kimura (1983) discussed how random drift in neutral genes could occasionally yield a new structure of high selective value (analogous to a useless poker hand being converted to a winning hand by the draw of a single card). Gould and Vrba (1982) also pointed out that the capture of current adaptations for new uses, or utilization of features that are of neutral selective value for new roles constitutes a crucial avenue of evolution. They termed such features exaptations, and understanding them requires a lineage paradigm. Gould (1989a) applies such ideas to the success of various phylogenies represented during the early Cambrian radiation of body plans. Wills (1989) concluded that molecular processes such as exon shuffling, gene duplication and transposition of regulatory recognition sites positively biases the likelihood of generating such exaptations (e.g. the invention of lactose in mammals (Gould and Vrba, 1982; Raff and Kaufman, 1983)).

Kauffman (1985, 1993) provides an epigenetic model that proposes that the inter-gene recognition sequences can be considered much like a wiring diagram. If various regulatory connections could be unplugged and reinserted in other circuits, considerable epigenetic potential would become available. Significantly, almost every control sequence known to be critical for regulation of eukaryote genes has been identified within the long terminal regions of retrovirus-like elements. These are the very pieces that are often left as 'footprints' during the movement of these transposons (reviewed by McDonald, 1990).

Confirmation of the evolutionary significance of transposons was recently revealed for the sex-limited protein gene (*Slp*) of mice. This gene was derived from a gene duplication in the histocompatibility complex, but has a novel regulatory response to androgen. This control was imported by the insertion of a retroviral-like element containing the relevant control sequences, and the remnants of the transposon are still evident (Stavenhagen and Robins, 1988). Syvanen (1984) and Finnegan (1989) point out that the kinds of genetic changes needed for the evolution of regulatory mechanisms would be difficult to obtain via anything other than the activities of transposons.

The picture that emerges is that of a regulatory organization that has considerable adaptive potential to re-wire itself. In all fairness, there is an alternative, non-adaptive scenario that also could be applied to transposon activity and stress. Axelrod and Hamilton (1981) suggested that parasitic elements that have co-evolved to minimize their impact on a long-term host might revert to full activity if they detect that the host might soon die. In this case, the increased activity of transposable elements during stress might not be an adaptation for evolvability, but a selfish strategy to flee a sinking ship. The occasional evolvability bestowed by this activity could be coincidental.

Such tactics would only be useful if increased transposition rates improve the chances of mobile elements to escape from the host. Of relevance here, however, Erwin and Valentine (1984) proposed that such transposons (including viruses) could mediate genetic exchanges among individuals, rapidly creating intrafertile subpopulations with novel characteristics. Thus, even the release of such elements from the genome could be relevant to the evolvability capabilities of a host lineage. Finally, there is strong inferential evidence that transposons that are normally incapable of escaping from a host lineage may do so via a vector such as a parasitic mite. Such transfers may even cross species borders (Houck *et al.*, 1991).

Evolution has a remarkable tendency to convert apparent handicaps into useful adaptations. It is quite conceivable that some small concessions by one or both parties could forge even selfish transposable elements into an organization with an optimal mix of short-term stability and longer-term evolvability (see also Orgel and Crick, 1980). If transposable elements represent a functional attribute of fluid organizations, then we might expect to find evidence for regulated control of their activity and restrictions on the sites that are available for insertion. Some areas may be favourably varied by transposons, but other areas that are easily disrupted might be protected.

There is increasing evidence that transposable elements are regulated by the host genome (e.g. Kubli, 1986; Csink and McDonald, 1990; Corces and Geyer, 1991; Shapiro, 1992). P elements in *Drosophila* are restricted to the germ line, and are disruptive to strains that are not coevolved to harbour them (Rio, 1991). There is also mounting evidence that the target areas available to transposons are constrained. A lineage paradigm predicts that transposable elements may be an integral feature for evolvability, and if so, molecular evidence should reveal some adaptive control of their levels of activity and localization of their target sites. For the *gypsy* retrotransposon, control elements in the transposon are recognized by transcription factors (e.g. suppressor of *Hairy-wing*), that may modulate the expression of the mutant phenotype within targeted genes (Corces and Geyer, 1991).

Thus, mutant phenotypes may be driven not simply by positional effects of transposon insertions that disrupt gene function, but by changes in the regulatory constitution of the gene. Such findings suggest that not only the activity of transposons, but the expression of mutant phenotypes driven by them may reside in a regulatory framework open to higher-order organizational control. Such mechanisms could, for example, allow cryptic accumulation of regulatory variation that can be selectively released, perhaps in response to stress (McDonald *et al.*, 1987; Travis, 1992). In fact, some mobile elements may be regulated by systems specifically attuned to stress, such as those involved in heat shock responses (McDonald *et al.*, 1987). In bacteria, functions relevant to evolvability may be activated by the so called SOS signal generated under stress by the *Rec A* gene (Echols, 1981). Echols (1981) interpreted this system as an adaptation for inducible evolution.

The regulated application of transposition in the immune system (Wills, 1989), and other examples, such as its functional role in mating-type determination in yeast (Klar, 1990), argues strongly for adaptive utilization of transposition. Besides a role in adaptive phenotypic processes, mobile elements could also secure a general evolvability function (Syvanen, 1984; McDonald, 1990). As Syvanen (1984) points out, an evolvability role for transposable elements has been dismissed by some authors as requiring group selection. If, however, there is occasionally strong selection for variability, then adaptive transposition could be selected during such periods, even if it proved neutral or slightly detrimental at other times. Thus, Syvanen (1984) has articulated a lineage-selection explanation for transposition, similar to that being developed here, and he recognized this as something other than group selection.

An evolvability role for both sex and transposition are potentially viable in a lineage paradigm. There is a paucity of theoretical discussion on transposition compared with sex, but the reader can apply nearly all of the discussion on sex (Chapter 5) to the evolution of transposons as well. It seems significant that the timing of sexual reproduction and increased transposition rates have similar ecological correlates. In a regulatory context, where complex integration may be involved, elevated mutation rates improve the chances of simultaneously capturing several components of a new regulatory circuit (Hall, 1991).

It might be expected that a model of subspecific lineage selection would yield genic selection given a uniform genome (i.e. variation in alleles but not in genome structure), populations with homogeneous spatial and temporal habitats and unbiased, complete sexual panmixia in large populations. Even given these classic assumptions, however, species selection and cladogenesis involving differential rates of extinction and speciation among species and clades would not be precluded (e.g. Stanley, 1979; Eldredge, 1985; Salthe, 1985). Thus, even in the best case, the genic view is insufficient for complete understanding, as even Williams (1988) has now argued. The model proposed here does not preclude genic selection, but relegates this to a single level of organization that may not have free rein. The dynamics of subspecific lineages emphasizes levels of selection that link the genic framework hierarchically to that of species selection, whereas these two frameworks are usually regarded as sharply dichotomous.

There will be strong interaction between unique forces and histories associated with the impact of environmental heterogeneity on the population structure of a species and the action of hierarchical selection pressure across levels of developmental organization within sublineages. As I will show, the entity forged by the interaction of these different processes is not simply the code for an individual organism adapted to its immediate environment, but a super-genome or dynamic genetic templet which may be capable of producing numerous types of organisms appropriately adapted to a wide range of temporally or spatially discrete environments (Chapters 5 and 6).

5

Metalineage selection and sexual reproduction

5.1 SEXUAL COMMUNICATION AND METALINEAGE ORGANIZATION

Recent models suggesting 'short-term' advantages of sex are actually transgenerational and, by considering genes, circumvent the key point made by Williams (1966a) and Dawkins (1976) that sexual reproduction is tantamount to organizational suicide on the part of individual organisms. Theoretically, parental identity is diluted at an exponential rate over successive generations. It seems most unlikely, however, that a key level of integration like that of the organism could be easily induced to commit organizational hara-kiri merely so that genes might act selfishly. Instead, organizational evolution appears to aspire to progressively higher levels (Chapter 1). Given that genomic organization is largely geared to the elaboration of phenotypes and that relatedness should probably be interpreted relative to organizational congruence rather than alleles at single loci, perhaps a shift to a genic view is not the appropriate vantage point for understanding sexual species and phenotypic evolution.

Wilson and Sober (1989) captured the quandary by pointing out that despite the ephemeral nature of individuals, the genetic organization that produces humans has not changed significantly in recorded history. They emphasize that this phenotype-making organization is as biologically relevant as any gene, and moreover, that the selfish gene framework may be fundamentally irrelevant to the question of functional organization. In the light of the general rule that increasing temporal scales are associated with larger and more complex organizations (Salthe, 1985), the correct answer instead may be that sex is an adaptation that improves the persistence of lineages. Williams (1975, 1988, 1992), a very strong advocate of genic selection, allowed for such a lineage persistence explanation of sex. Such a perspective suggests that the invention of sex yields a higher-order organization rather than a free field for selfish genes. Recognizable sublineages that otherwise might independently diverge are forged together into a cohesive entity (Figure 5.1). Moreover, such an organization is probably not simply some fortuitous consequence of the presence of sexual reproduction, but the main purpose for the existence of this phenomenon.

Each transition to a new level of biological organization is characterized by a number of key elements. There is a communication system or information transfer among lower-level components. Sex itself represents such a vehicle. The components then evolve complementary specialization. Male and female dimorphism represent such a criterion. In highly evolved/organized systems the components lose their independence and become non-dissociable. In most species, sexual reproduction is obligatory. In mammals, for example, failure to derive parthenogenic lineages is not related to the expression of recessive lethals associated

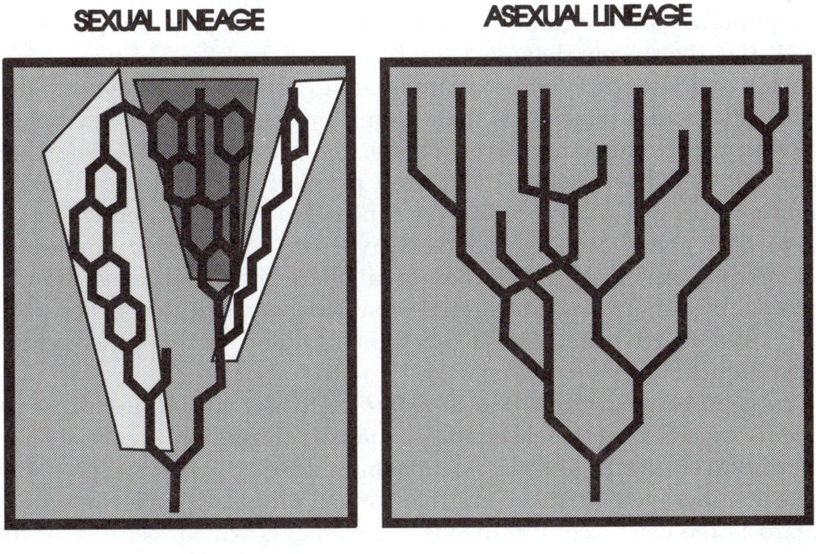

Figure 5.1 The structure of sexual and asexual lineages. In sexual species a hierarchy of recognizable sublineages may emerge on various temporal and spatial scales. At each level, recognizable branches are composed of lower levels with net-like integration. Sublineages may be differentiated by both variation in their regulatory organization and structural elements. In asexual lineages, the lineage structure is always that of a branching tree and net-like integration is absent at all levels.

with homozygosity as was previously thought (i.e. when a haploid egg doubles its genetic complement).

Instead, successful development requires both paternal and maternal imprinting of homologous chromosomes (Slack, 1991, p. 186). Specialization has proceeded so far that complete information cannot be derived simply by doubling the female contribution. The fact that a single organism does not represent a fundamental unit of reproduction in sexual species seems like something of the utmost importance, and yet this is largely glossed over in models of evolution considering individuals. Maynard Smith (1971b) observed that the apparent design of developmental mechanisms to resist asexuality needs some explanation.

A key problem appears to be how to reconcile the obvious state of high functional integration seen at the genomic and phenotypic levels with a system of recombination that theoretically disrupts linkages and the likelihood of genes with positive epistasis (in the population genetic sense) from remaining together. Part of the answer may be that the transition to the next higher state of organization requires a mechanism(s) to prevent takeover by selfish elements. Buss (1987) considered this idea with respect to cell lineages in organisms and concluded that consolidation of higher-level organizations involves defence of the bridge between levels (i.e. the system defends itself from rebels).

Haig and Grafen (1991) explored a relevant model of meiotic drive (where particular genes bias their own representation in the gametes at the expense of other genomic elements). Meiotic drive leads to short-term gains by selfish elements, but at the price of eventual organizational failure. Thus, it is a kind of lineage cancer. Haig and Grafen (1991) conclude that genes causing meiotic drive can only spread successfully where they can link themselves to a block of other genes. Random meiosis, by scrambling linkage relationships, ensures that genes potentially causing drive are just as likely to harm themselves as other genomic constituents.

Thus, global recombination and fair meiosis may be necessary constraints to create and perpetuate sexual organizations (Leigh, 1977). If so, wherever strong linkage relationships develop, they are potentially invasible by meiotic drive mechanisms. Particularly relevant in this regard is the existence of transposable elements which, because of their greater potential for direct selection at the genic level, could benefit by carrying elements capable of generating drive. Hickey and Rose (1988) point out that this may actually be the case with the segregation distorter system of *D. melanogaster*. Such elements could exaggerate the importance of drive phenomenon because they could possibly transport the relevant machinery for drive into any stable linkage groups that are protected from recombination. Thus, in two of the best known examples of drive, the t-haplotypes of mice and the segregation distorted system of

Drosophila, recombination is locally suppressed by inversions. Hickey and Rose (1988) even argue (incorrectly I think) that sexual reproduction may have originated as a mechanism for transposable elements to effect horizontal transfer. Haig and Grafen (1991) suggest that recombination acts as a scrambling mechanism to resist drive – clearly a feature beneficial to organizational persistence. Rather than selfish elements, genes then appear to be participants in a genomic democracy where all constituents must have fair and equal representation.

Haig and Grafen (1991) present their hypothesis in the popular context of immediate selective advantage, but drive mechanisms do not appear to arise with the necessary frequency. In a lineage framework, however, this could represent a key factor relevant to organizational transition and persistence. Given this kind of constraint on genome structure, a coordination of genes via intermediate signals (e.g. transcription factors), rather than physical proximity becomes a necessity (Chapters 2–4). Although batteries of important regulatory genes are strongly linked, they may be protected from invasion because their dense regulatory structure is easily disrupted, with likely lethal results.

A transition of sublineages to asexuality might also represent an organizational failure for a sexual lineage (Leigh, 1977; Nunney, 1989). Both these authors view sex as a feature promoting group persistence on long time scales, which may be overcome by selfish elements that can spread via short-term advantage. Nunney (1989) provides a model that shows that sexual lineages will be selected to reduce the likelihood of 'cheats'. If the frequency of cheaters is sufficiently low, the organization may have enough time to express its advantages to selection. Thus, cancers may result in organismal failures, but organisms have sufficient integrity to allow evolution at this level, and they also have evolved mechanisms to reduce cancer. Just as cancers do not mean that organisms are not real, the existence of short-lived asexual taxa does not mean that sexual species do not represent a higher level of organization. Nunney (1989) favours selection acting at the species level. In subdivided metapopulations, however, interdemic selection may act on relatively shorter time scales.

One feature of an adapted lineage would be the deployment of highly fit phenotypes. What other features might reflect the reality of such an organization other than the possible advantages of evolvability? Any such features would be most likely relevant to population scales of time and space rather than the organismal. Genetic correlation structures and reaction norms may well represent such features (Chapter 6). Thus, the evidence that sexual lineages constitute a higher level of organization is considerable. The necessary conceptual jump is that this organization is polyphenotypic. Components are not independent of the organization even though they have discrete spatial–temporal identity (e.g. individual males and females). Colonies of social insects or slime molds represent less controversial entities that reflect such distributed organization (Hull, 1976), and in the case of asexual lineages, most biologists recognize individuals as parts of a single genetic entity (e.g. Janzen, 1977).

If sexual lineages do represent higher-order organization rather than simply assemblies of independent organisms, then whether lineage selection should be considered to be group selection is debatable (although group selection undoubtedly does constitute strong lineage selection). The distinction is an important one and yet the issue remains unresolved. The idea that lineages may comprise units of selection is an old one, and yet this has only recently made the hurdle into mainstream theory. Thus, there have been convincing arguments that sexual species can be considered as individuals (Ghiselin, 1975, 1987; Hull, 1976, 1978, 1980, 1981; Salthe, 1985; Keller, 1987) but these ideas have not been developed at sub-specific levels.

Williams (1975), Jain (1979) and Brooks (1988) appear to regard subspecific lineage selection as something other than group selection, but Maynard Smith (1989) branded it as group selection with temporal separation of groups. How this definition applies to overlapping generations highlights the problem. The critical criterion must be to what extent a lineage represents a higher-order organization rather than simply a conduit of associated independent genes or individuals.

In a lineage framework, what constitutes group selection has a fairly restricted criterion. No biologists argue with the fact that phenotypes may incorporate features to promote offspring survival, even at the expense of the parental soma. This is allowed even though offspring are only half related. Extension of this idea (which represents the shortest temporal unit in lineages) to encompass larger lineages merely recognizes some degree of relationship among other individuals not conventionally regarded as kin. Maynard Smith (1976) placed considerable emphasis on physical separation as a criterion for distinguishing kin and group selection. It can be argued, however, that the difference between group and kin selection rests on whether families are considered kin or groups (Wade, 1980), because kin selection also requires structuring of populations into kin groups.

In a sexual species, there is a continuity of such identity, with no clear demarcation of where relatedness ends because strong reinforcement occurs over time. Thus, most models of 'group' selection in fact constitute selection of subspecific clades (Vrba, 1989a). Much of the controversy stems from the perception of some dichotomy where none actually exists. Wilson (1976, 1983) and Slatkin (1987) recognized such continuity while Dawkins (1989a) argues that kin and group selection are fundamentally different.

If the term group selection is to be applied to a process where entities are selected which have no internal organizational structure or relationship, then this process becomes restricted to the coevolution of interacting lineages that do not exchange genetic information. Otherwise the process is more correctly lineage selection. Even without genetic exchange, there is no fundamental barrier to deriving higher orders of functional organization. Presumably such associations could be concentrated at particular levels of hierarchical organization among the species concerned, and this could be of evolutionary importance.

Cladogenesis at phylogenetic levels where sexual communication has ceased may more appropriately be classified as a process of differential sorting (Vrba, 1989a), since there is no need for diverging lineages even to be spatially associated for the process to occur (Williams, 1992). Where phylogenetic lineages interact, then lineage selection may act at this level. Thus, I would classify all forms of selection within sexually communicating organizations as lineage selection and any form of selection where genetic information is not exchanged as true group selection. Even then, non-genetic channels of information transfer may be opened that allow organizational integration and species may evolve complementary genomic features, such as the gene-for-gene virulence/resistance structure found between many parasites and hosts (see also Goodnight, 1990a,b).

The search for some value of sexual reproduction that could offset the two-fold cost to organisms may be fruitless if some of the benefits reside with lineages. Where a new level of organization has been consolidated, it is inappropriate to seek its selective value among lower-level components (Buss, 1987). Rather than yielding a system where genes can be considered independent and 'selfish', the invention of sex instead lays down yet another overlay of constraining biological organization (Salthe, 1985) which includes the necessity of a system for sex determination and dimorphic epigenetic trajectories.

Ohno (1970, 1973), Wilson (1980, 1983), Kimura (1983) and Vrba (1989a,b) all provide a rather clear vision of this concept of higher-order constraints shaping lower-level structure. Lineages also have a historical component in that they represent biased and limited compartments of all possible genetic potentials or types of imposed genetic constraints. These concepts are sometimes related. Once an organization has selected one possible pathway out of many lower-level potentials, then subsequent organization is elaborated on this foundation. To change the lower-level component would then result in disintegration of the supported structure. Thus, we are becoming increasingly aware that much of biological organization represents 'frozen accidents' acquired during early lineage evolution. As organizations, however, sexual lineages constitute much more than simple accumulations of historical accidents.

There has been considerable difficulty in attempting to classify life in terms of either functional biological hierarchies or historical genealogical hierarchies (Salthe, 1985; Eldredge, 1985, 1986). The two appear mutually incompatible, but organisms come out as components of both. This nexus is predicted, however, if selection shifts at this point from acting on a single spatio-temporal organization in a single generation (organisms), to acting on a polyphenotypically distributed organization (Hull, 1980) over multigenerational time scales (sexual lineages). Significantly, in their attempt to classify the organism as the centrepiece in the various levels of biological organization, MacMahon *et al.* (1978) concluded that this selective level must include the ancestors of living individuals.

This highlights the fact that, if the hierarchy of life is mapped according to levels of organization, subspecific lineages follow naturally as a transition from organisms (Chapter 1). In this view, sex mainly serves as an integrator and communication channel for intra-lineage organization. Genetic polymorphism, the crucial fuel for recombination, may represent part of the system's global identity (Lerner, 1954; Lewontin, 1957). Changes in allele frequencies may represent the dynamics of the lineage's homeostatic balancing, a view foreshadowed by Lerner (1954) and Hull (1981). Hull (1981) saw that this would only be true if there was feedback between the organismal and lineage level. This follows if the organisms represent different faces turned to the world by a single, flexible, phenotype-making program, and the long-term persistence of such machinery is favoured by the success of locally deployed organisms.

5.2 HIERARCHIES AND LINEAGE STRUCTURE

There are two kinds of interacting hierarchies involved in a lineage paradigm. The first is the organizational hierarchy discussed earlier, which suggests that relevant units of genomic organization vary hierarchically, with circulating variants constrained by higher-order coadaptational constraints (Chapters 3 and 4). A key factor for the current thesis is that important variation may consist of differences in genome structure and wiring relevant to regulatory evolution. The second kind of hierarchy is derived from more traditional models relevant to the spatial and temporal fragmentation of populations.

Early theoretical treatments of sexual evolution usually assumed that the relevant evolutionary framework consisted of single large panmictic populations inhabiting ecologically uniform environments with low temporal variability. A non-equilibrium ecological framework suggests that spatial and temporal heterogeneity impose variable and/or uncertain selection pressures on subpopulations. Thompson (1976) reviews earlier literature pertinent to these ideas. The assumption that even low gene flow must lead to panmixia has been displaced by theoretical and empirical work showing that population differentiation and the lineage selection process can operate even with relatively high gene flow (Ehrlich and Raven, 1969; Endler, 1973, 1977; Wade, 1977; Wilson, 1989; West-Eberhard, 1989). Moreover, dispersal characteristics (and interactions among species) may themselves generate discontinuities highly relevant to the genetic structure of a species or the spatial heterogeneity of multi-species assemblages (e.g. Ehrlich and Raven, 1969; Endler, 1973, 1977; Hassell, Comins and May, 1991).

Such features will tend to fragment even continuously distributed sexual species into local subpopulations with various degrees of connectance. Local populations with some degree of relative discreteness constitute units known as demes. In a subspecific lineage paradigm, demes represent a key holon in hierarchical lineage differentiation. Wright (1964) came very close to capturing the subspecific lineage model proposed here when he described population structure as 'a loosely knit organism with protoplasmic continuity of a netlike pattern in space-time'. Even Fisher (1930) emphasized that the intimate manner in which organisms in sexual species are bound together tends to be overlooked in discussions of evolution of such entities. He suggested that variants in sexual species can be envisioned as a weaving of differently coloured threads into a 'single uniform fabric' (Fisher, 1930, p. 124). Arthur (1988, p. 61) observed that species are better regarded as systems of parallel lines rather than as single conduits.

Williams (1966a, pp. 113–114) suggested that if the overall genetic background changed relatively slowly, then a lineage and genic view could be reconciled where an allele was differentially favoured among gene pools. Although he felt that this was an unlikely circumstance, the evidence for a genomic hierarchy of regulatory genes with conserved upper holons brings this view more closely into line with Wright's model of interdemic selection outlined below. More recently Williams (1992) advocated the reality of clade selection and suggested its extension to subspecific levels. Thus, a view of sexual species as entities composed of recognizable subspecific lineages is not new.

Even across species, single loci are not suitable for recognizing phylogenetic relationships, but recognizable lineages can be discerned by statistical divergence among multiple loci (Hillis and Bull, 1991). Spatial discontinuities may further impose a hierarchy of diversifying lineages, each of which could have an internal hierarchy of sublineages on finer scales (Figure 5.1). Figure 5.1 assumes no elimination of phenotypes by natural selection, no phenotypic plasticity and equal mutation rates in a sexual *versus* clonal system. For simplicity, sexual sublineages are portrayed as having mutations (represented as bifurcations) and recombination (relevant to coadapted complexes) occurring at discrete times. Sex acts to maintain the identity of the metalineage even though individuals and sublineages may constantly vary. The rate and distance of sexual communication among sublineages defines the extent to which the inclusive lineage constitutes an individual (Hull, 1980). Harvey *et al.* (1992) suggest that all species represent metapopulations of this type at some appropriate scale.

Sex causes phenotypic regression on the mean for quantitative traits (i.e. the majority, Falconer (1981)) that helps maintain identity, promotes stasis and (in some cases) resists speciation (Felsenstein, 1981). Simultaneously, sex might allow a greater diversity of genotypic sublineages to interface to ecological subniches generated by multidimensional resource and constraint gradients (Jones, Leith and Rawlings, 1977; Michaels and Bazzaz, 1989). Case and Taper (1986), using models based on resource competition, calculated that sexual lineages could displace any single asexual clone if the niche width of the sexual organization was even 12% broader. Overall, the broader ecological niche would probably also be subjected to greater stabilizing selection than that of narrower clones (Figure 5.1). Thus, sexual lineages are more likely to exhibit stasis and diverge less rapidly than asexual clones (Stanley, 1979). Empirical support for this idea is provided by Lynch, Spitze and Crease (1989) and Schoen and Brown (1991).

For the asexual system, even though the range of phenotypic variability among sublineages may be large, each asexual sublineage has less adaptability and greater risk of extinction than the structurally more complex sexual lineage (assuming comparable genome ploidy) (Vrijenhoek, 1990). A swarm of asexual genotypes could occupy a significant portion of the range of a sexual lineage (the frozen niche hypothesis of Vrijenhoek). Furthermore, a sexual lineage would be displaced by an asexual swarm that occupied all available niches (Case and Taper, 1986). However, the likely persistence of such asexual diversity is unlikely over the long term, so asexuals are unlikely ever to win and hold enough of the potential niche to displace the sexual lineage. Thus, sex may be advantageous for lineage persistence, even though the price for individual organisms is attenuation of their transmitted identity.

Wright (1931, 1932) envisioned the adaptive landscape of a species as a surface with local fitness peaks with intervening valleys or saddles. Populations tend to climb local adaptive peaks. The question then becomes how can populations escape from such peaks to move to other, perhaps superior adaptive vantage points (shifting balance)? Wright's shifting balance theory mainly addressed changes in gene frequencies within and among loci in terms of their yields in fitness. This can be easily extended to a regulatory framework with the added proviso that changes in genome structure or organization may constitute important aspects of the adaptive landscape as well (i.e. gene frequencies may only represent foothills or merely the tips of fitness peaks).

A key component of Wright's (1980, 1982a,b, 1988) shifting balance theory is the action of local inbreeding and subsequent random drift to compartmentalize and accentuate variation within a species or meta-population (Schoen and Brown, 1991). The importance of inbreeding in contributing to evolution has been otherwise underemphasized in the modern synthesis, despite recognition of its essential role in artificial breeding programs (Lovtrup, 1974, 1987; Reid, 1985; Wright, 1988; Thornhill, 1993). The local reinforcement of genetic identity via in-breeding in demes (Shields, 1982, 1993) means that demes effectively represent subspecific lineages. The key factor determining the degree of local identity, even in species with mechanisms for selecting genetically different mates, is effective population size (Shields, 1993). In a regulatory genetic framework, this allows compartmentalization of a higher-order lineage into sublineages that can diverge in exploration of higher-order organization (Wade, 1992). Thus, exploration of the regulatory landscape takes place via a mechanism of parallel processing.

The number of lower-level sublineages, their relative size and their further subdivision will all influence the degree to which variance among subgroups is selectable (e.g. Wade and McCauley, 1980). A local deme that discovers a fitness jackpot, can export the discovery to adjacent demes. The increase in population size associated with success will also tend to reduce disruptive inflow of information from other demes (Wright, 1988). In this way, locally successful sublineages can convert others and ultimately higher order lineages to a new conformation. This last aspect has been seen as the weakest part of Wright's theory, but recent models show that locally discovered advantageous gene combinations can indeed be exported and incorporated into other sublineages. Moreover, this can proceed with relatively low rates of migration and in the face of two-way gene flow (Crow, Engels and Denniston, 1990). The significance of such population structuring is a dramatic increase in evolutionary rates (Bush *et al.*, 1977; Wright, 1988). Wade (1977) also makes the point that a process of group selection can exceed rates achieved by individual selection, whenever individual selection co-efficients are low, but variance among groups is high. Thus, processes of group selection need not be weak relative to individual selection, despite the common assumption to the contrary.

Both Wright (1982a,b) and Slatkin (1987) emphasized the importance of demography to population structure. Wright (1982a) suggested that under optimal conditions, populations may be large and continuous so that gene flow yields high stability. Fragmentation of the population under stress would greatly enhance evolution, perhaps accounting for the apparent periods of punctuated evolution seen in otherwise static species (Eldredge and Gould, 1972; Gould and Eldredge, 1977). Slatkin (1987) points out that the degree of demographic stability may be partially a species characteristic (e.g. weedy species are inherently unstable). He argues that 'group selection' in the sense of Wright's models will act much more strongly in highly fragmented species, particularly if there is considerable local extinction and recolonization.

Wright's models are generally recognized as a form of group selection (Lewontin, 1974; Wynne-Edwards, 1986; Slatkin, 1987) and have been harnessed to other relevant models of group selection. Wilson (1980, 1983) considered that local demes may represent genetic entities by virtue of relative panmixia within them. He extended the hierarchy of organization downwards to consider 'trait groups' that represent substructure within demes. Trait groups were envisioned as ecological groups of organisms whose interactions with one another (even if transient) may have evolutionary consequences. Wilson (1980) recognized that the fitness of an organism was partially relative to that of other genotypes present and partially related to the ecological success of the group as a whole.

The importance of this framework is that higher-order organization can evolve via differential selection of trait groups that improve the global fitness of the deme. In other words, the behaviour of an organism may indirectly enhance its own fitness via positive feedback through others. A simple way to think about this is that interactions that raise the local carrying capacity, and thus indirectly improve the fitness of participants, are selectable. Barton and Clark (1990) suggest that Wilson's trait group models and kin selection models can be reconciled if the associated organisms are viewed as either indirectly improving individual success (through resources or survivorship) or as representing indirect genetic resources (kin). Wilson (1983) also showed how kinship models could be recast as intrademic group selection.

Wilson's model is generally regarded as a form of group selection, but has gained widespread acceptance as an important evolutionary paradigm (e.g. Harvey, 1985; Williams, 1992). The central core of this model, however, resides in the group interactions indirectly reinforcing local individual selection. In this context it can be recast as considering individual selection where the environment of the individual includes interactions with others as part of a complex mosaic of selection pressures (Williams, 1992). However, it cannot be denied that interactions with others implies a group and Wilson's models pivot on some such groups being of selective advantage (Wilson, 1983).

Semantics aside, the model provides a viable framework for considering not only the evolution of features relevant to higher-order holons in hierarchically structured lineages, but also the coevolution of completely unrelated organisms into integrated ecological organizations. The concept of the 'superorganism', which entails genomic organization transcending that relevant to individual organisms, follows directly (Wilson and Sober, 1989). The fact that genic perspectives largely reject such a construct illustrates that the structured deme paradigm captures features that lie outside a framework of selfish genes.

Williams (1985) argues that group selection cannot be a strong force because organisms most often act selfishly rather than as team players. If lineage selection most often reinforced individual selection, however, the idea that such selection must lead to populations of universal altruists is an unfair expectation. Depending on the ecological circumstances, lineages might be best served by deploying highly competitive phenotypes. Evidence for the reality of lineage-level selection is probably better highlighted by adaptations for evolvability or persistence that might be irrelevant to or in conflict with maximized individual fitness. Lineage selection is likely to act most strongly where it effectively reinforces individual fitness, or allows features relevant to longer time scales than an organism's life-time. Thus organismal fitness may evolve in a context of maximization, whereas lineage fitness may impose a framework of geometric mean fitness.

In most cases, the persistence of lineages will be enhanced by achieving organisms of the highest individual fitness. The plastic expression of a cannibalistic morph in some salamanders may represent a feature that has been forged by both individual and lineage selection. Here the population contains a universal program to become a cannibal given particular environmental cues. It is perfectly reasonable for such nasty individual behaviour to be selected in lineages where such behaviour promotes long-term persistence or contributes in a boot-strapping way to evolution (e.g. by active elimination of low-fitness individuals). The deployment of a specialized cannibalistic phenotype in response to the environment implies the evolution of a plastic switching mechanism via lineage selection (Chapter 6).

Wynne-Edwards (1986) extrapolated Wilson's (1980, 1983) arguments upwards to consider the possible evolutionary importance of selection acting among demes. He viewed this model as a redemption of his earlier arguments for group selection acting to maximize local resource stability and supply (Wynne Edwards, 1962). His model incorporates the idea that interdemic selection may override the selfish interests of individuals. This formulation differs from Wilson's in that group selection is viewed as strongly antagonistic to short-term individual interests rather than being mainly reinforcing. This 'strong' form of the group-selectionist paradigm has been vigorously attacked over the last 30 years. In a lineage framework, however, it seems possible that the long-term fitness of sublineages may be indirectly enhanced by features selected relevant to higher levels. Wilson (1983), for example, shows clearly how a trait that is selected against in all environments may still spread via group selection if it enhances group performance and groups vary in genetic constitution. Such a model still involves a maximizing rather than a regulatory system. If a persistence criterion of fitness is accepted, then it seems premature to reject Wynne-Edward's arguments. The basis of his reasoning, that features promoting elevation and stability of the carrying capacity may be selectable, is essentially similar to Wilson's (1980, pp. 145–155). It is the stability component (reduction of fitness variance) that remains elusive. Regardless, convincing empirical support for Wynne Edward's (1986) hypothesis, that density regulation evolves as a population-level adaptation, remains scanty.

Individual underexploitation of resources to benefit others altruistically would not be expected if there is no longer-term lineage benefit deriving greater or more stable resources for descendants. Recall, however, that a lineage paradigm evokes a broader definition of relatedness and what constitutes descendants. Having said that, it remains that selective processes in various sublineages probably act most strongly to reinforce the immediate fitness of individuals or their more immediate descendants, a view strongly argued by Wright (1980). Wilson (1980) makes the point that the key criterion may be temporal scale, with the strength of selection attenuating as derived indirect benefits become more remote.

The evolution of sex may be one example of a feature relevant to long-term persistence that has negative fitness consequences to individuals. A nuance, however, is that adaptations for evolvability can have no relevance to organismal function, only to lineages. Adaptations to resources may be relevant to both levels so a conflict between levels favouring individuals may be accentuated. Features for evolvability, although imposing costs on organisms, may not conflict as directly with ecological adaptation and individual fitness.

Williams (1992) argued that clade selection is a viable paradigm but that it has suffered from an artificial emphasis on the species holon (e.g. Stanley, 1979; Vrba, 1989a). He advocates a hierarchical model of selection regimes that includes subspecific lineages (see also Vrba, 1989a,b). Wilson (1980, p. 42) also recognized hierarchies below the species level. For example, in populations with local dispersal, he viewed the population structure as resolving into a series of nested trait groups based on spatial gradients of relatedness. Such views are also implied in the models developed by Endler (1977) and Shields (1982).

In the light of the organizational framework developed here, it must be significant that the features of critical import addressed by Wright (1980, 1982a,b, 1988) and Wilson (1980, 1983) pertain to functional integration of higher-order features (gene interactions, coadapted genomes or co-evolved species) (see also Wade, 1992). Integration does not magically pop out via selection on individual selfish genes, but must be selected in its own right at the level relevant to its operation. Wright (1980, 1982, 1988) argued that coadaptation might not be expected under panmixia, but that coadaptive features mainly evolve via subspecific lineage selection.

Wilson (1980, 1983) arrived at similar conclusions relevant to selection of ecologically linked lineages (including inter-specific associates). Wilson (1990) also pointed out that the recombination of various genetic components into individual organisms may allow selection to differentiate higher-order interactive systems. This is equivalent to the other aspect of the lineage paradigm developed here – that selection may act on higher-order genomic features that may vary even within single populations (Chapter 4). Even large panmictic populations exhibit a species-specific epigenetic organization that implies that they represent coadapted genetic organizations and that the expressed variation may be related to homeostatic balancing selected at large temporal and spatial scales (Chapter 7).

Wright's idea that such selective regimes synergize selection of 'interaction systems' suggests these models are essential to understanding genomic organization and phenotypic adaptive suites. Wade and McCauley (1980) effectively addressed the same point when they noted that group selection may be highly relevant where the mapping of a genotype onto the expression of the phenotype is not simply a linear function of the component parts. In other words, when the whole is greater than the sum of its parts, lineage selection may act on those features responsible for the emergent organization. Wade and Goodnight (1991) experimentally confirmed that interdemic selection significantly accelerated evolution and that the key feature involved was gene interactions. A combination of subdivided population structure and interdemic genetic exchange may be necessary to maximize regulatory evolution (Wade, 1992). Although small, subdivided populations have been implicated in the rapid diversification of mammals (Bush *et al.*, 1977), Wake and Larson (1987) conjectured that the lack of significant phenotypic diversification among plethodontid salamanders may arise because their highly isolated populations rarely exchange genetic information.

A common reason for rejecting 'group selectionist' arguments in evolution has been the idea that this will always be a relatively weak force compared with the action of individual selection. Regardless of whether one considers a subspecific lineage paradigm as group selection or not, it remains that such dynamics are extremely important in the evolution of individual advantage based on coadapted features. As such, the action of this process in allowing populations to discover and consolidate local adaptive peaks is 'not fragile at all' (Wright, 1980) and such 'group selection' may represent 'one of the greatest creative forces for evolutionary change' (Wade, 1977).

It seems significant that the recent evolutionary framework has shifted from assumptions of single populations with global panmixia to one consistent with developments in non-equilibrium ecology. This view considers local ecological/evolutionary systems as 'metapopulations' congruent with a Wrightian framework of subspecific lineages (Olivieri, Couvet and Gouyon, 1990; Wilson, 1990; Hanski, 1991; Hanski and Gilpin, 1991; Hansson, 1991).

Simultaneously, this metapopulation framework has inspired hierarchical expansions of fitness and selection (Wade, 1985, 1992; Heisler and Damuth, 1987; Damuth and Heisler, 1988; Wilson, 1990; Goodnight, Schwartz and Stevens, 1992). Goodnight, Schwartz and Stevens (1992) explicitly recognized subspecific lineages (i.e. subpopulations with continuity across generations) as selectable entities.

Such models focus attention on dispersal rates as a key factor and recognize that this may evolve as a lineage metacharacter. Metapopulations are theoretically more stable than unstructured populations by virtue of their distributed risk, rescue of extinct demes and local adaptive diversification. Migration, besides governing rates of information exchange, may also determine the degree of metalineage fragmentation. An intriguing consideration is whether such factors could contribute to the evolution of species-specific migration rates.

The reality of recognizable levels of organization within lineages is the basis of taxonomy and a key factor in ecology (e.g. organisms, trait groups, demes, metapopulations, ecotypes, races, subspecies, species). Without sex, species would not exist (Ghiselin, 1988). Various levels may have different time scales of long-term fitness. Thus, the time scale for the selection of phenotypic plasticity is multigenerational, but the lineage paradigm can be extended further upwards where it merges directly with higher-order evolutionary processes concerned with species selection and cladogenesis (Stanley, 1979; Eldredge, 1985; Stearns, 1986; Vrba, 1989a,b).

Templeton (1989) points out that interbreeding 'species' (syngameons) are well known among plants and examples are being increasingly recognized among animals as well. Moreover, various levels of organization may sometimes remain discrete, despite some sexual communication (e.g. Van Valen, 1976). Thus, multiple 'entities' may be discerned at one spatial and temporal scale but they could still constitute a single higher-order organization. Among most animal lineages, the species is the upper echelon because communication stops there. In plants, hybridization remains a potent evolutionary force (Stebbins, 1950; Abbott, 1992). When a lineage paradigm is adopted, it becomes apparent that sex forges a higher-order organization that has multigenerational advantages and selective units of various temporal–spatial scales. Such a view has been largely overlooked by genic theorists but is increasingly recognized as a necessary extension by evolutionary ecologists (Harvey *et al.*, 1992).

5.3 LINEAGE SELECTION AND SEXUAL REPRODUCTION: AN OVERVIEW OF THE SELECTIVE ADVANTAGES OF SEX

The origin and maintenance of sexual reproduction is the Gordian Knot of evolutionary theory. Sex is not only crucial to any discussion of units of selection, it contributes directly to the rate and nature of evolution itself. In the subspecific lineage paradigm, it is the integrating channel of communication for higher levels of organization. For the genic paradigm, sex is shrouded in paradox. Consequently, sex occupies a nexus position in any discussion of phenotypic evolution.

Müller proudly announced in 1932 that genetics had 'solved' the problem of the evolution of sex: it was obviously an adaptation for evolvability. Fisher (1930) also proposed such a 'group selectionist' explanation of sexual evolution. Consolidation of the genic paradigm in the 1960s brought with it the realization that this problem was not only unresolved (Williams, 1966a), it was linked to a morass of difficult evolutionary questions (Williams, 1975). Consequently, the evolution of sex (outcrossing, segregation and recombination) has generated an avalanche of literature.

The genic view requires that adaptive features of organisms have immediate selective value at the level of a single generation (Williams, 1966a). The next obvious step along the way is then species selection: sex is maintained because sexual species become extinct less often or engage in cladogenesis (branch into several new species) more effectively (e.g. Stanley, 1979). The phylogenetic pattern of sexual *versus* asexual taxa in fact supports this interpretation, there being very few higher taxa that are primarily asexual (Maynard Smith, 1988a,b). In this respect Vrba (1989a) suggested that asexuality can be considered as a terminal organizational cancer. Species selection, however, operates on time scales of thousands and millions of years, which means that selection on organisms must dominate in power by many orders of magnitude (Maynard Smith, 1974; Stearns, 1987a). As discussed earlier, the degree to which this is true will depend on the degree of conservation of higher regulatory tiers.

This leads to a paradoxical problem that continues to defy resolution. Explanation of the near ubiquity of sexual reproduction in eukaryotes has proved difficult because the individual's genome is exponentially diluted with each passing generation and half of the individuals in a population (males) may contribute little to reproductive success (Williams, 1975; Maynard Smith, 1978a; Trivers, 1985). Furthermore, sexual selection may impose features upon the phenotype that are antagonistic to ecological adaptation (Thompson, 1976). The genic view thus requires powerful and immediate fitness returns to offset this theoretical two-fold disadvantage to individuals. Otherwise, in an equilibrium environment an asexual mutant would increase at twice the rate of (and rapidly eliminate) sexual females.

Species selection seems incapable of explaining the maintenance of sexual reproduction, unless the transition to this mode of reproduction, once attained, is extremely difficult to give up (i.e. it is a frozen accident). Although this could well be true in some cases (e.g. mammals), the widespread occurrence of some asexual forms in nearly all phylogenies, as well as species that adaptively regulate the occurrence of sexual generations, suggests that such an explanation is insufficient. Such genetic inertia is a possible contributing factor (Williams, 1975; Margulis and Sagan, 1988), but cannot itself explain the near ubiquity of sex. Recall, however, that the existence and even occasional success of cheats cannot be taken as evidence that real organization does not exist (Chapter 1).

The main effort of most evolutionary theorists over the last 30 years has been to find a genic, microevolutionary explanation for the prevalence and persistence of sexual reproduction and to find arguments weakening what are generally considered 'group selectionist' paradigms. Despite intense efforts and continued confidence in eventual success (e.g. Stearns, 1987a; Crow, 1988), this endeavour has failed to reach any convincing resolution (see books by Ghiselin, 1974; Williams, 1975; Maynard Smith, 1978a; Bell, 1982, 1988a; Shields, 1982; Stearns, 1987c; Margulis and Sagan, 1986; Bellig and Stevens, 1988; Michod and Levin, 1988). In fact some have concluded that there must be some long-term component after all (e.g. Gouyon, Gliddon and Couvert, 1988; Williams, 1988). Gouyon, Gliddon and Couvert (1988) suggest that group selection sorts lineages, such that those persisting obtain short-term benefits from sex, and this allows them eventually to harvest long-term benefits as well.

Theories pertaining to the evolution of sex are diverse and various proponents usually seek a single unifying explanation (e.g. Hamilton, Henderson and Moran, 1981). It is likely, however, that advantages accrue from several dimensions, and their summation is worth considering. Consequently, the view developed here will be a systems treatment. The key theories include the original lineage evolvability hypothesis which has two sides. On the one hand, sex accelerates the assembly of advantageous mutations (Fisher, 1930; Müller, 1932). Conversely, deleterious genes may be dumped into 'waste' individuals where they can be shed from the lineage, a hypothesis known as Müller's ratchet (Müller, 1964; Felsenstein, 1974; Bell, 1988a,b; Kondrashov, 1988; Chao, 1990, 1991, 1992). Both of these hypotheses have been variously classified as 'group selectionist' (Bell, 1982, 1988b). They are not exactly equivalent since deleterious mutations are much more common than favourable ones. Consequently, the possible evolutionary value of sex in combining advantageous genes at different loci is the dark side of the moon compared with the possibility of sex acting to eliminate deleterious mutations.

Kondrashov (1988) suggested that in well-adapted populations advantageous mutations only occur when the environment changes and so their assembly is really an environmental hypothesis. The evolvability hypothesis and Müller's ratchet, however, must be recognized as faces of the same moon. There is no necessity for organizational advance or genetic change to be associated with environmental change. For example, the evolution of homeothermic organisms might be advantageous in environments with fluctuating temperatures, but its evolution would not strictly require any change in the mean environment. Thus, the evolutionary hypothesis and the ratchet must be viewed as biased sides of the same coin.

The subspecific lineage model proposed here is only a slight modification of the classic evolvability hypothesis of species advantage for assembly of advantageous alleles. This mesoevolutionary hypothesis differs from evolutionary models seeking to establish a genic (single generation) microevolutionary advantage of sex, but also does not necessarily require species selection on very long time scales. If evolution, particularly regulatory evolution relevant to the epigenetics of phenotypes, has a transgenerational unit of selection, then the value of sex may well lie in the meso-evolutionary advantages conferred to subspecific lineages. Local lineage reinforcement of identity reduces the cost of sex (Bell, 1982; Shields, 1982).

The mesoevolutionary hypothesis includes both organismal selection and species selection, but more importantly, spans the gap between these extremes. Genic theorists often begin by recognizing individual selection as a key level, but then slide immediately to a genic framework as if these are somehow synonymous. In a lineage paradigm, alleles cannot be considered as independent entities resting against particular genetic backgrounds because they are themselves key determinants of that background, the background is highly organized and interactions may produce unique organisms with each bout of sexual reproduction (i.e. there is no constant background). In addition, the effective background constitutes the entire range of genotypic variants expressed within the lineage over the relevant time scale (i.e. adaptive variation can be an integral part of a dynamic strategy).

There are also genic evolvability hypotheses based on the immediate advantages of sex to parents in improved offspring survival. These hypotheses generally separate into biotic and environmental categories which emphasize temporal and spatial heterogeneity respectively (Bell, 1985). Models postulating the advantages of diversified offspring due to predation, interspecific competition, parasitism and disease fall under the heading of the Red Queen Hypothesis (Bell, 1982; Hamilton, 1988; Seger and Hamilton, 1988; Hamilton, Axelrod and Tanese, 1990; Ladle, 1992; Rennie, 1992). The name comes from Lewis Carroll's *Through the Looking Glass*, and was applied by Lerner (1954) and Van Valen (1973) to coevolutionary arms races between antagonistic species. The appellation refers to the idea that in an evolutionary sense species may be running as fast as they can simply to stay in the same place. Such theory considers the first linkage in a lineage framework and all that is being suggested here is that this may be extended to more than one generation.

Environmental evolvability hypotheses are mainly concerned with the advantages of diversified offspring in heterogeneous habitats. Reductions in sib-competition or parental–offspring competition may be achieved by differential resource utilization or compartmentalized habitat use in a shared environment (Ghiselin, 1974; Williams, 1975; Maynard Smith, 1978a; Young, 1981; Bell, 1985). Alternatively, a diversity of heterogenous habitats may be exploited by a species with variable genotypes (Vrijenhoek, 1990). Such scenarios are generally categorized as the 'Tangled Bank' hypothesis (Bell, 1982) after the famous paragraph in Darwin's *The Origin of Species* quoted earlier. The original hypothesis was contributed by Ghiselin (1974).

Shields (1982, 1988) put forward a powerful argument that the diversifying role of sex may be more theoretical than empirical. He argued that, in most low-fecundity species, inbreeding largely negates the two-fold cost postulated by mainstream theory. In this framework, sex is an optimal compromise for both removing deleterious mutations and faithfully transmitting parental genomes.

Variance models of sexual advantage are based on the idea that recombination increases the variance in the normal distribution of fitness by producing a few unusually high or low fitness genotypes. If there is very strong selection for only those genotypes of extremely high fitness (i.e. truncation selection), then sexual reproduction is more likely to produce and reinforce such genotypes (e.g. Williams and Mitton, 1973; Williams, 1975). Such models refer to the form of selection and distribution of fitness so they are not mutually exclusive of other hypotheses such as Tangled Bank (Young, 1981).

Until recently, nearly all theoretical discussions of sex focused on the value of recombination in bringing together alleles located at separate loci. Kirkpatrick and Jenkins (1989) demonstrated that, in the case of diploids, there is also a powerful potential advantage of segregation. This refers to the fact that sets of homologous chromosomes segregate independently to gametes in meiosis, so that chromosomes from parental genomes are effectively shuffled. Thus, if attaining homozygosity of a new mutation (or imported alleles) at a given locus is advantageous, a sexual diploid could do so more rapidly than an asexual diploid. The sexual species needs only to obtain a single mutation and (with a little help from inbreeding) homozygotes can then be obtained rapidly by segregation. An asexual diploid must obtain two separate and identical mutations at the same locus on homologous chromosomes. Thus, asexual diploids become stalled in the heterozygous state and cannot achieve fixation as rapidly. Although an important advantage for sexual species, it has been argued that this is still insufficient to offset the two-fold cost of sex (Wiener, Feldman and Otto, 1992).

Bernstein *et al.* (1985a,b, 1988) and Bernstein and Bernstein (1991) proposed that recombination is a mere side effect of DNA repair mechanisms and that outcrossing is a way of offsetting homozygosity produced by this process. Michod and Gayley (1992) recognized that DNA repair is served equally well by selfing as by outcrossing and concluded that recombination and outcrossing may serve two separate roles. Recombination acts to repair recognizable damage, while outcrossing serves to mask deleterious mutations (that cannot be repaired) in the heterozygous state. The evidence deployed by these authors leaves little doubt that mechanisms involved in recombination are intimately related to the evolution of DNA repair. Holliday (1984, 1988) added that repair may be relevant to information associated with chromatin structure. However, a litany of information suggests that sexuality has assumed an important secondary function related to evolvability, and thus, the questions pertaining to the function and maintenance of recombination and outcrossing lie in that sphere (Maynard Smith, 1988a). In particular, recent evidence suggests that meiotic gene conversion (repair) and crossing over are controlled by separate genetic systems (Carpenter, 1987; Engebrecht, Hirsch and Roeder, 1990; Hurst and Nurse, 1991).

Thus, there are several major categories of hypotheses related to the evolution and maintenance of sex. Any thorough discussion of sexual evolution would now require a multi-volume encyclopedia. Consequently, discussion here will be limited to aspects relevant to the lineage evolvability hypothesis, and the inadequacy of the genic paradigm. There is also a huge literature concerned with evolution of sexual selection in forging phenotypic adaptations and sex ratios (e.g. Charnov, 1982; Thornhill and Alcock, 1983) which must be passed over here. To begin, I will critically outline the general framework for the evolution of sex necessitated in a genic paradigm.

5.4 THE GENIC FRAMEWORK

5.4.1 Selection of linked modifiers

Firstly, modelling suggests that sex cannot be explained on the basis of single loci (Hamilton, Henderson and Moran, 1981), but involves the critical aspect of recombination among loci. Sex is also irrelevant to an organism (other than the imposition of being male or female), only impacting on the variation of offspring. Thus, sex necessarily applies to multi-locus benefits and is by definition not a phenomenon relevant to single genes or individuals. Although the results of Kirkpatrick and

Jenkins (1989) suggest that the benefits at single loci may be higher than previously thought (for diploids), most investigations have focused on recombination among loci.

All key models developed from the genic perspective rest on a foundation where genes controlling recombination are closely linked to the loci they influence. Selection of such recombinational modifiers occurs when they 'hitchhike' along with the successful combinations of target alleles that they create (e.g. Maynard Smith and Haigh, 1974; Felsenstein and Yokoyama, 1976; Strobeck, Maynard Smith and Charlesworth, 1976; Maynard Smith, 1988b,c; Brooks, 1988). The linked hitchhiker framework is necessary in a genic paradigm, otherwise selection (or recombination) is necessarily group selection (Felsenstein, 1974; Bell, 1982, pp. 123–125). To model a large genome requires numerous loci that modify local recombination rates and that these genes are rather tightly linked to their respective targets.

The final word on the reality of such organization is pending further molecular analysis. The lineage model does not preclude such hitchhiking but does predict that recombination, like other lineage features, may also evolve more global regulation. The meiotic drive hypothesis of Haig and Grafen (1991) also predicts more global or distant control of recombination. The current evidence supports the reality of fine-scale variation of recombination rate throughout the genome, some of which is associated with closely linked modifiers. However, modifiers of recombination may also be located distantly from their control sites and even on separate chromosomes (i.e. no linkage). Bell (1982, p. 410) suggested that lack of linkage, in fact, is more the general rule. Particular modifiers may also affect multiple targets. Thus, the evidence supports a view integrating both coarse and fine-scale regulation (Brooks, 1988; Crow, 1988; Maynard Smith, 1988a), and one consistent with a hierarchical model of regulatory organization. Differences in recombination rates between the sexes (Trivers, 1988; Lindahl, 1991) is further evidence of such higher-level control. If recombination evolves and is maintained only by hitchhiking modifiers, this should produce a mosaic pattern of independent recombination throughout the genome. Where there is selection favouring asexuality, loss of recombination would be a gradual, piecemeal process. Examination of various species should reveal some with high recombination at most loci and some whose recombination is reduced or absent in much of the genome or even restricted to only a few localized regions.

Unfortunately, this pattern is not inconsistent with a lineage model either. Single steps between sexual and asexual modes, however, is inconsistent with the genic model. Significantly, transitions to sexual or asexual reproduction are largely mediated by single genetic or environmental switches or events (i.e. in species with polyphenic expression of sexual forms), or where asexual lineages arise they usually do so in macroevolutionary steps, often associated with polyploidy. Such evidence (particularly regulated alternation of sexual and asexual generations in species such as aphids) is inconsistent with independent mosaic evolution of recombination rate (i.e. genic). Although such a model does not preclude adaptive local modulation of recombination, it allows the entire system to be switched on or off at a higher level. Significantly, global switches to asexuality that shut off meiosis also protect the system from·drive.

Genic models based on the selection of local modifiers of recombination are very difficult to apply to systems where sex is cyclical or facultative. As Kondrashov (1988) points out, the advantages of sex cannot simply accumulate in successive generations in such systems. Such switches are analogous to those associated with phenotypic plasticity. As will be shown, the best explanation of such systems is that the temporal unit of selection has a time frame encompassing not only both life-history strategies but, in particular, the switch mechanism itself. This will remain true, even if the genome is effectively compartmentalized to allow non-disruptive evolution of the asexual or sexual phenotypes (West-Eberhard, 1989). Brooks (1988) points out that the degree of linkage between recombinational modifiers and their target sites can be used as a measure of the intensity of selection at genic and higher levels of integration, and thus the lineage paradigm makes testable predictions. The tentative evidence supports a lineage, and not a genic paradigm.

5.4.2 Linkage disequilibria and epistasis

Theories of sex are all intimately connected to a phenomenon with the discouraging appellation of 'linkage disequilibrium'. It is useful to clarify this phenomenon because this is what recombination is all about (Felsenstein, 1974; Felsenstein and Yokoyama, 1976). In any large population with complete panmixia, random choice of mates, a homogeneous environment, and no selection or mutation, the frequency of given genotypes will come to a statistical equilibrium based simply on the relative frequency of the constituent alleles (Hardy Weinberg equilibrium). Imagine for example that we have two loci, each of which has two possible alleles (A and a or B and b). In a haploid organism, the possible genotypes are AB, Ab, aB or ab. In a population at equilibrium, the frequency of any particular genotype will be a constant proportion of the others due to their random statistical association. In fact, it is the very action of sex and recombination that achieves this statistical equilibrium. Now imagine that the only genotypes found in the population are AB and ab. The absence of Ab and aB genotypes means that there is a non-random distribution of genotypes or the system is in linkage disequilibrium.

Whether the system is at equilibrium or not can be determined by calculating the value of a quantity known as the linkage disequilibrium coefficient (D) where:

$$D = p(AB).p(ab) - p(Ab).p(aB)$$

In this case, D will equal zero whenever there is statistical equilibrium and increasing positive or negative values indicate disequilibrium. Disequilibrium may indeed arise due to linkage effects, as when selection at a particular locus causes alleles located nearby to hitchhike along. However, a major source of confusion is that disequilibrium may have nothing to do with linkage (Roughgarden, 1979, p. 113). For example, natural selection is a major source of disequilibrium and thus such states may be independent of physical linkage relationships.

This discussion underscores that in genic theories of sexual evolution, the degree of linkage disequilibrium is a key indicator of coadaptation or genetic epistasis among loci. Linkage disequilibrium is thus viewed as indicating epistasis, linkage or synergistic selection. As Hartung (1981) points out, however, although linkage disequilibrium may indicate that two alleles at different loci enhance one another's fitness, lack of this phenomenon indicates nothing at all about coadaptation. Very strong interactions may be obscured as the number of loci contributing to a feature increase. Clearly, the extremely integrated molecular and developmental organization described in Chapters 2 and 3 somehow exists in genetic systems expressing widespread linkage equilibrium.

Thus, the entire genic program of judging the value of recombination with measures of linkage disequilibrium at a few loci may be misdirected. If Falconer (1977) is right, coadaptation may actually be increasing as a species rebounds to its wild type format at linkage equilibrium. In other words, a population at linkage equilibrium may well represent the consolidated state of a highly advanced, coadapted genome. The idea that directional selection away from the optima (with associated linkage disequilibria) represents increasing coadaptation may thus be inappropriate in a global sense (see Chapters 7 and 8). This should be kept in mind during the following discussion.

Recombination is the great leveller and it can only act to restore linkage equilibrium. Once it has restored equilibrium, recombination then assembles and disassembles genotypes with equal frequencies (i.e. its effects disappear). Since natural selection is one of the main factors generating linkage disequilibrium, this suggests that recombination might actually be disadvantageous under some circumstances. The goal of the genic school has been to determine under what selective conditions recombination might be advantageous. This discussion suggests that the advantages of recombination must be associated with reducing linkage. The main factors generating linkage disequilibrium are: natural selection;

random drift associated with small populations and founder events; and mutations (a new mutation is in immediate disequilibrium because it has not had time to spread among various genetic backgrounds). Where real physical linkage exists such processes may amplify disequilibrium.

Hill and Robertson (1966) first pointed out that where selection is acting differentially among alleles at several loci, linkage effects can interfere with the response of either locus to selection. This occurs because the response of one gene to selection may be reduced because of linkage with a differently selected allele at another locus. Recombination, by reducing such linkage effects, can allow faster assembly of favourable combinations (Felsenstein, 1974). Maynard Smith (1988c) also showed how sexual reproduction might be advantageous under directional selection. The key factor is that recombination provides greater phenotypic variance so that selection is more effective.

Disequilibrium may also be associated with the action of natural selection in creating particular genotypes. Early theoretical studies based on haploid genomes showed that if all genotypes initially exist, then sex confers no advantages where selection is constant. In this case asexuals respond directly to selection (without the disadvantage of producing males), and recombination is just as likely to disassemble as to assemble useful combinations (Eschel and Feldman, 1970; Maynard Smith, 1971a; Thompson, 1976). It is important to stress that constant selection pressures on systems with existing genotypes may not favour recombination even if there is linkage disequilibrium and genetic polymorphism. Thus, constant disruptive selection for AB or ab genotypes in a haploid, two-locus system does not favour recombination. Recombination constantly acts to recreate the Ab and aB genotypes and consequently imposes a recombinational load (Maynard Smith, 1971a; Bell and Maynard Smith, 1987). If the sign of linkage disequilibrium changes frequently from one generation to the next, theory suggests an advantage to recombination. Thus, recombination may be advantageous if selection favours AB and ab genotype in one generation but Ab and aB genotypes in the next (Maynard Smith, 1971a). In this case the value of D changes sign across generations. Recombination may also be advantageous in systems subjected to directional selection where new genotypes are being generated from a previously optimal configuration.

The most obvious factors that could lead to changes in the sign of linkage disequilibrium are certain kinds of environmental cycles or fluctuations, population dynamics with periods of alternate high or low densities and spiteful coevolution of species with their predators, competitors, parasites and disease. This explains why these phenomena have been the main targets for genic explanations of sex. Although there is nothing intrinsically incorrect about the logic of these conclusions, I will argue that this framework is far too restrictive. In particular, the assumption that asexual lineages have equal access to all necessary geno-types, that none of these are locally lost and that theory based on very small haploid genomes can be extrapolated to large diploid organizations are all unrealistic. A realistic framework for sexual evolution must incor-porate large diploid genomes and heterogenous, non-equilibrium habi-tats where effective population sizes may be relatively small (compared with the classical assumptions of infinite size).

5.5 HYPOTHESES OF SEXUAL EVOLUTION

5.5.1 Mesoevolutionary advantage

My central thesis is that regulatory evolution occurs in sexual lineages, such that selection acts on hierarchical genomic organization on multi-generational time scales. This does not reject selection acting on organisms or immediate advantages to offspring, but it does suggest that features conferring longer-term advantages may also evolve as organizational features of the genome. Sexual reproduction is not only one aspect that might reflect such a feature, it is also the main vehicle creating the sexual organization itself. Both aspects will be considered here.

An immediate impact of adopting the subspecific lineage hypothesis is that it releases evolutionary theory from the necessity of showing single-generation advantages, but it does not require selection to act on restrictively long time scales either. If there really were a strong genic reason for having sex, one must question how so many asexual taxa and highly inbred species do so well for so long. The original explanations for the evolution of recombination and sexual reproduction were lineage-level (species) hypotheses. Fisher (1930) and Müller (1932) suggested that sexual species are capable of faster evolution than asexual species because favourable mutations arising at different loci can be rapidly assembled into a single individual via sex, whereas asexuals must obtain such mutations sequentially within a single line. Müller (1932) also emphasized that in asexual lineages, the two favourable mutations would compete with one another, which somehow has been lost in subsequent considerations of this process.

Crow and Kimura (1965) quantified the theoretical value of recombi-nation for assembling favourable mutations which is graphically illus-trated in Figure 5.2. Their paper was published in the midst of the genic revolution. Felsenstein (1974) added further strong support but the hypothesis was effectively declared dead by Bell in 1982. Others, although recognizing apparent weaknesses, were reluctant to administer last rites (e.g. Maynard Smith, 1984, 1988b; Williams, 1988; Hamilton, 1988) but the theory was definitely relegated to a home for the feeble. I will attempt resuscitation here.

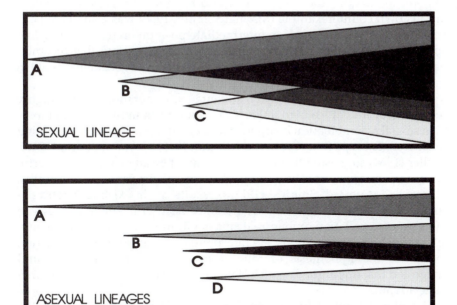

Figure 5.2 Müller's conceptual model illustrating the advantages of sexual reproduction for assembling beneficial mutations into single lineages. Redrawn after
Müller (1932) with permission of *American Naturalist*, University of Chicago Press.

Three major attacks have successfully weakened explanations of sex
based on long-term evolutionary advantages. Firstly, the strong supporting arguments presented by Crow and Kimura (1965) were successfully
attacked by Maynard Smith (1968) who showed that if mutations are
highly recurrent, then asexual lineages are not as disadvantaged as was
originally predicted. Even then the evolvability hypothesis was only
weakened, not rejected. An attack based on the idea of frequent adaptive
mutations is inconsistent with the common wisdom that advantageous
mutations are rare (Crow and Kimura, 1969). Even recent genic models
only work when such mutations are relatively uncommon (e.g. Hamilton,
Axelrod and Tanese, 1990).

In answer to this, Maynard Smith (1971a) further investigated the possible role of sex in contributing either to long-term evolution or immediate
advantage. For this he used two classes of models, those with recurrent
mutations or those examining the value of sex in shifting the frequency of
existing allelic combinations. Bell (1985) interpreted Maynard Smith's
(1971a) results as demonstrating fatal weaknesses in the Fisher/Müller
hypothesis. Maynard Smith's (1971a) own conclusion regarding this

hypothesis was that the case of Crow and Kimura (1965) was overstated, but his own counterexample (Maynard Smith, 1968) was also unrealistic. Felsenstein (1974) also took an intermediate position on this debate while supporting the Fisher/Müller mechanism. He pointed out that recombination was advantageous whenever populations were considered finite in size.

The essential conclusion is that if advantageous mutations are not highly recurrent, or if drift leads to linkage disequilibrium in finite populations, then sexual recombination is superior to asexual reproduction in obtaining the fittest genotype first (Crow and Kimura, 1965; Maynard Smith, 1971a). Müller (1964) suggested that this in itself might be an extremely powerful advantage because time lags may be a crucial determinate of success. Moreover, Maynard Smith's (1971a) results with existing genotypes suggested that recombination would only be useful where changes in the sign of correlation among relevant environmental factors frequently alternated across generations. He interpreted this rather strong restriction to favour the hypothesis of long term evolutionary advantage and suggested that this argued against Williams's (1966a) expectation of immediate advantage.

The idea advanced by Maynard Smith (1971a), that changes in the sign of linkage disequilibria are necessary to yield a recombinational advantage, has enjoyed widespread acceptance. It must be pointed out, however, that in the models deriving this conclusion, asexual lineages had access to all possible genotypes. In this case any potential advantage of recombination for assembling new combinations is largely neutralized. If particular alleles or combinations are lost by local fixation or drift, then sexual reproduction will be advantageous in rapidly restoring them when selection pressures change (e.g. Felsenstein, 1974).

The idea that sexual and asexual populations may have equal access to all necessary genotypes is unrealistic. In non-equilibrium habitats, founder events and populations bottlenecks will delete classes of genotypes and even available alleles. Sexual species only need restoration of the relevant alleles to rapidly regenerate all potential genotypes whereas asexuals would have to derive the specific genotypes as single or sequential events. Presumably the ratchet would also act to delete genotypes from the asexual menu. Just as phenotypic plasticity may be superior to circulating alleles where useful alleles may be lost, sexual reproduction may be superior to asexual reproduction where useful genotypes may be lost.

Although a genic view may justify considering only a couple of loci as the fundamental unit for sexual advantages, in a lineage paradigm, potential advantages must accrue across the number of loci relevant to determining fitness (Fisher, 1930; Maynard Smith, 1971a, 1974). As the number of relevant loci increases, the potential combinations that might be selectively advantageous increases exponentially. The possibility that a clonal lineage might maintain all of the combinations necessary to obtain an optimal combination among 100 loci is remote. This number of loci may not be unrealistic for many quantitative traits. Higher organisms such as humans may contain up to 100 000 genes, with considerable direct and indirect integration. Most genes, and particularly regulatory genes, are highly pleiotropic, interacting with potentially hundreds or more of downstream genes in regulatory hierarchies. In such genomes, if mutation is ignored, the advantage of recombination amounts to the difference between the range of variation that can be immediately housed in existing bodies *versus* the range of genotypes that can be potentially recombined from these existing bodies. In the case of mutations, sexual reproduction allows a search for a genetic background that best integrates the mutant to maximal advantage.

Thus, genome size alone may entail an advantage of sexual reproduction orders of magnitude larger than asexual reproduction if organisms represent adaptive suites with coadapted genomes (Chapter 7). Strong empirical support for these arguments was recently obtained via computer simulations (Findlay and Rowe, 1990). These simulations showed that if mutation rates are low, sexual lineages constitute a more effective algorithm for attaining a targeted optimum than do asexual lines. These simulations defined fitness as simply a linear function of similarity to an arbitrarily defined target genotype represented as a string of loci. In reality even a crude approximation to a given optimum may yield exponential returns in fitness. For example, a bacteria that can use a novel substrate, even inefficiently, may survive where others without this ability fail. Further increases in utilization efficiency and the acquisition of regulated expression over constitutive activity are likely to yield exponential rather than linear fitness benefits (Lerner, Wu and Lin, 1964; Clarke, 1983).

It is likely that where an optimum has shifted, or a novel fitness peak arises, even approximate shifts in the right direction will generally yield better than incremental benefits. Since most species probably generate only a small fraction of the genetic configurations that are possible, capturing useful allelic combinations or exploring the impact of new mutations on different genetic backgrounds would be particularly important if populations are relatively small (i.e. all eukaryotes). Consequently, the simulation models of Findlay and Rowe (1990) probably understate the potential value of sex as a adaptation for evolvability.

A second theoretical attack on the Müller/Fisher hypothesis was mounted by Eshel and Feldman (1970) who concluded that recombination is just as likely to disassemble useful genetic combinations as it is to assemble them in the first place. In their model, selection among asexual clones with the appropriate constitution would always reach an optimal peak faster and consequently they would hold their position better than sexual species. Thompson (1976) strongly supported this idea that sex does not accelerate evolution and Bell (1982) regarded this as one of the key coffin nails in the lineage evolvability hypothesis. Examination of Eshel and Feldman's (1970) assumptions, however, reveals that any possible advantages of sexual reproduction were excluded by design. Specifically, highly recurrent mutations are assumed to become available to both asexual and sexual lineages simultaneously. This then is another variant of the problem explored by Maynard Smith (1968, 1971a). The sexual lineages are saddled with the cost of breaking up some useful combinations, but are allowed no possible offsetting benefits. The conclusions follow as a mathematical necessity of these assumptions.

The assumption that adaptive mutations are highly recurrent, originally introduced by Maynard Smith (1968, 1971a), needs to be re-evaluated in the light of the mounting evidence that morphogenesis in eukaryotes is modulated by regulatory genes and that phenotypic evolution involves reorganization or modification of regulatory hierarchies rather than simply point mutations in structural genes. For example, White (1978) convincingly argued that the evolution of genome structure and many regulatory features involves rare or even effectively unique events. These arguments dovetail nicely with the insight of Kauffman (1985, 1993) that regulatory evolution involves a re-wiring of existing circuitry that would be best carried out in a fluid genome (Wills, 1989; McDonald, 1990). Kimura (1983) also argued that the classic assumptions of recurrent mutation and back mutation (particularly to a useful state) are highly unrealistic. Although recurrent adaptive mutations could occur, their frequency in effective population sizes in nature may be low, particularly in relevant regulatory features. If so, sex could be valuable for capturing and assembling variants relevant to adaptive morphogenesis, most of which may effectively represent unique or rare evolutionary opportunities. Such ideas recourse back to discussions of progress in evolution (Chapter 1). There may be a largely unpredictable unfolding of life, partially determined by what potentials from the generative holon are realized.

Classic arguments point out that if mutations are sufficiently rare, but are recurrent then they may be fixed by selection in either sexual or asexual lines before the next mutations occurs (Crow and Kimura, 1965). In this case, sex confers no advantage. In the case of effectively unique events, the same reasoning suggests that different lineages may accumulate unique strings of regulatory changes. Sexuals can bring these together (section 5.5.3), but asexuals will simply continuously diverge (Figure 5.1). Given that sexual species tend to show considerable stasis, if adaptive mutations are all that recurrent, then they should already have been discovered, refined and incorporated (either as loci or as circulating alleles). Certainly recurrent null mutations in regulatory genes are prevalent, but loss of function is much different from obtaining adaptive genes. Even most 'gain of function' mutations documented in development result from loss of function in a inhibitory regulator, rather than adaptational advances (Slack, 1991). The assumption of recurrent adaptive mutations, particularly with the associated restrictive assumptions of fixed loci in non-variant karyotypes, is inconsistent with the modern reality of phenotypic evolution.

Sexual lineages are more likely to 'discover' appropriate combinations sooner. If these are rare regulatory mutations or novel combinations, asexual lineages may never discover or incorporate them. Thus, at least in the initial stages, sexual lineages will evolve faster because they have a head start (Müller, 1964). Secondly, there is no reason why the sexual lineage cannot achieve and hold an optimal state. Presumably selection would reinforce the frequency of the relevant alleles among loci in the population so these combinations would increase in frequency over time. If different organisms contain alternative large chunks after some period, recombination can also assemble larger subunits instead of simply isolated genes. Strong selection would eventually ensure fixation. Thus, the Eschel and Feldman (1970) argument only pertains to the rate of evolution near an adaptive peak when both asexual and sexual lineages have access to the same combinations of mutations.

Once consolidated on a peak, the asexual lineage has very little chance of tracking a peak shift. It may be doomed by its invariant specialization (the classic explanation for the demise of asexual taxa). The sexual lineage would be better capable of tracking an unstable peak (probably a common phenomenon in a non-equilibrium world) and has a greater potential to utilize mutations to escape from a peak that disappears or radically shifts. This insight was actually a key conclusion of Eshel and Feldman (1970).

Finally, most arguments postulating the advantages of asexuality in responding to selection have been based on theoretical studies or simulations of haploid genomes. In the case of diploids (i.e. most higher organisms), segregation must also be considered. Because diploid asexuals have no segregation, they can only move from one homozygous state to another via two mutations at the same locus on homologous chromosomes (Bull and Harvey, 1989). Sexual species need only one mutation and segregation can then quickly produce homozygotes (Hedrick and Whittam, 1989; Kirkpatrick and Jenkins, 1989). Moreover, sexuals need not even wait for mutations; they can import novel, functional information from other local populations. Thus, in the case of diploids, sexual species can both escape faster from genetic fixation of any particular state and they can achieve a new state of fixation faster than asexuals. The disadvantages of recombination in sexual species which exists when asexuals have the same initial genetic variation is more than offset by the segregational advantage of sexuality.

The full implications of Kirkpatrick and Jenkins's (1989) results have not been considered in a regulatory lineage paradigm because they focus only on alleles. Returning to the insight of White (1978) that structural changes in the genome may be essentially regarded as unique events (with high regulatory implications), then the second mutation needed for asexual lineages to escape from heterozygosity may never come. Essentially this means that capturing any significant structural changes in the genome can only be accomplished by sexual lineages. Since changes in karyotype might be major factors contributing to outcrossing depression (and subsequent evolution of isolating mechanisms), sexual species may be better adapted to consolidate novel karyotypes and speciate in this dimension. Wright (1980) emphasized that the demic structure of sexual species would greatly facilitate such evolution.

To whatever extent karyotypic evolution is seen as advantageous, then these chips must all be raked to the sexual side of the gaming table, because asexuals cannot even place a bet. Furthermore, interchromosomal recombination may be a major mechanism generating gene duplications via unequal crossing over (Ohta, 1988; Maynard Smith, 1989). Examination of the genome structure of various clades reveals that numerous structural changes have been successively fixed in different species (e.g. Dobzhansky, 1970; White, 1978). White (1978) estimates that at least 90% of speciation events are accompanied by karyotypic changes. There can be no doubt that such changes are sometimes associated with adaptive evolution (e.g. Nevo and Shkolnik, 1974). In the long term, the advantages of participating in such tinkering must contribute to selection among subspecific lineages and among species.

A third blow to the Müller/Fisher hypothesis was a masterful compilation of empirical evidence by Bell (1982). He showed that the frequency of asexual reproduction increased as taxa were examined along a spatial scale proceeding from oceans to lakes to ponds to water films. Since environmental variation also increases along this scale, the evolvability hypothesis predicts that the frequency of sex should increase (and conversely parthenogenesis should decrease) from oceans to water films. Glesener and Tilman (1978) made similar observations. Thus, the data appear diametrically opposed to the theory. These data are convincing, but cannot be used to reject the evolvability hypothesis because sex is not the only mechanism available to organisms to deal with environmental variability.

Alternative strategies include phenotypic plasticity or general purpose genomes with high dispersal and/or selective habitat choice. Such tactics are often associated with parthenogenesis (Lynch, 1984; Harvell, 1990; Rollo and Shibata, 1991). In fact, Bell's (1982) choice of gradient descends both spatial and temporal environmental scales but does not consider the scale of generation time. With very high variability (i.e. much shorter than the generation time), or intermittent (but predictable) recurrence of particular conditions (i.e. equal to or longer than generation time), phenotypic plasticity or other homeostatic mechanisms may be a superior strategy to that of circulating alleles (Lewontin, 1957). A case in point is that the ecological success of aquatic snails occupying unstable pond environments is attributed to their renowned phenotypic plasticity (Rollo and Hawryluk, 1988). Such organisms may also be suited for asexuality or hermaphrodism to ensure the success of founder events in highly fractured and unstable habitats (Tomlinson, 1966). When considering environmental variation, predictable variation that can be met plastically (Chapter 6) should probably be separated from that which is not easily predicted.

Bierzychudek (1985), Lynch (1984) and Hebert (1987) drew attention to the fact that organisms adapted to extreme environments or reproducing asexually are often polyploids and/or species hybrids. Loss of sexuality in such cases may simply be a requirement for stability of the genotype. This adds a further confounding overlay onto this entire question. The immediate point is that Bell's (1982) evidence is strongly confounded by the existence of such alternative tactics so his results are not necessarily inconsistent with some function of sex in meeting environmental uncertainty.

Finally, studies with aphids indicate that a great deal of care must be taken in interpreting sex in the context of life histories (Moran, 1992a). Aphids are of temperate zone origin and transitions among various specialized morphs (including sexual generations) are environmentally cued. Photoperiod is most commonly used with regards to sex. Aphids that have colonized the tropics are mostly asexual, perhaps because they no longer experience the requisite photoperiod (Moran, 1992a). So here we have a taxa where sexuality is associated with temporal heterogeneity and diversified niches (shifts among alternative hosts). Asexuality, in direct contrast to Bell's (1982) analysis, is associated with less variable tropical environments of high biotic diversity. These aphids also show loss of dispersal and host switching and consequently have lower niche variation as well. Moran (1992a) makes the interesting observation that tropical aphids usually have highly developed mutualistic relationships with ants, that would clearly protect them from predators. Is asexuality in aphids a genetic constraint? Can mutualism substitute for short-term advantages of sex? Does sex serve a role in adapting to temporal and spatial environmental/niche diversification in temperate regions? The aphids illustrate the complexities involved in interpreting life-cycle changes. Despite the current lack of enthusiasm for a lineage paradigm, the hypothesis that sex improves long-term survival in an uncertain world remains a conceptually parsimonious explanation of the evidence (e.g. Williams, 1988).

5.5.2 Centre court advantage

Since recombination may act against the products of selection (i.e. it reduces linkage disequilibria), it is usually seen as a disadvantage in a genic framework. Chapter 7 argues that sexual species have coadapted genomes, long-term integrated elements of which represent the wild phenotype. This phenotype may exist at a state where linkage disequilibrium is minimal (Lerner, 1954), but coadaptation is high. In this context selection producing disequilibrium may represent a reduction in coadaptation rather than vice versa. For lineages, selection may vary temporally on multigenerational time scales. Thompson (1976) pointed out that it is important to distinguish whether selection is acting on new mutations or old genetic elements. Recombination may provide a relatively unexplored benefit where variation is based on elements that are consolidated components of the dynamic organization (centre court advantage).

Recombinational load refers to the fact that linkage disequilibrium produced by natural selection may be disassembled by recombination. Asexual species lack such a process resisting selection. It is important, however, to consider the joint action of all forces influencing evolutionary rates rather than each individually. Discussions of recombinational load often imply that, rather than resisting some small proportion of selection's work, recombination is actually reversing the process. As long as selection is strong enough, sexual organisms can proceed towards fixation, regardless of recombination. Inbreeding, in the sense of genetic reinforcement, will accentuate this process, reducing recombinational load (Shields, 1982).

Falconer (1977) suggested that antagonism among higher-order fitness features may be responsible for the most optimal phenotype residing at mean values for a species (e.g. small mice have lower fecundity, but large mice have poorer survival). Selection away from the mean wild type is possible, but must be maintained to offset such internally driven stabilizing selection until genetic variation is reduced. Wade (1992) argued on theoretical grounds that, in sexual species, interactions may generally be opposite in sign to the main additive effects of genes. Such a situation might also contribute to internally driven stabilizing selection. Presumably, such coadapted systems can transcend their selection limits if a genetic reorganization occurs at a more distant fitness peak. However, barring a genetic revolution associated with the discovery of a new adaptive peak, maximal fitness in a coadapted genome may reside at a highly balanced mean phenotype (Manning, 1976). Coadaptation may drive its own internal stabilizing selection that may resist directional selection in any given direction (Lerner, 1954).

Only where selection changes direction or is relaxed will its products be disassembled by recombination. This in itself may actually be an advantage. On average, the best genetic position for responding to new selective regimes of unknown direction will likely be from linkage equilibrium. Consider a situation where selection occurs in discontinuous bouts. A good analogy might be a game of squash. An asexual species playing in an evolutionary court might respond quickly to a drop shot in the front right corner, but must then wait there for the next volley. A lob to the opposite back corner would leave such a player badly out of position. A sexual species, after making the front corner shot, would return to centre court (i.e. the imprints left by selection would be restored to linkage equilibrium). Several mechanisms would be involved in doing so. Firstly, recombination will tend to restore linkage equilibrium (i.e. random distribution of all possible genotypes) under neutral selection. Intermittent communication or continuous introgression of alleles from neighbouring sublineages will tend to restore those lost by fixation or drift, so that equilibrium frequencies of alleles will also tend to be restored. If the wild type is maintained by coadaptation, then stabilizing selection may also act to restore balance. Thus, in sexual organisms, the imprint of previous selection regimes may gradually be erased. This process is of adaptive significance because a population at genetic equilibrium has the maximum available degrees of freedom for rapidly responding to the next volley of selection. Centre court advantage is a crucial tactic for a winning game in any racket sport where volleys are intermittent and of unpredictable direction. Any player who does not use it will lose against one who does. In a lineage paradigm, centre court advantage could be a crucial advantage bestowed by segregation and recombination via sexual reproduction.

Because asexual species cannot erase the imprint of the last selective regime, they will always be beaten in the long term by a sexual player, once selection coefficients exceed the recombinational disadvantage carried by sexuals. This will be true of both haploids and diploids. Only where the same selection pressures recur repeatedly would an asexual player have any long-term advantage. In fact, where nature makes the same shots in predictable cycles, asexuals could be favoured (Chapter 6).

Clearly the advantages of sexuality would be greatest in a 'game' situation where the opposing player spitefully searches for the next most disruptive shot (i.e. Red Queen coevolution). However, centre court advantage would accrue in any situation where there is wide scope and uncertainty with respect to the future direction of selection. This model is thus very similar to that proposed by Maynard Smith (1980) where selection for the optimal phenotype constantly shifts (see also Felsenstein, 1988). A species using the tactics of centre court advantage would exhibit a typical wild type and would likely have considerable evolutionary stasis, both of which are ubiquitous characteristics of sexual species

(Lerner, 1954). Recent evidence suggests that recombination might be favoured in a regime of normalizing selection with a shifting optimum, a situation possible conforming to the proposed model (Bergman and Feldman, 1990).

A related advantage of sexuality may be to confer a more complex genetic response to environments defined by numerous niche dimensions. In ecological jargon this might mean a finer grained response is possible. In our racket court analogy one can imagine the floor digitized into squares that may be resolved to various levels. The sexual species, using recombination, may be capable of resolving a finer grain and a more precise ecological response (Michaels and Bazzaz, 1989). Thus, sex may at once contribute to genetic uniformity at upper tiers of a lineage and ecological diversity at lower tiers (Bell, 1988c). Based on general models of interactive networks, Kauffman (1993) showed that the highest adaptive peaks will tend to be clustered in the central region of a rugged fitness landscape. Such a situation would accentuate centre court advantage because most of the best positions are all within a few steps of centre court.

An asexual species might be hard put to compete within the circumference defined by the selection limits of sexual entities (all other things being equal). Where resource competition occurs, the sexual lineage may win by virtue of its broader niche alone (Case and Taper, 1986). If centre court advantage is important as an evolutionary explanation of sex, it is clearly a tactic relative to global organization and long time scales. (i.e. to mesoevolution of higher-order genetic organization).

Sexual reproduction and recombination appear to be adaptations that provide evolvability in uncertain and complex environments. Centre court advantage would allow sexual species to track environmental fluctuations better and make optimal use of available carrying capacity where fluctuations necessitate time lags. With greater uncertainty, a species might increase recombination rates. A reduction in generation time would also indirectly increase recombination rates within a lineage, and the increased intrinsic rate of increase would further improve environmental tracking. Thus, increasing uncertainty relative to the organism should tend to select for smaller body sizes. In many cases, the decrease in generation time would probably provide both the ecological and evolutionary flexibility needed without changing individual recombination rates by much. These ideas are related to the selection arena hypothesis of Stearns (1987b) that postulates that species may have excess production geared to allow natural selection to sieve out variants of high fitness. The data of Burt and Bell (1987), showing that recombination rate increases with generation time, are not inconsistent with this interpretation. All other things being equal, any factor selecting independently for increased generation time might require offsetting increases in recombination rates.

In many cases such species are more advanced organizationally and have smaller populations. Exploration of the regulatory field and damage control might thus require more individual recombination.

Maynard Smith (1980) reasoned that in a lineage under strong normalizing selection, recombination would be selected if the optimum frequently shifted in an uncertain way. Moreover, directional selection in such a system should select for increasing recombination rates (Maynard Smith, 1988a,c). Empirical evidence for shifts in the predicted direction is provided by Zuchenko, Korol and Kovtyukh (1985).

In an organized lineage adapted for evolvability, we might also expect to find that the degree of recombination in the genome may not be a fixed attribute. Just as the activity of transposons may be regulated to vary according to environmental conditions (e.g. stress), the likelihood that particular regions of the genome may recombine may also come under regulated control. There is certainly ample evidence that sexual reproduction as a whole is often controlled by functional switches cued to the environment (Maynard Smith, 1988a). It is important to distinguish, however, between switches that simply shut the entire meiotic machinery off or on from the rate of recombination that is expressed during the sexual mode.

Strong support for a lineage paradigm would be obtained if the degree of recombination itself was sometimes an adaptively modulated feature. In other words, a lineage paradigm predicts that the genetic organization might facultatively modulate individual recombination rates according to cues indicative of individual stress or impending stressful environments (e.g. realized growth or reproduction rates). Recombination rate itself might be phenotypically plastic in which case not all changes observed in selection experiments need represent genetic changes in the sense of hitchhiking modifiers. Early empirical support for this idea was provided by Bergner (1928). Parsons (1988) reviews considerable evidence that recombination rates may be modulated in response to various sources of environmental stress including temperature and nutrition. Such an anticipatory response requires multigenerational selective scales.

For sexual species with a highly coadapted wild type, selection may most often be stabilizing. Maynard Smith (1979) suggests that a long period of such selection will select for reduced recombination, but Manning (1976), Shields (1982) and Bernstein and Bernstein (1991), point out that sex would still be advantageous for sexual species in terms of reducing DNA damage and the mutation load or assembling compensatory mutations (Wagner and Gabriel, 1990). Thus, sexual lineages may at once maintain a coadapted genome in good repair and an ability to respond in numerous directions to unpredictable and multidimensional selection regimes.

This topic will be revisited in Chapter 7 after considering the role of reaction norms and genetic correlations in a lineage framework. This discussion suggests that a considerable portion of genetic variation may be bound up in epistatic linkages that normally resist selection and lead to decay of changes brought about by short-term selection. Such variation may be released in response to particular kinds of stress or under inbreeding, allowing variants to be selected or genetic reorganizations to be facilitated under extreme conditions. A crucial question then arises as to whether such attributes are accidental, or perhaps reflect adaptive organization of a higher order entity that requires both flexible local adaptation as well as an ability to reorganize if necessary.

5.5.3 Crosstalk among subspecific lineages

There are three kinds of possible adaptive variation that may be available for mesoevolution. These are: (1) new mutations; (2) variation within demes; and (3) variation imported from other demes. Most genic theories presume that faster incorporation of new mutations is fortuitous but not necessarily of selective significance because these are relatively rare. With respect to adaptive aspects most recent discussions pertain to recombination of existing variation in single panmictic populations. In a lineage paradigm, communication among sublineages becomes important.

If bacteria and other lower organisms are considered, the original value of sex may have involved transfer of genetic information. In such cases exchanges may be unequal and may not involve recombination (Margulis and Sagan, 1984, 1986, 1988). In modern bacteria, exchange of extrachromosomal plasmids allows rapid shuffling of important genes such as those involved in antibody resistance (Maynard Smith, 1990). Such evidence suggests an important role for horizontal gene exchange, consideration of which leads immediately to a lineage perspective. For example, in considering to what extent bacteria represent clonal lineages, Hull (1980) and Maynard Smith (1990) concluded that sexual lineages may have a net-like rather than tree-like internal structure, but that these two views may represent poles of a continuum formed by degrees of recombination.

In Wright's model, individual demes might have such a net-like internal structure, but if subpopulations are not linked by constant exchanges, the higher-order structure may again appear tree-like (e.g. the branches of sexual lineages may have net-like substructure). Williams (1992) also considered this same problem and placed considerable weight on the idea that units of selection only exist if their history can be drawn on a dendrogram. It is clear, however, that net-like substructures can derive higher-order lineages and that sexual species may have compartmentalized structure that allows a hierarchy of recognizable sublineages to emerge.

Gene exchange could constitute relatively constant introgression of information as occurs in clines or established hybrid zones, or it could involve intermittent crosstalk among sublineages that otherwise may be isolated for varying periods. Hamilton (1993) recognized that successful incorporation of useful adaptations derived from hybrid crosses would be synergized by inbreeding. This aspect of local sublineages forging coadapted fitness jackpots and exporting them is the cornerstone of Wright's shifting balance theory (Wright, 1980, 1982a,b, 1988; Crow, Engels and Denniston, 1990). Wright (1982a) recognized the importance of inbreeding in producing a diversity of sublineages of high phenotypic variability. Furthermore, he recognized that not only was such inbreeding a crucial aspect of successful artificial breeding programs, but that a further applied practice was the exporting of locally derived superior forms to other inbred backgrounds (e.g. the use of a few stud bulls to sire offspring in numerous herds). This represents artificial application of crosstalk. Wright (in Crow, Engels and Denniston, 1990) explicitly recognized that local inbreeding provided advantages similar to those obtained by asexuality (see also Shields (1982) and Hamilton (1993)), while crosstalk yielded the full advantages of sex. Given that Wright (1988) estimates the shifting balance process in subdivided populations can enhance evolutionary rates in the order of a million times, the elusive advantages of sexual reproduction could well reside with such dynamics. This is especially so if phenotypic evolution is largely concerned with regulatory organization. Information in this case is not confined to single genes, but includes structural variants and higher order epistatic circuitry.

The theoretical support for the mesoevolutionary hypothesis has been developed only as far as the value of recombination for assembling a few useful combinations (Crow and Kimura, 1965). The spatial and temporal diversification of lineages, however, occurs across a broad range of scales reflected by local structured demes, metapopulations, ecotypes, races, subspecies, species and syngameons (= groups of interbreeding species). Occasional sexual crosstalk between such lineages may vary in frequency and introgression, but its impact is to introduce novel regulatory information and structural genes that (as far as the recipient lineage is concerned) could be viewed as new mutations. A key difference is that such genes have been sieved and polished in another lineage and are consequently highly biased towards functional utility. Moreover, considerable information may arrive at once, in large organizational blocks. Because regulatory changes can produce novel, unexpected results (Dickinson, 1991), the mixing of regulatory variants from separate demes could generate novelty that neither deme could obtain otherwise.

By analogy, we could imagine two companies independently developing high-technology robots. A merger could well provide advances in efficiency, production rate or product standards, provided a reorganization incorporated the best features of both firms. Unions between sublineages attach one set of coadapted genes to a different coadapted complex. Extensive segregation and recombination (and selection) would be the only way to achieve any useful synthesis from such mergers. Alternatively, local lineages may maintain some autonomy, while importing useful information for otherwise independent evolution. An analogy might be the development of weapons technology by different nations.

This mechanism has not been fully explored in arguments of sexual evolution, possibly because most theories assume that populations approach local optimal adaptation and gene flow is thus largely disruptive. Maynard Smith *et al.* (1985), however, did recognize the importance of crosstalk with respect to genetic transfer of adaptive information among bacteria via plasmids. They clearly articulated the idea being put forward here when they stated that sexual reproduction in diploids 'sorts the genes in a lineage into a gene pool in a way that allows transfer of advantageous gene combinations into sublineages...'.

There are really two kinds of potentially advantageous genetic interactions – one associated with heterozygosity at single loci (positive heterosis) and the other associated with positive associations among loci (epistasis). Hybrid vigour is often obtained by wide outcrossing due to increased heterozygosity. However, the F2 generation generally has reduced fitness, presumably because of disassembly of positive epistatic associations by recombination (Lerner, 1954; Endler, 1977; Shields, 1982). It may be that a truly coadapted genome maximizes both kinds of advantages. There may also be different weightings associated with the degree of outcrossing (i.e. inbreeding stressing epistasis, outbreeding stressing positive heterosis).

It is not necessary, however, that these be mutually antagonistic. Although epistasis is usually considered to be a measure of gene interaction, this is only true in a relative genetic sense. Although they are often discussed in the same breath, coadaptation and epistasis are not completely overlapping concepts. Functionally interactive genes (i.e. coadapted organizational components) will show no epistasis in a genetic sense if the relevant loci are both fixed. Similarly, functionally linked genes may also have allelic variants that may show both positive heterosis (e.g. where heterozygotes can enzymatically span a wider range of conditions) or positive epistasis that varies in value according to the environmental variation spanned by the effective population. A genome can be highly coadapted (i.e. integrated and organized) even where little genetic epistasis is apparent and heterozygosity is prevalent.

Having said this, it still remains that evolution proceeds by selecting variants that have higher relative fitness. Outbreeding depression associated with disrupting epistatic associations will depend on the degree of outcrossing and we are led to conclude that there may be an optimal level of communication to maximize advantages. In a lineage paradigm, however, the advantages must be viewed across generations. Outcrossing may improve performance of F1 progeny but may decrease the average fitness of F2 progeny. But what about the next 10 generations? Bell (1988b) and Mitton (1993) suggest that low fitness of F2 phenotypes may occur due to expression of deleterious genes in the homozygous state that are then subject to removal (à la Müller ratchet). Certainly, recombination might not be expected to forge a superior constellation of attributes immediately, and even if it does, selection must act long enough to sift and amplify the relevant constituents. The question then becomes whether some F2 or F3 variants are superior, even if overall mean fitness of the population is temporarily lower. Empirical data appear sparse. Hamilton (1993) suggests that superior forms may indeed arise as new epistatic combinations are created by recombination.

Some authors have rejected the idea of genetic revolutions based on the argument that they cannot occur in a few generations (Chapter 7). This is an unfair burden to place on the concept. Recombination may require several generations to dismantle coadapted gene complexes contributed by different lineages. The potential value of crosstalk is vastly amplified where regulatory evolution is occurring because there is a finite number of sub-lineages available to explore the adaptive landscape, and some may make useful discoveries that differ from those of others. Sex provides a vehicle allowing such crosstalk, the ability to sift useful grain from non-adaptive chaff, and a mechanism to derive a new recipe from the mix of available ingredients. Non-equilibrium ecological models suggest that competition does not necessarily act rapidly to eliminate all but the few fittest genotypes, which may allow time for such a process to occur.

Although gene flow may well be disruptive if precise local adaptation is necessary, presumably this will vary according to the degree of divergent evolution in the alternative sublineages, as well as the actual differences in their local habitats. Recent evidence shows that relatively high levels of gene flow (even 40% between demes) are not necessarily disruptive to local adaptation and differentiation (e.g. Endler, 1973, 1977; Wade, 1977; Templeton, 1989; Harrison, 1991). It is also not necessarily true that demes in different locations are adapting to radically different selective regimes. If the sublineages have essentially similar niches, crosstalk may offset loss of useful alleles that occurs due to founder events, random drift in small populations or associated with population bottlenecks.

It is often assumed that demes derived from the same parental stock and experiencing the same selection regime will converge on the same genetic solutions such that uniformity would result without any gene flow (e.g. Slatkin, 1987). The experiment of Rutledge, Eisen and Legates (1974) on tail length in mice dramatically shows that this is not necessarily so, and this is only one example of problems with replication in quantitative genetic studies (Riska, 1989). Rutledge, Eisen and Legates (1974) interpret their results to support the framework of Wright where selection and random forces interact. The diversity of solutions possible in a fragmented population is a fundamental aspect of Wright's theory. Useful invention may be imported from lineages that constitute polygenic features of greater significance than single alleles. Where mice are selected for large size, for example, they reach a selection limit presumably associated with loss of genetic variation. If separate lineages that have both reached selection limits are crossed, they often reach even larger body sizes and show renewed responses to selection (Falconer, 1977). Such results are commonly obtained in artificial selection regimes, and although inbreeding depression may also be a contributing factor, the fact remains that superior responses can often be obtained by selecting separate lineages and then crossing them.

Crosstalk also is essential in maintaining higher-order lineage identity and evolution. For example, the evolution of phenotypic plasticity may be associated with this process (Chapter 6), and phenotypic canalization is another likely candidate. In this context, the entire question of whether even divergent selective regimes are necessarily associated with disruption of local adaptation is questionable (West-Eberhard, 1989). Lineages without outcrossing cannot engage in crosstalk at all and those without recombination and segregation could not synthesize new regulatory and structural information into a useful configuration.

Lineage crosstalk, unlike mutation, involves the acquisition of genetic information that has already been adaptively honed by another sublineage. Thus, the mesoevolutionary value of sex is not simply limited to the capture and consolidation of useful combinations of crude new mutants. In a biotically diverse and environmentally heterogenous world, opportunities for crosstalk abound and the potential evolutionary rewards seem high. Even in the simplest case, restoration of alleles lost via random drift could be a powerful force maintaining sex in populations fractured into many small demes by ecological heterogeneity. This hypothesis seems closely related to Müller's ratchet, and could contribute to the maintenance of sex even in relic species exhibiting evolutionary stasis (Kondrashov, 1988). As in other features of relevance to a lineage paradigm, it is important that fitness be considered on multigenerational time scales.

Although Wright masterfully explored the evolutionary implications of demic population structures, he viewed this structure as largely a fortuitous consequence of environmental heterogeneity. Shields (1982) arrived at a model essentially identical to that of Wright, but he obtained this perspective by considering the adaptive value of inbreeding for reinforcing local genetic identity. Shields (1982) also applied genic arguments but his theory is extremely relevant to the lineage paradigm. In particular, Shields (1982) recognized that aspects promoting lineage identity (including mate selection, habitat choice and dispersal characteristics) were likely to be adaptively selectable. He recognized that adaptive mechanisms involved in isolating species (to prevent disruption of coadapted genomes) are likely also to be active at other levels in the hierarchy of population subdivision. Thus, local demes may themselves have mechanisms promoting their own subspecific lineage identity (my terminology).

Shields's (1982) ideas, however, are unidirectional. Selection always acts to promote the highest possible fidelity in parental information (balanced against other trade offs such as offsetting Müller's ratchet). In fact, Shields (1982) viewed the entire process as a positive feedback system promoting selection for greater identity which then selects for even greater isolating mechanisms. Thus the system drives its own differentiation. This works as long as coadapted genomes are always disrupted at every higher level by communication. Shields's (1982) theory needs only to be elaborated to consider the balance among the three advantages of reinforcing (i.e. inbreeding), importing and exporting information. Systems with genetic reinforcement derive benefits largely ascribed to asexuality and may apply these benefits to occasionally imported novelties. In such systems, gene exchange may be the primary function of sex (Templeton, 1989). This expansion provides a general framework for considering the forces promoting lineage cohesion and diversification that might act across the hierarchy of spatial and temporal scales and associated selective units in the regulatory hierarchy.

The maintenance of bimodal dispersal phenotypes (i.e. mainly residential, with some high-dispersal forms), would be an ideal way to achieve both ends. Such a leptokurtic distribution of dispersal phenotypes appears widespread (Endler, 1977; Shields, 1982; Wynne-Edwards, 1986). When high-dispersal phenotypes are produced at intermittent intervals or in localized patches, they are also invariably associated with sexual reproduction and outcrossing (Maynard Smith, 1974, 1984; Williams, 1975). Such association is strong support for the idea that crosstalk is an adaptive attribute of higher-order lineages. Ghiselin (1974) also suggests that more widespread and ecologically diverse species have higher recombination rates than those with restricted ranges and this would also act to allow effective integration of transmitted information. Thus, the lineage paradigm suggests that there may be conflicting selection pressures to minimize disruption of local adaptation and identity while maintaining a smaller channel of potentially useful information.

It will be necessary to reinterpret the huge literature on the evolution of isolating mechanisms and dispersal in light of this paradigm. The hypothesis to be tested is clear: lineage isolating mechanisms (and these could occur at numerous levels) should only evolve where the disadvantages of local disruption exceed the advantages of maintained information exchange. Adaptive variation in dispersal rates and in mate and habitat selection should reflect this as well. A lineage paradigm suggests that even where isolating mechanisms are operating strongly, windows of potential communication may be adaptively opened in response to particular temporal or environmental conditions. There is no doubt that intermittent crosstalk has been a major source of evolutionary innovation and advance. Even at the highest levels of divergence (i.e. species), hybridization events and associated reorganizations (including polyploidy) are significant sources of both new species and organizational advances (Stebbins, 1950; Ohno, 1970; White, 1978; Abbott, 1992). 'Leaky' species abound, and this problem is becoming of practical concern, because this provides escape routes for genetically engineered genes (Ellstrand and Hoffman, 1990).

Introgression between species can lead to at least four outcomes (Abbott, 1992): (1) increased genetic diversity in a species; (2) either reduction or strengthening of reproductive barriers; (3) transfer or origin of adaptations; or (4) the origin of a new taxa. In the British Isles alone, 715 plant hybridizations have been documented, of which about half show some degree of fertility (Abbott, 1992).

There is no doubt that disruptive selection can resist hybridization among diverging lineages (White, 1978). However, the phenomenon of introgressive hybridization illustrates that separate lineages may also exchange useful information without necessarily losing their individuality (Van Valen, 1976; Templeton, 1989; Arnold, Buckner and Robinson, 1991; Whittemore and Schaal, 1991; Abbott, 1992). In such cases, different genetic elements from one population may penetrate into another at very different rates. The degree of disruptive selection on these various elements might be assessed by comparing the relative introgression of mitochondrial lineages relative to that of other genetic elements.

The crucial question is to what extent such communication is encouraged by design *versus* the likelihood that it represents fortuitous accidents. How high up the organizational scale might adaptive communication be found? If we consider syngameons (interbreeding species swarms) as the upper extreme where communication generally stops, it seems unlikely at first glance that the 'species' involved represent sublineages in the sense of a higher-order adaptive organization. However, we must ask why such species do not develop isolating mechanisms, even where the frequency of hybridization events is rather common (Van Valen, 1976; Templeton, 1989). We may be missing something important if we simply assume that there has not been enough time. There is currently insufficient information to say where organized lineages stop and fortuitous assembly or unavoidable disruption begins.

Parallel discussions have emerged with respect to bacteria. The best evidence is that bacteria are largely clonal with only occasional recombination. Thus, taxonomists distinguish good 'species' and these are composed of recognizable sublineages (Maynard Smith, 1990; Maynard Smith, Dowson and Spratt, 1991). However, such entities are so leaky that gene exchange continues among even remotely related forms, and the organization of evolutionary relevance may in fact reside at the genus level (Maynard Smith, 1990).

The possibility that a species swarm represents a real ecological and evolutionary organization cannot be entirely dismissed. For example, burdock occurs as two recognized species, great burdock and common burdock, which have remarkably different dispersal characteristics and seed-packaging tactics (Rollo, MacFarlane and Smith, 1984). These species readily hybridize. Great burdock has large, low dispersal burs which open and drop most of their seeds close to the parent. This results in local dense patches that are competitive in developing successional habitats. This strategy also exposes seed predators (that are dumped out with seeds) to the rigors of winter and consequently this species is not highly predated (<15% of seeds eaten by moths). Common burdock makes many small burs with higher dispersal potential and which remain tightly closed. These burs are much more likely to discover newly disturbed habitats favourable to these weeds. Incidentally, such a packaging strategy also promotes local inbreeding in new sites. This species routinely suffers seed predation exceeding 80% (probably because their burs stay closed) but dispersal allows escape from the moths. Thus, the two species are adapted to different ecological pressures associated with successional conditions and seed predators. In any given site, a genetic succession may occur as the early arriving common burdock are eventually replaced by hybrids and great burdock. Eventually great burdock dominates as succession proceeds (Rollo, MacFarlane and Smith, 1984).

There is insufficient information to say how much genetic introgression occurs here or how the two phenotypes are genetically maintained. The example does suggest how two 'species' may form an apparently integrated complex in which communication may not be entirely selected against. The organizational unit might sometimes comprise a single 'species', but it could also encompass both. West-Eberhard (1989) has explored how species with genomes organized to produce two genotypes may consolidate one or both to derive new species. Certainly, new species are known to be derived via hybridization, but a question that has not been addressed is the possibility that such mergers could yield a single organization with adaptive phenotypic plasticity.

5.5.4 Hierarchical hopscotch

The lineage paradigm is based on two kinds of hierarchies: the regulatory hierarchy (some of which is probably a fixed organizational templet and some of which is dynamically distributed), and the lineage hierarchy (species, races, ecotypes, demes, organisms). The interaction of these hierarchies allows sexual lineages to explore the potential epigenetic landscape available in their existing organization. The wild type phenotype is only the tip of the iceberg in terms of the variation available for expression. In sexual lineages the range of existing developmental possibilities can be exposed, either by selection acting progressively upwards on levels of modifiers, by inbreeding that generally compartmentalizes coadapted complexes or by environmental factors that alter the phenotypic expression of the developmental organization (Waddington, 1957, 1975; West-Eberhard, 1989). I will refer to this exploration of the regulatory hierarchy as hierarchical hopscotch.

The first case is illustrated by breeding programs where directional selection for a given trait is imposed for numerous generations. Sexual lineages are capable of generating phenotypes that deviate from the original wild type by a factor of three or more standard deviations before a selection limit is reached (Falconer, 1981). Directional selection acting on a sexual population will mainly act on one dimension of the organization, progressively exposing variation for further selection. For asexual lineages, barring mutations, what you see is what you get (Crow, 1988). Thus, one huge advantage of sexual reproduction is to open the regulatory hierarchy to selection. Considerable early literature was aimed at documenting the release of genetic variation by recombination (Thompson, 1976). Although this appears to be one of the major potential advantages of sex, it is rarely specifically addressed by recent genic theorists, even though considerable microevolution is mainly concerned with just this process.

I termed this process hierarchical hopscotch for the explicit reason that it entails a lineage-level feature that is entirely inconsistent with the genic view. Over evolutionary time, selection might well traverse particular developmental configurations numerous times. In such cases, the lineage organization may retain the imprint of past selection regimes (i.e. derive an organization pre-adapted for particular shifts). In other words, although species may not inherit acquired characteristics, they could possibly re-acquire previously selected features with surprising rapidity. Unless one had knowledge of the historical selection regimes shaping a particular lineage, such phenomena would have a definite Lamarkian appearance. It would be most useful to review the tenacity of the Lamarkian paradigm from this perspective. Quantitative proof of this would require a compilation of plastic and genetic responses of genomes to various environmental regimes, and a test of whether there is bias in developmental shifts in adaptive directions. This seems like a difficult but testable hypothesis.

Finally, there is possibly a synergistic link among the adaptive contributions of centre court advantage, crosstalk and hierarchical hopscotch. A sexual lineage at wild type equilibrium can respond to selection immediately with no need of new adaptive mutations. Asexuals cannot move at all except by mutation. Sexuals can then return to equilibrium via recombination, crosstalk and stabilizing selection. If selection frequently uses particular 'volleys', the sexual lineage may be primed to respond to them appropriately via adaptive genetic correlation structure or plastic norms of reaction (Chapter 6). The entire picture appears to be congruent with known properties of sexual species, but would not be arrived at by a genic paradigm.

5.6 MÜLLER'S RATCHET

Müller (1964) pointed out that asexual lineages can only maintain their fittest genotypes by selection against a constant barrage of deleterious mutations. In small populations, the fittest class may be accidentally lost, resulting in a global reduction of fitness. In theory, this process could be reiterated across generations, the genotype being progressively polluted with accumulating deleterious mutations. This would ultimately lead to extinction. Like a ratchet, each transition to a new worse state is irreversible unless fitness can be restored be recurrent adaptive mutations (see also Felsenstein, 1974; Felsenstein and Yokoyama, 1976; Manning, 1976; Maynard Smith, 1978a; Kondrashov, 1988; Chao, 1990, 1991, 1992). Genic theories suggest that asexual populations may establish an equilibrium balance between recurrent mutation and stabilizing selection. As in the mesoevolutionary arguments above, a regulatory hierarchy controlling epigenesis means that functional recurrent back mutations are unlikely (especially in small populations), which could imply that the impact of the ratchet is likely to be very high. Wagner and Gabriel (1990) suggested, however, that in complex genomes where phenotypic features are polygenically determined compensatory mutations at other loci could offset loss of function at any particular site. This reduces the immediate impact of the ratchet and extends the time course for deterioration. Lynch and Gabriel (1983) also pointed out that in such circumstances, asexuals may mask accumulations of mutations. The high redundancy of epigenetic systems is congruent with these ideas.

Molecular evidence suggests that the evolution of epigenetic mechanisms does not necessarily follow the most optimal route, and inefficient contrivances can be seen, even relatively high up in the epigenetic hierarchy (Chapter 3). Loss of function at one locus can sometimes be offset by compensatory loss of function at others (e.g. Lawrence, 1992). It is interesting to note that such a process could lead to a reduction in complexity and a more parsimonious genetic organization. Thus, this process of compensatory adjustment may actually lead to design improvements, although the yield in fitness of this process is unknown. The store of possible compensatory mutations is finite, however, so that this process may slow the impact of the ratchet, but ultimately may not prevent its advance.

In sexual species, mutation-free genotypes can be restored by recombination. Simultaneously, several deleterious mutations may be combined in 'waste' individuals whose demise improves the shedding of harmful genes from the lineage (Kondrashov, 1988). Thus, sexual reproduction may be an important adaptation for offsetting the accumulation of deleterious genes. This will be most effective where deleterious mutations decrease fitness more than predicted by their independent contributions (Kondrashov and Crow, 1991).

The function of sexual reproduction in eliminating harmful mutations is clearly a feature relevant to transgenerational (lineage) fitness. Bell (1982, p. 101) classified this as group selection and rejected it on several grounds (Bell, 1982, p. 387). Despite this, the high probability of deleterious mutations has made this particular lineage aspect palatable even to genic theorists, particularly where it can be argued that there are at least some deleterious mutations per genome per generation (Kondrashov, 1988). This seems like an unnecessary restriction and points out the rather small transition required to convert a genic view to a mesoevolutionary framework.

In non-equilibrium environments, fluctuating selection pressures associated with density and environmental changes could well allow the assimilation of deleterious mutations during periods of relaxed selection. The ratchet could be accentuated in asexual species under such conditions and the value of occasional sexual recombination might similarly be enhanced. The basic theory is no different from the current genic arguments, but is merely extended to multigenerational time scales where selection does not necessarily have to act strongly in any one generation or short-term sequence of generations. The advantage of sexuality for offsetting the ratchet is now widely recognized (Shields, 1982, 1988; Kondroshov, 1988). Even Bell (1988a,b) now argues strongly that sex may function to offset Müller's ratchet, although he also points out that it does not fully explain the major ecological correlates of sex. Alternatively, Kondrashov (1988) argues that the strength of this hypothesis is its generality. Given the near ubiquity of sex in eukaryotes, a mechanism 'must' exist to explain its maintenance even in relic species showing little evolutionary dynamics or in relatively constant conditions.

Nearly all discussions of the ratchet have been theoretical or based on phylogenetic interpretations. Vrijenhoek (1984) observed significantly higher loads of deleterious mutations in clonal fish compared with their sexual relatives. Experimental evidence for the operation of the ratchet has also been documented in segmental RNA viruses (Chao, 1991, 1992; Chao, Tran and Matthews, 1992). These small genomes are susceptible because RNA has higher rates of mutation than DNA. Such results suggest that organizational advances associated with larger genomes may have required sexuality to offset mutational degradation.

Bell (1988b) provides theoretical estimates that the ratchet may impact strongly on asexual unicells with fewer than 10^5 members, and if correct, is highly relevant to the relatively small populations maintained by larger eukaryotic organisms. He argues that the ratchet would have strongly resisted any transition to multicellularity in the absence of sexual reproduction. Moreover, the ratchet would strongly oppose increases in genome size associated with organizational advances unless accompanied by sex. Lynch and Gabriel (1990) added that the accumulation of deleterious mutations would itself reduce population sizes, allowing homozygous expression of these and existing recessive genes (Mitton, 1993), in a 'mutational meltdown'. Lynch and Gabriel (1990) suggest that asexual species are unlikely to persist more than 10^5 generations.

Bergner (1928) extended the duration of various stages of *Drosophila* using a wide variety of environmental and dietary manipulations. A similar effect was obtained in all cases. As the physiological age of germ cells increased, the frequency of crossing over also increased. Selection for greater longevity (longer generations and less frequent recombination) also yields increased recombination rates as a correlated effect (Wattiaux, 1968). Such responses might be relevant to the ratchet since more mutations are likely to accumulate between recombinational events. Parsons (1988) reviews some literature suggesting that recombination rates may decline with age within species. The ratchet clearly predicts the opposite and more data are needed to draw any firm conclusions, especially since trends with age within organisms may be confounded by the action of senescence.

Associated features of the genome relevant to a function of repair in sexual species are the fragmentation of genes into discrete functional domains (intra-locus recombination can restore wild type functions from two mutated alleles (Ohno, 1970)), redundant gene copies and diploidy. The segregation advantage of sexual recombination also becomes important (Kirkpatrick and Jenkins, 1989). These ideas are consistent with the intimate association of sexual reproduction with major organizational advances and in particular the transition to diploidy.

The evolution of diploidy is intimately linked to the problem of Müller's ratchet. In most higher eukaryotes, the haploid phase is highly transient (Charlesworth, 1991). An immediate advantage of transitions to diploidy is the masking of recessive deleterious mutations. However, this advantage may be transitory as the number of such mutations then increases until a new, worse equilibrium is reached. In terms of compensatory mutations diploidy is also advantageous. Since many asexual species are derived via polyploidy, such species may have exceptional abilities to mask recessive mutations or obtain compensatory adjustments (i.e. slow the ratchet).

Recent theoretical treatments suggest that diploidy may be a favoured state given certain assumptions. Kondrashov and Crow (1991) showed that if there are many mutations of non-additive effect, truncation selection will yield superior reduction of mutation load in diploids whenever the degree of dominance of deleterious mutations is less than a quarter. Such assumptions appear consistent with the empirical evidence (Charlesworth, 1991). Perrot, Richerd and Valero (1991) considered the case where diploids and haploids freely interbreed. In this case, diploidy displaces haploidy whenever dominance is less than a half. This occurs because both genotypes will obtain similar frequencies of the mutant, but diploids can mask them in the heterozygous state and obtain higher mean fitness. The available evidence suggests that dominance of the wild type is obtained via developmental mechanisms and that diploidy will be increasingly favoured in larger genomes if the mutation rate is relatively high (Charlesworth, 1991). Thus, diploidy may be particularly relevant to genomes that deploy complex multicellular phenotypes.

The phylogenetic distribution of asexuality shows that such lineages rarely establish new species and almost never taxonomic units higher than genera. This suggests that asexuality represents dead end twigs on the tree of sexual life (Müller, 1932; Maynard Smith, 1978a, 1984). A major consternation for the ratchet hypothesis is the existence of a few notable exceptions. These include the bdelloid rotifers, many Protozoa, oribatid mites and a few others (Hurst, Hamilton and Ladle, 1992). Large population sizes and small genomes are two factors that might alleviate the ratchet in such species.

Recent evidence also suggests that sex may be 'covert', in such species, occurring intermittently or in non-obvious ways. Thus, Pernin, Ataya and Cariou (1992) documented strong evidence that the amoeba, *Naegleria lovaniensis*, is occasionally sexual and Hebert (1987) mentions a genus of asexual cynipid wasps in which one species is now known occasionally to express sexuality. Hurst, Hamilton and Ladle (1992) suggest that large swarms of dead-end asexual lineages could be continually spun off a few core sexual lineages in such organisms. Vrijenhoek (1984, 1990) believes that a diversity of many specialized clones may be derived from sexual ancestors by freezing ecologically specialized sublineages with asexuality. In such cases, the diversity of asexual sublineages still requires generation by a sexually reproducing lineage. Proof of this would be strong inference for the universality of the ratchet. We badly need genetic studies of the bdelloid rotifers to determine if genetic exchanges and recombination are occurring at least intermittently.

The advantages of diploidy are also relevant for somatic repair systems which can then use homologous chromosomes as information to repair damage and replace lost information more precisely (Berstein *et al.*, 1985a,b, 1988; Bernstein and Bernstein (1991); Bell, 1988a,b). This advantage, however, adds new problems associated with maintaining universal symmetry among homologous chromosomes. In sexual diploids, recombination and gene conversion work to maintain universal organizational integrity. Thus, the evolution of diploidy may have required sexual reproduction to provide a mechanism for repair of DNA damage exceeding the abilities of the somatic repair system, and also as a mechanism for maintaining a congruent organizational structure for exchange of complementary information.

Because the constraints are different, primary asexuality in haploid protozoans or bacteria may be a very different situation from secondary asexuality in diploid organisms. The advantages of diploid repair may eventually wind down as homologs diverge via mutation. The segregational advantages of sex would allow chromosomes that are complementary, as far as repair function is concerned, to be subjected to paired selection in various organisms. It would also restore evolvability to a diploid organization (Kirkpatrick and Jenkins, 1989). Presumably, the repair system of the haploid asexual, although not being as precise as a diploid, is not constrained by the need for homology. Thus, the evolution of diploidy and of sexual outcrossing are undoubtedly closely related. It may be significant that the only known sexually reproducing amoeba is also unusual because it is diploid (Pernin, Ataya and Cariou, 1992). Together, recombination and gene conversion forge sexual diploids into unified organizations with symmetry for functional repair. The repair advantages are both somatic and via inter-generational selection.

Asexual haploids, as long as they reproduce before significant somatic decay occurs, may not have as great a need for sex. The immortality of such organisms and their lack of senescence may both derive from their ability to reproduce and rapidly push new phenotypes through the sieve of selection before any significant DNA damage accumulates (Chapter 11). To do so requires a very brief somatic existence and consequently small size. The correlates of asexuality with unicellular organisms with largely haploid genomes is consistent with such an interpretation.

In a lineage paradigm, the evolvability advantage of sexual reproduction (both segregation and recombination) will also be enhanced by integration with transposable elements. Segregation will help to restore matched homologous chromosomes following breaks and rearrangements mediated by transposons. Recombination could act in an intra-locus manner to assemble novel or optimal combinations of regulatory recognitions sequences and/or exons. Both mesoevolutionary advantage and ratchet reduction would be served by sex in such cases. It is tantalizing to note that both the activation of transposable elements and the timing of sexual reproduction are often mutually associated with ecologically stressful conditions. A wealth of theory now supports the idea that genome deterioration may be unavoidable in small eukaryotic populations without sexual recombination. This perspective will resurface in consideration of senescence in Chapter 11.

5.7 THE RED QUEEN

The search for a single-generation selective regime that could explain the ubiquitous maintenance of sex in eukaryotes has increasingly emphasized the Red Queen hypothesis (Levin, 1975; Glesener and Tilman, 1978; Jaenike, 1978; Hamilton, 1980, 1988; Bell, 1982; Rice, 1983; Seger and Hamilton, 1988; Hamilton, Henderson and Moran, 1981; Hamilton, Axelrod and Tanese, 1990; Bell and Maynard Smith, 1987; Bremermann, 1987; Ladle, 1992; Rennie, 1992). The main attraction has been that the most powerful selection for recombination is expected where the sign of linkage disequilibrium changes spatially or temporally (e.g. Maynard Smith, 1971a; Bell and Maynard Smith, 1987). Environmental factors are unlikely to have such properties except in restrictive cases that cannot support a general explanation (e.g. Bell and Maynard Smith, 1987). Interactions with parasites may result in a negative correlation between frequency and fitness (Bell, 1985; Seger and Hamilton, 1988). Coevolution with other biological entities that have antagonistic fitness interactions is wonderfully congruent with theoretical predictions for sexual advantage.

Of all the possible sources of coevolutionary disequilibria, the most powerful would be the interaction of multicellular eukaryotes with their pathogens and parasites. Because of their much shorter generation times and larger numbers, diseases have the potential to evolve much faster than their hosts (Hamilton, Axelrod and Tanese, 1990). Models of recombinational advantage in disease environments substantiate the potential value of sex in the evolutionary arms race between host and exploiters (Rice, 1983; Bell and Maynard Smith, 1987; Hamilton, 1988; Hamilton, Axelrod and Tanese, 1990; Chao, 1992), particularly for large, slower-growing species with small populations. The frequency of chiasmata, which are indicators of recombinational frequency, are positively associated with generation time, consistent with the increasing need of organisms with less frequent sex to compensate via increased recombination intensity (Burt and Bell, 1987).

With respect to the coadaptation of the genome and possible guided genetic trajectories, it seems highly relevant that one of the correlated responses to selection for greater longevity in *Drosophila subobscura* was an increase in recombination (Wattiaux, 1968). Presumably recombination would also enhance relevant variation for evolving positive coevolutionary relationships (Wilson, 1980, 1983). In this case earlier considerations of coadaptations and their possible disruption by recombination are also relevant. It appears that, once consolidated, mutualistic symbioses may select for reduced recombination associated with a much more stable biotic environment (Law and Lewis, 1983).

There can be no doubt that Red Queen evolution is a key selective regime explaining the maintenance of sexual reproduction. Having said that, I would argue that this hypothesis is still insufficient to support a general genic theory for the evolution and maintenance of recombination. Even in its strongest form, it does not provide general explanation and it cannot reject a lineage-level model.

Firstly, if sex evolved as a mechanism principally to offset coevolutionary arms races, then recombination, particularly in a genic model, should be restricted to those attributes specifically adapted to such interactions. Instead, recombination is deployed throughout the genome and impacts on nearly all phenotypic features. In the genic Red Queen model, only interspecific competition could contribute to such general recombination. Widespread recombination is more consistent with a general evolutionary function of sex. This is also fully congruent with the empirical association of recombination and generation time (Burt and Bell, 1987).

Although there may indeed be high variation in features associated with disease resistance (such as variation in the histocompatibility complex), and recombination can certainly contribute to variation, the immune system is the supreme example of an adaptively plastic mechanism. A range of antibodies sufficient to recognize nearly any new invader can be generated by the immune system of a single organism. This involves a complex integration of regulated transposition, compartmentalized variation and differential cell proliferation, the basic workings of which are a remarkably conserved and integrated mechanism (Wills, 1989). All of the models postulating immediate Red Queen advantages of recombination in host–parasite systems largely ignore the existence of the immune system. Given the extreme power of the immune system, the question then becomes, how much residual value can recombination overlay on this mechanism?

Genomic variation is a crucial requirement for sex to operate against parasites and there is good evidence that polymorphism may be maintained as a defensive feature (Rice, 1983; Seger and Hamilton, 1988). However, asexual species, which presumably have little defence against Red Queen onslaughts, do not sicken and die with the rapidity expected from known rates of microbial evolution. Nor does the rapidity with which pathogens overtake lineages that convert to asexuality appear sufficient to offset the two-fold cost of sex via immediate individual advantage in every generation. Among plants, clones may live over a thousand years in the same locality without being overrun. Similarly, although conservationists are concerned about species like the cheetah which have little genetic variation, the immediate risk to them appears to be human habitat destruction and alien diseases in zoos. Disease does not appear to be quickly 'winning' in their natural environment, and even highly polymorphic species may be badly affected by exposure to new diseases.

Although asexual or genetically depauperate species must be at a disadvantage against new disease, the long-term persistence of those with genetic variation reduced by inbreeding, asexuality or founder effects suggests that the contribution of sexual reproduction to immunological power may be overstated. Shields (1982) echoed similar sentiments. Hamilton (1980) suggested that a crucial factor in immunological responses is self-recognition. The ultimate danger for a species would be a parasite that breaks this code. Sex could serve constantly to change the locks. The above discussion applies equally well to this idea. Both these hypotheses of varied immunology and self-recognition appear highly relevant to the maintenance of sex, but their full impact is more likely to be reflected on multigenerational time scales.

In our enthusiasm to explain the maintenance of sex via immediate advantage, the vastly more relevant question of how this marvellous plastic immune system could have evolved has been overlooked. Given that this is a sophisticated invention utilizing regulated transposition to instantly create novel 'alleles', the entire mechanism lies outside the genic paradigm. Furthermore, this is an integrated regulatory system which generates novelties of adaptive potential. As such, it does not appear amenable to a genic process. Instead, the properties of the immune system appear to have been forged by long-term selection involving interaction among many known players and a need for robust responses to new ones. Intermittent frequency in exposures and variation in the particular cast of contenders could favour phenotypically plastic responses where the costs of maintaining fixed defences are high. The reality of plastic mechanisms deployed against specific Red Queen protagonists is now well documented for predators and competitors (Harvell, 1990), but was virtually unknown 20 years ago. Given the evolutionary stability of many prokaryotes (i.e. *Escherichia coli* consists of only a few clones) such arguments can also be extended to numerous diseases to a limited extent. Such phenotypically plastic features are best understood in a lineage-level framework (Chapter 6).

Coevolution between most known protagonists tends towards reduced fitness impacts on the host and stabilization of fitness in both players. Variation among individuals in disease responses may be due to genetic variation, but other factors like age or stress may be equally important in the case of coevolved pathogens. This may involve long-term lineage coevolution, not local changes in genic balances. In many cases, this involves evolution of the immune system rather than changes in the frequency of any particular allele. As in other plastic systems, the disadvantage of circulating alleles recombined by sex is that crucial alleles may be lost by fixation of alternatives by selection, during founder events, or via random drift in small populations. The immune system retains the full degree of freedom needed for producing most potential antibodies, and this potential is protected from loss. Seger and Hamilton (1988) provide models suggesting that highly polymorphic species may exhibit chaotic trajectories of gene frequencies that might protect rare alleles from loss. It seems highly unlikely, even with maximal recombination rates, that populations of large, long-lived eukaryotes could meet pathogen onslaughts with only the genetic variation circulating in their relatively small populations.

One mechanism that could protect immunological variation from loss might involve mechanisms that actually recognize variants and actively select them. In mice, for example, animals can detect variation in the histocompatibility complex, and females prefer males that have different variants. This might be of high selective value if heterozygotes derive superior immunological responses. Such a system is consistent with a view of organisms as complementary components in a lineage-level organization adapted to maintain variation and scramble resistance characteristics.

The most dire threats to a lineage arise when a disease coevolved to a different lineage escapes from that system and invades a naive host (e.g. bubonic plague, yellow fever, AIDS, chestnut blight, myxomatosis virus of rabbits, Dutch elm disease). Although such events may be rare, they have very strong impacts. Our own species has experienced many such events. This is analogous to the discussion of lineage crosstalk in that opportunity for such events abounds and only systems with high evolvability can hope to respond. Thus, some characteristics of immune systems may indeed be driven by selection for evolvability, but strong selection from really novel opponents might also contribute to the maintenance of sex. Given that the immune system is attuned to handle most commonly encountered pathogens, the most important evolutionary sphere may involve changes or extensions in the regulatory apparatus itself. It seems likely that the nature and power of the immune system may derive, not simply from the coevolutionary arms races between commonly encountered agencies, but from unpredictable and intermittent invasions (often by diseases honed by attacking related members of a clade). In this context lineage crosstalk might be a crucial mechanism for importing and incorporating defensive features.

Those features promoting better survival in hosts and reduced pathogenicity in disease agents will be simultaneously favoured in local sublineages, whereas demes with more severe interactions will become extinct or have reduced export. This form of group selection is believed to explain the coevolution of myxomatosis virus and European rabbits in Australia (Lewontin, 1970; Wilson, 1983) but, combined with recombination during crosstalk, may reflect a general mechanism for the coevolution of inclusive (i.e. higher order) antagonistic lineages. Wynne-Edwards (1986) provides a good review of examples. It should be noted that in the case of pathogens and small herbivores a 'deme' may reside with a single host (Thompson, 1982; Rice, 1983; Wilson, 1983; Karban, 1989). Thus, individual hosts may be matched against entire pathogen lineages. This again points out the small transition necessary to extend models of individual selection to a workable lineage paradigm.

Although lineages are not immediately overrun by pathogens on converting to asexuality, recent evidence shows that they do accumulate higher parasite loads than their sexual relatives (Lively, Craddock and Vrijenhock, 1990; Moritz *et al.*, 1991; Ladle, 1992; Lively, 1992). Such trends are also expected to apply to highly inbred populations (Hamilton, 1993). Such evidence directly favours the Red Queen, as does the general distribution of asexual forms in habitats with low biotic diversity. However, several of the examples pertain to parthenogenetic hybrids. The Red Queen predicts that outcrossed individuals should be more resistant to parasites and species hybrids are extremes of outcrossing (Rice, 1983). In some cases such hybrids may have more parasites, a factor that could act as an isolating barrier (Whitham, 1989).

Clearly the problem may be thornier than first anticipated. Where hosts and parasites are highly coevolved, outcrossing of higher-order lineages could lead to destabilization of interactions and increased parasite loads. For example, transposable elements can be considered in a framework of coevolved genetic parasites. Hybrid dysgenesis in *Drosophila* is caused by the introduction of mobile P elements into systems with no coevolved regulatory control. The P elements perform a hatchet job on the naive host genome and greatly reduce fitness.

The widespread existence of coevolved genetic parasites may impose barriers to gene flow at higher levels (Hickey and Rose, 1988). In this case the value of outcrossing and recombination in terms of Red Queen advantage may actually change sign at different tiers in the lineage hierarchy. Lineages may be envisioned as towing unique parasitic lineages as they course through time (e.g. Timm, 1982). The strong coevolution of these biota with their host means that in some sense they represent part of a coevolved organization. A merger that allows sexual communication by hosts will also be accompanied by potentially dangerous mergers of parasite lineages. Whitham (1989) also pointed out a possible advantage of hybridization: a few susceptible hybrids may act as the focus of attention for parasites and may disrupt the ability of the parasites to overcome the resistance of either contributing parental lineage. Alternatively, a lineage harbouring a coadapted parasite may have a competitive advantage against a naive competitor (Sarkar, 1992). The coevolution of hosts and disease is an extremely important aspect that is relevant to both ecology and evolution. To capture it fully, a lineage paradigm with multi-generational time scales is required.

All of the models investigating Red Queen advantages of recombination are based on haploid genomes. In the case of diploids, the transition from heterozygosity to homozygosity is a severe bottleneck for asexuals (Kirkpatrick and Jenkins, 1989). Where hosts or prey derive maximum benefits by obtaining extreme genotypes (i.e. disruptive selection for various combinations of homozygotes at numerous loci), or shifting rapidly between these states, then asexuality would be highly disadvantageous. The segregational advantage of sex in Red Queen evolution may be extremely high. The interaction of recombination and segregation may be even more crucial.

The above arguments pertaining to immune responses do not extend directly to Red Queen evolution associated with competitors or predators. Here too, however, plastic adaptations may evolve where antagonists are well acquainted (Harvell, 1990). My intention here is not to reject Red Queen advantages to recombination. They are undoubtedly a key factor in the maintenance of strategies for variation. However, the existence of these other response modes may offset the short-term need for recombination and they also suggest that units of selection with multi-generational time scales may also be important.

Van Valen (1973) considered the key features of fitness to be the ability to make a living in an adaptive zone, and the ability to resist invasion of that zone by antagonistic lineages. Any lineage that failed to evolve would see the available environmental resources decline and losses to antagonists increase. Glesener and Tilman (1978) also recognized that the evolvability of the surrounding sexual biota would create uncertainty favouring retention of sex in any particular species. Although the Red Queen is currently seen as salvation by the genic theorists, it should be pointed out that Van Valen (1973) derived his hypothesis as one of group selection largely at the species level. The rate of branching of lineages was considered the key factor impinging on a given lineage. New lineages invading an occupied adaptive zone could arise in distant phylogenies, or the lineage itself could branch and replace other sublineages. Thus, the Red Queen model is as relevant to a lineage as to a genic paradigm.

5.8 TANGLED BANK

The idea that sex is advantageous for adaptive diversification because it produces variable genomes is one facet of the Tangled Bank hypothesis. The present discussion differs from the genic view, however, in that most genic arguments are cast in a light viewing competition among close relatives as the main driving force (e.g. Bell, 1982, 1985). There is a meso-evolutionary aspect of this hypothesis, which is largely ignored in the current synthesis. This is related to the old problem of whether diversification is largely driven by competition (niche specialization) or whether it is driven by lack of competition (adaptive radiations). Life tends to explore all potential avenues that are unconstrained. Most major adaptive radiations occur where a founding lineage encounters a nirvana of empty niches (i.e. cichlid fish in Lake Victoria; *Drosophila* in Hawaii; the opossum's ancestor in South America) or a catastrophe wipes out the current dominant tenants (Benton, 1987). Interspecific competition may constrain such radiations, although it may also drive specialization on finer scales (Rensch, 1960). Thus, there are two important evolutionary aspects associated with Tangled Bank. Diversification potential is an important evolutionary attribute in its own right (Simpson, 1944; Stanley, 1979; Arthur, 1988), but is clearly a mesoevolutionary aspect that can be independent of local competitive interactions. The ability to generate a diversity of forms so that available evolutionary passageways are discovered and explored appears to be a fundamental attribute of life that may be of equal import to fine-scale compartmentalization of such diversity by competition (i.e. competition acts on existing diversity, and deletes rather than generates). Thus, Moran (1992a) observed that asexual aphids show little diversification from ancestral sexual forms and are generally regarded as conspecifics or races. Moran took this as evidence that such asexual forms are ultimately evolutionary dead ends.

It is crucial to separate these two aspects. Arguments that sex is advantageous because it allows an individual to ameliorate competition seem to ignore the equally relevant fact that sex may produce a combination of attributes that can better survive in a different habitat, and that this survival may be completely oblivious to competitive situations that may exist elsewhere. Although improved features (relevant to environment) may be important in local competition, they may also be of selective advantage even where competition is weak or intermittent.

A prodigious effort related to the advantages of sex to plants growing with neighbours of varying relationship has been carried out by J. Antonovics and associates. They used the grass *Anthoxanthum odoratum* which reproduces sexually via seed or asexually via tillers. Several of these studies have demonstrated strong advantages for sexually diversified genotypes. However, no support for the sib-competition aspect has been derived. Sexually derived offspring exhibit minority advantage (Antonovics and Ellstand, 1984; Kelly, Antonovics and Schmitt, 1988), but are not favoured as density increased which is required by sib-competition models (Antonovics and Ellstrand, 1984; Schmitt and Antonovics, 1986; Kelly, 1989a,b). Failure to substantiate the sib-competition component needed to derive immediate sexual advantage has focused attention on the alternative frequency-dependent mechanism relevant to the Red Queen (Antonovics and Ellstrand, 1984; Ellstrand and Antovics, 1985; Schmitt and Antonovics, 1986; Kelly, Antonovics and Schmitt, 1988). Aphid and rust infections appear to act in a frequency-dependent manner at the seedling stage (Schmidt and Antonovics, 1986). Kelly (1989a), however, rejected both Red Queen advantage and sib-competition as explanations of sexual advantage in the field. McCall, Mitchell-Olds and Waller (1989) compared the performance of self-fertilized *versus* outcrossed seed in the herb *Impatiens capensis*. They also found no evidence that sex reduced sib-competition or that frequency-dependent selection was important in maintaining the outcrossing mode.

All of the above studies addressed the importance of sex where only local environmental variability was considered, and the overall conclusion appears to be that no advantage is obtained with relatively low environmental heterogeneity (e.g. Kelly, 1989b). On larger scales, however, sufficient genotype × environment has been detected for *I. capensis* to maintain genetic polymorphism (Bell and Lechowicz, 1991; Bell, Lechowicz and Schoen, 1991; Lechowicz and Bell, 1991; Mitchell-Olds, 1992). This suggests that the diversifying aspect of Tangled Bank may be relevant on lineage-level temporal and spatial scales. It also seems possible that reductions in intraspecific competition might still emerge at degrees of relatedness differing more than that among sibs or half sibs (all that has been considered to date).

The plant species that have been used are also a somewhat biased sample because they grow in patches which may be largely derived by selfing or vegetative propagation. Such species may be relatively unaffected by inbreeding depression and may derive benefits from group patch dynamics that outweigh local intraspecific competition. One must ask why the grass species is sexually self-incompatible, and why *I. capensis* invests at all in very expensive outcrossing flowers that are designed to prevent self-fertilization. Such systems suggest that there is a bimodal distribution of advantages relevant to strong local reinforcement of lineage identity and larger-scale sexual diversification. Seed dispersal must be a key factor in understanding the value of sex. For *I. capensis* for example, it is assumed that explosive seed dispersal obtains only short-distance movement of seeds away from parents. My own observations suggest that this is generally true, but that spring flooding and runoff also result in long-distance movement of seeds in the moist habitats favoured by this species (e.g. floodplains). Thus, the best evidence suggests that selfing or vegetative reproduction may be favoured for local adaptation, but that there may be considerable interpopulational diversification in heterogenous environments that favours the retention of sex (Mitchell-Olds, 1992).

Significantly, Tangled Bank rests on a foundation of spatial heterogeneity that leads directly to a subspecific lineage framework. Support for the hypothesis (reviewed by Bell, 1982, 1985) includes the reality of local adaptation of plants to their habitat, and the general finding that mixtures of genotypes have higher yields than monocultures (i.e. sexual lineages may have higher local carrying capacities). The potential temporal aspect of Tangled Bank is widely ignored. If Tangled Bank is provided with temporal dynamics (e.g. a can of worms), then the ability of a lineage to maintain its footing would be greatly enhanced by having numerous, widespread, flexible feet rather than a single inflexible leg to stand on.

5.9 CONCLUSION

If sexual reproduction is considered to reflect an important transition in organizational advance, then the effective units of selection become lineages with multigenerational scales of selection. Such entities may be selected, not only to deploy functionally adaptive phenotypes, but also for features relevant to their integrity and evolvability. If all populations are heterogenous at some scale, then a Wrightian model of meta-population evolution allows lineage selection (which encompasses both processes of kin and group selection) to act on relatively short time scales.

Such evolution could ultimately forge a genetic templet that incorporates sexual reproduction and outcrossing simultaneously to maintain cohesion in a polyphenotypically distributed organization. Such entities would have considerable potential for local genetic adaptation in the sense of homeostatic balancing while retaining the potential for rapid reorganizational evolution in novel or stressful conditions.

6

Phenotypic plasticity, allelic switches and canalization: a metalineage perspective

6.1 INTRODUCTION

Polymorphism refers to the existence of multiple phenotypes within a single species. Such variation may be achieved by circulating alleles or batteries of genes that derive markedly different organisms. Graded responses may also be achieved genetically by variation in numerous alleles of small effect, usually associated with environmental gradients. There is a considerable literature concerned with factors such as spatial heterogeneity that maintain genetic variation (Hedrick, Ginevan and Ewing, 1976; Roughgarden, 1979; Hedrick, 1986). Single genomes may achieve similar results, however, without any genetic variation, a phenomenon termed phenotypic plasticity.

Whereas development was simply ignored by the genic paradigm, the associated phenomenon of phenotypic plasticity was viewed as downright inconvenient (Sultan, 1987). Until the 1980s there were only a few major treatments, but interest is now compounding (Gause, 1942; Schmalhausen, 1949; Bradshaw, 1965; Shapiro, 1976; Jain, 1979; Caswell, 1983; Smith-Gill, 1983; Lloyd, 1984; Lynch, 1984; Via and Lande, 1985, 1987; Lively, 1986b; Schlichting, 1986, 1989a; Fagen, 1987; Matsuda, 1987; Sulton, 1987; Via, 1987; Levin, 1988; West Eberhard, 1989; Stearns, 1989, 1992; Harvell, 1990; Cheplick, 1991; Thompson, 1991; Hall, 1992; Moran, 1992a,b; Newman, 1992; Scheiner, 1993). A related phenomenon, canalization, occurs where developmental systems converge on a single phenotype that resists disruption by either genetic or environmental variation (Waddington, 1957; Rendel, 1979; Levin, 1988; Scharloo, 1991).

The significance of plasticity and canalization for a lineage paradigm is multifold. Both phenomena uncouple the genotype from the phenotype and promote evolutionary stasis. Both processes also serve to increase the integration of the higher-order lineage (Levin, 1988). Recall that a genic view of recombination suggests that the disassembly of advantageous gene combinations and mixing of genes relevant to alternative selection regimes would make the differentiation of multiple adaptive phenotypes from a single genome problematic. The widespread ability of single genomes to differentiate several coadapted phenotypes confirms that different coadapted gene complexes can be obtained without mutual disruption. Secondly, by their nature, plastic mechanisms usually have time frames only visible on multigenerational scales, or in spatially heterogeneous habitats of relatively coarse grain (which amounts to the same thing from an organismal perspective). Plasticity among individuals of phenotypically fixed traits is a characteristic of the genome (i.e. the genetic templet), not of individuals (Scheiner and Lyman, 1991; Scheiner, 1993).

Non-equilibrium ecology emphasizes the importance of both spatial and temporal heterogeneity. Modern genetic models are also adopting a structured format that recognizes various important units (e.g. individuals, trait-groups, demes, metapopulations). Evolution occurs via the interaction of ecological variation with genetic variation through developmental filters. Genic models usually assume homogeneous environments and uniform genetic backgrounds, which represent utterly unrealistic assumptions (Wilson, 1980). Such models preclude the expression of adaptive plasticity (Gomalkiewicz and Kirkpatrick, 1992).

A lineage perspective is relevant to plasticity because genetic structure does not necessarily map in a one-to-one manner onto ecological structure. Thus, a single lineage may straddle a range of temporal and spatial environments or even niches that require considerably different local adaptations. In such cases plasticity represents a metacharacter relevant to the heterogeneity experienced by a lineage, and the relevant evolutionary framework conforms to a Wrightian population structure where sublineages may occupy patches with different ecological contingencies (Gomalkiewicz and Kirkpatrick, 1992). Scheiner (1993) recognizes the importance of considering structured populations and points out that all previous models are based on panmixia.

Plasticity is deployed in response to nearly every major ecological problem faced by organisms. It may involve single features or coordinated phenotypic modulation. Below I will outline some of the major categories of phenotypic flexibility associated with lineage organization and ecological contingencies. Some of these, such as ontogenetic diversification and sexual polymorphism, are not usually considered in the category of phenotypic plasticity. I include them here because they all involve developmental processes for deploying divergent phenotypes from a single genetic templet.

Although population genetics recognizes the reality of adaptive plasticity, in practice evolution has been largely viewed through a microevolutionary lens that holds that an individual genotype expresses one phenotype and evolution proceeds via differential survival and reproduction of alleles at a single locus (Roughgarden, 1979; Sibly, 1989). The initial paucity of research on adaptive plasticity is due to neglect rather than prevalence, since even that model of population genetics, *Drosophila melanogaster*, shows strong shifts in phenotypic organization with environmental factors like temperature or density (Giesel, Murphy and Manlove, 1982a,b). Significantly, plasticity has now moved to the forefront of evolutionary ecology (Stearns, 1992). Relevant aspects for the current thesis include maternal influence, polymorphic switches, norms of reaction, plastic developmental trajectories, genetic assimilation and canalization.

6.2 SOME EXAMPLES OF ADAPTIVE PHENOTYPIC FLEXIBILITY

6.2.1 Diversification of phenotypes in response to niche variation

An appreciation that single species may show significant social and eco-logical niche specialization is relatively recent (Williams, 1992, p. 98). The specialized castes of social insects are the quintessential example (e.g. Oster and Wilson, 1978; Holldobler and Wilson, 1990; Robinson, 1992). Smith (1990a,b) documented that dimorphism (and even trimorphism) for size and shape of bills occurs in finches of the genus *Pyrenestes*. This genetic polymorphism appears related to niche segregation associated with seeds of various size and hardness. The possible contribution of plasticity or maternal influence to this situation appears to be unexplored. Similarly, parasitic wasps may show marked variation in body size and ovipositor characteristics among alternative hosts (West-Eberhard, 1989). In the tiny wasp, *Trichogramma semblidis*, individuals emerging from the eggs of butterflies are fully winged whereas those that emerge from the eggs of alder flies are wingless and show other morphological modifications as well (Gottlieb, 1992, p. 153). Moran (1992a) reviewed the plasticity associated with host alternation that commonly occurs in aphids.

Modification of the feeding apparatus relevant to changes in diet are widespread in animals (reviewed by Johnston and Gottlieb, 1990). Such changes may occur during development due to exposure to diets that alter the muscular and skeletal conformation of the head (e.g. Bernays, 1986; Meyer, 1987, 1990; Wainwright *et al.*, 1991; Thompson, 1992), or they may be cued to environmental or host-derived signals (Greene, 1989). Collins and Cheek (1983) documented the expression of specialized cannibalistic morphs in a salamander in response to high densities.

Several divergent phenotypes adapted to differing ecological niches have been identified among fish species (Liem and Kaufman, 1984; Wilson, 1989; Meyer, 1987, 1990; Wainwright *et al.*, 1991). In one case this may be genetically determined, but otherwise the diversification appears plastic. Among plants, their modular structure and relative immobility of parts allows the diversification of parts within single individuals. Lloyd (1984) reviewed such multiple strategies, which are common in leaf morphologies, flower types and dispersal characteristics of seeds. Thus, plastic modifications in response to resource acquisition are widespread and may be striking. The recognition of ecological diversification within single species has been confounded by the fact that such features are often secondarily superimposed upon ontogenetic or sexual dimensionalities. Furthermore, the complexity of responses (involving the entire adaptive suite) and continuous mapping of forms to environmental gradients makes such phenomena difficult to identify and study.

6.2.2 Plasticity and predictable environmental cycles

Numerous phenotypic elements show altered expression that anticipates predictable environmental cycles, notably those associated with annual periodicities. For example, numerous insects have darkly pigmented forms in one season or environment and lighter coloured forms in others that are not obtained by local selection (Shapiro, 1976). Seasonal dispersal is an especially important feature modulated via phenotypic plasticity. Responses may involve numerous integrated traits, including the presence of wings, their size and the development or histolysis of associated flight musculature (Harrison, 1980; Rankin and Burchsted, 1992). Diapause is yet another feature largely modulated by seasonal plastic responses (Masaki, 1980). Although seasonal responses are particularly important, it is a general rule that plasticity requires clear environmental signals that are correlated with the key selective factors (Moran, 1992a,b; Scheiner, 1993). Most seasonal polymorphisms utilize temperature and photoperiod.

Where the same genotype spans several temporally distinct environments, dichotomous characters may be evident in single stages of development at different times. However, the mapping of appropriate ontogenetic stages onto optimal temporal periods can also result in ontogenetic synchronization with no evidence of dichotomy within any particular stage.

6.2.3 Sequential ontogenetic diversification

The extremely divergent phenotypes represented by an aquatic dragonfly larva and the superb aerodynamic adult are obtained without disruption by the same genome. Of course mechanisms to facilitate transition are required. In this case the larva must have a behavioural program to crawl out of the water. In *Drosophila*, the same initial epigenetic programming is used, but cell lineages destined for adult structures are sequestered in localized compartments (imaginal discs and abdominal nests). These are later elaborated into the fly phenotype plasticity and represent a mechanism allowing a single genome to express divergent phenotypes without genetic disruption. Moreover, in many cases, the transitions among developmental stages, as well as timing of the life-cycle and diapause events, are environmentally cued (Smith-Gill, 1983; Newman, 1988a,b, 1989, 1992).

6.2.4 Inducible defences: anticipating the Red Queen

Inducible defences are fast becoming the model paradigm for under-standing the evolution of plasticity and consequently are considered in some detail here. The significance of the vertebrate immune system as a plastic mechanism was stressed earlier. Although other defensive charac-teristics of the phenotype are often genetically fixed, a barrage of recent evidence attests that defensive adaptations relevant to both predation and competition may be plastically deployed in both plants and animals. Reviews of such 'inducible defences' include Lively (1986b), Havel (1987), Sih (1987), Schultz (1988), Dodson (1989), Karban and Myers (1989), Stearns (1989), Adler and Harvell (1990), Harvell (1990) and Spitze (1992). In social or semi-social species, the regulated deployment of soldiers at the demographic level, such as the seasonal modulation of defensive phenotypes in aphids (Akimoto, 1992), also conforms to a framework of inducible defences. Some of the effects would be akin to converting rabbits into porcupines, if wolves are smelled in the woods.

Gilbert (1966, 1980) led the way with the finding that the rotifer, *Branchionus calyciflorus*, develops long posterolateral spines if chemicals from predacious rotifers (*Asplanchna* spp.) are detected. *Daphnia ambigua* responds to specific chemicals that signal the presence of predatory *Chaoborus* by developing spines and spiked head capsules (Hebert and Grewe, 1985). In the absence of predators such features are not expressed. Harvell (1984) showed that the bryozoan *Membranipora membranacea* produces spines when attacked by the nudibranch *Doridella steinbergae*. Spine formation was coordinated so that spines were induced around the entire perimeter of attacked colonies. Lively (1986a) documented induced shell dimorphism in the barnacle *Chthamalus anisopoma* in response to carnivorous gastropods. Numerous such examples attest to the ability of a single genotype to alter morphological features that improve survivorship.

Also relevant to predation are cryptic phenotypes. Greene (1989) found that caterpillars of *Nemoria arizonaria* mimic either twigs or catkins at different times of the year, a phenotypic plasticity tied to cues from the host plant. Harvell (1984) made the point, of relevance here, that such adaptations are not individual-level features, the genome being the appropriate evolutionary unit.

Life-history features may also be modified. In the snail, *Biomphalaria*, animals exposed to parasitic *Schistosoma* (but not necessarily infected) increase their reproductive effort. This is adaptive because infected snails may eventually be castrated by the parasite (Minchella and Loverde, 1981), a tactic that increases the body size of the snail. Washburn *et al.* (1988) found that free-living ciliates that are prey to mosquito larvae are induced by chemicals associated with mosquitoes to convert from prey to potentially lethal parasites. Crowl and Covich (1990) assessed the impact of water that had been conditioned by crayfish on the life history of the aquatic snail *Physella virgata*. Snails that detected crayfish matured at body sizes double that of controls, and they also lived twice as long.

Among plants, the main response mode is via the production of defensive chemicals (reviewed by Karban and Myers, 1989; Havel, 1987; Schultz, 1988). Baldwin and Schultz (1983) even found evidence that attacked trees may release chemicals that induce defensive adjustments in unaffected neighbours. Inducible defences may have very devious complexity. Turlings, Tumlinson and Lewis (1990) showed that corn plants release volatile chemicals in response to damage sites associated with caterpillar saliva. The signals attract wasps parasitic on the caterpillars. Other defences may be similarly herbivore-specific, although most plant responses represent generalized defence (Karban and Myers, 1989).

Induction of plastic defences is not limited to invertebrates or plants. Skelly and Werner (1990) found that tadpoles exposed to dragonfly predators reduced their size at metamorphosis. This effect was mediated largely by reduced levels of activity. Wilbur and Fauth (1990) found a similar response to the newt, *Notophthalmus viridescens*. Bronmark and Miner (1992) demonstrated that carp changed their body morphology to a deeper conformation in the presence of predacious pike. Such changes reduced the risk of predation, but increased drag while swimming. This would probably increase foraging costs. This may represent a common cost to aquatic animals via its impact on net acquisition of resources.

These examples document a broad range of tactics related to chemistry, morphology, behaviour and even life histories that may be modulated by specific information from the surrounding biota. Responses may be complex. Stemberger and Gilbert (1987) documented that the defences may be induced by a plethora of divergent predators and that the particular phenotypes expressed may vary with the kind of predator. Parejko and Dodson (1991) found that the degree of induction may scale to predator density. Brett (1992) explored the implications where not all predator species (e.g. larger fish) are deterred by the expressed defence.

Weider and Pijanowska (1993) reported and reviewed evidence that *Daphnia* may respond to the presence of fish by maturing at smaller sizes, but to invertebrate predators by maturing at larger sizes. Thus, the mix of predators and their relative impacts may be important. Responses may be relevant to competitors as well as predators. Buss and Grosberg (1990) documented the deployment of specialized fighting organs used for intraspecific competition in the hydroid, *Hydractinia* sp. Harvell and Padilla (1990) found similar responses in the bryozoan, *Membranipora*.

Theory suggests that the evolution of plasticity is favoured over constitutive deployment where the threat from specific antagonists varies in time and/or space, and the defensive features represent an expense that is better allocated elsewhere when it is not needed (Lively, 1986a; Havel, 1987; Havel and Dodson, 1987; Dodson, 1989; Harvell, 1990; Riessen and Sprules, 1990; Spitze, 1992). The relative cost of induced defences may vary with resource levels (Black and Dodson, 1990). Among plants, plasticity may be favoured where the generation time of herbivores is much shorter than the generation time of the host (Karban and Myers, 1989), and due to their modular construction, responses may be highly localized (Haukioja, 1990).

Fixed defences may be favoured where clear information about predators is unavailable, or where there is a high risk involved in obtaining it (Sih, 1987). Parejko and Dodson (1991), for example, found that *Daphnia* may show induction of defensive features in response to non-predatory mosquito larvae which are distantly related to predatory *Chaoborus*. Such miscues would impose a cost. These studies document that trade offs may be a crucial aspect in the evolution of polymorphisms, the fitness of the different forms being divergent under different conditions. Bradshaw (1965) refers to species that lack plasticity as analogous to being sewn into one's winter underwear. Alternatively, the fact that populations of *Daphnia* that have been isolated from fish for numerous generations still respond appropriately to fish signals (Weider and Pijanowska, 1993) suggests that genetic inertia may allow such species to keep their winter underwear in a drawer for a very long time.

6.2.5 Sexual polymorphism

Sexual dimorphism is another example of divergent phenotypes that are part of a single genetic organization. Although the dimorphism of males and females is well known, further differentiation may occur to the extent that species exist that have three kinds of males and one female (Shuster and Wade, 1991). In many cases such forms are generated by circulating genetic switches, but in many others, divergent phenotypes are controlled by environmental cues like temperature (Bull, 1983; Stearns, 1989). The flexibility of the genome in this regard is exemplified by the bluehead wrasse, *Thalassoma bifasciatum*. In this fish, individuals may change from females into males when they reach sizes large enough to defend a territory. Removal of the dominant males results in larger females changing sex (Warner, Robertson and Leigh, 1975). Shapiro (1980) obtained similar sexual transformations of females when males were removed in another fish, *Anthias* sp.

Even within a single sex there may be multiple phenotypes that are evolutionarily successful. Thus, satellite males who intercept females attracted to calling males have evolved numerous times where callers incur costs associated with parasites or predators (Thornhill and Alcock, 1983). Where males have differential success at obtaining resources for growth, large individuals may invest relatively more in weaponry and aggressiveness while smaller individuals may opt for mobility or stealth (Eberhard, 1980, 1982; Gross, 1985; Eberhard and Gutierrez, 1991). There is a relatively large literature pertaining to theories of sex allocation (Charnov, 1982; Cockburn, 1991, pp. 203–229), but like ontogenetically sequential phenotypes, this aspect is often considered separately from other categories of phenotypic plasticity.

Where divergent males occur, however, they are often expressed according to the resources that are available (e.g. Eberhard, 1980, 1982; Eberhard and Gutierrez, 1991) rather than by circulating alleles (Figure 6.1). In this case, the development of large horns used in competition for mates is favoured only in male beetles that reach large sizes. In such cases, alternative coadapted constellations appear to be adjusted to maximize the fitness of individuals across the variability existing in resource acquisition.

6.3 MATERNAL AND PATERNAL INFLUENCE

If lineage selection is an evolutionary reality, design elements that transcend generations or specific environments should be expected. Routes for adaptive parental manipulation of offspring, for example, should be prevalent. In fact, maternal gene products not only act extensively in early embryogenesis, they are indispensable for most major formative events prior to gastrulation (Raff and Kaufman, 1983; Arthur, 1984; De Pomerai, 1990; Slack, 1991; Lawrence, 1992). The development of an organism requires the activity of the genome of both the progeny and its mother. Genetically, organisms are lineage products, not discrete entities at all. This machinery for transgenerational integration is not limited to early formative events, but is a major factor influencing numerous aspects of life-history tactics and environmental fine tuning (e.g. Wellington, 1965; Shibata and Rollo, 1988; Mousseau and Dingle, 1991; Moran, 1992a).

Even relatively fundamental attributes such as rates of metabolism,

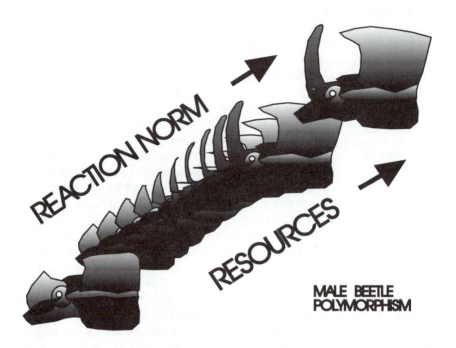

Figure 6.1 Variation in the size of horns among scarab beetles. Males display a reaction norm of horn development modulated by the growth performance of individuals. Original based on a figure of *Xylorectes lobicollis* from Eberhard and Gutierrez (1991), with permission of *Evolution*, Allen Press.

growth and maturation may be modulated via this mechanism (Capinera, 1979; Shibata and Rollo, 1988; Giesel, Lanciani and Anderson, 1989; Rossiter, 1991). In some cases transgenerational effects extend back to grandmothers (Mosseau and Dingle, 1991). In a hierarchical context this means the highest levels of epigenetic control may not even reside in an organism, but instead resides in an ancestor. Kirkpatrick and Lande (1989) pointed out that maternal inheritance can produce evolutionary phenomena not predicted by simple Mendelian evolution in single generations (e.g. responses in the opposite direction to immediate selection, continued evolution after removal of selection and oscillations in evolutionary responses). Where maternal effects act strongly on a particular trait (e.g. propagule size), the expression of the paternal genome on offspring phenotypes may be delayed for one generation (Reznick, 1981).

Maternal influence constitutes a transgenerational phenotypic mechanism of great evolutionary importance. For example, different developmental results could be invoked via simple changes in the amount of yolk supplied to eggs or via specific chemical signals associated with maternal mRNA or proteins, such as sequestered transcription factors. The direction of spiral shell coiling in snails like *Lymnaea* is determined by the maternal genotype in this way (reviewed by Arthur, 1984, pp. 132–136). Other key characteristics of offspring influenced maternally include diapause, migration and reproductive investment (Groeters and Dingle, 1988). Simple changes in yolk content can alter major ecological correlates, such as competitive ability, predator defence, stress resistance, growth rates and developmental schedules.

Recent advances in this area include the direct manipulation of egg yolk in lizards and other organisms (Sinervo, 1990, 1993; Sinervo and Huey, 1990; Bernardo, 1991; Sinervo *et al.*, 1992). Larger eggs produce longer incubation periods, larger hatchlings, slower juvenile growth rates and faster sprint speeds. Such variation in egg quality has been shown to have consequences related to larval activity and feeding behaviour, growth rates, body sizes and adult dispersal patterns in moths (Wellington, 1965, 1980; Capinera, 1979; Barbosa, Cranshaw and Greenblatt, 1981; Rossiter, 1991). Shibata and Rollo (1988) also found that egg size was associated with developmental consequences in terrestrial slugs, larger eggs producing slower growing individuals with altered maturation times. Such mechanisms may allow mothers to deploy offspring with variable developmental characteristics that would allow spreading of risk with respect to seasonally variable resources (Shibata and Rollo, 1988; Rossiter, 1991). Even local adaptation of populations to their environment may fulcrum on variation in yolk allocation to eggs (Sinervo, 1993).

Although larger offspring may be produced by organisms that have higher levels of resources, evidence for adaptive modulation of this characteristic comes from species that do the opposite. Thus the pillbug, *Armadillidium*, produces larger offspring when food is scarce, presumably to improve offspring survival (Brody and Lawlor, 1984). Sinervo *et al.* (1992) manipulated egg size to generate lizards of extreme small and large size. Survival was maximized by larger eggs in females and intermediately sized eggs in males. Parental fecundity was favoured by small eggs. Thus, optimal egg size was balanced by alternative selection pressures for fecundity or survival.

The genetic templet of social insects may generate four developmental stages, two sexes, a worker-queen dichotomy and several specialized sterile phenotypes. Caste differentiation represents an aspect of parental or sibling manipulation of the offspring phenotype (Oster and Wilson, 1978; Holldobler and Wilson, 1990; Wheeler, 1991) of remarkable evolutionary significance. Stearns (1992) and Roff (1993) reviewed the trade off between maternal investment in offspring size *versus* offspring number, but in most considerations, the possible impact of this trade off on progeny quality has been virtually ignored.

Among plants, maternal influence may impinge upon seed characteristics related to dormancy, dispersal, germination, size, competitive ability, cold resistance and defence (Capinera, 1979; Roach and Wulff, 1987). Cytoplasmic contributions from the mother are also responsible for transmitting lineages of cell organelles (e.g. mitochondria and chloroplasts) as well as numerous cytoplasmic symbionts such as the bacteroids of cockroaches.

Of particular relevance to parental influence is the rapidly expanding knowledge suggesting a developmental role for the phenomenon of genomic imprinting (Jablonka and Lamb, 1989; Peterson and Sapienza, 1993). In this case, the pattern of gene expression inherited from a parental genome is partially a function of genetic constitution, but is also dependent on the structural patterning of chromatin which may vary among parents (Chapter 2). Genomic imprinting thus represents a secondary signal of inherited developmental information that is not directly associated with genetic constitution. This information can remain stable over numerous generations. Because this information system is non-cytoplasmic, it provides a mechanism for both eggs and sperm to influence progeny in significant ways. The small cytoplasm of spermatozoa has previously excluded them from serious consideration in terms of non-genetic information transfer.

The role of genetic imprinting in deriving sexual dimorphism is now well established (Chandra and Nanjundiah, 1990; Peterson and Sapienza, 1993). Whether it might also be deployed to alter the phenotypes of offspring in response to ecological contingencies has not been considered at all. It seems highly likely, however, that such a clear route for short-term direct modification of offspring by parents has been thoroughly explored by natural selection. All that would be required is that the regulatory machinery controlling local chromatin structure be linked in a specific way to environmentally derived signals. Jablonka and Lamb (1989) have explored this theoretical idea to some degree. Paternal effects are also known in plants but have been assumed to act via the cytoplasm (Roach and Wulff, 1987).

The possible importance of genomic imprinting is underscored by evidence emerging from hybridization studies. Genomic imprinting may be of nearly equal significance to actual genetic constitution in the expression of phenotypes derived from interspecific crosses (Heslop-Harrison, 1990). This is clearly of relevance to the earlier discussion of cross-talk in sexual lineages. The phenotype of one parental race or species may dominate that of the other (sometimes via repression of entire chromosomes) and genomic imprinting is revealed where divergent results are obtained depending on species-by-sex interactions. For example, a male donkey and female horse yield a mule phenotype, but a male horse and female donkey yield a hinny, a remarkably different outcome. Imprinting might also be important in regulating degree of expression in tandems of duplicated genes or with respect to movements of transposable elements.

Of particular importance to lineage selection is the asymmetry between parental and offspring control of epigenetics. Parents have the potential to manipulate offspring in ways that may decrease the fitness of some, in order to maximize global reproductive success. Thus, the evolution of eusociality can be viewed from a standpoint of manipulation rather than kin theory. Similarly, where environments vary, parents may produce offspring of different types, some of which have lower chances of success but larger potential rewards (e.g. dispersal *versus* residential forms).

Among plants, the tissues surrounding the seed embryo and endosperm are maternally derived. Thus, the parent has free reign to provide dispersal-enhancing structures, and these tissues may also control diapause. In many species, dichotomous types of seeds are produced (Capinera, 1979; Westoby, 1981; Roach and Wulff, 1987). Without such manipulation, no individual might choose to engage in a high risk or delayed fitness behaviour as long as safer alternatives were available (Westoby, 1981). Such diversification of offspring may reduce the variance in long-term fitness of lineages (Westoby, 1981; Kaplan and Cooper, 1984).

6.4 POLYMORPHIC SWITCHES

A prediction of the lineage paradigm is that some adaptations may be expressed only in certain environments (e.g. individual plants that have different leaves or other structures in water or in air), or only in some generations (e.g. locusts and aphids). The phenomenon of divergent phenotypes is widespread and suggests some possible ways that lineage selection may operate. Imagine a species that exhibits very different phenotypes in dichotomous environments, a phenomenon termed developmental conversion by Smith-Gill (1983). For example, a single species of waterstrider may have either winged migratory forms that exploit temporary, uncertain habitats or wingless residential forms that exploit more permanent environments (Shapiro, 1976).

Population geneticists examine how selection in these environments favours appropriate alleles and eliminates inappropriate ones to achieve the fittest phenotype and in many species this model is applicable. What is ignored by this perspective, however, is that numerous other species, even close relatives in the same groups, achieve similar polymorphism without microevolution (Shapiro, 1976). Such species contain the necessary code for both phenotypes and use environmental cues to direct the relevant epigenetic trajectory. Secondly, the genic view fails to appreciate that the circulating allele may only constitute a switch for extensive epigenetic organization for either phenotype that is shared in either genotype.

The real problem of the evolution of bifurcating epigenetic systems, which include such switches, is largely ignored by a genic focus which only considers changes in frequency of the switch itself. Thus, Roff (1986), in his excellent review of the evolution of wing dimorphism in insects, eliminated all those examples where environmental factors controlled the observed polymorphism. Even though such control is via a genetic mechanism (Schlichting, 1986; West-Eberhard, 1989), such systems are often not considered genetic.

Bacteria provide a model illustrating the evolution of genetic switches to obtain plasticity. Many bacteria are capable of utilizing specific nutritional substrates, but production of the requisite enzymes is repressed unless the substrate is actually present as in the classic example of the 'lac' operon in *Escherichia coli* (Jacob and Monad, 1961a,b). Clarke (1983) reviewed experiments with *Pseudomonas aeruginosa* where strains were selected to metabolize novel amides with progressively more complex structure. Not only were strains obtained that produced the new required enzymes, but regulatory genes specific to the new substrate also evolved to modulate constitutive expression. Thus, the evolution of novel plasticity, at least for biochemical phenomena, occurs on scales only slightly longer than the conventional microevolutionary model. It seems likely that such models can be extrapolated to the more complex epigenetic hierarchies of eukaryote regulatory systems. In such systems the possibility that a switch can be effected via genomic imprinting should also be considered.

One crucial difference between plastic and genic mechanisms is that even the simplest plastic system is multi-locus whereas genic systems might involve alternate alleles at a single locus (if only the switch is considered). There are probably many different ways to obtain plasticity, but models based either on pleiotropy (changes in the expression of one gene) or epistasis (changes associated with interactions among genes) are best supported by empirical studies (Scheiner, 1993). Pleiotropy might derive plasticity where the expression at one locus is modulated by input from the environment. For epistasis, two loci linked to an environmentally sensitive regulatory gene is conceptually simple. In fact, the hierarchical nature of epigenesis suggests that there are likely to be complex alternative trains of epigenetic machinery downstream from any given switch. Survival of the plastic lineage is a consequence of among-locus organization rather than of any particular locus or of a specific allele. Both lineages may have exactly the same phenotypic capabilities but use different mechanisms to achieve them (Figure 6.2). Clearly, the relevant organization for plasticity is selected over a longer time frame than that of an organism in one generation. In both lineages, each organism expresses only one phenotype in its life-time and these may be the same in each generation. The mechanisms are invisible in single

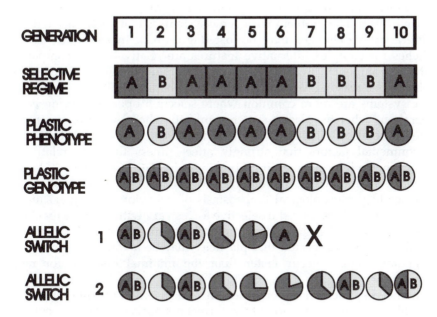

Figure 6.2 Genetic structure of a plastic switch mechanism compared with a system employing circulating alleles.

generations, their impact only being revealed on multigenerational time scales or across varying environments.

For adaptive plasticity to evolve and be maintained, natural selection must assemble or scrutinize the entire mechanism by integrating across generations expressing particular aspects. The evolution of a switch compartmentalizes the genome, allowing accumulation of coadapted modifiers to be selected for each divergent phenotype during different generations (West-Eberhard, 1989). Bacterial studies demonstrate that the evolution of plasticity is multigenerational since modification of regulatory components occurs across successive individuals (e.g. Wilson, 1976; Clarke, 1983). The presence of pre-existing loci both for making functional enzymes and for their regulation, however, pre-adapts bacteria for rapid shifts in regulated enzyme capabilities, since the requisite structural framework is already present. The evolution of the original switching mechanism itself involves structural modifications that would require even longer time scales (i.e. it may be easier to modify existing operons than evolve completely new ones).

This discussion begs the question as to why some organisms have their adaptive responses organized intra-genomically on the basis of phenotypic flexibility (among loci) whereas others realize similar capabilities via inter-genomically circulating alleles (Bradshaw, 1965; Caswell, 1983). The problem with genic systems is that alleles needed for future adaptive changes may be lost (Figure 6.2). In fact, such allelic switches for complex alternative features are relatively rare compared with environmental switches, and the instability of allelic switches is probably one reason (West-Eberhard, 1989). Shapiro (1976) observed that genic systems are more common where selection operates during each generation, whereas plastic systems prevail where selection is inter-mittent (see also Bradshaw and Hardwick, 1989). It is also likely that if environmental factors that strongly affect fitness fluctuate relatively rapidly, then a lineage obtaining fitness via circulating alleles would also be relatively inefficient. Developmental plasticity or mechanisms for resilience (e.g. physiological homeostasis or behavioural compensation) would be strongly favoured if selective forces vary with frequencies close to or less than the generation time. In terms of spatial variation, Bradshaw (1965) suggested that plasticity may be favoured where environmental changes occur faster than the minimal distance for race formation.

Another key factor is information. If a clear environmental signal is available to ensure expression of the correct phenotype at the correct time and place, then phenotypic plasticity is favoured (Harvell, 1990; Scheiner, 1993). If a clear signal is unavailable or has not been discovered the systems may be forced to rely on microevolutionary responses or adaptive 'coin flipping' (Kaplan and Cooper, 1984). In this case, alternative epigenetic trajectories are produced without direct environmental linkages and selection is allowed to determine their success. Where some information is available but uncertain, the coin may be biased in different directions.

An obvious factor that is ignored in many reviews is the mobility of the organism. Plants, colonial invertebrates and other organisms of low mobility (such as snails, or some pond animals) do not have a high degree of habitat choice or the ability to move if local conditions change. Intertidal organisms that settle from planktonic larvae also have relatively few degrees of freedom for habitat selection or avoiding changes. Such organisms may be selected for high physiological and morphological plasticity because they do not have a sufficient behavioural buffer. Organisms that stay in a single local habitat for most of their lives might favour developmental deployment of diverse adaptive suites, but mobile organisms that move around numerous habitats, or that precisely track a narrow niche, have less need for such capabilities.

Clearly, discrete polymorphism obtained by allelic switches or environmental signals are closely related phenomena. Both involve the diversification of the genetic templet to obtain two divergent adaptive suites. This idea underscores that circulating alleles may not contain all the relevant information for alternative forms, but may often constitute signals that switch developmental trajectories down different arms of a hierarchy that is present in all individuals. The significance of this phenomenon is that alternative adaptive phenotypes may be evolved simultaneously on either side of a switch. Given appropriate conditions, or perhaps a critical degree of divergence, one of the alternative phenotypes may be abandoned, or the two phenotypes may consolidate into separate species (Matsuda, 1987; West-Eberhard, 1989; Wilson, 1989; Meyer, 1990; Moran, 1991, 1992a). West-Eberhard (1989) has assembled very convincing evidence that such polyphenic speciation may be a widespread process.

Polymorphic speciation does not require the existence of any intermediate phenotypes of low fitness. Rather than Wright's (1980, 1982a,b, 1988) view of evolution taking place on an adaptive landscape, or even Simpson's (1944) modification of this concept to that of a choppy sea, the evolution of lineages may be more analogous to adaptive whirlpools that spin off new eddies that are already full-blown adaptive suites. That is, genetic templets may occasionally give birth to new adaptive templets that constitute one arm of a bifurcating developmental hierarchy.

6.5 NORMS OF REACTION AND PLASTIC DEVELOPMENTAL TRAJECTORIES

Plasticity is not limited to cases of extreme phenotypes or intermittent selection. Where an environmental factor has a continuous rather than discontinuous character, a single genotype may show gradual phenotypic adjustments (Figure 6.3A). Such a response is termed a norm of reaction (Schmalhausen, 1949; Stearns, 1989, 1992). Polymorphic switches are sometimes viewed as an extreme case of such a framework. It is also important to differentiate between what might be regarded as adaptive plasticity and variation associated simply with constraints on the function of the phenotype. This is not as easy as it may sound. For example, growth rate may vary according to the quantity and quality of available food, so variation associated with such constraint might not constitute adaptive plasticity, a point emphasized by Smith-Gill (1983). However, if the expressed level of growth is part of an integrated strategy (e.g. reduced growth and reproduction but increased investments in longevity with food shortages: see Chapter 10) then some element of adaptation may be involved.

In aquatic snails, for example, dietary restriction reduces productivity, but larger species may defer reproduction to support growth while smaller species may support reproduction at the expense of growth (Rollo and Hawryluk, 1988). Such results are consistent with the general hypothesis that iteroparous species may reduce reproductive effort under poor rations whereas semelparous species may not (Baird, Linton and

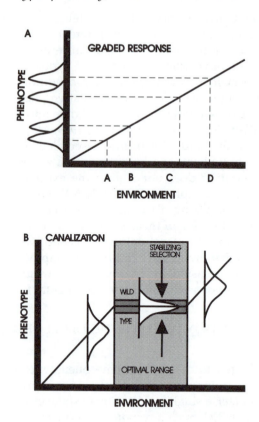

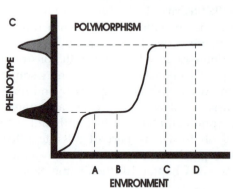

Figure 6.3 Various norms of reaction deriving (A) graded responses, (B) canalization and (C) polymorphism. The graded response is modelled after Gupta and Lewontin (1982) with permission of *Evolution*, Allen Press. The concept of canalization was adopted after Rendel (1967) with permission of Academic Press. The representation of polymorphism is similar to an empirical example in Eberhard and Gutierrez (1991).

Davies, 1987). Adaptive plasticity to offset the impact of inadequate diets has already been mentioned. Baird, Linton and Davies (1987) found evidence supporting adaptive plastic responses to resource variation in a leech, but allocation also varied according to mortality risk associated with temperature. Thus, the plasticity of traits may respond simultaneously to more than one environmental factor.

Norms of reaction are commonly associated with environmental factors, such as temperature, salinity, pH, light intensity, photoperiod, moisture, aspects of nutrition or seasonal cycles. It is often convenient to restrict the definition of a reaction norm to a single aspect of the phenotype. Thus, maturation time may display a norm of reaction to temperature or the concentration of some chemical associated with a known predator. Size of reproductive propagules may likewise display a reaction norm in response to a factor such as maternal nutrition. Clearly, where reaction norms exist for many such factors, heterogenous environments may elicit numerous plastic phenotypes, differentially exposing associated genetic variation to the action of natural selection or perhaps genetic assimilation (Stearns, de Jong and Newman, 1991).

Norms of reaction are not limited to simple morphological or physiological features, however, but may also involve the complex interplay of life-history attributes such as maturation times, body sizes and longevity. Stearns and Crandall (1984) documented that the entire constellation of life-history attributes associated with an organism's adaptive suite may vary along plastic developmental trajectories according to the realized growth potential of the organism (see also Calow, 1984; Rollo and Shibata, 1991). Such trajectories represent the mutual integration of several reaction norms governing different features (Schlichting, 1986, 1989a) and it is convenient to retain this term.

In such cases, reaction norms are sometimes represented as covariation between life-history features (e.g. age *versus* size at maturity) divorced from the environmental signal influencing each component. Patterns in such relationships reflect the integration of life-history features at higher levels of the phenotype. To take a specific example, the optimal age at maturity is both a function of growth rate (larger sizes = greater productivity) and mortality rates (Stearns and Koella, 1986; Perrin and Rubin, 1990). Because these rates vary environmentally, the evolution of plastic responses is favoured, although very different tactics may be expected depending on ecological contingencies and the nature of the trade off structure. Resources are likely to play a dominant role in the plastic modulation of life-history tactics via impacts on productivity (Rollo and Hawryluk, 1988; Ford and Seigel, 1989). Rollo and Shibata (1991) documented such plastic developmental trajectories in a terrestrial slug with low genetic variation. Phenotypes differed so remarkably with food quality that they could be mistaken for different species. Shifts in phenotypic features relevant to dispersal, sexual weaponry, defence, reproductive effort and diapause may also be tied to individual foraging success or performance (Goodman, 1979; Wellington, 1980; Eberhard, 1982; Pettifor, Perrins and McCleery, 1988; Parejko and Dodson, 1991). In other cases, the response may be directly linked to the density of conspecifics rather than resources (e.g. Collins and Cheek, 1983).

Bradshaw (1965) and Scheiner (1993) stressed that plasticity might be limited to only a single feature of an organism, and that it can be selected independently of other features or even of the mean of the particular feature under consideration. This may be true to some extent, but Schlichting (1986, 1989a) convincingly argued that a single reaction norm does not evolve in a vacuum. Modification of one norm may enhance or interfere with the adaptive expression of others. If individual genes are not independent of their genetic background, it is less likely that larger organizational pieces like reaction norms would be either. Consequently, selection must also involve higher-order integration among norms as well as with fixed or canalized phenotypic features. Such integration is fundamental to both the basic coordination of development and the realization of an adaptive suite of features being expressed under appropriate conditions.

The form of plastic developmental trajectories may be radically different depending on the environmental factor involved. An excellent example was described by Lints and Lints (1971) who found that the rate of development slowed and adult body size declined in *Drosophila* reared at increasing larval densities. Similar results are obtained by manipulating food concentrations (Gebhardt and Stearns, 1988). An inverse relationship was obtained across a range of temperatures, however, larger body sizes being associated with slower development at colder temperature (Lints and Lints, 1971). Longevity was most closely related to development time, and consequently was associated with larger flies in the cold or smaller flies at high densities. Similar results were obtained in bruchid beetles (Moller, Smith and Sibly, 1989b). Given that the complexity of multifactor environments may yield unique selection regimes in any locality (Jones, Leith and Rawlings, 1977), organisms may show complex adaptive adjustments that have been generally ignored by a genic perspective.

There is rapidly accumulating evidence that the expression of life-history constellations in particular environments may largely represent a plastic adjustment of multivariate fitness features to local environmental conditions (e.g. Schmalhausen, 1949; Eberhard, 1982; Marshall, Levin and Fowler, 1986; Trewavas, 1986; Kaitala, 1987, 1991; Sultan, 1987; Holliday, 1989b; Rollo and Shibata, 1991; Chapin, 1991; Groeters and Dingle, 1988; Ford and Seigel, 1989; Newman, 1988a,b, 1989, 1992; Schlichting, 1989a; Van Noordwijk, 1989; Moller, Smith and Sibly, 1990; Perrin and Ruban,

1990; Sinervo, 1990; Stearns, 1989, 1992). In plants, allocation strategies tend to be simply proportional to size, but Hickman (1975) found that *Polygonum cascadense* allocated relatively more to reproduction under increasing stress.

Unlike discrete switches, many of the same genes may likely be involved in phenotypic expression along norms of reaction (Hillesheim and Stearns, 1991). In a regulatory framework, the degrees of expression of various genes may lead to coordinated changes in their penetrance along environmental gradients. Coordinating a multiplicity of traits along a plastic developmental trajectory is clearly a difficult problem. It is possible, however, to obtain integrated flexible responses by linking most secondary features to a few key endocrine signals (Ketterson and Nolan, 1992). For example, the adaptive shifts in numerous life-history tactics do not require that each feature interprets the ecological complexity realizing foraging success. Instead, independent aspects could be coupled to the individual growth performance of the organism. Dangerfield and Hassall (1992), for example, found that variation of life histories in the pillbug, *Armadillidium vulgare*, arose as plastic responses to individual growth conditions. Such organization would have the advantage of automatically accounting for numerous impacts on growth rate (such as differences in genetic growth potential, food quality, temperature or variations in symbiotic populations that help digest food). Numerous such constraints would be converted to one or a few physiological signals. Integration based on fundamental aspects, such as growth performance, age or body size could account for many of the very consistent trends in organism design that emerge from allometric analyses (Peters, 1983; Calder, 1984; Schmidt-Nielsen, 1984). Most major life-history trends, for example those associated with the r–k continuum, are largely related to body size.

An obvious way for such integration to be achieve epigenetically is shifting development according to titers of key hormones (Ketterson and Nolan, 1992). Trewavas (1986) and Sachs (1988) developed this idea with respect to plants. In social insects, caste determination is associated with levels of juvenile hormone (Wheeler and Nijhout, 1981; Matsuda, 1987; Robinson, 1992), and in some species, various castes conform predictably to allometric exaggerations of particular features with body size (Huxley, 1932; Wheeler, 1991). Nijhout and Wheeler (1982), Hardie and Lees (1985) and Moran (1992a) document the hormonal coordination of numerous phenotypic trajectories in insects (e.g. associated with metamorphosis, phase determination in aphids and locusts, colour polymorphisms and caste determination). Some maternal effects also involve hormones (Mousseau and Dingle, 1991). Our work with the Supermouse documents a similar situation for growth hormone (Chapter 8).

It would be interesting to know whether caste differentiation in naked mole rats is linked to growth hormone. The fact that small body sizes are associated with increased behavioural activity and decreased fecundity in mice suggests that such reaction norms might serve as a pre-adaptation for differentiation of worker and reproductive castes in social animals. We tested this idea in carpenter ants as well, and found that smaller workers were dramatically more active than larger workers or queens (M. Ham and C.D. Rollo, unpublished). Interestingly, Weider and Pijanowska (1993) noted that *Daphnia* expressing smaller sizes due to the detection of fish were notably more reactive and alert. Thus, such associations between smaller sizes and increased activity may be rather pervasive. Hormonally coordinated responses to stress have also been documented in plants (Trewavas, 1986; Chapin, 1991).

A significant feature of coordination derived via endocrine control is that the expression of a given feature, although perhaps responding to some environmental constraint, is not simply the imposition of a corresponding developmental constraint. Instead, the epigenetic or physiological response may vary according to the long-term fitness of the organism, and particular strategies may be expressed in a feedforward way that anticipates environmental changes, and allows amelioration of their impact (Rollo and Hawryluk, 1988). In other words, stress is met with some specific and adaptive phenotypic expression (Trewavas, 1986; Chapin, 1991). Such capability will be important whenever responses require time to enact, and where the strategy would be ineffective if engaged only in response to direct immediate impacts. Matsuda (1987) and McKinney and McNamara (1991) have emphasized how variation in the mix or timing of hormones provides corridors for coordinated phenotypic evolution.

6.6 CANALIZATION

The phenomena of plasticity and canalization are essentially opposite sides of the same coin. In fact, graded responses, canalization and discrete polymorphism may all be conceptualized as variants of complex norms (Figure 6.3). Waddington (1957) envisioned development as a ball rolling down a landscape of bifurcating hills and parallel valleys. Not only could the ball make alternative choices at particular points, but the steepness and height of surrounding hills could limit the trajectory to a narrow path or prevent lateral movements to alternative pathways to various degrees. These last aspects Waddington (1957) termed canalization – the precise tracking of a particular epigenetic pathway. Where expression of a particular feature within certain specifications is advantageous, then selection should act to minimize environmental disturbance of the epigenetic course, minimize disruptions of required resources and ensure that alternative genotypes converge on the same phenotypic endpoint.

Thus, canalization may minimize the phenotypic impact of both environmental and genetic variation (Waddington, 1957; Rendel, 1979; Levin, 1988). Physiological homeostasis, compensatory responses to nutritional deprivation and developmental canalization are thus widespread adaptive aspects of organism design; canalized monophenism appears to be a more pervasive phenomenon than polymorphism (Lloyd, 1984).

Waddington (1957, 1959) envisioned dominance relations, heterozygosity and numerous regulatory cross linkages as contributing to this phenomenon. At a higher level, the requirement for integration of particular features means that canalization involving the entire adaptive suite is likely to be pervasive. That is, canalization of a particular epigenetic trajectory, or canalization of alternative plastic developmental trajectories involves the integration and modulated expression of reaction norms. Selection for canalized developmental systems may arise via stabilizing selection which tends to eliminate any phenotypic expression outside the range of the wild type (Bradshaw, 1965; Falconer, 1981; Zeng, 1988). This results in a normal distribution of phenotypes. Alternatively, requirements of adaptive integration (particularly key life-history trade offs) may limit the phenotype to a narrow range of normally distributed features (Salthe, 1975; Schluter and Smith, 1986). On short-time scales, stabilizing selection may dominate over directional selection (Zeng, 1988).

Where there is selective advantage converging on a particular phenotypic optimum, one obvious mechanism of canalization might be via conservation of precise propagule sizes. For example, it is well established that the seed sizes of numerous plants are relatively refractory to environmental or maternal variations. It is generally true that offspring size is less flexible than offspring number (Bradshaw, 1965; Lloyd, 1984; Stearns, 1992). Waddington (1975, pp. 256–259) recognized that relatively invariant wild type might be realized by canalization. Moreover, the mapping of phenotypes onto fitness may allow a relatively broad range of equal fitness and much greater genetic heterogeneity than might otherwise be expected (see also Levin, 1988). Such phenomena might well explain the extended periods of phenotypic stasis exhibited by many species (Kerszberg, 1989; Levin, 1988). In fact, Rendel (1967, p. 153) suggested that canalization should result in short periods of rapid evolutionary change interspersed with long periods of stability, exactly anticipating Eldredge and Gould's (1972) thesis of punctuated equilibria.

Rendel (1967) proposed a quantitative genetics model for canalization, and also considered the genetic regulatory mechanisms that might contribution to this phenomenon. He considered some increasing distribution of a developmental gene product which could perhaps be some morphogen or an endocrine hormone. Phenotypic responses to this substance involve thresholds such that above a given lower threshold and below another upper threshold the same phenotype is always produced. Rendel (1967) considered that modification of regulatory thresholds could also vary dominance relations among genes, one mechanism of masking genetic variation in phenotypes. Rendel's (1967) model can be represented as mapping across a range of environments (Figure 6.3B). Phenotypes expressed across the range of the intervening plateau are relatively invariant (the wild type) even though a large reserve of low penetrance genetic variation may be present (Schmalhausen, 1949). Consequently, considerable existing genetic variation may not be exposed to natural selection (Sultan, 1987). Canalization is associated with the phenotypic plateau which Rendel (1967) considered to reflect the integration of features defining species. A system may be pushed off the plateau by an unusual environmental factor, or an environmental extreme that affects the developmental system. Usually some genetic variation is expressed across the plateau so that strong directional selection could still eventually push the system above or below the thresholds for canalization.

Thompson (1991) stresses the idea that plasticity may continually be adjusted by selection rather than resisting it. Mutations may sometimes also have a destabilizing effect (reviewed by Scharloo, 1991). However, developmental systems typically have considerable resilience and robustness that can compensate or mask the expression of new mutations. This is particularly true in higher vertebrates, where progress in molecular genetics has been impeded by difficulties in detecting relevant mutations. Kerszberg (1989) developed models showing how such features improve the fitness of lineages.

Outside the range of canalized development, previously unexposed genetic variation may be fully expressed and organizational integration may be weaker (Schmalhausen, 1949; Rendel, 1967, 1979; Scharloo, 1991). Responses of the developmental system to environmental factors may be greatly increased, but such responses may be non-adaptive morphoses rather than functional responses (Schmalhausen, 1949). In bacteria, analogous shifts occur when inducible enzymes escape regulatory control and are expressed constitutively (Lerner, Wu and Lin, 1964). In general, alternative adaptive phenotypes tend to be fully expressed in response to some particular threshold associated with an environmental factor whereas morphoses tend to respond incrementally to environmental variation.

In the context of a regulatory hierarchy, what this means is that potential alternative pathways are revealed to selection when a system falls off the canalized plateau. If any of these variants are superior in the new environment or against the new genetic background, a new organization could be consolidated. This constitutes genetic assimilation in the case of a novel environment, or genetic coadaptation in the case of a new mutation. The expression of hidden genetic variation following decanalization could well fuel genetic revolutions or speciation events (Levin, 1988). If there is no selective advantage of a new organization, the system may be selected to adjust for the environmental or genetic disturbance and restore the wild type (Lerner, 1954). In this case, the breadth of the canalized plateau may still be extended. Waddington (1957, pp. 185–186) suggested that if canalization resists selection, then applying stress to a phenotype might be a valuable adjunct to improve selection programs. The proof of this has been dramatically illustrated in *Drosophila* where selection for increased longevity can only be obtained if flies are reared at high density (Chapter 10). Clearly, canalization is an adaptive aspect forged on a multigenerational scale. Scharloo (1991) points out that the model proposed here may be too simplistic in that more than one epigenetic response may be involved, or particular subunits of a morphological attribute may have independent thresholds.

A key component of phenotypic fitness will be the environment influencing it. In most considerations, the ability of an animal to assess the favourability of a habitat and select appropriate conditions is ignored. Clearly in sexual species with both genetic variation and phenotypic plasticity, habitat choices will be a major determinant of success. Rausher and Englander (1987) make the point that such choice may be a relatively sophisticated and plastic attribute in higher vertebrates, but may have a simpler basis in lower forms like insects. Because behaviour represents a plastic mechanism that might buffer the genotype in this way, this area is probably best considered in the context of plasticity and canalization (see also Scheiner, 1993).

There may be behavioural equivalents to developmental canalization and decanalization. Thus, Geissler and Rollo (1987) found that cockroaches fed high quality diets linked to particular olfactory cues, showed strong avoidance of diets associated with novel cues. However, if the animals had nutritional deficiencies, they readily added novel foods to their menu. Reduced host discrimination is commonly observed in animals subjected to starvation (Hoffmann and Parsons, 1991). Behavioural shifts, perhaps driven by food shortages or competition, may be a major factor exposing phenotypes to novel selection regimes (Gottlieb, 1992).

It is perhaps a general rule that constancy is one feature that requires flexibility to maintain it in others (the Law of Requisite Variation). The evolution of perceptual and cognitive systems must represent the ultimate in plastic systems in organisms, and these probably yield fitness benefits largely by stabilizing the phenotypic milieu with respect to resources, risks and operating environment (Real, 1991). There will be costs and constraints associated with features designed to acquire and process information (e.g. Orians, 1981). In this regard, higher organisms appear to be allocating relatively larger amounts of resources to perceptual and cognitive features and may derive benefits in terms of both performance and persistence.

One rather general reflection of canalization in nature is species-specific body size. Genetic, internal and environmental sources of variance may be reduced as body sizes converge on a relatively conserved target. Riska, Atchley and Rutledge (1984), in an analysis of targeted growth in mice, found that an important factor was antagonism between parameters of rates *versus* duration (e.g. animals with faster early growth might mature earlier). Decanalization in such cases may separate important aspects of rates and their timing, allowing changes in life-history evolution via heterochrony (McKinney and McNamara, 1991).

When environment constrains performance, animals may make compensatory adjustments to maintain relatively constant trajectories. Key adjustments include increased food intake on low quality diets, and changes in respiration rates that may effect savings (Rollo and Hawryluk, 1988; Surbey and Rollo, 1991). Following periods of food shortage such adjustments may be effected to yield compensatory catch up growth that offsets the period of retarded performance (Rollo, 1984; Riska, Atchley and Rutledge, 1984). Such phenomena argue that the developmental program has phenotypic targets that are determined as long-term evolutionary set points. In this framework, the hypothesis that all elements related to fitness are maximized may be mistaken. For example, the optimal foraging hypothesis that organisms maximize their net intake of energy (Pyke, 1984) may be incorrect if feeding and metabolism are regulated to achieve growth trajectories set to long-term evolutionary targets.

A crucial aspect of canalization relevant to evolution is that the greatest degree of canalization is obtained in optimal habitats. Rendel (1967) suggests that the fittest form is an intermediate in which various developmental factors are balanced. As any one aspect of the environment exceeds the optimal range (i.e. becomes stressful), canalization may decline and increasing phenotypic variability may be expressed. Levin (1970) expressed similar ideas some time ago. He also suggested that canalization may be stronger in heterozygotes and consequently inbreeding or population bottlenecks would tend to disrupt canalization and reveal hidden selectable variation. A similar result might be obtained by directional selection on any one feature. As the phenotype exceeds the wild type range, canalization may be increasingly disrupted and greater genetic variation may be expressed. Within the optimal range of environments, canalization and other homeostatic systems may largely neutralize minor environmental and genetic impacts on phenotypic variation. Within a single optimal environment or narrow range, plasticity and genotype–environment interactions cannot be detected (Geisel *et al.*, 1982b).

6.7 GENETIC ASSIMILATION

A key aspect of phenotypic evolution is the tug of war between environmental and intrinsic factors shaping epigenesis (McKinney and McNamara, 1991). Genetic assimilation is perhaps the key process associated with such dynamics. As such it represents a potentially important mechanism for the evolution of regulatory organization that warrants much greater attention than it has been awarded (e.g. Williams (1966a) suggested it was unimportant). The idea that the environment might be involved in genetic changes, in any manner other than direct selection, has been generally misunderstood or rejected because it smacks of Lamarkianism. While it is true that phenotypic modifications imposed by the environment (e.g. cutting off the tails of mice) cannot be inherited, environmental impacts that induce the genotype to express variant epigenetic pathways do represent a route with important evolutionary implications. Hypothetically for example, selection of mice that express variation in reduced lengths of tails in response to some drug (perhaps low doses of thalidomide during gestation) could lead eventually to mice that express no tails to even small amounts of thalidomide.

In fact, the literature on genetic assimilation predicts that at this point, the drug could be discarded, and the mice would remain tailless in subsequent generations. Thus, the environment acts, not as a selective agent, but as a crucible that determines the type and amount of genetic variation that is exposed. In this case, the selection of tailless mice might be very difficult without the coaction of the drug. Selection, however, is not acting on some new feature that the environment imposes on the phenotype. Instead it acts on a feature that the genotype itself is expressing. The key is that the environment modulates this expression by interacting with the genome via epigenesis. Like any other Darwinian process, there must be genetic variation (Waddington, 1957). There is no mystery associated with genetic assimilation as soon as it is understood that not all genetic variation is expressed in all environments.

There are actually two closely related phenomena of relevance here. The 'Baldwin effect' (reviewed by Simpson, 1953b; Gottlieb, 1992) also holds that characters acquired by individuals may be reinforced or replaced by hereditary factors following selection. Waddington (1957) distinguished this from genetic assimilation on the basis of the source of variation. He envisioned genetic assimilation as acting on existing variation that was simply revealed to selection by environmental factors. The Baldwin effect differs in that it relies on incorporation of new mutations into phenotypes expressed in novel habitats.

The real significance of genetic assimilation is that new mutations are not necessary for achieving major genetic reorganizations. If an environmental factor can be found that exposes previously hidden epigenetic potentials, major phenotypic modifications can be achieved with existing regulatory variation. New mutations may similarly expose novel regulatory variation.

To give a more realistic scenario relevant to plasticity, consider a population of mice that are exposed to a frigid environment. The tails of mice are naked because they function as a heat radiator and are valuable for thermoregulation (Falconer, 1981). In cold environments, selection may act to reduce tail length to conserve heat. However, mice are already adaptively organized plastically to grow shorter tails when raised in the cold, and longer tails in hot environments (Falconer, 1981). Thus, selection for reduced tail size in the cold will select those individuals that intrinsically have the shortest tails but also those with the greatest epigenetic ability to reduce tail size in response to cold (i.e. selection is on a specific aspect of phenotypic plasticity). Thus, to select mice for short tails, a cold environment might expose hidden genetic variation in temperature-sensitive epigenetic programs for tail length, greatly facilitating the selection response. Waddington (1959) demonstrated such an effect with respect to salt tolerance in *Drosophila*.

If tail size remained shorter than the norm on return to a warm environment, then this would represent a case of genetic assimilation of an epigenetic response to low temperatures. This kind of selection regime is not as controversial as the thalidomide example and points out how this process could be widespread in nature. The evolution of short-tailed northern rodents (e.g. lemmings, *Microtus*) might have been greatly facilitated by existing mechanisms that expose the requisite variation in response to temperature (i.e. adaptive norms might reinforce or steer microevolutionary responses).

Gause (1942) made the interesting observation that where phenotypic plasticity in responsiveness was high, such 'adaptable' types responded faster and to a greater extent to natural selection than less environmentally flexible forms. Moreover, the plastic types tended to have initial conformations that were further from that finally realized following successful selection (i.e. the most successful types were often the opposite of those predicted from their initial phenotypic distribution). This phenomenon has been virtually ignored since.

In other cases, the phenotype is obtained by a canalized epigenetic trajectory. Thus, if mice showed no genetic variation in tail length, or there was no variation in the plastic response to temperature, then selection could not act on it. Some potential pathways may only become visible to selection under unusual conditions. A change in habitat enforced by dispersal to a new region, a local change in environment, or a mutation in the genes relevant to habitat choice (e.g. host selection in phytophagous insects), will expose organisms to novel selective forces in a new developmental environment. Numerous factors such as food quality, temperature and photoperiod may have profound developmental impacts. In many cases, the reason for this is that the factors are already linked directly to epigenetic switches. Matsuda (1987) provides an excellent review showing that endocrine hormones are functionally linked to a broad range of major phenotypic features and developmental pathways and that these hormones are responsive to environmental variation. He suggests that genetic assimilation could be a very strong evolutionary force acting via this pathway.

Novel environments or epigenetic shocks may serve to expose novel development constellations to selection. This will be especially true where a shift destabilizes canalization. In particular, higher-level variation that is normally buffered by lower-level modifiers may be exposed. It was partially the existence of environmentally induced phenotypes that exactly mimicked the products of certain gene mutations (phenocopies) that contributed to Goldschmidt's (1940) radical ideas about evolution. Waddington (1957, 1975) experimentally examined the phenocopy phenomenon and concluded that genetic assimilation is a major pathway for genomic reorganization. Selection on phenotypes showing increased response to particular environmental factors may progressively increase the penetrance of hidden developmental potentials to the point where the environmental stimulus is no longer necessary for expression.

Probably the most astonishing demonstration of the potential power of genetic assimilation was Waddington's (1956) genetic fixation of phenocopies of the *bithorax* mutation. In the entire order Diptera, all families of flies contain only two wings, whereas most other insects have four. Waddington (1956, 1957) found that occasional four-winged flies could be produced by exposure of *Drosophila* eggs to ether vapours at a critical time in development. Waddington (1956) selected those flies showing strong response to the ether, and the frequency of such forms increased rapidly in each generation. Finally, a lineage of four-winged flies was obtained that was fully expressed with no need for exposure to ether at all (see Scharloo, 1991 for other examples). In an evolutionary sense, this major reorganization was fixed in a very short evolutionary time frame. Ho *et al.* (1983) found that penetrance of the *bithorax* phenotype in response to ether may increase even without accompanying positive selection. Such a result might be explained if expression resulted from direct modification of chromatin structure by the ether (Jablonka and Lamb, 1989). Environmental impacts on chromatin structure could alter the genome so as to effect stable phenotypic changes.

The existence of considerable genetic variation, allowing sexual species to obtain phenotypes well outside the range normally expressed in the wild type, has been well explored in population and quantitative genetics (see section 5.5.4). Such studies have tended to examine gradual changes in specific traits in response to directional selection and they have tended to play down tendencies for phenotypes to show discontinuities or unusual linkages to other traits. Phenocopies demonstrate that such gradual, independent shifting of specific features is indeed not the only relevant variation stored in sexual lineages. In fact, the necessary genetic variation to obtain macroevolutionary reorganizations may exist without recourse to new mutations. Reconfigurations of existing epigenetic systems may not be restricted to gradual continuous shifts in particular characters as the genic view suggests.

The reality of this was recently suggested by our research on transgenic Supermice that contain extra rat growth hormone genes. Although most morphological and behavioural aspects of the mice were not obviously discontinuous (except colour, sex and size associated with the extra growth hormone genes contained on one chromosome), mice occasionally appeared with highly deviant phenotypes. These spinner mice were extremely hyperactive, abnormally reactive to disturbance, highly aggressive, small in size, showed poor parental care and were likely to run rapidly in circles. We even obtained a variant that did high flying back flips instead of running in circles. When disturbed it appeared to be running on a non-existent wheel. Similar dwarfs or pygmies have appeared after about nine generations of selection for small body sizes, even though the original stocks showed no evidence of such 'mutants' (Falconer, 1953).

Spinner mice provide two lessons. Firstly, they demonstrate that inbreeding or strong directional selection may yield unexpected novelties or discontinuities rather than smooth incremental variation. Moreover, in the course of our experiments, we obtained several hundred of these spinner mice against the normal genetic background. However, only three spinners were obtained in mice with extra growth hormone genes. This strongly suggests that the aberrant phenotype is normally masked in faster-growing mice. The existence of *spinner* genes, in fact, might well be advantageous in larger mice, to offset the general association of larger size with reduced levels of activity and low reactivity (Chapter 8). Consequently, *spinner* genes may constitute an adaptive component of the normal mouse genetic template. In the case of pygmy or spinner mice, penetrance might be altered as selection changes the genetic background.

Falconer (1981, p. 279) interpreted this phenomenon in terms of spontaneous thresholds for expression where some genes have thresholds outside the range of wild type variation. The idea that the penetrance of a gene may vary according to environmental and spontaneous thresholds is relevant to all aspects of plasticity and canalization. For example, Parejko and Dodson (1991) found that the sensitivity of *Daphnia* clones to express inducible defences was higher among those from ponds inhabited by predators. Thus, there may be genetic variation in the thresholds of switches (see also Collins and Cheek, 1983). Parejko and Dodson (1991) also found that the switch threshold was sensitive to food availability, induction being greater with poorer resources.

An unexplored source of potential regulatory evolution involves the Red Queen coevolution of plants and the herbivores (notably insects). Plants mainly defend themselves chemically from various herbivores via a broad range of compounds that are not otherwise involved in primary metabolism (thus they are sometimes termed secondary substances). Some of these materials mimic important regulatory hormones and disrupt the development of herbivores (e.g. juvenile and ecdysone hormone analogues which impact on insects). Presumably insects coevolving with a particular host may develop more specific or protected hormonal signals to outwit the plants, but there may also be a greater exposure of regulatory variants due to such an arms race. Switches to new hosts may have particularly disruptive epigenetic consequences, potentially fuelling macroevolutionary changes. Even host water variation can impact on insect phenology (Wood, Olmstead and Guttman, 1990).

Given that plants have evolved a metabolically targeted biochemical sophistication sufficient to support our entire pharmacology industry, it is likely that these 'ecological designer drugs' could derive both the initial exposure of genetic variation in target (or incidental) species, and the subsequent action of genetic assimilation if any of these variants are of value. In this regard, it may be significant that the signal used by caterpillars of *Nemoria arizonaria* to switch from phenotypes adapted to catkins to those adapted to leaves may be the level of tannins in their food (Greene, 1989). Tannins are a common defensive compound produced by plants, which may have developmental impacts on herbivores. I am not aware of any literature exploring such possibilities, but the potential importance of such specific interaction is likely. To the extent that symbiosis and positive aspects of coevolutionary integration also involve regulatory integration, we must also consider that such evolution may initially involve mutual disruption of individual canalization and consequent genetic assimilation of integrative regulatory organization.

The key problem from an evolutionary perspective is understanding how a regulatory hierarchy may be exposed to selection since variation may be buried in canalized epigenetic systems. Clearly, recombination, mutation and hybridization are relevant to this problem, as well as environmental impacts. Genetic assimilation is usually considered only from the environmental perspective, but clearly it is only part of a larger framework that includes other destabilizing forces.

One of the most intriguing evolutionary problems has been the rapid speciation of organisms that are not constrained by occupants of adjacent niches. Thus, the diversification of cichlid fish in Lake Victoria is a classic case where cladogenesis has occurred at an enormous rate (Meyer, 1987; Dorit, 1990). Schmalhausen (1949) suggests that absence of endemic predators was a key factor. Molecular evidence reveals that not only did this entire diversification involve the branching of a single lineage, but the current genetic variation across all of these species is less than that observed within many other single species such as among human beings (Meyer *et al.*, 1990). Thus, all of this radical behavioural, ecological and morphological change has proceeded with relatively little genetic change. This suggests that most of this speciation has involved regulatory re-organizations that have not yet accumulated large amounts of genic divergence in structural genes. How could this occur?

Cichlid fish are prone to show highly plastic modifications in the structure of the mouth and feeding musculature in response to particular diets. Even the eyes may change in size (Meyer, 1987, 1990; Dorit, 1990; Wimberger, 1991). Such plasticity would expose potential regulatory variation associated with feeding on various foods and presumably could be genetically assimilated with considerable rapidity. In fact, the entire vertebrate musculoskeletal system responds plastically, and in adaptive ways, to use-related stress (Gordon, 1989). Maynard Smith (1983) refers to a goat that was born without forelimbs and whose skeletal and muscular system adapted to walking on its hind legs.

Given a range of different feeding niches that represent alternative adaptive peaks, a generalized cichlid ancestor could chase its own plasticity across the regulatory maze of epigenetic organization. Possibly, the new compartmentalization of genes could itself destabilize the phenotype, revealing further variation available to natural selection. When I first thought of this, it seemed like a rather radical idea. Subsequently, however, I found that Meyer (1990), Wimberger (1991) and Hall (1992) independently had the same thought. It seems unlikely that a single ancestral lineage could contain sufficient variation to diverge into so many remarkable divergent phenotypes, so some new mutations are likely involved as well. However, the genic view that divergent pheno-typic evolution must involve a massive divergence in numerous alleles of small effect is not supported by the molecular evidence for this system, and although conjectural, the potential role of genetic assimilation does not seem far fetched.

It is tantalizing to note that most speciation events, even those involv-ing allopatric models, are usually also associated with an ecological niche shift (Mayr, 1963). One reason why this should be so may be that such shifts expose higher-order variation needed to fuel genetic reorga-nizations (i.e. isolation without niche shifts should otherwise just as frequently derive new species). Gottlieb (1992) has stressed that changes in behaviour may be involved in subjecting development to new envi-ronments. Although completely conjectural, the paradigm developed here also suggests that species with evolutionary 'experience' of particu-lar selective regimes (e.g. salt tolerance in *Drosophila*) may be organized to express and rapidly assimilate particular adaptive conformations in response to the relevant environmental factor (Scharloo, 1991). Schmalhausen (1949) made the interesting point that new mutations that release genetic variation at the ends of a canalized plateau would have the appearance of being in the direction of selection. Clearly this is relevant to the possible reality of the Baldwin effect.

6.8 MESOEVOLUTION OF PLASTIC ORGANIZATIONS

Until recently, evolutionary theorists largely neglected the possible selective important of temporal variation and instead emphasized spatial heterogeneity to explain features like sex, plasticity or maintenance of polymorphism (e.g. Bell, 1982; Via, 1987). Molecular biologists have also documented the presence of latent genes, some of which have the potential for expression following mutation, recombination or movements of transposable elements. Maintenance of such sequences has also been attributed mainly to spatial heterogeneity (Moody and Basten, 1990). A lineage paradigm suggests that temporal heterogeneity (which may favour plasticity) may be of considerable importance. Thus, Scheiner and Goodnight (1984) estimated that up to 96% of phenotypic variance in a grass was plastic, whereas only 1–10% was genetic. Significantly, spatial heterogeneity is often associated with intermittent selection on scales exceeding the life-time of the relevant organisms.

The emphasis on spatial heterogeneity and allelic variation was reinforced by early models of speciation. The allopatric model assumed that differentiation of a species into races, subspecies and eventually new species required geographic isolation and restricted flow of alleles (Mayr, 1963). In the hierarchical lineage model, such a selection regime (which is probably common) would be associated with divergence in the regulatory hierarchy at numerous levels. As part of his early thesis to reject the evolvability and mutation accumulation hypotheses for the evolution of sex, Bell (1982) explicitly argued that temporal variation was relatively unimportant. He reasoned that spatial heterogeneity was associated with potentially larger and more discontinuous variation than temporal aspects.

Spatial heterogeneity is unquestionably an important aspect influencing diversification, both with respect to sex and plasticity (Wade, 1990). In these cases, however, selection is also often multigenerational and so still constitutes a lineage framework. Moreover, there is widespread prevalence of seasonal polyphenism (Shapiro, 1976), phenotypic responses to fluctuating population densities and adaptations to predictable periodic stress or disturbance associated with various factors (e.g. fire, flooding, drought). Seed banks in plants represent an example of a feature adapted to long-term temporal uncertainty. Such evidence leaves no doubt that both spatial and temporal variation are important in the evolution of sex and plasticity, and often spatial and temporal variation will be inseparable as far as the lineage paradigm is concerned. Even Bell now emphasizes the importance of the Red Queen hypothesis with respect to sex, a view that requires a multigenerational time frame (i.e. parents and offspring) (Burt and Bell, 1987). Temporal variation may impact on the entire metalineage at once, and so may exert a stronger effect than spatial heterogeneity that might affect only part of the population (Roff, 1993).

There is now widespread acceptance that life-history evolution is largely consistent with the mechanisms providing adaptive 'bet-hedging'. Plasticity is one aspect of this in that the flexible deployment of adaptive variation reduces the impact of selection across environments and consequently reduces variance in fitness of the lineage over time and space (Moran, 1992b). Genetic tracking of environments entails a cost, in that inappropriate genotypes must be eliminated. Although genetic tracking would be necessary if there were no information available about the habitat, where such information exists, plastic genomes can use it to adjust local phenotypes without loss; that is, they suffer less 'ecological load' (Hickman, 1975; Hoffman, 1978).

It is highly significant to a lineage framework that several authors have concluded that spatial heterogeneity may be incapable of yielding bet-hedging features (Seger and Brockman, 1987; Philippi and Seger, 1989; Moran, 1992b). Fitness in spatial heterogeneity adds arithmetically since all members are contemporaneous. Fitness in a temporal framework, however, is multiplicative and bet-hedging features are based on the geometric mean of fitness associated with temporal variance (Slatkin, 1974; Gabriel and Lynch, 1992; Moran, 1992b). Seger and Brockman (1987) suggested that genetic polymorphism maintained in an evolutionary stable equilibrium by spatial heterogeneity could not yield a bet-hedging feature, although frequency dependent selection is one exception (Lloyd, 1984; Moran, 1992b). Frequency selection can yield genetic polymorphisms, particularly with respect to sexual dimorphism and predation. Genetic polymorphisms in butterflies that serve in mimicry are also good examples. My own feeling is that this view is too dichotomous. The key factor may be to what extent plasticity in one genotype can span a range of environments. Within certain boundaries, a single plastic genotype might suffice. Well beyond these, different locally adapted genotypes must be deployed (Sultan, 1987). However, there is a middle ground where both plasticity and genetic diversity maintained by environmental heterogeneity can act to yield lineage homeostasis and shifting balance in the sense of Lerner (1954) and Wright (1988). Gabriel and Lynch (1992) concluded that plasticity may be favoured by increasing spatial heterogeneity and increasing variance across generations combined with decreasing within-generation temporal variance.

There is some evidence that spatial heterogeneity may favour genetic polymorphisms, whereas temporal heterogeneity may favour plasticity (Hedrick, 1986; Bull, 1987; Seger and Brockman, 1987). Alleles that are locally absent in space may still constitute a part of the organization of metapopulations, but alleles lost in time must be reforged from mutations. Thus, Schmalhausen (1949) provides an example of a praying mantis in which alternative green and brown forms are genetically determined and choose vegetation with appropriate background colours. Spatial variation can stably maintain the genetic polymorphism. The change to a white coat colour among mammals that experience seasonally snow-covered environments could not be obtained by such circulating alleles, but requires plasticity. Plasticity in response to spatial variation is not ruled out, however, as is illustrated by chameleons that can flexibly alter their colour over very short time scales. Scheiner (1993) points out that selection for plasticity could take place at the individual level for traits that are labile in single phenotypes. Evolution of traits that are plastic among individuals, but fixed within phenotypes, however, requires lineage selection acting at the level of the genome.

There are two important aspects relevant to the evolution of polymorphic switches and norms of reaction. Firstly, we wish to know what are the epigenetic mechanisms that yield such switches or a single reaction norm. This aspect, although crucial for understanding, is effectively absent in the literature (Via, 1987). This is not surprising given that most advances relevant to even single phenotypes have been achieved only recently. The evolution of phenotypic plasticity must be a more complex version of the evolution of morphogenesis in single phenotypes described earlier. Ultimately, the actual mechanisms that mediate epigenetic pathways must be applied to understand this problem.

Secondly, we need to understand how genetic variation is related to the evolution of plasticity and how they interact. These are important questions related to genetic organization. For example, how can plastic mechanisms be adjusted, canalization ranges altered, and what happens if genotypes with different plastic responses cross (do they blend or maintain some discrete aspects)? Current theories are largely based on quantitative genetics, a field that assumes that most characters are normally distributed, and continuous responses to selection can be obtained via segregation of numerous genes that make incremental contributions to the trait (Barton and Turelli, 1989). Whereas the genic view largely ignores specific functional integration among loci, quantitative genetics ignores even the Mendelian discreteness of particularly alleles. Despite this, quantitative genetics has been remarkably successful when applied to artificial breeding programs (Falconer, 1981) and has made increasingly greater contributions since the 1970s to evolutionary ecology (Lande, 1979; Mitchell-Olds and Rutledge, 1986; Clark, 1987b; Van Noordwijk, 1990; Rose, 1991; Stearns, 1992; Roff, 1993).

Via and Lande (1985, 1987) and Via (1987) developed quantitative genetic models of plasticity based on the idea that a particular trait expressed in two different environments can be considered as two separate traits. This conforms to a model of a linear reaction norm connecting two environments (Figure 6.4). Assuming that there is genetic variation in reaction norms (i.e. different genotypes are present), they then asked what are the consequences of traits being genetically correlated. This amounts to considering the degree and sign of pleiotropy in interaction among genes expressing these alternatives. The genetic correlation is obtained by examining how similar responses are among genotypes. A genetic correlation of +1 indicates that reaction norms among genotypes are parallel (slopes are equal), and selection for a different degree of response between traits (a new slope) is not possible. Increases in one trait are invariably linked to increases in the other.

A genetic correlation of −1 indicates that crossing reaction norms inter-

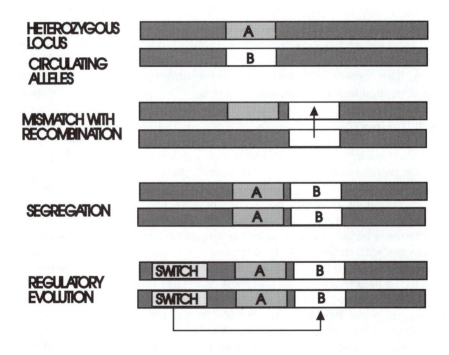

Figure 6.4 A possible mechanism for converting systems of circulating alleles to a system of phenotypically plastic switches. Another possible way of deriving the same thing could be gene duplication followed by divergent specialization. A simple pleiotropic model can be envisioned if the expression of a single locus was modulated by environmental signals, rather than effecting a switch.

sect perfectly at a single point. Although there may be considerable variation among slopes, and this is selectable (Stearns, 1992, p. 62), the genotype with the highest ranking in one environment will be lowest in the other and vice versa. There is no variation available to select responses in the same direction or to hold one trait constant while varying the other.

High positive or negative correlations reflect a low amount of scope for independent adjustment of a trait in two environments, so selection for the optimal phenotype in each environment will be slowed by disruptive selection in the other. Plasticity can evolve most readily where there is no correlation (norms have high variation in slopes and a network of intersections). The optimal norms of reaction can be achieved unless correlations are perfect, polygenic variation is limited or plasticity has an additional cost. In this model reduced migration and population subdivision will favour local genetic adaptation rather than plasticity. The last point does not apply to features derived in response to temporal heterogeneity which points out the importance of assumptions in deriving predictions. Many theoretical models considering the evolution of plasticity have been formulated, perhaps inappropriately, in a framework of spatial heterogeneity (reviewed by Scheiner, 1993).

Via's (1987) model predicts that shifts in the mean plastic responses of a population may be selected most easily where there is lack of pleiotropy (low genetic correlation or lack of shared genetic elements) between traits expressed in different environments. This amounts to a high degree of variation in slopes of reaction norms that allows obtaining any combination of alternative phenotypes. The original norms of reaction necessary for the model to work may simply be non-adaptive 'morphoses', that are then selected to forge adaptive plasticity. It seems questionable, however, that a large amount of such non-adaptive variation might be maintained in populations.

Via (1987) recognized that selection to achieve a new norm of reaction can be compartmentalized into two components: direct and correlated responses. Selection in one environment will be affected by correlated responses in the other. Of particular importance are correlations that are in the opposite direction to selection (i.e. where both characters are selected to increase, but they are negatively correlated, or the characters are selected in opposite directions but are positively correlated). Evolution of plasticity is predicted to be slowed where one environment is relatively rare or selection is much stronger in one environment than the other.

De Jong (1989, 1990a,b) also considered the case where reaction norms are assumed to be linear. In this case, the evolutionary fuel available to selection was determined to be the variation in the slopes of norms among genotypes, specifically the additive genetic variance in slopes divided by the total variation in slopes. The conceptual difference in de Jong's model is that it is based on the idea that genes may act on plasticity as a trait in itself. Alleles may affect both the slope and intercept of the reaction norm. In a narrow environmental range, selection will act most intensely on the local value of the expressed norms. Any slope will do as long as the appropriate phenotype is derived. As the environmental range expands, the slope of the general norm becomes more important, even if this conflicts with local adaptation.

Such an interpretation derives a view of plasticity as a constraint rather than an adaptation. Antagonistic pleiotropy among common genetic elements prevents perfect local adaptation due to stabilizing selection on the reaction norm at the level of the metapopulation (= sexual lineage). Presumably this would mean that greater fitness of the metalineage is achieved at some cost to local sublineages. West-Eberhard (1989) suggests that where one environment disappears, or different plastic phenotypes are allopatrically separated in different environments, consolidation of one alternative adaptive suite could proceed faster and to a greater extent than when the organism must maintain the machinery to deploy both alternative phenotypes. This amounts to release from antagonistic pleiotropy associated with shared genetic elements. Moran (1992b) reviews evidence that prolonged selection of a polymorphic species in one environment may reduce fitness in others.

Via's model considers only two environments, so that a single adaptive norm is always possible if appropriate variation is available (a single straight line will always connect two points). De Jong's model considers the case of more environments so the chance that a single linear norm passes through all optimal phenotypes declines. Of course, the existence of such constraint hinges on the assumed inability of a reaction norm to deviate from a single simple function (in this case linear) across environments. Scheiner (1993) recognized that norms need not be constrained to linearity, and suggested describing them with polynomial expansions. In West-Eberhard's case, the requirement to integrate numerous norms in a adaptive plastic trajectory could be the source of constraint.

We have very little empirical understanding of the epigenetic mechanisms involved in the evolution of plasticity. However, the regulatory hierarchy described earlier must define the framework within which such evolution operates. The molecular evidence suggests that the evolution of regulated development, epistasis and pleiotropy may involve actual changes in genome structure as well as genic constitution (e.g. changes in the DNA or protein binding domains of transcription factors, changes or movements of DNA control elements, changes in enhancer regions, exon shuffling). Mutations of particular alleles in the traditional sense (point mutations in structural genes) are not ruled out either, but they may be of relatively minor importance (Wilson, 1976). Such structural changes will still lend themselves to traditional quantitative genetic analyses when lower-level variants of small effect are segregating, but a different conceptual model than one of the mass balance of alleles in the genic sense is needed. In fact, if specific regulatory integration proves to be the case, the model that is required is selection of alternative genomic structure (i.e. lineages), where regulatory control involves assembly across generations. Moran (1992a) suggests that major evolutionary transitions in aphids are mainly associated with shifts in developmental switches that derive divergent phenotypes and that the requisite varia-tion exists among sublineages in metapopulations.

A crucial problem with the evolution of plasticity, as was also true of sex, is the idea that polymorphic specialization or local adaptation may be disrupted by recombination among individuals emerging from alter-native selection regimes. The crucial factor is antagonistic pleiotropy among common genetic elements. The hierarchy of developmental organization will confer nearly universal pleiotropy among genes (Atchley and Hall, 1991) where single phenotypic targets are involved. Models to date also assume that pleiotropy is a relatively fixed property of the genotype. To the degree that antagonistic pleiotropy disrupts attaining the optimal phenotype among varying environments, it may be that regulatory evolution will act to reduce it. This could be achieved by incorporation of separate genetic elements into one genome and the evolution of switches.

An extremely simple way in which a system of circulating alleles can be converted to plasticity illustrates how fundamental changes in genome structure could act. Imagine a single locus with two circulating alleles. A frameshift mismatch during crossing over could result in one chromo-some obtaining both alleles in adjacent positions. Alternatively a gene duplication followed by recombination could achieve the same end (Figure 6.4). Ohno (1970) recognized the importance of such mechanisms, particularly with respect to the evolution of gene families. In the present case, the adaptive function of the two genes has been honed during a prior period of circulation as competing alleles. Once consolidated within a single genome, the further linkage of expression to environmental or developmental contingencies could easily follow. Hamilton (1993) sug-gested that such a solution as an alternative to segregation of alternative alleles may cost too much code. This question was addressed in Chapter 4.

There is widespread recognition that the evolution of phenotypic plasticity likely involves changes in genome structure associated with duplications, rearrangements and regulatory integration (e.g. Shapiro, 1976, 1984; Lande, 1980; Smith-Gill, 1983; Schlichting, 1989a). Thus, systems of circulating alleles could eventually be converted into plastic systems via structural modification of the genome. In this context, plasticity may represent a more advanced form of epigenetic organization than circulating alleles (Lewontin, 1957; Sultan, 1987). In organizations achieving high plasticity, some of the advantages of sexual reproduction might be lost (Case and Taper, 1986; Sulton, 1987). Significantly, inducible defences are largely associated with species that are clonal or largely asexual (Havel, 1987; Harvell, 1990). Rendel (1967) also suggested that more advanced genetic organizations may reflect greater degrees of canalization, a view that is not inconsistent with polymorphism (each morph may be canalized). Both ideas are based on modification of the epigenetic machinery to maximize local fitness and decouple the genotype from local selection pressure.

Within single genes the level of transcription could be modulated via differential activity of transcription factors generated via environmentally sensitive genes, or single genes could be differentially used in separate gene nets via incorporation of new regulatory recognition sequences. Alternatively, completely separate control could be relegated to separate promoters within the same gene. The existence of such regulatory structure is widespread. Models that view the evolution of plasticity as limited to a world of competing alleles that interact in a framework of continuous and potentially antagonistic blending may be missing much of the real story. In a hierarchically structured regulatory net, a crude modulation of the phenotype could be derived at higher levels of control, with varying diversification at lower levels to obtain phenotypes with relatively precise meshing to local environments. Reliable environmental information may be more the limiting factor than epigenetic capabilities. Maynard Smith (1983) refers to a relevant example in the plant genus *Geum*. Hybrids between one species that produces plumed, wind-dispersed seeds, and another that produces hooked, animal-dispersed seeds, express both kinds of seeds on a single plant. The quantitative genetic prediction of an intermediate intergrade is not realized.

Models of plasticity based on continuous phenotypic responses to the environment (e.g. linear or simple curvilinear norms) are particularly amenable to quantitative genetics and the prevalence of such responses may justify the application of such models. However, one must ask whether such responses occur simply because a gradual modulation of the phenotype is the best adaptive strategy in response to continuous environmental variation. In other words, such responses may well represent the best way of responding to a continuously varying environment rather than reflecting any intrinsic limitation of the epigenetic organization to achieve locally fit phenotypes. Although mathematically attractive, linear models of plasticity ignore the rather powerful regulatory complexity available for phenotypic deployment.

Developmental systems are not limited to simple linear incremental responses (Figure 6.3C). Greene's (1989) caterpillars are a good example, as are the radical shifts in leaf design in aerial *versus* aquatic leaves on the same plant. Social insects also illustrate that multiple and relatively discontinuous phenotypes may be diversified by a single genome. Such complex alternative phenotypes imply that the reaction norms linking them are not much of a constraint, at least in the long term. Viewing norms as straight lines or smooth curves is a useful conceptual tool, but epigenetic regulatory cascades may have significantly different properties than those that can be captured by such simple representations. Plasticity can respond to discontinuities or span more than one adaptive peak. Several studies have documented that evolution of local phenotypic expression and the degree of plasticity of traits may be relatively independent of one another (e.g. MacDonald and Chinnappa, 1989; Schlichting and Levin, 1990; Scheiner and Lyman; 1991). If so, plasticity may not be limiting local adaptation, but instead may facilitate it where environments vary (within limits). For example, it seems quite likely that alternative morphs derived via phenotypic plasticity could well be independently canalized to obtain local optimal conformations.

Single locus models (i.e. genic) are undoubtedly insufficient to address the phenomenon of phenotypic plasticity (Via, 1992). Adaptive modulation of phenotypic expression most likely involves epistatic integration among genes (Scheiner and Lyman, 1991; Scheiner, Caplan and Lyman, 1991; Scheiner, 1993). Most likely a hierarchical cascade of regulatory genes, functionally integrated via a specific recognition mechanism, controls canalized epigenesis, bifurcating polymorphic genotypes and continuous norms of reaction. This would involve both pleiotropy and epistasis in a complex response of regulatory machinery coordinated via the neuroendocrine system (e.g. Nijhout and Wheeler, 1982; Matsuda, 1987; Robinson, 1992). Bargmann and Horvitz (1991) have shown that the plastic control of development into either normal or diapausing phenotypes in the nematode, *Caenorhabditis elegans*, is mediated by chemosensory responses of particular neurons to food, or a pheromone signalling high population density. Such studies underscore that although plasticity can be analyzed as a genetic phenomenon, the functional mechanism may emerge at higher levels of phenotypic organization, in this case the cellular organization of the animal.

Although responses to novel situations may be neutral or inappropriate (morphoses), norms in established populations are probably largely adaptive (Schmalhausen, 1949; Sultan, 1987). Thus, norms may be higher-order lineage attributes associated with a particular regulatory structure. The idea that genotypes can be characterized by particular norms of reaction is prevalent in evolutionary ecology (Stearns, 1989, 1992; Hillesheim and Stearns, 1991; Roff, 1993). This essentially derives a framework where various sublineages are considered as discrete entities with particular plasticities. Wade (1990) recognizes this explicitly. Gomalkiewicz and Kirkpatrick (1992) propose a model for the evolution of reaction norms that is essentially congruent with selection acting among subspecific lineages.

There are two aspects to consider. Firstly, a genotype is not characterized by a single norm of reaction, but may express numerous norms relevant to different phenotypic features and different environmental variables. Achieving a dynamic adaptive suite across a range of environments requires coadaptation among such norms (Schlichting, 1989a). If selection acts on reaction norms or complexes of norms as recognizable features, then very large components of lineage organization must be involved and variation is likely to be seen at the level of sublineages. If there are conflicts between achieving the optimal expression of a particular trait, and evolution of plasticity in this trait, it most likely may derive from requirements for integration where different features are linked by trade offs. Where complex modulation of phenotypes is required, there may be limits on the ability of epigenetic systems to achieve this in a coordinated way, across a broad range of environments. The regulatory machinery to do so may not yet be available, it may be too expensive to maintain, or appropriate environmental information may be lacking.

The second aspect is related to the distribution of genetic variation in the metapopulation. Van Noordwijk (1989) goes so far as to suggest that population can be considered to be 'bundles' of reaction norms associated with different genotypes. This might reflect the existence of lower-order sublineages. The models of Via and Lande (1985, 1987) and Via (1987) suggest that for two environments, a single adaptive norm of reaction may be achieved as long as there is not complete overlap in the genes contributing to the alternative phenotypes. This suggests that genetic variance will be decreased as plasticity evolves. Disruptive selection resulting from conflicting demands upon shared genetic elements might maintain genetic variability. These models, however, assume that a single genotype is capable of unrestricted plasticity, that is, a single genotype can be most fit in all environments.

Gillespie and Turelli (1989) reconsidered Lerner's (1954) ideas that heterozygosity may help derive phenotypes with greater stability and fitness. Moreover, they assume that no single genotype can derive optimal phenotypes across all environments. Balancing selection acting at the lineage level recognizes overdominance of heterozygotes expressed across the range of experienced environmental heterogeneity. Such a model predicts maintenance of genetic variation and genotype–environment interaction at equilibrium. Scheiner (1993) points out that there is little empirical support linking plasticity to genetic heterozygosity, but the entire question is complex. For example, heterozygosity might have little influence on the plasticity of individual genotypes, but might considerably extend a reaction norm that emerges at the metapopulation level.

The range of plasticity achieved within a single genotype may be limited. To span the full range of environments encountered, or complex ecological dimensionalities, no single genotype may be available that can provide the optimal phenotype in every environment (Newman, 1988b; Gomalkiewicz and Kirkpatrick, 1992). For example, imagine that optimal body size changes across an environmental gradient of temperature, such that the genotype expresses a plastic response for larger body sizes at lower temperatures. However, if there are limits to the breadth of plasticity, then it may be that the optimum body size at lowest temperatures may only be achieved by a different genotype than that achieving smallest body sizes at highest temperature.

This amounts to asking whether for any mean genotype, there is a limit on the possible variance achievable. It seems very likely that there are such limits (e.g. Schlichting, 1986; Gabriel and Lynch, 1992). Moreover, where adaptive plastic trajectories are concerned, it is the flexibility among traits that may limit the range of plastic adaptability. Plasticity is only expected where there are trade offs in the fitness of alternative phenotypes in different environments (Moran, 1992b). Achieving the optimal phenotype involves modulating the trade offs structure among traits. The range of adaptive plasticity of any genotype will be limited by the least flexible element it expresses. Thus, genotypes with different means and degrees of plasticity among traits may constitute parts of a global adaptive system incorporating both plastic and genetic elements (Figure 6.5).

Gabriel and Lynch (1992) have approached this problem by considering that genotypes express a tolerance curve in response to stress. They then consider the implications of reaction norms that shift this curve or increase its breadth. They assume that maintaining breadth in a single phenotype is expensive and imposes a fitness cost in the optimal environment. In this case, a reaction norm that shifts a narrowly adapted tolerance curve allows high local adaption.

There is good evidence that the ability of a given phenotype to sustain an increase in its performance may be limited to a scope of two to four (maximum of six or seven) times over that expressed in optimal conditions (Chapter 8). Limitations on sustainable scope appear to span phylogenies, suggesting a rather fundamental constraint on phenotypes. Such limitations may also apply to the adaptive range of particular reaction norms.

This question of what range of adaptability can be achieved by one genetic program is exactly the same problem that plagues those interested in questions of speciation and biodiversity. In a communicating sexual organization, a combination of plasticity and genetic variation in plasticity may be maintained and deployed to obtain appropriate environmental responses. Such mixed control of features via both genetic

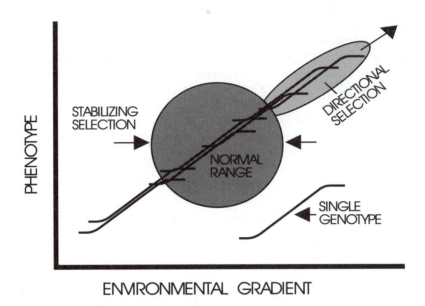

ENVIRONMENTAL GRADIENT

Figure 6.5 Integration of genetic variation and plastic reaction norms in species adapted to a broad ecological niche. The full range of a species' norm may be compartmentalized across genotypes, any one of which may not span the full range of response. The shape of the reaction norms for individual genotypes is modelled after Gabriel and Lynch (1992).

variation and plasticity is widespread (Lively, 1986b). In the above example, stabilizing selection for an intermediate body size might define the wild type but periodic environmental shifts or spatial heterogeneity may maintain both plasticity and genetic variation as part of the overall flexibility of the metapopulation (Figure 6.6).

Sultan (1987) articulated this clearly: genetic variation may allow spanning a greater environmental range than could be achieved by any single genotype and plasticity then allows refined adjustments to variability in smaller scales. Similar ideas were expressed by Levins (1963, 1968). The fact that norms may broadly overlap and each span a range of environments would tend to maintain variation in variable habitats. Prolonged local directional selection on either extreme would tend to eliminate certain genotypes, while the metapopulation might maintain the full range.

Figure 6.6 depicts a possible mix of genetic and plastic variation that might reflect the adaptive structure of a lineage. Unexpected support for

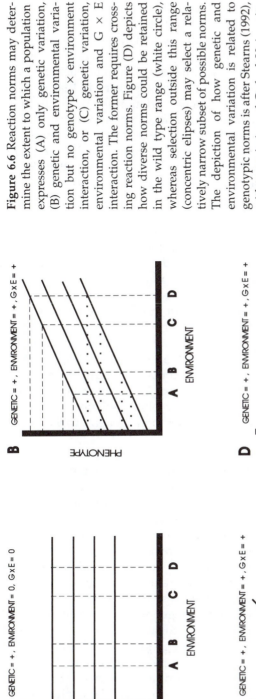

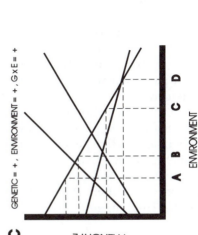

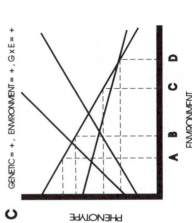

Figure 6.6 Reaction norms may determine the extent to which a population expresses (A) only genetic variation, (B) genetic and environmental variation but no genotype × environment interaction, or (C) genetic variation, environmental variation and G × E interaction. The former requires crossing reaction norms. Figure (D) depicts how diverse norms could be retained in the wild type range (white circle), whereas selection outside this range (concentric elipses) may select a relatively narrow subset of possible norms. The depiction of how genetic and environmental variation is related to genotypic norms is after Stearns (1992), with permission of Oxford University Press.

such an interpretation has recently emerged from the study of caste differentiation in eusocial insects (Moritz and Southwick, 1992; Robinson, 1992). In honey-bees, for example, queens may mate with as many as 17 drones, and sperm is mixed in the spermatheca. This results in colonies with numerous subfamilies, workers within single families being 'super-sisters' in terms of relatedness. The emerging picture is that division of labour is mediated partially by genetic variation (resulting from re-combination in queens and polyandry). Various genotypes vary both in their predilections to do various jobs and in their thresholds of response to switch to new tasks. Highly specialized workers may be rare geno-types. Genetic variation may be required to ensure that all tasks are adequately met and may also improve the flexibility of the colony to adjust ergonomic allocations rapidly under shifting demands (Moritz and Southwick, 1992; Robinson, 1992).

Such a situation suggests that the transition from coadapted meta-populations (i.e. most sexual species) to true Superorganisms sometimes incorporated genetic variation as part of the transition. Alternatively, limits on phenotypic plasticity may have required importing genetic variation as the complexity of the social system expanded. A similar advantage might accrue in ant colonies with multiple queens, since even closely related queens would provide sperm from different males. It would be most interesting to know whether polymatrial colonies are ergonomically more complex or efficient than those with single queens. This hypothesis also suggests that recombination might be higher in eusocial species with greater social complexity. The relevance to sexual metalineages is clear. Metapopulations may obtain greater centre court advantage via a structure that incorporates both genetic variation and plastic extensions of these variants.

The expression of genetic variation, environmental variation and variation associated with genotype–environment interaction is expected to be influenced by the kinds of reaction norms expressed among geno-types and across environments (Figure 6.6). Thus, relations among norms might result in only genetic variation (Figure 6.6A), only genetic and environmental variation (Figure 6.6B), or a mix of genetic, environmental and genotype–environment variation (Figure 6.6C).

Of particular relevance to the question of genetic variation is the case where reaction norms expressed by different genotypes cross (Figure 6.6C). Where reaction norms cross (i.e. norms vary in slope), selection for a higher phenotypic value may favour one genotype in one environment, and an alternative genotype in another because their phenotypic ranking switches. Such a change in ranking is called genotype–environment inter-action. Crossing reaction norms and genotype–environment interactions are pervasive (Bell, 1985; Via, 1987; Gebhardt and Stearns, 1988; Stearns, 1992). This suggests that variation in norms may be maintained in metapopulations because there are trade offs in the ability of a single norm to produce phenotypes spanning the entire range of experienced temporal and spatial variation.

Crossing reaction norms are also involved in another important phenomenon. Where reaction norms intersect, the same phenotype is expressed by both genotypes. Natural selection cannot see them (Gupta and Lewontin, 1982). Stabilizing selection for a wild type phenotype might result in an organization where most norms cross in the optimal or intermediate environment. Clearly we are addressing a potentially important aspect of canalization (Figure 6.6D). Canalization need only dampen norms in the vicinity of the intersection point to yield a single phenotype relative refractive to both genetic and environmental impacts. Such a situation predicts that additive genetic variance may be lowest in the optimal range and the various genotypes may conform to a model of neutral evolution under those conditions. Genetic variation could then be maintained by lack of selection.

As the environment shifts to extremes, the reaction norms among genotypes diverge, revealing high phenotypic variation visible to selection (Figure 6.6D). If the environmental perturbation is novel, the expressed variation is likely to be random (morphoses). However, if such environmental shifts have been commonly experienced then selection will have tended to eliminate genotypes with inappropriate norms and the phenotypic shift may be adaptive. The genotypes expressed in the wild type in this case reflect the previous history of environmental variation. The divergence of norms associated with different genotypes is predicted to increase steadily with the degree of environmental stress. Whether selection differentiates among norms (and eliminates those diverging too far from the optimal trajectory) will depend on the fitness consequences of deviation and on how frequently environments that elicit such divergence occur (Figure 6.6D). Norms that may be regarded as adaptive plasticity within the normal range of environmental variation may lose precision in infrequent or environmental regimes due to the attenuation of selective scrutiny.

Models of genetic variation in reaction norms in single traits also suggest that plasticity may modulate the amount of expressed genetic variation (Stearns, 1992). Where the adaptive suite is concerned, numerous key features need to be modulated in an integrated way to achieve a plastic developmental trajectory. Thus, later maturation at small sizes but producing a few well endowed eggs might represent a suite of tactics in a low quality environment. In a better environment, rapid maturation at larger, more fecund sizes and the production of numerous smaller eggs might be favoured. Numerous other design elements might also be tied to these alternative life-history strategies.

The linkage among such traits may be via pleiotropy, competition for a shared resource pool (physiological trade offs), or via a common responsiveness to a particular environmental variable (Stearns, 1992, p. 65). Where plastic developmental trajectories are concerned, one environment may favour a correlated increase in trait A when trait B increases. In another environment, a correlated decrease in trait A may be associated with increases in B.

Stearns (1992) illustrates this for the common case where the fastest growing genotype matures latest in a good environment, but earliest in a poor environment (Figure 6.7). The result is that the genetic correlation between traits shifts from positive to negative in the different environments because the reaction norms for the alternative genotypes cross. Empirical support for these ideas has been obtained by Giesel (1986), Gebhardt and Stearns (1988), and Newman (1989, 1992).

This discussion of plasticity points out that both the amount of genetic variation in particular traits, as well as the interrelationship among traits (positively or negatively covarying), may be altered. In a lineage paradigm, the implication is that not only may the kind of phenotype deployed be adaptively altered, but there is potential control over the degree and direction of microevolutionary responses. The degree of selection on a trait may be altered by plasticity in its expression (e.g.

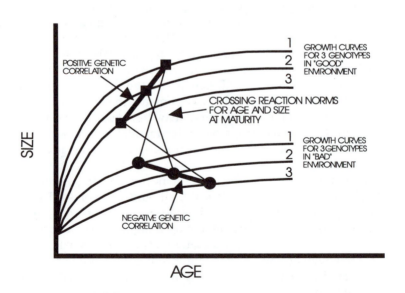

Figure 6.7 Crossing reaction norms can lead to reversals in the sign of genetic correlations. Redrawn from Stearns (1989) with permission of *Functional Ecology*.

inducible defences are only visible in some environments) and the way that this trait is integrated with other phenotypic features. Thus, adaptive plastic trajectories change phenotypic correlations and the indirect effects of selection on particular traits (Stearns, de Jong and Newman, 1991; Stearns, 1992).

Plastic modulation of the phenotype can alter selection intensities (Schlichting, 1989b). Such modulation also can influence how much, if any, genetic variation is revealed (decanalization) and whether genetic correlations among traits are positive (facilitating selection) or negative (inhibiting selection). It is also likely that the interface of phenotypic and environmental complexity will yield conflicting selection pressures on the integration of traits. Where the capacity of the system is limited by complexity, potential costs of plastic mechanisms or insufficient information, the lineage must rely on natural selection of circulating genetic elements to polish the fit of the phenotype to the local environment.

Finally, in systems where cognitive abilities have evolved sufficiently, the interfacing to the environment may be achieved behaviourally. In this case instincts can probably be considered as reaction norms. Innate responses may be finally tuned to local contingencies via cognition. In simple animals with limited cognitive flexibility, norms of behaviour may be more easily discerned. Both instincts and cognition, however, are different aspects lending organismal plasticity and are likely subject to the selective regime outlined here. The possession of instincts, transgenerational transmission of information at the cognitive level, and cognitive adjustment of local responses are also likely to buffer phenotypes from selection to varying degrees.

6.9 GENETIC CORRELATIONS AND GUIDED EPIGENETIC TRAJECTORIES

Discussions of phenotypic plasticity, reaction norms and polymorphic switches suggest that the crucial framework for understanding phenotypic evolution is the epigenetic regulatory landscape. Developmental capabilities may be modified by either environmental influences or the impact of circulating genetic elements. Thus, numerous cases of environmentally induced 'phenocopies' have been documented that mimic the impact of mutations. Similarly, in polymorphism, the divergent phenotype may be either environmentally induced, genetically determined, or achieved by a mix of mechanisms.

In the case of genetic determination, a circulating allele may act as a switch (West-Eberhand, 1989), or a block of genes constituting a supergene may be exchanged (Wills, 1989). For example, inversion polymorphisms in *Drosophila* may be a mechanism for protecting blocks of coadapted genes from recombination (e.g. Dobzhansky, 1970; Lewontin, 1974). In the land snail, *Cepaea nemoralis*, various combinations of genes that control the colour of the shell and body as well as degrees of banding on the shell are circulated in supergenes that create large linkage disequilibrium (Jones, Leith and Rawlings, 1977; Cain, 1983).

In an epigenetic framework, for any genetic change that might modify the phenotype, appropriate environmental influences may achieve similar ends (Cheverud, 1988; Van Noordwijk, 1990). In such a framework, at least some circulating alleles may be considered as constituting exchange of developmental information. In fact, in a regulatory framework, the exchange of genetic elements with no direct contribution of structural protein should be common. In a sexually organized lineage, such information is often ancient in nature, and the system may not simply change its expression relevant to the actual functional capability conferred by the allele, it may also modulate the expression of other phenotypic elements to accommodate the allele to maximum adaptive purpose.

Schmalhausen (1949) pointed out examples of genes that yield different attributes (e.g. eye colour) in different genic backgrounds. Similarly he documented how epigenetic organizations may adjust to accommodate mutations. Thus, strains of *Drosophila* containing a mutation that resulted in lack of eyes, gradually re-express eyes despite the continued presence of this gene. The accommodation of genomes to offset the deleterious impact of mutations is a well established phenomenon (e.g. Lerner, 1954) and attests to the integrative nature of genetic organization.

This may be extended to remarkable changes in the expression patterns of genes according to their genetic background. A particularly interesting example from the standpoint of the present focus is wing-form determination in the planthopper, *Laodelphrax striatellus*. In this species there is strong phenotypic plasticity for long or short wings in response to several environmental factors. However, this plastic response is restricted to the female sex (Mori and Nakasuji, 1990). Relating this system to what we know about insect epigenetics, such a result could occur if the system that establishes sexual status disconnects the switch allowing wing polymorphism so that males (except in rare instances) always have long wings. The occasional short-winged males are significant because they show that the male genome still carries both pathways.

Berg (1960) called attention to the fact that some phenotypic features were selected in correlated pleiades or blocks whereas others were relatively independent of one another. He was among the first to recognize that such associations might reflect adaptive phenotypic integration. The pervasive integration of the epigenetic program is reflected in the ubiquitous expression of genetic correlations (Falconer, 1981). A genetic correlation refers to an inherited association among traits that is apparent in breeding programs, and in the indirect changes in numerous features that come about when selection is applied to any single trait. Darwin (1859) referred to this phenomenon as the 'mysterious laws of corre-lation', and the main interest in them has been restricted to artificial breeding programs where they have generally been regarded as an inconvenience that interferes with obtaining desirable results.

A major interest in genetic correlations then has been the recognition that they may alter responses to selection on any given traits (Lande, 1979, 1980, 1982; Lande and Arnold, 1983; Reznick, 1985). If traits show strong negative correlations, for example, this may oppose directional selection for both features to increase. Alternatively, positive correlations may facilitate achieving a common goal, but may resist selection on traits in opposite directions. Dickerson (1955) called such phenomena genetic slippage. Scheiner and Istock (1991) obtained a particularly clear demon-stration of this phenomenon with respect to the relationship between development time and diapause in mosquitoes. Such phenomena may also be involved in producing asymmetrical responses to selection in various directions.

Genetic correlations are believed to arise due to linkage disequilibrium and pleiotropy in the action of shared genetic elements (Hegmann and Dingle, 1982; Clark, 1987a,b; Stearns, 1992). In turn, these aspects may derive from developmental constraints, functional integration and energetic trade offs among phenotypic features (Chapters 7 and 8). Genetic correlations are also sensitive to changes in gene frequencies (Mitchell-Olds and Rutledge, 1986; Clark, 1987b; Stearns, 1992). Some evolutionary ecologists have consequently ascribed paramount importance to such correlations based on the logic that they define the directions available to microevolution (e.g. Reznick, 1985; Bell and Koufopanou, 1986). The advent of evolutionary ecology has also recast interest in genetic correlations to consider their relevance to phenotypic integration and to ask how genetic correlations themselves might evolve (Schluter and Smith, 1986; Clark, 1987a,b; Fairbairn and Roff, 1990; Stearns, 1992). Like plasticity, genetic correlations could arise via stabilizing selection among fitness components at the upper lineage-level of organization and they may involve higher tiers of regulatory genes (Falconer, 1977; Zeng, 1988).

Just as some sorts of plastic responses may be non-adaptive morphoses, some genetic correlations may arise due to unavoidable epigenetic linkages, or may represent non-adaptive expression of the existing organization to novel selection pressures. Genetic correlations are generally regarded as arising from both developmental constraints and functional coadaptation of components (Kingsolver and Wiernasz, 1991). A crucial question is whether there is any evidence that genetic correlations, like phenotypic plasticity, reflect adaptations of metapopulations to heterogenous environ-ments. Only in the last decade has this been seriously considered. Several studies suggest, however, that the answer is emphatically yes.

Good evidence that genetic correlations may serve an adaptive function is revealed by research on the milkweed bug, *Oncopeltus fasciatus*. Milkweed bugs display phenotypes that vary according to whether selection favours local maximization of reproduction (residential forms) or movement to exploit new environments (migratory forms). The two extremes mirror the adaptive suites commonly expressed by other insects with clearer demarkation of residential and migratory adaptive suites. Significantly, selection for a single feature of the phenotype relevant to these alternative strategies (i.e. either short or long wings) yields phenotypes expressing the entire assemblage of appropriate morphological, behavioural and reproductive traits (Palmer, 1985; Palmer and Dingle, 1986). Moreover, selection for flight propensity also carries with it appropriate correlated divergence in wing length, body size and maturation time (Palmer and Dingle, 1989). Very similar results were obtained by Fairbairn and Roff (1990) with sand crickets. That such features represent meta-adaptations, rather than an unavoidable or fortuitous consequence of genetic organization, is suggested by the fact that populations of *O. fasciatus* that are not selected for occasional migration (e.g. island forms) do not show this suite of correlated responses (Dingle and Evans, 1987; Dingle, Evans and Palmer, 1988). Denno, Olmstead and McCloud (1989) suggest that the costs of flight and associated morphology may impose costs on fecundity, development time and longevity that underlie such shifts in life-history tactics.

Rather than being exceptional, it appears to be a general rule that selection on a single phenotypic feature shifts a complex assemblage of genetically correlated traits (Falconer, 1981). The house mouse, for example, is a model system for vertebrate development and evolution (Slack, 1991). Two-way selection for large or small body sizes selects an entire assemblage of traits. Large mice have allometrically smaller extremities, grow and mature faster, produce larger, more frequent litters and are more efficient at converting food into tissues than smaller mice. Smaller mice are remarkably more excitable and active than large ones (e.g. MacArthur, 1949; Falconer, 1953; Falconer, 1981). Later I will show how this is consistent with adaptive adjustment to levels of available resources.

Drosophila is the model organism for genetics, and it too shows pervasive genetic correlation structure (e.g. Tantawy and El-Helw, 1966; Giesel, 1979, 1986; Giesel, Murphy and Manlove, 1982a,b; Clark, 1987a; Rose, 1991; Stearns, 1992). The significance of this genetic organization will be considered in more detail later. Schluter and Smith (1986) suggested that genetic correlations may themselves evolve. They interpreted the correlation structure of song sparrows, which have relatively canalized phenotypes, as most likely arising from selection for joint optima among traits that are functionally interdependent. Moreover, stabilizing selection might jointly determine not only the correlations themselves, but also the phenotypic variance in the species. Scheiner, Caplan and Lyman (1991) also stressed that stable correlations may reflect functional integration among traits. Tucic, Cvetkovic and Milanovic (1988) emphasized that such organization may tend to resist selection in some situations, a circumstance that could contribute to evolutionary stasis.

These examples suggest that the genome may be organized appropriately to alter the expression of many characters given the genetic status at various relevant loci. Both structural and regulatory organization could be involved, but the emphasis is probably on regulatory aspects. In a lineage framework, if sex and circulating alleles are properties of a higher-order organization, then it would be expected that features of the templet and communication system have evolved further than the genic prediction of random segregation of alleles within randomly organized slots. The above discussion suggests that, at least in some cases, the organization is adapted to facilitate large phenotypic shifts in response to familiar selection regimes. Scheiner and Istock (1991) drew a similar conclusion with respect to the coevolution of development time and diapause in mosquitos. Similarly, Gilbert (1986) concluded from selection studies on developmental rates and pupal sizes of *Pieris* butterflies that the release of genetic variation may be under genetic control, that it is released specifically in response to ecological contingencies, and that such dynamics imply a level of organization that transcends the individual.

The existence of such 'guided epigenetic trajectories' does not imply that evolution in general is directed, but only that regulatory organization forged in a multigenerational framework may adaptively steer microevolution to a greater degree than has been previously considered (Dingle, 1984). That is to say, the dice may be loaded so as to steer microevolution in particular directions or maintain integrated responses between known adaptive peaks. To the extent that localized control of recombination is possible (Brooks, 1988), areas containing coadapted genes with some linkage could evolve locally reduced recombination rates. If the reality of this phenomenon is accepted, it reflects a degree of epistasis, linkage and pleiotropy that is entirely inconsistent with a genic paradigm.

Just as adaptive phenotypic plasticity may be limited to the range of environments frequently experienced by the organism, guided genetic trajectories may only represent adaptive responses within a relatively narrow range of selection pressures. The potential range of phenotypes that can be achieved by hierarchical hopscotch is much greater than any form expressed in wild type populations in their normal range of environments. Genetic correlation structures may change across environments or in response to selection (Lande, 1980; Service and Rose, 1985; Bell and Koufopanou, 1986; Wilkinson, Fowler and Partridge, 1990). Alternatively, strong conservation of genetic correlation structure is found in some systems (Schluter and Smith, 1986; Mitchell-Olds and Rutledge, 1986; Spitze, Burnson and Lynch, 1991; Scheiner, Caplan and Lyman, 1991).

The conservation of such structure, if adaptive, would presumably be related to the frequency of selection pressures. In addition, the possible strategy embodied in certain kinds of correlational responses may only be effective across a particular range. Outside that range, a reorganization may be required to achieve an optimal phenotypic conformation. It is important to emphasize that the phenomenon of phenotypic plasticity refers to the adjustment of phenotypes in response to differing environmental cues by a single genotype. Genetic correlations refer to the adjustment of phenotypes in response to changes in genetic information and may involve no change in the environment. Clearly, however, wherever selective regimes are changing, both aspects will be involved in responses by metapopulations (Schlichting, 1986). This frameworks suggests that if particular phenotypic combinations represent adaptive trade off structures that vary in fitness among environments, then phenotypic correlations derived via adaptive plastic trajectories and genetic correlations associated with guided genetic trajectories should tend to be similar (Van Noordwijk, 1990; de Jong and Van Noordwijk, 1992).

Cheverud (1988) argued that this is probably the case in most instances. Schmalhausen (1949, p. 75) observed that the adjustment of plants to local ecological conditions converged on similar features regardless of whether this was achieved via plasticity or local genotypic fixation. Bradshaw (1965) reiterated this point. Thus, natural selection may tend to shape plastic and genetic responses in similar ways (see also Falconer, 1981; Schluter and Smith, 1986). It is also clear, however, that strong directional or stabilizing selection will tend to eliminate genetic variation, and temporarily at least, genetic correlations.

Moller, Smith and Sibly (1989b) suggested that phenotypic plasticity and genetic correlations may be complementary. In particular, they found that a key life-history trade off between adult fecundity and longevity was evident in plastic responses of beetles to environmental variation. However, there was no evidence of this trade off when examining genetic correlations because there was no genetic variation. Newman (1988a,b) also found that a lack of genetic variation in a plastic response is a key life-history feature of toads. This suggests that modulation in some feature may be mainly or entirely plastic, while others are adjusted genetically. In this case, understanding the local adaption of phenotypes would require examining both genetic and plastic correlations. Where a particular strategy is robust, it follows that the genetic templet could deploy phenotypes biased towards local adaptation, even in novel environmental ranges. In other cases, epigenetic organization relevant to the normal range might act as constraints, temporarily retarding appropriate evolutionary responses.

Stearns (1992) pointed out how phenotypic plasticity can alter the intensity of selection on phenotypes, the amount of genetic variation expressed in a trait, and the genetic covariance among traits. Genetic correlations may further alter selection responses in their own right. A question left unconsidered is whether changes in genetic constitution could be integrated with the expression of plasticity as part of global lineage adaptability.

Schlichting (1986) observed that the organization of traits as revealed by genetic or plastic correlations may vary among environments, and suggested that this must reflect changes in the functioning of regulatory and modifier genes (i.e. modification of modifiers). The integration of adaptive epigenetic trajectories via hormones was reviewed earlier. Fairbairn and Roff (1990) suggest that neuroendocrine signals may also be involved in the expression of genetic correlations. In any case, there is strong evidence that epigenetic organizations have a potential to modulate phenotypic expression adaptively in response to both environmental and internal signals. The earlier discussion of potential lineage-level advantages of sexual reproduction should be overlaid with this framework. In particular, such mechanisms add considerable power to the arguments developed relevant to centre court advantage, crosstalk, and hierarchical hopscotch.

7

Adaptive suites, coadapted genomes and dynamic genetic templets

7.1 INTRODUCTION: COADAPTED GENOMES AND PHENOTYPIC ADAPTIVE SUITES

If genomes are integrated via specific regulatory interactions and organizational levels of integration are selected on various temporal scales, then the idea that particular combinations of genes or phenotypic features are coadapted comes to the fore. The basic idea can be traced back to Darwin (Wallace, 1991). In a genetic sense, coadaptation is as old as Wright's (1931, 1932) adaptive landscape model, with its intrinsic implication that certain combinations of genes form adaptive peaks, whereas others may be associated with depressions or valleys in fitness. That the genome represents a coadapted complex of genes and that phenotypes constitute highly integrated systems are ideas that have widespread recognition (e.g. Wright, 1931, 1932; Dobzhansky, 1937, 1970; Goldschmidt, 1940; Schmalhausen, 1949; Mayr, 1954, 1955, 1963, 1975, 1982, 1983, 1992; Waddington, 1957, 1975; Rendel, 1967; Lewontin, 1974; Alexander, 1975; Antonovics, 1976; Endler, 1977; Carson, 1982, 1987a, 1990a,b, 1991; Templeton, 1982b, 1989; Carson and Templeton, 1984; Reid, 1985; Wallace, 1991).

Despite this, there has been relatively little synthesis regarding the source of such coadaptation, or even what coadaptation really means. Ecologists have focused on the integrated features of the phenotype and their interface with the environment. Geneticists have focused on the association of genes that epistatically enhance fitness (Wallace, 1991). Phenotypic aspects of coadaptation are recognized as constituting the adaptive suite; genetic aspects as the coadapted genome. These two concepts are obviously closely related, and much can be gained by forging their linkage.

There are at least five kinds of constraint relevant to the concepts of coadapted genomes and adaptive suites: historical, genetic, metabolic, developmental and ecological.

Historical constraints reflect that the current genetic format and variations available in extant individuals represent a limited subset of all possible configurations. Barring mutation, further change is limited to pathways defined by this local subset rather than global potentialities. Historical constraints also pertain to solutions that were consolidated in early evolution or development and which have considerable subsequent organization resting on them (Gould, 1989b). In such cases, the fundamental features represent 'frozen accidents' that are difficult to bypass or change (Ohno, 1970, 1973). Historical constraints can occur at numerous levels of organization. The genetic code was mentioned earlier. A higher-level example might include the concept of fundamental 'Bauplans' associated with various phylogenetic body designs (Rensch, 1960; Gould, 1989a; Hall, 1992; Stearns, 1992). Examples might include vertebrate tetrapods, insect hexapods or molluscan spirals, each of which is associated with radically different evolutionary potentials and restrictions.

Genetic constraints such as linkage groups that oppose recombination, local regulation of recombination, genetic correlations or genetic architectures that limit compatible sexual communication may affect genotypic variation among sublineages. The organization of the genome as a hierarchically structured regulatory system with downstream and cross-branching linkages must also impose limitations on subcomponent assembly and likely evolutionary pathways (Chapters 3 and 4).

Metabolic constraints limit the adaptive system to particular potentialities based on such factors as the maximum rates and yields of biochemical pathways, availability and range of suitable substrates, competition among pathways for resources and accumulation of inhibitory or damaging byproducts. Beaumount (1988) suggests that coadaptation may be maintained via selection against extreme phenotypes associated with disrupting metabolic flux. The Balance Hypothesis outlined in section 8.7 suggests that metabolic rates must be tuned to levels of resources.

It is significant that key researchers concerned with negative genetic correlations have explicitly recognized that these probably reflect trade offs in resource allocation occurring at higher levels of metabolic and physiological function (e.g. Falconer, 1977; Rose and Charlesworth, 1981b; Rose *et al.*, 1984; Sibly and Calow, 1986; Stearns, 1992). This means that coadaptive synergisms and trade offs constituting the genetic templet may reside at higher levels of integrated gene interaction, so that epistasis and pleiotropy are not necessarily visible at lower levels (Falconer, 1977; Hartung, 1981; Rose, 1982). In fact, these interactions may emerge more strongly as larger genetic fields (some of which may use some common genetic elements) are considered (e.g. growth *versus* reproduction, defence *versus* growth). Franklin and Lewontin (1970) argued strongly that two locus theory was inadequate to assess the true degree of correlated linkage among higher-order genetic units and concluded that large, correlated blocks of genes may be the normal state. Any argument rejecting coadaptation on the grounds of a lack of evidence for pleiotropy, linkage and epistasis at the level of a few loci must be offset by the widespread evidence for such associations in higher-order phenotypic features programmed by hierarchies of such genes (Chapter 8).

Developmental constraints arise from the need to elaborate a complexly integrated morphology based on the simple rules of cell differentiation, movement, secretion and adhesion. Thus, although the wheel has proved to be an extremely useful technological device, it is conspicuously absent from developmental systems (Gilbert, 1991). Paley was quite correct that the existence of a watch implies a designer. The nature of evolution and development have made the discovery of wheels or intermeshing gears exceedingly unlikely. Levinton (1988) and Hall (1992) suggested that interlocking developmental processes may lead to progressively greater resistance to disruption (i.e. increasing constraint).

Maynard Smith *et al.* (1985) view developmental constraints as an adjunct to stabilizing selection in maintaining species integrity. Physiological studies are revealing very strong aspects of coadaptation that are unavoidably required by changes in body size or metabolic rate (Weibel, Taylor and Hoppeler, 1991). Such requirements may limit the allowable range of variation within bodies of particular dimensions. A positive aspect of developmental constraints is that they define channels for change that may be explored by a lineage, even in the absence of selection (Gould, 1989b; Kauffman, 1993).

Lande (1986) argues that developmental features usually have selectable genetic variation so are unlikely to constrain evolution. The presence of genetic variation, however, does not preclude nearby selection limits or the impact of genetic correlations. Thus, metabolic rates of similarly sized animals may display selectable variation, but there is likely a maximum metabolism allowable for any given body size. As Antonovics (1976) observed, the key question is not whether genetic variation exists, but how it is constrained by coadapted gene complexes.

Ecological constraints represent the integration of forces imposing stabilizing, balancing, disruptive or directional selection on the genome. It must be reiterated, however, that lineages are not simply genetic ships adrift in an ecological sea, they are tenacious overlays on their habitat that have attributes specifically adapted to environmental and biotic contingencies. Thus, evolved characteristics such as dispersal, habitat recognition, or behavioural, physiological and developmental flexibility actively filter the habitat. Furthermore, many species act to modify environments to suit their own needs better. Environmental impacts and interactions with development will also be important. The prevailing view is that most selection on species is globally stabilizing with only locally limited shifts in directional or disruptive selection (Schmalhausen, 1949; Kimura, 1983; Maynard Smith, 1983; Lande, 1986; Zeng, 1988; Carson, 1991; Stearns, 1992).

Organisms are constructs elaborated across numerous organizational levels, and thus reflect selection acting simultaneously or intermittently on all of these levels. We are particularly concerned with the ultimate function of phenotypes, their ecological competence and success, since this is what shapes the genome (Haukioja, 1982). Given that there are numerous problems that the organismal phenotype must adequately solve, it is not usually possible to maximize the adaptive value of any one particular feature. The current conformation of a given feature reflects six key criteria:

- Constraints that prevent global optimization, as described above.
- The degree to which the optimal conformation obtainable has been realized (evolutionary lags occur).
- The extent that natural selection acts globally to maximize organismal rates and efficiencies will be reflected in features that are synergistic (one feature promotes another), overlapping (one feature does two jobs) or redundant (one feature backs up another in case of failure or overload). Such integrative requirements may prevent particular features from being optimal for their specific function to optimize system-level fitness.
- The resources available to allocate to a feature may be limited. Trade offs among features occur because there is an upper limit to quantities and flow rates of material and energy. Thus, if all features draw from a single limiting pool, increases in one may incur costs in another. External variations in resources may further constrain intrinsic potentials. Most life-history theory considers that important components of fitness are traded off against one another (e.g. Gadgil and Bossert, 1970; Charlesworth, 1980, 1990; Rose, 1984b; Sibly and Calow, 1986; Rollo, 1986; Cockburn, 1991; Lessells, 1991; Stearns, 1976, 1992; Roff, 1993) to achieve some optimal compromise.
- Related to this is the problem of specialization. Optimization of the phenotype for one purpose may simultaneously make it less effective outside the immediate range of specialization.
- Finally, besides the need to be adapted to the environment in terms of simple chemical and physical criteria, organisms are subject to ultimate failure and risk. The design must ensure perpetuation. In the genic view this simply boils down to maximization of successful reproduction in each generation. In a lineage view it entails maximizing

longer-term persistence. Persistence can be maximized by minimizing failure as well as maximizing success. Clearly the correct criterion should be one that balances these requirements (which may not always be in conflict). Maximization of any particular aspect may be curtailed, if this incurs offsetting risks.

Thus, a species' phenotype reflects a constellation of attributes that derive from historical aspects, selection for internal organizational features associated with elaborating and supporting the phenotype (genetics, metabolism and development), and ecological forces that determine the competence and selection of these organisms. There are undoubtedly a limited number of possible configurations that can accommodate all of the above forces in an optimal way. The wisdom of the early naturalists, that organisms are intricately balanced so as to carry out a diverse spectrum of different and specific functions effectively has become formalized in the ecological concept of the adaptive suite. The essence of this appellation is that each feature of the phenotype may be adaptively honed by evolution for maximal competence, but the integration of various aspects may be of even greater importance (i.e. the whole is greater than the sum of its parts).

To link the ecological and genetic concepts, it is necessary to extend the concept of the adaptive suite to encompass all of those levels of the phenotype from cell metabolism, to physiology and morphology. All of this higher-order organization can be envisioned as reflecting downwards to impose constraints on genetic structure (e.g. Wilson, 1980, 1988; Vrba, 1989b). The entire adaptationist program in evolutionary ecology is concerned with holding organisms up to the light of optimality models to see how well features of interest conform to theoretical expectations (Maynard Smith, 1978b; Pyke, 1984; Krebs and Houston, 1989; Parker and Maynard Smith, 1990; Schmid-Hempel, 1990). In many cases organisms do remarkably well (Alexander, 1982; Pyke, 1984), although not all aspects of the phenotype may function as adaptations (Gould and Lewontin, 1979; Kauffman, 1993).

Clearly the genetic reflection of the adaptive suite is the coadapted genome. Since DNA ultimately represents simply the order of bases to spell out triplet codons, there are few strong intrinsic constraints on the potential arrangements of unselected DNA. Consequently, any constraints on DNA imposing order must arise via selection at various levels associated with functions of DNA replication or transcription, karyotype, metabolic machinery, cell function, epigenesis, ecological success or evolvability. Thus, the concepts of adaptive suites and coadapted genomes are different sides of a common coin. The natural selection of adaptive suites produces coadapted genomes by downward constraint (Chapter 1). Because adaptive suites are hierarchically organized, this should impose a similar organization on coadapted genomes, although this might only be visible dynamically and not structurally.

A crucial problem for the lineage paradigm is how genetic polymorphism is linked to the coadapted genome and Lerner's (1954) model of genetic homeostasis. The idea that heterozygote advantage could generate such balance has been thoroughly explored (Dobzhansky, 1970; Carson, 1982; Wallace, 1991; Mitton, 1993). Heterozygotes appear to have less environmental variance (i.e. greater canalization) which may be of selective advantage under stabilizing selection. Heterozygotes may also have better performance or broader functional ranges than homozygotes. Such heterozygote advantage (or overdominance) may explain the maintenance of genetic polymorphism in sexual species (Mitton, 1993), although others remain unconvinced (e.g. Lewontin, 1974; Hamilton, 1993).

The fact that trade offs generating pleiotropy may arise at higher levels is a possible escape from this restriction. Balance maintained by overdominance could emerge even at the level of thousands of genes (Mitton, 1993). Thus, Falconer (1977) suggested that additive genetic variability in fitness components coupled with antagonistic pleiotropy among them could give rise to heterozygote superiority for global fitness. In this case, individual components might maintain genetic variation even though additive genetic variation for global fitness would be minimal. The factors associated with key life-history trade offs generating antagonistic pleiotropy (e.g. longevity, reproduction, growth, storage, activity and maintenance) are all polygenically determined traits that could feasibly conform to Falconer's hypothesis (Rose, 1982; Rose, Service and Hutchinson, 1987). Very similar ideas were articulated by Hartung (1981). Mayr (1954, 1955) and Carson (1982, 1987a) viewed co-adapted balances as arising from complex interactions among heterozygous loci.

The potential utility, in terms of flexibility or performance, that might be derived from a heterozygous locus or interactions among heterozygous loci has not yet been fully explored in a regulatory genetic framework, but clearly this is a possibility that is unlikely to have been bypassed in evolutionary explorations. The complexity of regulatory networks is not only consistent with the emergence of particular trade off watersheds at higher levels of genetic organization; it suggests that such trade offs are more likely to be found there than at the level of one or a few loci.

Wallace (1991) proposes a model where the genome is globally co-adapted, but genes are capable of integrating in various combinations to achieve different local coadapted complexes. In this case, the coadaptation is present, but it is loose. Selection can easily disrupt one coadapted complex, but the result may be a shift to an adjacent coadapted complex.

Such a model suggests that measures of linkage disequilibrium at the level of a few loci may not easily detect coadaptation. Carson (1990a) has a similar view, suggesting that local adaptation does not usually involve fixation of homozygous alleles, but rather shifts in polymorphic states. In a species that occupies a non-equilibrium environment, the value of sex in mixing sets of potentially synergistic genetic elements then becomes consistent with the maintenance of a broadly based heterozygous optimum capable of adaptive excursions in response to fluctuations in directional selection. Hamilton (1993) suggests that the improved fitness obtained in offspring from heterozygous crosses may not be due to heterozygosity itself, but to the homozygous fixation of superior multi-locus gene complexes. In all these interpretations, the concepts of a global but variable organization linked by sexual communication is consistent with centre court advantage (Chapter 5). Such a model only applies, of course, to time-tested alleles that are coadapted to the global organization, not new mutations.

7.2 COADAPTATION AND GENETIC REVOLUTIONS

The reality of adaptive suites and coadapted genomes is a nexus for numerous key questions, such as the effective units of selection, micro-*versus* macroevolution, punctuated *versus* gradual speciation, the evolution of biological organization and the understanding of organism form, function and development (to name only a few). It is disconcerting that, despite the central importance of the adaptive suite paradigm in ecology, its reality and implications have rarely been specifically addressed other than in a piecemeal fashion. This strange state of affairs is probably due partially to the fundamental reductionism associated with the prevailing genic view. Models of population genetics do not even consider organisms, only gene frequencies.

The adaptive suite concept is such a cornerstone for evolutionary ecology that there has been little debate about the concept in that literature. To deny the reality of coadapted genomes necessarily denies the existence of adaptive suites and yet this has indirectly become an interesting debate in genetic circles. Disagreement arises because if there are shifts from one adaptive peak to another, a coadapted genome might be expected to undergo a relatively rapid reorganization and this is at odds with classic genic theory which predicts very gradual shifts over long periods of time and the relative independence of selection among traits.

The observation that species tend to remain relatively unchanged for long periods of time, punctuated by periods of rapid genetic reorganization, is one plank in the theory of punctuated equilibria (Eldredge and Gould, 1972; Gould and Eldredge, 1977, 1986), and the reality of species stasis is supported by empirical evidence. A possible mechanism for punctuated equilibria might be the disruption of coadapted genomes and their consolidation on new adaptive peaks which may occur following shifts to new environments or population bottlenecks (Rendel, 1967; Carson, 1991; Mayr, 1992). Gould and Eldredge (1986) also recognized the relevance of such a Wrightian model.

The key debate specifically related to the coadapted genome concept has centred on founder events and how they may have contributed to the explosive evolution of the Hawaiian drosophilid flies. Clearly this debate has general evolutionary implications. One camp maintains that founder events may destabilize coadapted genomes, allowing drift, genetic reorganization and perhaps altered sexual selection to consolidate new species in rapid genetic revolutions (Mayr, 1954, 1982, 1992; Templeton, 1980, 1982a; Carson, 1982, 1989, 1991; Carson and Templeton, 1984; Provine, 1989).

Three basic models have been suggested. Mayr (1954, 1982) proposed that peripheral populations (or those established by low numbers of founders), might accumulate larger degrees of homozygosity, and that this might lead to destabilization of coadaptation maintained by hetero-zygotic interactions. Genetic revolutions and shifts to a new organization might follow. Such a model would be applicable to any population subjected to conditions promoting local inbreeding. Mayr (1954) saw inbreeding as particularly important because it affects all loci at once, and brings recessive genes into homozygous condition where they can be viewed by selection. In the case of metapopulations, genetic variation is compartmentalized among demes and local combinations may be unique. The major criticism of this model has been that the genetic variation needed for reorganization is eliminated by increasing homo-zygosity, but as discussed below, this criticism has now been largely relaxed (Lewin, 1987).

Templeton (1980, 1982b) proposed the genetic transilience model, largely based on the compartmentalization or fixation of genes of large phenotypic effect by founder events. Templeton (1980) intended tran-silience to mean a genetic revolution restricted to only one part of the genome. Reorganization is required to accommodate the altered balance of fitness associated with segregation of a few major modifiers. Such a model is remarkably congruent with the epigenetic hierarchy (Chapters 3 and 4).

Carson's model (Carson, 1975, 1982; Carson and Templeton, 1984) pro-poses that founder events disrupt coadapted gene complexes, probably via compartmentalization of the ancestral genome, and that release of selective constraints in the new environment allows a population flush during which the population may continue to drift away from the ancestral equilibrium. Coadaptation would eventually be restored, but this might differ from the older conformation.

Opponents of these models argue that genetic and demographic factors of themselves cannot generate such revolutions, and that new selective forces must be involved in effecting changes (e.g. Barton and Charlesworth, 1984; Charlesworth and Rouhani, 1988; Barton, 1989). Associated with this dissention is the argument that evolution must be gradual (Barton and Charlesworth, 1984) or at best, may involve a series of weak peak shifts (Barton, 1989). Indirectly such arguments imply that coadaptation of itself is not of strong evolutionary importance (i.e. it will not impede selection in any significant way, or its disruption will not of itself drive further selection or lead to speciation). The crucial question is how strongly may coadaptation be expected to contribute to stasis, resist selection or require reorganization given that disruption occurs? Thus, Barton and Charlesworth (1984) conclude that 'absolute coadaptation between closed systems of alleles is unlikely'. Similarly, Barton (1989) states: 'There is no reason to suppose that coadaptation and homeostasis bind a species into a coherent unit, resistant to change.' Such ideas tend to emerge wherever a genic focus is emphasized. Such views somehow expect the high integration of epigenetic systems to somehow just 'pop out' of selection on genes; not so. Such a perspective leads to the idea that species are not adapted units, have no special significance (e.g. Williams, 1966, p. 252) and consequently integration imposes little constraint on evolutionary potentials.

Given that my central thesis is that lineages, including species, instead represent rather coherent organizations, it is clear that this debate is of critical importance. Thus Mayr (1954) made the point that '. . . a well integrated coadapted gene complex constitutes an evolutionary unit in spite of its intrinsic variability'. Even then, the disagreement boils down to one of degrees. Even those arguing against founder-event speciation employ Wright's adaptive landscape model, with its inherent recognition of coadapted assemblages of genes (e.g. Barton, 1989). Key arguments put forward to deny the importance of genetic revolutions include the following:

- Selection can be obtained in nearly any direction in most species, and even canalized traits can be selected.
- Reproductive isolation between species usually involves many genes rather than just a few. This excludes macroevolution as a major force and favours the idea that isolation occurs gradually in small steps.
- There are no theoretical reasons to expect any strong genetic constraints that will inhibit evolutionary change. Even strong linkage or epistatic interactions do not impede selection.
- The importance of inbreeding is greater on multigenerational time scales, and is not as important in single founder events followed by population flushes.

- Reduction in genic variability associated with founder events does not intrinsically suggest that adaptive peaks will shift.

These arguments suggest that the key problem is in defining how genomes might be constrained by coadaptation. Barton and Charlesworth (1984) certainly are right in that genome structure can ultimately evolve in any direction.

However, genome evolution relevant to eukaryotes is not selected directly. Selection acts on the epigenesis, form and function of phenotypes and it is the existence of phenotypic adaptive suites that imposes the coadapted genome. Discussion of coadapted genomes cannot be divorced from the ecology, development and physiology of phenotypes and this is a key factor largely ignored in genetic discussions. Coadaptation of the genome can arise from requirements for internal organizational features needed to deploy and operate bodies as well as external forces requiring adaptive integration.

The main genetic argument pertains to whether founder events or population bottlenecks could lead to speciation without the co-action of disruptive or directional selection. This amounts to asking whether selection may act on the internal organization itself or whether externally imposed forces are more important. In the hierarchical view, the phenotypic constraint structure results from a combination of environmental and epigenetic selective forces, which ultimately dictate that certain combinations of traits are mutually compatible and others are disadvantageous. This is then reflected in a coadapted genome. In this light, the debate does not have an either/or solution. Wright (1980) emphasized that fluctuating selection may be just as important in local peak shifts as random drift. During a population bottleneck both elements are also likely to be involved. Carson (1989) emphasized the stochastic nature of destabilization, but also held that a new coadaptive balance would arise under directional selection.

The existence of coadapted genomes does not exclude the possibility that directional selection on any one component will nearly always be effective. Consequently, this cannot be used as an argument to imply that coadaptation is weak. What coadaptation does predict, however, is that selection may be slowed or modified by features such as dominance, canalization, plasticity, epistasis, linkage and pleiotropy. Other predictions include the existence of selection limits, genetic correlation structure, selectable rates of recombination and mutation, and selection that is easier in one direction than another. All of these phenomena are well documented (Lerner, 1954; Falconer, 1981; Brooks, 1988).

The evolution of phenotypes to buffer themselves from environmental factors (physiological homeostasis, behavioural flexibility, developmental stability), to deploy plastic alternatives in response to heterogeneity, to rebound resiliently from short-term stress, or to integrate robustly across environmental perturbations, will all reduce the impact of natural selection on a coadapted genome (Schmalhausen, 1949; Thoday, 1953; Lewontin, 1957; Sultan, 1987; Rollo and Shibata, 1991) and should promote species stasis. Such stasis was not predicted by the classical models (Gould and Eldredge, 1986). The unexpected tendency for populations to restore their original phenotype after directional selection is released (reviewed by Lerner, 1954) is strong evidence that some optimal balance exists, and that internal forces apply sufficient selection to drive such changes.

A general lesson from artificial selection experiments is that whenever strong directional selection is applied to bias a particular phenotypic aspect, the general fitness of individuals declines as selection limits are approached (e.g. they become less fertile or viability is reduced). This is undoubtedly a crucial aspect driving the system to return to wild type. Wright (1980) points out that this phenomenon arises from pleiotropic integration rather than inbreeding depression. Such a situation is not predicted by genic theory. The fact that a coadapted balance is involved is also suggested by the fact that further selection on other phenotypic features can sometimes restore fertility and viability.

Another phenomenon relevant to coadaptation is related to the interbreeding of distantly related lineages. Although the F1 generation often displays hybrid vigour or phenotypic luxuriance, this is usually followed in the F2 generation by marked reductions in mean performance (Shields, 1982). Such results suggest that regulatory integration is disrupted in the F2 generation by recombination of elements between the two genomes whereas integration is maintained in the F1 because each genome is represented fully (Endler, 1977; Wallace, 1991). In support of this, F1 hybrids are relatively uniform in phenotype whereas subsequent generations display an explosion of phenotypic variability (e.g. Stebbins, 1950) mostly with lowered fitness. The hybridization event essentially destabilizes the genome.

Genetic mechanisms involving the segregation of coadapted gene complexes protected from recombination are well known. The chromosome inversions of *Drosophila* spp. are one example (Prakash and Lewontin, 1968; Dobzhansky, 1970; Anderson *et al.*, 1991). The evidence suggests that polymorphisms in such complexes persist for millions of years and are associated with ecological adaptation. Such results suggest that coadaptation resides at higher levels of the genome. Elements may be coadapted within chromosomes and complement one another in heterozygous chromosome associations without disruption (i.e. heterozygote advantage pertains to larger genetic elements than single loci). Destabilization appears to involve reorganization within chromosomes, which then disrupts complementation, achieving positive heterosis.

The hierarchical integration of regulatory genes also allows compartmentalization of genomic features. Presumably, if adaptive suites are not of evolutionary importance, we would not expect to find single genomes organized to produce divergent metamorphic, sexual or ecotypic polymorphisms. Instead, continuous phenotypic distributions would always be the rule. It must be highly significant that speciation in the Hawaiian *Drosophila* is usually associated with the fixation of particular inversions formed in ancestral species. Inversion polymorphisms rarely transcend speciation events, whereas isozyme variation is not strongly impacted (Templeton, 1980, 1981; Carson and Templeton, 1984). This situation appears to be genetically analogous to speciation by consolidation of alternative plastic polymorphisms as envisioned by West-Eberhard (1989).

To prove the existence of adaptive suites requires only that we document the existence of mutually incompatible constellations of phenotypic characters. An elegant example mentioned earlier was Kingsolver's work on thermoregulation in pierid butterflies. Recall that two thermoregulatory strategies have evolved, one involving direct absorption of thermal radiation via dark pigmentation, and another involving reflection of thermal radiation via white pigments. There is a dichotomous divergence of coadapted colour and behavioural programs associated with different thermoregulatory tactics. In fact, the integration of features relevant to thermoregulation and energetic power in butterflies includes aspects of behaviour, colour, insulation, morphology and enzymatic coadaptation (Watt, 1985; Kingsolver and Wiernasz, 1991).

Pianka (1978, pp. 92–94) gives a nice example of adaptive suites associated with the quality and availability of prey in lizards. Animals eating abundant but low quality food (e.g. horned toads eating ants) require large digestive systems and tend to have tank-line bodies and low mobility. Those that hunt high quality but widely dispersed prey tend to be sleek and mobile. A similar dichotomy occurs in the residential *versus* dispersive phenotypes of numerous species (Chapter 6).

Adaptive suites may also be discerned in organisms with unique ways of life. Naked mole-rats are remarkable as the only known eusocial vertebrate (Sherman, Jarvis and Alexander, 1991; Cohn, 1992). Their odd constellation of attributes including nakedness, slow growth, ectothermy, burrowing life-style, eusociality and specialization on underground tubers 'are unlikely to be understood unless they are considered *together*' (Alexander, 1991, my italics).

Subterranean rodents have much lower basal metabolic rates than rodents generally, and show a convergence of metabolic and morphological attributes (Lovegrove, 1986). Such convergence may extend to other phylogenies, evidenced by the remarkable morphological similarities between moles and their insect equivalent, the mole cricket.

Phenotypically integrated features may not be visible at the genic level, but must represent an important level of organization contributing to the stasis and cohesion of coadapted genomes.

The reality of coadapted genomes suggests that genetic reorganizations of a revolutionary nature may occasionally be favoured but this does not impose the restriction that such reorganizations occur in a single or very few generations. Although revolutions may be fast, they still require multigenerational periods (Carson, 1991). Mayr (1992) also emphasizes this point and suggests that the term revolution was unfortunate in this regards. In particular, the acquisition of new coadapted balance may span hundreds or thousands of generations and proceed via microevolution (Carson, 1982). The hierarchical regulatory model also does not require that numerous genes are always involved in speciation. The study of Meyer *et al.* (1990) demonstrating low genetic variation in cichlid fish is good support for this contention. Regardless, the model also does not preclude that many (perhaps most) species are obtained gradually by the accumulation of numerous divergent alleles. Species may still have to climb to adaptive peaks gradually once they reach the foothills. The argument that genetic revolutions cannot occur because numerous genes must be involved over relatively long time frames can be dismissed by documenting one counter example (e.g. Meyer *et al.*, 1990).

Finally, the evidence that most species presently vary at numerous loci does not address the issue of the particular genetic variation associated with the speciation event itself. Nor does it consider that some variants may be of paramount importance while others are effectively neutral. If phenotypic evolution is mainly via regulatory integration, then the actual genic content may be relatively less important than the degree of 'rewiring' (Carson, 1991; Kauffman, 1993). In this respect it is significant that some species of Hawaiian *Drosophila* and some Lepidoptera show no fixed electrophoretic differences even though they vary widely in behaviour, morphology and inversion polymorphisms (Templeton, 1981; Carson, 1982; Mayr, 1982). With reference to parthenogenetic strains of *Drosophila mercatorum*, Templeton (1979) observed that coadaptation involved changes in loci regulating fundamental aspects of development, rather than changes in loci coding isozymes.

Somehow, the hierarchical regulatory model must be incorporated into the concept of the coadapted genome. Given that epigenetic hierarchies may involve heterogenous alleles in obtaining balanced phenotypes, simplification of the organization could lead to destabilization or increased phenotypic variance. Founder events or the fracturing of large populations into locally inbred demes during population bottlenecks is theoretically one avenue for exposing potential regulatory variants to prevailing selection. Inbred populations effectively represent potentially unique constellations of regulatory structure, associated with homozygous fixation of different components across the entire regulatory network or localization of structural variants.

The idea that such a process may increase selectable variance is counter-intuitive for genic models, and consequently, recent evidence that moderate to strong inbreeding *increases* selectable genetic variance is of considerable interest (e.g. Bryant, McCommas and Combs, 1986a,b; Goodnight, 1987, 1988; Bryant and Meffert, 1988, 1990; Bryant, 1989; Carson and Wisotzkey, 1989; Carson, 1990a,b). Current explanations for the increased variance associated with bottlenecks or populations sub-divided into small demes invoke fixation of rare recessive alleles or changes in selectable variance associated with changes in epistatic inte-gration (Robertson, 1952; Goodnight, 1987, 1988). Epistatic genetic varia-tion is converted into additive genetic variation. Levin (1970) anticipated these results in an early article where he reviewed evidence that stress in peripherally isolated populations might disrupt canalization and release previously hidden genetic and phenotypic variation. The epigenetic hierarchy outlined earlier suggests that genes of major impact do exist and that the hierarchy is intermeshed by pleiotropic interactions. Such genetic architecture is capable of radical phenotypic shifts (Carson and Templeton, 1984). Wade (1992) argues that genetic interactions will generally evolve to be opposite in sign to the main additive effects of genes. This might also promote coadaptation but allow release of selectable genetic variation on destabilization.

Templeton (1982b) emphasized that such systems will be more likely to drift during population bottlenecks. Carson and Templeton (1984) also recognized that such systems may show release of genetic variation following destabilization. Although there is little data directly relevant to genetic variation in higher-order epigenetic holons, it is possible that antagonistic pleiotropy among genes for major fitness components could maintain variation as discussed above (Carson and Templeton, 1984; Rose, Service and Hutchinson, 1987). In this case, genetic revolutions could be a rather pervasive phenomenon. Presumably, mutations in higher-order genes could also generate revolutions in metapopulations that are sufficiently fractured.

Carson's (1975, 1982) suggestion that the genome could be subdivided into open and closed components is also of relevance here. He proposed that genetic revolutions must involve reorganizations in mechanisms governing key developmental systems and life-history components, whereas those affecting isozymes or structural genes may be relatively unimportant. Clearly this is a recognition of the importance of regulatory *versus* structural aspects (Chapter 3) and implies that some genes (e.g. higher in the epigenetic hierarchy) are more relevant to variation in phenotypes than other elements (structural genes or lower-order regula-tory modifiers). Mayr (1975, 1982) and Templeton (1981) also recognized the possible importance of regulatory integration in coadaptation and genetic revolutions.

Carson's (1975) model suggests that considerable higher-order variation exists but is essentially locked up in coadapted balances. Disruption, as occurs in bottlenecks, could then release it to selection. Bryant and Meffert (1990) showed that this appears to hold for flies passing through bottlenecks, established populations diverging in both shape (strong bottlenecks) and size (weaker bottlenecks). Such phenomena are associated with changes in genetic covariance structure and the release of selectable genetic variation (Bryant, 1989).

Mayr (1954, 1963, 1982) proposed that inbreeding of itself could generate new selective forces, but until recently the potential importance of this idea was questioned because, according to genic theory, inbreeding should always reduce the genetic variation needed to respond. Certainly any founder event may release a species from local high density, competitors or parasites (Cockburn, 1991). To the extent that these forces impose constraining coadaptation on the genome, founder events represent a release of constraint and alteration of forces shaping co-adapted integration. This might well fuel internally driven selection to obtain a new balance and would also allow a greater degree of drift as envisioned by Templeton (1982b), Carson and Templeton (1984) and Carson (1991). Bryant, Combs and McCommas (1986a) suggest that larger blocks of genes may constitute selective units following population bottlenecks which might contribute to possible shifting.

A coadapted genome produces adaptive suites geared to particular environments. Carson's (1975) closed genetic system may be relatively protected in optimal habitats but could be destabilized by novel or extreme environmental factors. Consequently, it is highly significant that some fitness components of *Drosophila melanogaster* have proved to have little response to artificial selection in optimal conditions, but considerable selectable variance is released under various forms of stress such as temperature extremes, high density, or food limitations (Giesel, Murphy and Manlove, 1982a; Luckinbill and Clare, 1985, 1986; Clare and Luckinbill, 1985; Arking and Clare, 1986; Gebhardt and Stearns, 1988). Similar increases in phenotypic and expressed genetic variance occur under dietary restriction or crowding in mice (Totter, 1985; Hoffmann and Parsons, 1991, p. 120). Schmalhausen (1949, p. 74), proposed that directional selection would also destabilize the genic balance of the wild type form, leading to the release of previously hidden genetic variance (changes in penetrance), an intuition supported by the study of Gilbert (1986).

Regardless of the source of destabilization (i.e. selection, drift, mutations, simplification or compartmentalization) an organized system would be strongly selected to restore balance subsequently or improve associated decreases in fertility and viability (Carson and Templeton, 1984). Following such restructuring, a lineage might not readily fuse with the old wild type, even if communication with ancestral populations was restored (i.e. speciation has been at least initiated). This represents a genetic revolution in every sense of the definition. Thus, recent evidence provides strong exceptions to both the genic predictions of high genetic variation associated with species diversification, or low heritable variation associated with inbreeding. There does not seem sufficient evidence to reject revolutions fueled by founder events, although their occurrence would clearly be greatly synergized by locally divergent selection pressures. Overall, the lineage model suggests that revolutions should be important, the main limitation being availability or penetrance of regulatory variation.

7.3 GENETIC REVOLUTIONS AND METAPOPULATION DYNAMICS

Although the main focus of this debate has been founder events, such a situation is only one reflection of numerous models that predict that small, relatively isolated populations are most prone to speciation (Mayr, 1954, 1982; Templeton, 1982b). Peripheral isolates are another contender. Possibly the strongest candidate for such events, however, occurs when a relatively large panmictic population is subjected to a global stress that impacts across a broad geographic range. Rather than contracting into a single small population, such a situation will lead to the fracturing of the population into numerous small demes with varying degrees of isolation (Figure 7.1). Some local habitats might also become vacant, allowing local founder events to occur within the species' range. Aspects of the global genetic templet may be relatively compartmentalized among demes, and local drift and inbreeding would tend to accentuate disruption of former coadaptation. Each deme has the capacity to explore differing gene interaction systems (Wade, 1992). Such population structure is most likely to allow escape from stabilizing selection (Lande, 1986; Mayr, 1992).

Such circumstances not only allow the operation of group selection, but more importantly, the potential for local genetic revolutions is greatly accentuated (Carson, 1982) which will synergize group selection. Thus, we return to Wright's (1980, 1982a,b, 1988) model for adaptive peak shifts. The really important insight here is that any new adaptation that emerges is relevant to the organization of the population as a coadapted complex (Carson, 1982; Mayr, 1975, 1982; Templeton, 1989). That is, lineage organization is the relevant feature to consider, and this emerges at the population rather than individual level.

Intraspecifically, such processes allow species to explore their adaptive landscape (hierarchical hopscotch, Chapter 5), sequentially shifting onto

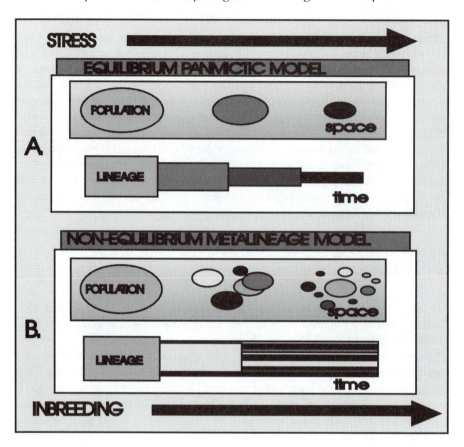

Figure 7.1 Two models of genetic and ecological dynamics. (A) In an equilibrium world with high homogeneity, populations may be considered as large, continuous and panmictic (i.e. Fisher's framework for evolution and that assumed in ecological models based on simple logistic equations). Under stress such populations may decrease in size and be subjected to inbreeding. (B) In a non-equilibrium world with high heterogeneity, large populations are rarely uniform or panmictic. Stress results in fracturing the population into a metapopulation structure where demes may reside in ecologically divergent patches. Even in a uniform environment, however, demes would be expected to diverge in regulatory structure and allelic constitution.

progressively higher adaptive peaks (Templeton, 1982b; Charlesworth and Rouhani, 1988). As the process continues, it may become increasingly more difficult to descend such peaks and highly coadapted complexes may be relatively resistant to shifts (Mayr, 1975, 1978; Carson, 1982; Charlesworth and Rouhani, 1988). Destabilization may also allow access to a more distant peak which may enforce some degree of genetic isolation or impose reduced fitness on hybrids formed during inter-breeding of differently adapted populations. This represents incipient speciation. Thus, shifting balance may be involved in the attainment of increasing lineage coadaptation within species, as well as escape and consolidation of sublineages on new peaks representing novel species. Kauffman (1993) explores ideas concerning destabilization of populations climbing adaptive peaks as their number of components and connectance increase. Such factors may prevent species from reaching pinnacles of adaptation, or under some conditions, destabilization may allow escape from local peaks.

Carson and Templeton (1984) were careful to point out that their ideas were specific to founder events, and as such could not be extended as a general basis for macroevolution. The central ideas of intrinsic coadapta-tion, destabilization and shifts to new coadapted conformations do, however, represent such a model (Carson, 1982). Radical shifts in the environment, marginal populations, founder events, population bottle-necks, hybridization events and metapopulation dynamics can all be fitted into such a framework (Templeton, 1980, 1981; Carson, 1982).

The coadaptation between the sexes may be a particularly important aspect subject to disruption in species with complex mating behaviour, something that was of particular interest to Goldschmidt (1940). Galiana, Ayala and Moya (1989) support the contention that population bottle-necks may also generate some degree of sexual isolation in the desired lines, a result clearly important to the genetic reinforcement of a local coadapted constellation.

The genetic templet is an extension of the coadapted genome concept to a lineage paradigm. Multigenerational selection on a hierarchically structured genetic organization can evolve an entity capable of deploying numerous adaptive phenotypes, and one which may also incorporate features promoting evolvability. Rather than rejecting the idea that co-adaptation is important, the evidence suggests that it is rather crucial to understanding the evolution of phenotypic organization. The templet concept expands the horizon within which coadaptive integration might be considered. We should not reject the idea that higher-order organi-zation exists just because it is distributed among individuals, it is only expressed in some environments or because some features are circulating. Rather, we need to enlarge our concept of organization to recognize that a single system may be dynamic, and that local genetic and phenotypic variation may be an integral part of long-term, larger scale adaptation (e.g. Lerner, 1954; Carson, 1990a; Mrosovsky, 1990). As Carson (1982) points out, evolutionary stasis may emerge once a species has attained a state of stable variation.

Of great significance to the possible selective value of sexual reproduc-tion to lineages is Goodnight's (1988) theoretical conclusion that high

recombination would greatly facilitate the conversion of epistatic variation to additive genetic variance. Epistatically derived features decay
when selection is relaxed (Chapter 5) because they may be associated
with linkage disequilibrium. A conversion process would allow various
players to be deployed more permanently to various positions in the
field.

The discussion of phenotypic plasticity and genetic correlations is also
highly relevant to the debate regarding genetic revolutions. The coadaptation debate that destabilization and conversion of epistatic variance into
heritable additive genetic variation contains an inherent assumption that
such processes represent non-adaptive disruption of a globally balanced
initial optimum. The discussion in Chapter 6 suggests that norms of
reaction may release genetic variation in particular ways and contribute
to genetic correlations as well. Genetic correlations may effectively steer
microevolution in particular directions or resist selection in others
(Antonovics, 1976; Lande, 1986). Plasticity may allow a genotype to
adjust its phenotypes to avoid selection. Both of these phenomena may
be embedded in a deeper genetic structure (Stearns, 1992) that might
conform to Carson's (1975) locked portion of the genome. A higher-order
organization could potentially exert some adaptive control over such
mechanisms.

In a hierarchically organized sexual lineage, stabilizing selection yielding lineage identity may act at different levels. Thus, local sublineages
may be subject to particular local forces, but part of their nature may also
be derived in stabilizing selection at the level of the metalineage to which
they belong. In the previous discussion, the possibility that norms of
reaction themselves are subject to stabilizing selection that might conflict
with local adaptation is one example.

Goldschmidt's (1940) support for the importance of macromutations
and the reality of genetic revolutions was sufficient to earn his rejection
from the genic community. A less well-known heresy was his contention
that the formation of local races may not represent incipient speciation,
but simply the local ecological deployment of optimal interfaces by
a globally adapted species. This conclusion was based on a rather
admirable review of geographic distribution patterns, although Mayr
(1992) complains that Goldschmidt selectively ignored available information.

The allopatric model of speciation via local divergence of races is well
established (Mayr, 1963), so Goldschmidt's (1940) conclusions were
clearly too extreme. Mayr (1954) actually accepted Goldschmidt's (1940)
interpretation, holding that isolation was the crucial aspect pertinent
to speciation events, and as long as some communication continued,
speciation was unlikely. The increasing evidence that local specialization
can be maintained within sexual lineages, even in the face of considerable

genetic exchange (Endler, 1973, 1977; Antonovics, 1976), however, highlights Goldschmidt's conclusions.

Significantly, in his exploration of the reality of coadaptation and genetic revolutions, Carson (1982) was led to the same conclusions as Goldschmidt (1940). At least in some cases, differentiation may well represent local adaptation of a higher-order genetic entity and not necessarily disruptive fracturing of the lineage into incipient species. Carson (1982) suggested that speciation required modification of the deeper balanced portion of the genome whereas most subspecies or races did not meet that criterion. The relevant factors are the frequency of reunions, the compartmentalization of relevant genes within the hierarchical templet and changes in regulatory integration.

Recently, Barnett and Dickson (1989) reviewed the literature on cold adaptation in populations of the house mouse. The authors were unable to differentiate whether divergence among populations with respect to this factor represented true evolutionary divergence, or simply local adaptive deployment of the fundamental mouse design to local contingencies. Temperature is a pervasive selection factor for lineages of small rodents, and in *Mus* the optimal phenotype is not only temperature-specific; relevant genotypes appear available at short notice. The existence of reaction norms for body size and tail lengths in response to temperature may well be relevant here.

A really interesting question is whether stabilizing selection at the level of metalineages may impart some organizational features at a rather deep genetic level that might adaptively steer the release of genetic variation and direction of microevolutionary responses among sublineages. Genetic inertia and genomic memory (Chapter 4) could also contribute to such an organization. Although the reality of such a situation would be distasteful to those hoping to explain evolution via the differential segregation of independent selfish genes, the question as to whether such mechanisms exist and may be adaptive is now beginning to be asked (Gilbert, 1986; Clark, 1987b; Stearns, 1989, 1992; Price and Langen, 1992). The range of potentially adaptive response in such a system would be limited to the global umbrella of stabilizing selection. Outside this range, fortuitous adaptive conformations might still be derived by extrapolation, but non-adaptive morphoses or true disrupted coadaptation are more likely. In such a model, speciation events would only be expected outside the range defined by metalineage organization and coherence.

8

Trade offs and synergisms in phenotypic organization: an evolutionary framework

8.1 INTRODUCTION

The key to understanding the genetic organization and evolution of phenotypes lies in the realm of their epigenetics and ecology (i.e. organismal design). We have already considered the genetic and developmental spheres. To understand how ecology shapes epigenetics it is best to consider organisms as complex machines, in fact, the most complicated machines in the universe (Dawkins, 1989a).

Persistence of lineages is ultimately obtained by reproduction of phenotypes, but also by features improving the chances that reproduction will be achieved and by features improving offspring survival and adult competence. At the basic level, DNA templets could simply make copies of themselves without elaborating phenotypes. In fact, such simple molecular evolution might be considered the most fit state in an unlimited soup of nutrients with no competition (Dawkins, 1989a), because replication would be most rapid and efficient with shorter DNA strands.

It is instructive to note that, in a truly unlimited and unconstraining environment, every organism that has ever existed, and many, many more that we can or cannot imagine, would eventually be realized. This follows because evolution is thought to be blind, acting only on random mutations that occur by chance in existing organisms. Because natural selection is merely an editor, the existing end points of the diversifying primordial lineage in some sense represent those pathways that natural selection has not acted against. Natural selection can be imagined to act like the cyclops, Polyphemus, carefully checking each sheep for hidden riders as it emerges from his cave. Only a sheep, a man hiding under one, or a wolf in sheep's clothing might pass.

With a little reflection, the value of sex to facilitate the assembly of sets of building blocks that have variously passed this test of invisibility also becomes very clear. The oozing, crawling, swimming, shuffling, running and flying emergence of primitive DNA from the primordial soup, has depended upon organizational overlays that allow sequestered templets to persist in otherwise hostile habitats, such as thermal vents, brine pools or deserts. With respect to organizational advance this represents the acquisition of further cloaks of invisibility, each of which allows access to a further recess of the universe.

It is most startling that what effectively represents cloaks of invisibility from a gene's-eye view turn out to be graceful and elegant robots from an ecological perspective. The emperor's invisible cloak is actually of a most complicated weave, and the designer's hallmark is organization (Chapter 1). It may be artificial to try to separate the genetic program, that ultimately elaborates and runs the phenotype, from the phenotype itself. It is possible, for example, to consider the robots as the unit of selection and the DNA simply as one component of organization – the memory. A computer is nothing without a program to run it, but a program without a computer is just as useless. The point is that the organizational evolution that has fostered lumbering robots must also be reflected in the on-board genetic organization. Besides being replicators, genes have functions, and themselves have structures relevant to the timing and location of their contributions (i.e. genes have phenotypes too). Phenotypic overlays improve the fidelity and uniformity of success (at some cost in production rates).

Phenotypes may also improve or expand avenues of resource acquisition, improve defence and offset competition. In fact, there is a range of functions that are nearly universal to all phenotypes, and this provides a framework for considering organismal design. In essence, an organism can be considered as a resource allocation system that has a number of universal demands (Figure 8.1). Because of limitations imposed by features such as biochemical pathways, cell surface areas, size of the digestive tract, and DNA fidelity, there are immediate and life-time limitations on available energy, resources and time. This necessitates competition among demands from different features. The way that resources are allocated is not random, but is a physiologically controlled process termed the resource-allocation strategy (Sibly and Calow, 1986). The strategy must be adapted to the existing range of environmental constraints and opportunities (Figure 8.1). The environment, then, can be considered as a kind of templet that shapes the potential solutions on the part of the organism (Southwood, 1977, 1988). Thus, the genetic templet concept is meant as the evolutionary reflection of Southwood's habitat templet.

In animals, behaviour may act as an integrative filter between the

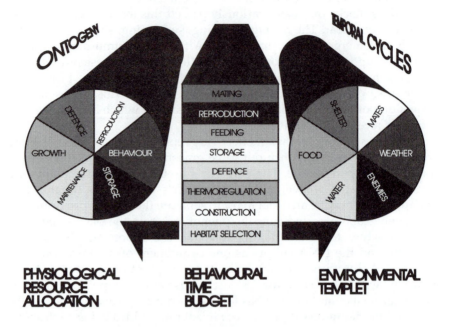

Figure 8.1 A conceptual model for holistically considering animal phenotypes. All animals must allocate resources among a common suite of physiological demands. The particular resource allocation strategy involves regulated deployment rather than a simple free-for-all among these requirements. Behaviour is one aspect of the program that requires investments, but it also determines the amount of resources obtained, and may have other benefits associated with avoiding predators, or selecting environments for optimal performance. Thus, behaviour acts as an interface linking the physiological sphere to the environmental templet. A holistic consideration of physiological resource-allocation strategies and behavioural time-budgeting tactics provides a framework appropriate for considering phenotypic evolution.

physiological sphere and the environment (Boggs, 1992). Since plants largely lack such a filter, they must be correspondingly more flexible physiologically and developmentally. The kind of behaviour exhibited by animals, its intensity and its duration, are also under specific control. The overall strategy is termed the behavioural time budget (Figure 8.1).

Behavioural features specific to one function (e.g. foraging) are nearly always mutually exclusive of others (e.g. mating or sleeping). Thus, the allocation of time to specific behaviours may take the form of a time budget (McFarland, 1977). Behaviour is linked to resource allocation via its energetic costs, and because this is the main avenue for acquiring resources such as water, sexual partners, food and temperature. Thus, in animals, resource allocation and behavioural budgets are functionally interconnected (Boggs, 1992). Resource allocation strategies dictate the energy available for sustaining rates and durations of different behaviours but such strategies must take into account both returns and costs of behavioural investments.

This framework of resource allocation is capable of explaining considerable variation in phenotypic form and function, but there are several important overlays that alter the realization of this simple economic approach. The main ones include morphology and associated physical aspects of phenotypic function, as well as specialization and associated limits on the allocation system. Risk is particularly important and may apply to a broad range of sources including external forces such as microclimate or predation, as well as internally driven factors such as failure to meet metabolic requirements or failure associated with senescence.

8.2 SOURCES OF SYNERGISM AND TRADE OFFS: SPECIALIZATION

The ultimate question for both ecology and evolution is why there are so many kinds of organisms (Hutchinson, 1959). One factor pertains to the general principles of morphological, behavioural and physiological specialization as reviewed extensively by Futuyma and Moreno (1988) and Cockburn (1991). Almost invariably, specialization to improve performance (rates and/or efficiencies) in one aspect is associated with reduced performance in others. Such a principle is perhaps most obvious with morphological attributes. Thus, bees with short broad tongues are well adapted to mop up nectar from plants with many, tightly clustered and open flowers (e.g. goldenrod), but bees with long narrow tongues are better on flowers with nectaries recessed in deep corollas. Similarly, the beaks of birds may be most efficient with seeds within particular size ranges (Smith, 1990a,b).

Futuyma and Moreno (1988) review specialization in the context of within individual niche breadth or breadth across varying individuals. Thus, in constant environments a single specialized phenotype may be deployed. In variable environments a single generalized phenotype may be favoured, or alternatively a genetic or phenotypically plastic polymorphism might evolve (Levins, 1968).

The ability of genomes to harbour code for divergent specialized phenotypes may be limited by disruption caused by differential utilization of common elements or by selection pressures that are too weak, infrequent or locally rare to leave a lasting imprint on a templet that is subject to constant erasure via mutations (Chapter 6). Thus a general purpose genome capable of generating phenotypes for all habitats and niches on earth has not evolved. Instead, even the initial Cambrian radiation was highly diverse and surviving lineages have bifurcated into compartmentalized phylogenetic lines, each with limited and specialized ecological attributes.

Despite the failure of life to evolve as a truly general purpose genome, the question of generalist *versus* specialist organisms remains a central issue on smaller scales (e.g. insects that eat numerous plant hosts *versus* those restricted to single species or genera). Species adapted to narrow, constant niches generally display low ranges of physiological tolerance. Such cases may reflect a lack of selection outside the observed range, leading to loss of response similar to the atrophy commonly observed in cave animals or parasites. Alternatively, exposure to a very broad range of ecological variables and values may limit the ability of a single phenotype to optimize performance within this range (a jack of all trades is a master of none). The idea that generalist strategies entail a cost in specialization is pervasive in the literature, but clear demonstrations are difficult to obtain.

Competition is relevant to specialization in three ways (Futuyma and Moreno, 1988): specialists may exclude generalists; competitors may shift facultatively to alternative, less utilized resources; or evolutionary character displacement may occur. Predators and parasites may also impact on niche width if utilization of enemy-free space is possible (reviewed by Cockburn, 1991). It is important to place such features in a non-equilibrium context. Although specialists may outperform generalists in any given locality, generalists may persist in a non-equilibrium world because specialists cannot globally saturate their niches. Moreover, by integrating across numerous subniches, generalists may obtain greater long-term stability/ persistence whereas specialization may be associated with greater susceptibility to local or even global extinction. Finally, there is a tendency to view specialization as an irreversible dead-end process in an evolutionary framework, the idea being that a species with numerous adaptations to a particular niche is constrained by its own specificity. Although there may be some truth to this idea, it must be viewed with extreme caution. There are numerous examples of relatively specialized organisms that have undergone extensive evolutionary radiations. In some cases, a specialization may be the gateway to a new evolutionary universe.

8.3 SOURCES OF SYNERGISM AND TRADE OFFS: MORPHOLOGY

A crucial overlap on this framework of allocation is the morphological structure of the phenotype. Thompson (1942) pioneered this area which is only now entering a real renaissance. Hall (1992) reviews the rapid advances in understanding the genetic control of morphology in the skull and mandible of rodents. Although the connection is obvious, the actual linkage between resource-allocation strategies and morphological evolution remains in its relative infancy. The greatest progress is being made with plants, largely because their modular format and relatively simple designs (i.e. stems, branches, leaves and flowers) are amenable to analysis and simulation. Among animals, a limited number of fundamental body plans (Bauplan) prevail. According to Margulis and Schwartz (1988, in Gilbert, 1991), there are 33 extant Bauplan and no new ones have been added in the last half billion years. In fact the trend shows a general reduction following an initial proliferation of perhaps 100 initial designs (Gould, 1989a; McKinney and McNamara, 1991).

Morphology is probably the most neglected of the areas crucial for a complete development of evolutionary ecology, and one that may contain many interesting surprises. To some extent, for example, resource allocation and morphological evolution may be decoupled. Several unpublished studies by my students are illustrative. Firstly, in the plant genus *Impatiens*, the touch-me-nots, the escaped ornamental *I. glandulifera* has a radically different branching pattern and leaf design than the native species, *I. pallida* and *I. capensis*. Flower size and number vary widely among all of these species. Despite their radically different morphologies, and considerable evolutionary distance, the allocation of mass to reproduction, stems and leaves is nearly identical in all of these species. Resource allocation may form a constraint structure within which morphology is somewhat free to vary.

In burdock, the two introduced species, *Arctium lappa* and *A. minus*, have radically different reproductive packaging. *Arctium lappa* has burs that are twice as large as those of *A. minus*. This is apparently achieved by a mechanism that may be highly relevant to resource allocation: *A. minus* has roughly twice as many branches and thus, twice as many sinks for reproductive allocation. Despite differences in both number and size of reproductive packages, the total mass of burs is a relatively constant proportion of above ground plant mass in both species (unpublished data).

Alexander (1982) analyzed the allocation of materials to skeletons in relation to the needs for resisting strain and predictable stress. The evolution of the Warren truss in bird's bones is a nice illustration of the need for maximizing strength while minimizing mass (Welty, 1952). Allometric studies also reveal considerable insight into changes in allocation among features associated with changing body size (Peters, 1983; Calder, 1984; Schmidt-Nielsen, 1984).

Behaviour is also intimately connected to morphology. It is striking, in fact, how often organisms converge on common solutions of morphological and behavioural integration. Rensch (1960) pointed out, for example, how saber-toothed cats that spear their prey with their greatly enlarged incisors evolved with remarkable fidelity, and complete independence, in both eutherian and marsupial lineages. Most readers will be familiar with the predatory elegance elaborated in the design of the praying mantis (an orthopteran insect) but few may be aware that adults of the unrelated Mantispidae (a neuropteran) are nearly identical in form and function, even though the larval stages of this insect are specialized parasites of spiders' egg cases. In this case, the two adult phenotypes represent different coadapted genomes that are convergent reflections of the same ecological solution (Chapter 1).

By its nature, however, morphological evolution will be one of those areas contributing the most to phenotypic diversity. Thus, raptorial front legs that can be used for capturing prey have evolved in numerous lineages including stomatopod shrimps, ambush bugs, water scorpions, assassin bugs, mantids and mantispids to name only a few. Although the examples of convergent evolution are remarkable, a global view suggests that unity amidst diversity is more the rule, with the emphasis strongly on diversity. There may be some fundamental generalities that may be ultimately extracted concerning morphological evolution (e.g. allometry is already finding some), but there will always be a need for specific examination of particular features. For a delightful example, explore the analysis carried out on the design of winged propellers as seed dispersal mechanisms in plants (Azuma and Yasuda, 1989). Such studies point out clearly how different solutions will of themselves require different adaptive constellations of features which cannot be directly compared. Thus, plants that rely on parachuting seeds will have quite different selective frameworks.

Resource allocation appears to be a more fundamental level of organization and is consequently more amenable to comparative generalities. In the discussions that follow, the reader should retain the idea that considerations of allocation underlie a superstructure of morphology, even though consideration of that complexity cannot be elaborated further here.

A final important dimension in the form and function of phenotypes is their development. Phenotypes usually derive from a single cell (except for asexual somatic budding) and during all stages of growth and maturation, the organism must be fully competent. Thus, the epigenetic trajectory of an organism must be integrated to the temporal changes occurring in the environment, or to those arising from movements of the organism (e.g. differentiated life cycles of insects or animal parasites with multiple hosts). If selection in early stages would disrupt adaptive aspects of adults, then a metamorphosis may be required. Otherwise, all features expressed at one stage in life must not only meet the needs of the current environment, but also the need to elaborate modified features in later phenotypes. In animals, of course, behaviour must change so as to interface the current phenotype in the best way with the current environment.

If we put all of this together, we obtain a relatively comprehensive framework for considering phenotypic adaptive suites. The adaptive suite may then be defined as the epigenetic integration of physiological resource allocation, morphology and behavioural time budgets in response to a given dynamic habitat templet. This ecological templet defines the constraints, opportunities and risks that limit phenotypic form, function and life histories. Because there are a limited number of fundamental problems, it is possible to explore whether a general theory of organism design is a possible goal for evolutionary ecology.

8.4 TRADE OFFS IN PHYSIOLOGICAL AND LIFE-HISTORY ATTRIBUTES

8.4.1 Overview

Stearns (1992, p. 79) lists 45 key trade offs expected in life-history evolution based on possible interactions among current and future reproduction, parental survival, growth and condition, the size and number of propagules and the growth, condition and survival of offspring. The emphasis of the present discussion will centre on physiological trade offs relevant to phenotypic design, since others have been well reviewed by Cockburn (1991), Stearns (1992) and Roff (1993). A relatively complete trade off matrix is obtained based on 10 fundamental allocation pathways (Table 8.1). This suggests 45 basic physiological trade offs highly relevant to the evolution of organism design. Others could be added but these constitute the most fundamental attributes.

Evolutionary ecology is primarily concerned with understanding the holistic integration of form and function of organisms. Life-history theory rests on the assumption that key fitness features are traded off against one another. This is known as the principle of allocation. This is interpreted in a framework of age-specific schedules of reproductive investment and mortality (Fisher, 1930; Cody, 1966; Williams, 1966b; Gadgil and Bossert, 1970; Stearns, 1976, 1992; Charlesworth, 1980, 1990; Partridge and Harvey, 1985, 1988; Sibly and Calow, 1986; Caswell, 1989; Dingle, 1990; Charnov, 1991; Cockburn, 1991; Rose, 1991; Lessells, 1991; Partridge and Sibly, 1991; Reznick, 1992a; Roff, 1993).

If increases in one feature can only be obtained at the expense of another, then variations among key fitness components are predicted to show negative correlations. Such inverse relationships might be expected to be visible in the structure of phenotypes. This could be approached by comparing organisms of closely similar design, or by broader analyses of key features across phylogenetic boundaries.

Experimental manipulations that force organisms to vary investments in one feature allow the result to be compared with controls. The rationale for

Table 8.1 The resource allocation matrix for phenotypic features. Longevity is incorporated as a resource sink involving various repair, replacement and protection costs. Information processing refers to all investments relative to detecting, storing and computing information

Trait 1		Trait 2									
	Code	L	R	G	T	B	W	S	D	I
Maintenance	M	1	2	3	4	5	6	7	8	9
Longevity assurance	L		10	11	12	13	14	15	16	17
Reproduction	R			18	19	20	21	22	23	24
Growth	G				25	26	27	28	29	30
Thermoregulation	T					31	32	33	34	35
Behavioural activity	B						36	37	38	39
Water/ion balance	W							40	41	42
Storage	S								43	44
Defence/immunity	D									45
Information processing	I									

such approaches has been that if there are constraints on phenotypic design, they should be visible by comparing organisms, independent of their genetic underpinnings (Partridge and Sibly, 1991; Partridge, 1992). Such a framework constitutes what is termed the adaptationist program, and a major approach has been the application of optimality models that measure how closely observed features conform to mathematical predictions (e.g. Maynard Smith, 1978b; Pyke, 1984; Parker and Maynard Smith, 1990).

An alternative approach is based on the idea that natural selection tends to consolidate positive interactions among key fitness features (either by fixation of portions of the genome that contribute positively to all features, or by some sort of heterozygous balance allowing persistence of variation locked into coadapted, overlapping complexes). Genes selected for their positive impact on one or more features might also be incorporated, even if they have some negative impacts on others. Universally positive genes should be fixed, and universally negative genes should be eliminated. Given the extreme degree of pleiotropy in epigenetic systems, however, genes with universal impacts in either direction may be relatively rare. Most genes may have positive impacts on some aspects, but negative effects on others. It is the impact on global fitness (that of the lineage) that determines the ultimate selection pressure on genes and higher-order complexes. At this level the fitness of a gene may also vary widely with the particular genetic background.

In the end, genetic variation promoting positive interactions among fitness components should be exhausted so the remaining genetic covariation is of a largely negative sign. Such negative correlations are then indicative of trade offs among components of the adaptive suite. If all genetic correlations are positive, then a single fittest genotype should displace all others (Moller, Smith and Sibly, 1989a). Consequently, where variation exists, genetic correlations should be predominately negative, a state referred to as antagonistic pleiotropy (Falconer, 1981; Rose, 1984a; Reznick, Perry and Travis, 1986; Clark, 1987a; Rose, Service and Hutchinson, 1987; Caswell, 1989; Charlesworth, 1990; Wade, 1992). Estimates of genetic correlations are obtained either in breeding regimes in particular environments with no selection, or by examining correlated responses under selection regimes (Falconer, 1981; Stearns, 1992).

A schism has developed among evolutionary ecologists with respect to which kind of measurement is most appropriate for revealing trade offs – phenotypic or genetic? The genetic camp holds that since microevolution requires genetic variation, the genetic correlation structure defines the direction of short-term evolutionary responses. Scheiner and Istock (1991) provide an example. Thus, the only relevant information can be argued to be genetic (Reznick, 1985, 1992a,b; Bell and Koufopanou, 1986; Reznick, Perry and Travis, 1986; Clark, 1987b; Rose, Service and Hutchinson, 1987; Caswell, 1989; Price and Langen, 1992).

Problems with the genetic approach include the difficulty of accurately estimating genetic correlations, and that these may change during selection or vary with the environment. Different selective factors may derive different correlations, and correlations may also vary with ontogeny (Mayer and Baker, 1984; Service and Rose, 1985; Bell and Koufopanou, 1986; Mitchell-Olds and Rutledge, 1986; Clark, 1987a,b; Riska, 1989; Dingle, 1990; Stearns, de Jong and Newman, 1991; Cowley and Atchley, 1992; Stearns, 1992; Roff, 1993). Furthermore, genetic correlations are very sensitive to local changes in gene frequencies, and they are likely to be altered by the degree of inbreeding (Stearns, 1992). The release of genetic variation under inbreeding, strong directional selection and environmental stress may have unpredictable impacts on genetic correlations which we are only beginning to appreciate. Release of such variation may allow evolutionary responses that would not be predicted from the genetic correlation structure estimated in optimal environments (Charlesworth, 1988).

Problems with the phenotypic approach include controlling for genetic variation while avoiding inbreeding. Phenotypic correlations may reflect adaptive plastic responses that may display no genetic variation in the wild-type range. Although such systems may be somewhat refractory to selection, they presumably still reflect trade offs and so represent a valid test of theory. However, phenotypic correlations may change across environments (Schlichting, 1989b), so a metapopulation framework is required.

Phenotypic correlations and genetic correlations are intimately connected because both probably reflect functional integration of the phenotype (Stearns, de Jong and Newman, 1991). The likely resolution is that both approaches are relevant. Cheverud (1988) concluded that phenotypic correlations may generally reflect underlying genetic correlations and they will tend to be similar (see also Falconer, 1981, p. 283). In a framework of resource allocation among conflicting demands, genetic and phenotypic correlations are also predicted to converge (de Jong and Van Noordwijk, 1992). In other cases the two approaches may reveal different aspects of the trade off structure (Moller, Smith and Sibly, 1989b; Reznick, 1992a), and consequently they are complementary.

In identifying trade offs, lack of genetic variation does not mean that trade offs do not exist (Partridge and Sibly, 1991). Trade offs may still be expressed and documented by experimental manipulations of plastic phenotypes. Given that trade offs derive from physiological and life-history processes (Partridge, 1992), genetic correlations are necessary, but not sufficient for testing theories of organismal integration (Sibly and Calow, 1986, pp. 63–64). Consequently, holistic understanding of phenotypic design and evolution may ultimately require knowledge of both aspects (Stearns, de Jong and Newman, 1991).

Despite the simplicity of the assumptions, it has proved extremely difficult to test theory based on the principle of allocation, let alone develop any general understanding of trade off structures. Of the various approaches outlined above, the expected negative correlations are only obtained consistently where experimental manipulations of phenotypes are applied, or genetically, under selection regimes or in extreme habitats that emphasize one aspect of the allocation network (Table 8.1).

It is crucial to understand the kinds of problems that can potentially obscure detecting trade offs relevant to adaptive suites and coadapted genomes. These are outlined below.

8.4.2 Holism

It is a general rule that holding one factor constant in variable environments requires variation in others. This law of requisite variation is one reflection of coadaptation that extends to the global trade off structure. The principle of allocation does not exclude the possibility that two features may show positive correlations if a third feature is traded off. Thus, in sibling species of cockroaches, one showed superior rates of growth, reproduction and better food utilization efficiency than another

(Rollo, 1986). The higher rate species, however, had considerably reduced longevity, which could represent a compensatory dimension.

Unless a holistic framework is considered, the expected negative associations may be missed. The framework provided in Table 8.1 should be considered as an integrated network in which the trade off structure emerges only in the global balance of the entire system. Approaches that consider only pairs of features can be expected to fail whenever potential trade offs can be absorbed by another factor, or a number of other systems can jointly absorb imbalances (Pease and Bull, 1988).

Complex responses by phenotypes to varying environments were emphasized in Chapter 6. Such dynamics can also generate confusion unless a holistic framework is adopted. Thus, Lints and Lints (1971) showed that the association of large body size with longevity was reversed depending on whether temperature or rearing density was varied. Across both treatments, however, greater longevity was associated with slower development.

Finally, phenotypic adaptations that serve more than one master should be expected in adaptive suites. For example, materials deployed against herbivores by plants may also serve in structural support, water conservation, protection from ultraviolet radiation or fire resistance (e.g. Bazzaz *et al.*, 1987). Such synergisms should result in positive correlations in response to such problems via a single avenue of resource allocation, so expected trade offs may not emerge. Integration most often means that all components of the phenotype have more than one function and all functions map onto several components. The impact on any single component will most often represent trade offs or compromise (Darlington, 1958, p. 223).

8.4.3 Genetic and phenotypic variation

In sexual species inhabiting non-equilibrium environments, the idea that there is a single superior genotype is exceedingly unlikely (Hedrick, 1986). In a lineage framework, some aspects of coadapted balance may only be visible at higher resolutions of temporal and spatial scale. Recombination and adaptive release of genetic and phenotypic variation will generate phenotypes varying in fitness within and across environments (i.e. fitness is never a fixed attribute for a single gene, coadapted complex or individual).

In any single environment, some individuals will be genetically superior to others. Recombination may be one feature that contributes to such variance in fitness among individuals (Chapter 5). In such cases a positive association of key fitness components may emerge across genotypes (i.e. some individuals are better at everything). Such a situation does not mean that there are no ultimate trade offs.

An inverse argument has been made with respect to inbreeding. Rose (1984b) proposed that many examples of positive genetic correlations obtained by Giesel and coworkers (Giesel, 1979; Giesel and Zettler, 1980; Giesel, Murphy and Manlove, 1982a,b) arose because they used inbred stock. The expression of deleterious recessive genes in such lineages might create fitness differentials where the fittest individuals are better at everything. This would then realize positive correlations of features among individuals (e.g. the fastest growing individuals might also have greater longevity and greater reproduction). Again, such positive correlations would be evidence of variance in fitness, but do not mean that trade offs are not operating.

Mutations are also likely to change the relative fitness of their bearers and lead to positive correlations of features within populations. A key theory explaining the maintenance of genetic variation in populations is that a balance is struck between mutations and selection (Kimura, 1983). Mutations are likely to be pervasive in populations and controlling for their impact in simple comparisons may not be possible.

It might be expected that problems could be avoided by examining clonally reproducing species, or those in which strong inbreeding has eliminated deleterious recessives. Even then, however, individual variation in quality may be generated by mechanisms such as maternal influence. Thus, Rollo and Shibata (1991) found extremely high variation in individual performance in a population of slugs believed to be genetically uniform. The source of this variation was likely differential quality of the eggs (Shibata and Rollo, 1988). In this case, positive correlations among features emerge via a non-genetic mechanism, and such phenomena are likely to be widespread.

8.4.4 Variation in resources and storage

If resources vary among individuals, relevant trade offs may not be detected. This idea was captured by Van Noordwijk and de Jong (1986) in an analogy based on cars and houses. Although every household is faced with a trade off in terms of allocating income to expensive houses or cars, across a city, expensive cars will tend to be associated with expensive houses. Only if everyone had exactly the same income would the trade off become globally apparent. Variation in income may be stochastic (e.g. lotteries), or possibly may reflect individual quality. In either case, a positive correlation between cars and houses emerges despite the known trade off.

A particularly troublesome source of variation related to resources is density, and comparisons require strict control (Law, 1979). Populations typically show cycles in density, and selection can act on features under crowded conditions that is antagonistic to performance in sparse populations (Mueller, Guo and Ayala, 1991). Most species plastically alter expression of life-history traits in response to density (Chapter 6).

The principle of allocation predicts that trade offs will be evident because there are limitations on the ability to acquire and process resources. If an organism is forced to increase allocation to a particular process, no trade off will be required if acquisition and processing rates can be increased to pay the extra cost. Possible limits on such compensatory ability will be discussed in section 8.5. Within limits, however, such responses may be successfully engaged and consequently expected trade offs may be avoided or ameliorated.

It has been argued that features associated with resource acquisition should be subject to strong directional selection for maximization, whereas resource allocation strategies may arise mainly via stabilizing selection (Houle, 1991; Spitze, Burnson and Lynch, 1991). The question then emerges as to why compensatory resource acquisition is possible at all (it should be fixed at maximum). We will return to this under discussion of the Phenotypic Balance Hypothesis. Regardless, the evidence for compensatory adjustments in rates of feeding and growth is widespread (e.g. Surbey and Rollo, 1991) and such responses could well alleviate trade offs to a considerable extent.

Most organisms have considerable ability to store resources that can be deployed during times of high demand. In such cases trade offs may be avoided in the short term (Tuomi, Hakala and Haukioja, 1983). Sibly and Calow (1986) placed considerable emphasis on whether costs were expressed immediately (direct costing) or whether they emerge later (absorptive costing). Clearly the degree of reserves may have an important bearing on this. Storage allows organisms to synchronize acquisition and allocation tactics to environmental variation in resource levels and other relevant conditions (e.g. favourable conditions for reproduction). Storage does not necessarily eliminate trade offs, but may change the time frame for their resolution. In addition, storage is likely to have costs associated with maintenance, transport and the need to mobilize them metabolically for use (i.e. they are themselves part of the trade off structure).

Given heterogenous environments and pervasive competition for resources, even a clonal population with identical individuals may display individual variation in foraging success. Stochastic variation in resource acquisition will affect both immediate costs and accumulation of stores. Vepsalainen and Patama (1983) recognized that even if resource levels are constant, the efficiency of using them might vary. Thus, comparisons of several species or subpopulations under identical conditions (e.g. a single temperature) may favour some and disfavour others, depending on how far from their optima these conditions deviate and how plastic the organisms are with respect to acclimation. Such sources of variation may have an immediate impact on all fitness features leading to positive correlations among measured components.

In complex environments the performance of a genotype in one environment may not predict its response in others (Bell, 1984a,b). The relative ranking of genotypes may then change across environments (Stearns, 1992). Such phenomena will vary expressed correlations even leading to reversals in genetic correlations where norms of reaction cross (Stearns, de Jong and Newman, 1991). Where traits are fundamentally integrated, genetic correlations may maintain stability (Scheiner, Caplan and Lyman, 1991; Stearns, de Jong and Newman, 1991), although lack of genetic correlation does not mean that there is no functional integration (Mitchell-Olds and Rutledge, 1986). If there are trade offs among features constituting the best responses in a heterogenous environment, then genetic correlations may vary. This arises as a developmentally based mechanism, and the crucial question is whether such changes themselves represent an adaptive modulation of the genetic organization.

8.4.5 Disease

Wherever disease or pathogens exist, their impact will increase the variance in performance among hosts or among infected sublineages. Disease may both compromise the general functioning of a phenotype and require increased allocation of resources to defence and repair. Disease is frequently cited as a possible confounding factor in trade off studies of rodents, and may be expected to generate positive correlations in key life-history traits that will obscure trade offs.

The other side of this coin is the possible differential distribution of symbionts (such as the fat body bacteroids of cockroaches that allow utilization of stored nitrogen). Such organisms could also generate variation in the quality of otherwise genetically identical hosts.

8.4.6 Senescence

Senescence is regarded as a general and continuous deterioration in phenotypic performance. Consequently, all else being equal, senescence will generate positive correlations among fitness features if the immediate performance of differently aged individuals is compared (i.e. young adults are generally better at everything compared with older ones). Learning, increased size or suicidal end-of-life breeding will obscure such a generality.

The discussion of Müller's ratchet suggests that the concept of senescence may extend to the lineage level. Small populations or those reproducing asexually may display reductions in fitness associated with accumulation of deleterious mutations. Thus, variation in general fitness that could generate positive correlations might be generated both within and among populations.

8.4.7 Conclusion

This discussion underscores that there are a plethora of factors that are likely to generate positive correlations among fitness components when-ever simple comparisons are made among individuals or populations. In fact, it seems most likely that such approaches should most often find positive correlations (Bell, 1984a,b; Bell and Koufopanou, 1986; Spitze, Burnson and Lynch, 1991; Santos *et al.*, 1992). Such positive correlations, however, cannot be interpreted to mean that physiological or life-history trade offs do not exist. So how can trade offs be measured?

Negative phenotypic or genetic correlations supporting the principle of allocation have been consistently obtained in comparisons utilizing experimental manipulations of phenotypes or applications of artificial selection regimes upon key fitness components (Partridge and Andrews, 1985; Partridge and Harvey, 1985, 1988; Bell and Koufopanou, 1986; Partridge, 1986, 1989, 1992; Charlesworth, 1990; Lessells, 1991; Reznick, 1992a; Stearns, 1992; Roff, 1993). Most studies based on unmanipulated comparisons have yielded a mixture of results, but positive correlations predominate.

Only well-designed experiments allow for control of factors that produce variation in individual quality. In the following discussion the reader should apply this discrimination to various studies purporting to dismiss the reality of trade offs based on the finding of positive corre-lations. The literature is extremely confused from this perspective. For example, one cannot claim that there is no trade off between longevity and behavioural activity based on the finding that in a given population there was a positive or weak correlation between these variables. No control for individual quality is present.

Similarly, one cannot claim that rodents under dietary restriction obtain enhanced longevity at no cost because basal metabolic rate did not decline. Trade offs may emerge in other dimensions such as thermo-regulation, reduced activity, slower growth or lowered reproduction. Holism is lacking. Even in experimental studies, negative associations across treatments (the relevant measure of trade offs) are often dismissed because positive correlations are observed within one or both treatments. Clearly, there are numerous reasons why such positive correlations might emerge within treatments, but these are not relevant to the existence of the trade off of interest. Trade offs must exist in any system that has competing demands but limits on maximal performance. Evidence for such limits is explored below.

8.5 SUSTAINABLE SCOPE

Peterson, Nagy and Diamond (1990) documented that for organisms of any given basal metabolic rate, maximal metabolic output is inversely related to duration of performance. Thus, in humans, energy outputs close to 200 times above resting can be achieved for a few seconds, but as the duration of performance is increased, the upper limit of sustainable output declines. For example Tour de France bicyclists sustain an elevated scope of only five times basal over the 22 day race. Moreover, very brief, intense outputs depend on stored energy. When metabolic scope is measured over intervals such that it is sustained directly by food, most mammals, birds and lizards show magnitudes of elevation only 1.5–6 times above resting rates (Drent and Daan, 1980; McNab, 1980; Kirkwood, 1983; Peterson, Nagy and Diamond, 1990; Weiner, 1992).

Estimates of field metabolic rates of birds are similarly 2.5–3.2 times that of basal rates (Kirkwood, 1983; Nagy, 1987; Root, 1988; Masman *et al.*, 1989). In fact, many species may be operating at close to maximal capacity (Weiner, 1992), in which case, further costs may have stronger impacts. Recent data on hummingbirds, which have the highest active metabolisms among vertebrates, suggest an activity scope of no more than 2–4 times basal rates (Suarez *et al.*, 1990). Some adaptive adjustment occurs. Hunting species generally have higher sustainable scopes than ambush predators (Karasov, 1986). Athletic species such as dogs and horses have maximal capacities as much as 2.5 times those of sedentary species like goats and cows (Weibel, Taylor and Hoppeler, 1991). Presumably such differences in VO_2 max would be correlated with maximal sustainable scope. Seasonal adjustments also occur (Bozinovic and Rosenmann, 1989).

What factors could result in limitations on scope? Karasov (1986) suggests thermal constraints, available foraging time, digestive constraints, cellular metabolism and circulatory supply. In terms of maximal metabolism, features associated with respiration span the lungs, heart, circulatory system, blood and mitochondria (Taylor, 1987; Weibel, 1987; Weibel, Taylor and Hoppeler, 1991). These all appear to be coadapted relative to maximal performance requirements for a given organism (which means that the system most often has considerable slack). Many of these features also scale with body size in relatively invariant ways which probably reflects the need to maintain coadaptation as metabolic rates and body sizes covary. Realization of maximal metabolic capabilities or scope is often only possible with rigorous exercise regimes. Regardless, for any particular body size, there are undoubtedly limitations on maximal metabolism which may reside at numerous levels from the enzymatic to that of organs and organ systems.

In terms of sustainable scope rather than maximal performance, limitations may derive from resource acquisition and processing. Drent and Daan (1980) suggested that constraints associated with food supply may limit birds to sustainable performances no more than 3–4 times basal. As with respiration itself, resource acquisition and processing involves a broad assembly of coadapted features ranging from the digestive system, blood, circulatory system and digestive enzymes. Some adaptive adjustment is possible. For example, the efficiency of digestion may be improved in response to shortages or decreased food quality (Calow, 1975a,b), or the length of the gut may be increased on low quality diets (Sibly, 1981). Seasonal modification of gut morphology may sometimes involve reaction norms cued by photoperiod (Weiner, 1992). Despite such flexibility, there will be upper limits to resource acquisition and processing for organisms of any particular size. There is overlap among the systems serving respiratory and resource spheres, requiring that these two aspects are coadapted. For small, high-rate species like hummingbirds and shrews, there is very good evidence that digestive bottlenecks are the key factor limiting scope (Karasov *et al.*, 1986; Saarikko and Hanski, 1990).

The fact that both respiratory and resource processing systems show considerable accommodation under stress, but such elevated performance is normally lacking, suggests that there are either costs associated with such configurations, or the system has a limited range of performance within any given conformation (e.g. basal metabolic rates may need to be increased to sustain greater muscle mass, intestine size or mitochondrial density). Kirkwood (1983) noted that increasing gut size to accommodate food processing is offset by the high metabolism of digestive tissues.

Species are limited in their maximal and sustainable scopes, and this can have considerable ecological impact, such as restricting species from habitats with costs exceeding these limits (e.g. where it is too cold or foraging costs are too high) (Hoffmann and Parsons, 1991). Masman *et al.* (1989) provide evidence that the brood size in kestrels is limited by the maximal sustainable scope of the foraging parent (3–4 times basal rates). Vermeij (1987) suggests that species with restricted metabolism may have little energetic flexibility after meeting essential demands. Given the likely high fitness consequences of sustainable scope, it is significant that higher scope as well as growth may be associated with greater heterozygosity (Hoffmann and Parsons, 1991, p. 199; Mitton, 1993). If so, this may represent an important aspect deriving coadapted complexes that maintain genetic variation.

A further consequence of limited sustainable scope is the elevation of immediate mortality associated with stress. The literature on reproductive costs clearly documents that immediate and deferred costing may contribute separately to age-specific mortality rates and potential longevity. The highest demands on animals likely occur during periods of lactation or feeding of young, when metabolic costs may commonly be elevated by up to five times (Karasov, 1986). Presumably, the likelihood of immediate failure is related to the configuration of demands (including somatic support), available resources (including storage) and energetic efficiency. Stearns (1992) provides a relevant review of the impact of reproduction on the condition of parents, and Weiner (1992) observed that reproducing birds and mammals frequently enter states of negative energy balance. A rather tight linkage of active metabolism to basal rates opens the door for integrating theory relevant to environmental costs, metabolic rates, body sizes and mortality schedules.

This discussion suggests that sustainable scope is rather invariant with respect to basal metabolism. Although scope may be relatively constant when measured as a ratio of basal to active rates, basal rates vary enormously with body mass. The most relevant measure is energy flow at the cellular level, and mass-specific metabolism increases rapidly with decreasing body sizes (Chapter 10). Consequently, the absolute value of scope is greater in smaller animals.

However, if there are limits on metabolic output, and smaller species operate closer to these limits, then they may have less ability to compensate than larger species. Alternatively, larger species could have lower mass-specific metabolic rates, but may be capable of greater relative elevations. In this case, larger, lower-rate species might derive greater ability to compensate for stress.

Rollo and Hawryluk (1988) reviewed the literature and tested the hypothesis that organisms operating at intrinsically higher specific metabolisms would have relatively less scope to compensate for stress such as via enhanced feeding to offset reduced food quality. In other words, organisms operating closer to their maximal sustainable scope should have less ability for further short-term outputs or adjustments. This idea was strongly supported. Larger, slower-rate snails had much greater flexibility to adjust for food stress than a smaller, higher-rate species. There can be no doubt that additional scope must be related to that which is already being used for other purposes. In this way, numerous costs compete for resources within a relatively narrow sustainable framework. A more fundamental question is whether there is some intrinsic relationship between the magnitude of basal rates and sustainable scope.

Gause (1942) made a number of relevant observations related to plasticity and responses to selection. For example, mice with high metabolic rates supported by high levels of circulating erythrocytes were relatively inflexible in accommodating to low-oxygen altitudes. Mice with initially lower metabolic rates were able to adapt to high elevations by increasing their initially lower erythrocyte concentrations.

Growth in *Drosophila melanogaster* also provides possible evidence for such a scenario. This species is one of the fastest growing Metazoa (Lints and Gruwez, 1972) and selection to utilize temporary food resources has resulted in extremely rapid metabolic enzyme systems (Gordon, 1959). Significantly, of more than 40 known mutations affecting rate of development, all slow it down. Selection for shorter duration of development is largely ineffective (Lints and Gruwez, 1972; Partridge and Fowler, 1992), and it may be that this species could only improve performance by decreasing body size. Alternatively, obtaining larger adult sizes may require longer larval durations (Santos *et al.*, 1992). Tucic, Cvetkovic and Milanovic (1988) did obtain additive genetic variance for developmental duration with flies reared at high density. Their data do not allow evaluation of this idea, however.

Although limits to cellular and biochemical function must exist, the data of Peterson, Nagy and Diamond (1990) suggest that sustainable scope is a remarkably constant ratio of basal to sustainable metabolism across a very wide range of body sizes and resting metabolic rates within phylogenies. Thus, most species apparently operate within a relatively similar framework of flexibility and such a constraint undoubtedly imposes a rather tight bottleneck on immediate maximal function (see also Schmidt-Nielsen, 1984). A second question relates to how much sustainable scope must be diverted to somatic support to enhance longevity (Chapter 9). These are relatively new questions so there is a paucity of data. It is tantalizing to note, however, that the mammal with the greatest recorded sustainable scope was the marsupial mouse, *Sminthopsis crassicaudata* (scope 6.9 times resting). This is a very unusual species in that males engage in intense sexual activity that represents a form of programmed early death (Arking, 1991, p. 85). Could such elevated scope be achieved at the cost of prolonged life?

Unfortunately, the important question as to whether scope varies with body size must await further analysis. Calder (1987) reviews literatures suggesting that maximal oxygen consumption scales as $M^{0.79}$ and daily field metabolism as $M^{0.81}$. Weibel, Taylor and Hoppeler (1991) estimate that maximal respiration scales as $M^{0.86}$. If Hayssen and Lacy's (1985) value of basal rate is correct ($M^{0.70}$), then large animals will have lower specific metabolic rates, but their relative scope will be increasing as Rollo and Hawryluk (1988) originally proposed. Unfortunately, the mass exponents obtained in other studies are so variable that no clear interpretation is yet possible. Kirkwood (1983) estimated that maximal energy intake scaled as $M^{0.72}$, very similar to Hayssen and Lacy's (1985) exponent for basal rates. Moreover, several studies explicitly addressing maximal metabolic rates obtained values more in keeping with the surface rule of $M^{0.66}$ (Bozinovic and Rosenmann, 1989; Masman *et al.*, 1989; Weiner, 1992). Bozinovic and Rosenmann (1989) concluded, however, that larger animals appeared to have disproportionately higher maximal capacities. If relative scope increased with size, it might allow a greater range of accommodation and greater flexibility relative to environmental conditions. Other salient aspects are that small animals use proportionally more energy for muscular contractions than larger ones (Calder, 1987), and that larger body size derives decreased transport costs for terrestrial and some aquatic species (Schmidt-Nielsen, 1984; Wieser, 1991). Similarly, endothermy reflects a cost that increases exponentially with smaller size in endotherms.

Bennett and Harvey (1987) observed that the active metabolism of birds scales as $M^{0.65}$. They concluded that smaller species require a proportionally greater elevation of metabolism above resting for normal activity. The allometric equation for lactation in mammals suggests that the investment by a 5g shrew is equivalent to 28% of its body weight, whereas that for an ox is only about 2% (Roff, 1993, p. 138). Such data suggest that relative costs may be higher for smaller species, and larger ones may consequently have more flexibility for compensatory adjustments, even if sustainable elevations of metabolic rate are a constant proportion of basal rates.

Root (1988) analyzed the metabolic rates of birds along isotherms associated with their northern winter limits. He argued that the metabolic rate at these northern boundaries represented an elevation of 2.45 times their normal basal rates. Allometric equations for basal metabolism or that at the northern winter limits (i.e. maximal scope) against body mass had slopes of 0.80 and 0.92, respectively. The constant of 2.45 was obtained by ignoring these differences in slope (that were not statistically resolved), and adjusting that for maximal scope to the same slope as basal rates. Another derivation based on data obtained in another way, however, obtained slopes for basal and maximal sustainable metabolic rates against body mass of 0.81 and 0.98, respectively. These data argue that larger species may indeed have greater scopes. Although tantalizing, more detailed analyses are required to derive firm conclusions.

Calder (1987) suggested that basal metabolic rates may be minimized to reduce the costs of sitting in idle. However, if idle involves processes like repair, or is required for growth or accumulating reserves for otherwise unsustainable metabolic outputs later, then basal rates may correlate to active rates. Clearly, basal metabolism includes a number of energetic sinks that could be profitably separated for future theory.

Basal metabolic rates correlate with active rates rather well (McNab, 1986a), which would argue for (relatively) constant sustainable scope. McNab (1980, 1986a) argued that the linkage of basal metabolic rates to the intrinsic rate of increase (Hennemann, 1983) should generally select for maximal basal rates. Hoffmann and Parsons (1991) documented, however, that persistence in stressful environments usually involves reducing metabolism to effect global energetic savings. In this case, rates could still be locally maximized subject to environmental constraints.

Wieser (1991) also postulated that fish are selected for maximal metabolic flow through, and that constraints on acquisition and scope define a framework dictating the allocation of resources during stress or ontogeny. In particular, high costs of transport and growth in very early stages may tightly constrain juvenile fish. Older fish have reduced metabolic demands, slower growth and cheaper costs. This provides more flexibility for metabolic and behavioural adjustments.

Sacher and Duffy (1979) observed that among strains of mice, those of intermediate size had the greatest ratio of average to resting metabolism, which suggests that scope may decline in animals selected for extremes of size (Figure 8.2). If so, stabilizing selection for greater scope could contribute to coadaptation of the wild type. These data are of further interest because the resting metabolic rates of larger and smaller mice were less than those of intermediately sized mice, and yet mice of extreme sizes had reduced longevities. Sacher and Duffy (1979) concluded that the way that organisms partition respiration between resting and active rates may be the key factor impacting longevity.

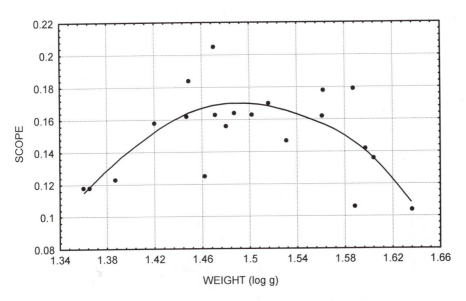

Figure 8.2 Comparisons among 21 genotypes of mice indicate that the operational metabolic scope (average minus the minimum metabolic rate measured as oxygen consumption per gram per hour) shows a parabolic relationship to body weight, with both larger and smaller mice having reduced scope. Such a situation may be significant with respect to stabilization of optimal body sizes. Calculated from data provided in Sacher and Duffy (1979), $r^2 = 0.49$, $p < 0.05$.

8.6 THE SAFE TUNING HYPOTHESIS

These results suggest that there must be some rather fundamental constraints restricting organisms to a particular range of operational metabolism. If resting metabolism reflects a wide range of actual hidden functions as well (e.g. growth, storage, repair, defence, reproductive provisioning), then it might also be worth considering whether there is a lower sustainable scope. In particular, what are the consequences of failures in resource supply?

A very large amount of literature suggests that organisms subjected to short-term deficiencies are capable of rebounding via rates of growth, respiration and feeding that exceed normal ranges (Sibly and Calow, 1986; Wieser, 1991). The degree of compensatory responses is remarkably similar to the magnitude of observed sustainable scopes (i.e. 2–5 times). The widespread existence of such capabilities suggests that normal rates are in fact submaximal. Organisms appear to be operating at rates that are safely tuned to environmental opportunities and constraints. Such safe tuning allows some slack for compensatory adjustment when shortages occur. How can this be reconciled with the idea that basal rates should logically be maximized?

A likely explanation is that basal metabolism is associated with fairly strong restrictions in upward or downward adjustment. It is not too difficult, for example, to understand why shrews starve to death with such renowned rapidity. Metabolic demands cannot easily be shut off unless the organism changes gear entirely (e.g. numerous small endotherms engage torpor during shortages). The lesson here is that high basal metabolisms may be associated with risk if resources sufficient to sustain them fail. Selection could then act against higher-rate individuals because they would fail or be uncompetitive in low-resource situations. This is consistent with selection for geometric mean fitness of lineages (Seger and Brockmann, 1987; Philippi and Seger, 1989).

A clear demonstration of this hypothesis was provided by Derting (1989) who artificially elevated the metabolic rate of cotton rats using implanted thyroxine pellets. With *ad libitum* food, the higher rate individuals expressed growth rates elevated by 25–34% over normal rats, and appeared to incur no associated costs. Thus, the results bear out the conjectures of McNab (1980) and Vermeij (1987, p. 82), that high metabolic rates are advantageous in good environments because they generally elevate performance features. Derting (1989) also illustrated why such rates may not be normally expressed. With a slight reduction in food availability (90% *ad libitum* levels), the high-metabolism individuals did not grow or lost weight while normal rats continued to grow, albeit at a reduced rate. The likely mechanism was greatly increased foraging costs associated with the need to meet high basal demands. Roberts (1981) obtained similar results with mice artificially selected for enhanced growth. On *ad libitum* diets the large strain grew faster and more efficiently than normal mice. With restricted diets, however, the large strain was less efficient and had slower growth than controls. Totter (1985) suggested that the increased variance in survivorship among rodents on dietary restriction might arise if individuals configured for high productivity were impacted more strongly by food shortages.

Several authors have suggested that selection for lower, safely tuned metabolic rates may occur if high-rate individuals are occasionally uncompetitive or fail (Karasov, 1986; Rollo, 1986; Vermeij, 1987, p. 21; Giesel, Lanciani and Anderson, 1989; Lanciani *et al.*, 1990). Hoffmann and Parsons (1991, p. 155) suggested that selection for such submaximal rates may require group selection, but the failure of high rate individuals in resource-poor environments allows for selection of geometric mean fitness in lineages. In a lineage paradigm, even occasional resource bottlenecks could select for rates of growth, reproduction and performance that are considerably below theoretical maxima.

Millar and Hickling (1990) observed that the mass exponent of fat deposition in mammals scales positively with body mass ($M^{1.45}$). Calder (1984) also reviewed this generally, and unpublished data from my laboratory found the same trend among species of cockroaches. Thus, this trend may be relatively general and might be synergized by the reduced costs of transport in larger organisms. Based on such considerations, Millar and Hickling (1990) concluded that unpredictable environments may select for larger body sizes and associated fasting endurance (see also Millar and Hickling (1992) and Speakman (1992) for debate).

These authors also postulated that low food supplies would select for smaller body sizes because such organisms consume less food on an absolute basis. Costs, however, must be interpreted at the level of individuals. The high metabolism of shrews, for example, is unlikely to reflect low costs or relatively low food supplies. Shrews probably require relatively predictable and high supplies of resources.

Energy constraints on body size associated with the costs of endothermy in small animals, or problems with heat dissipation in larger forms (which would impede scope), have also been frequently suggested (Kendeigh, 1972). Such models and those of Millar and Hickling (1990, 1992) suggest that optimal body sizes and ultimate constraints on size will vary with environments, and they successfully explain variation in body size associated with altitude or latitude. Arguments like these suggest that there may be an energetic linkage between body size and environmental resources. Small organisms may be adapted to relatively fine spatial and temporal scales which would also predict the existing associations between body size, physiological rates and population densities (Brown and Maurer, 1986; Damuth, 1987; Nee, Harvey and Cotgreave, 1992).

Sibly and Calow (1986, p. 119) interpreted various allometric curves to imply that large animals showed reproductive investments well below physiological maxima. Others have argued that the clutch sizes of birds may be submaximal (e.g. Boyce and Perrins, 1987). Bad years may impact more heavily on large clutches than small and selection may again act on geometric mean fitness. Stearns (1992) reasoned that the best strategy would be to overproduce in any given breeding attempt, and then trim the clutch to a sustainable size to achieve a local maximum. The correct interpretation is likely to fulcrum on the costs of overproduction to parental survival, and the flexibility of the reproductive strategy. Thus, Perrigo (1987) found that house mice with high foraging costs adjust litter sizes by selective cannibalism. Deer mice under the same conditions do not do so and are more likely to defer reproduction or experience brood failure. Although these rodents are very similar in size and morphology, deer mice typically live twice as long as house mice. This is consistent with the suggestion of Masman *et al.* (1989) that longer-lived species may be more conservative in energy expenditures.

Several selection studies on mice are also relevant. Falconer and Latyszewski (1952) and Falconer (1960) selected mice for growth on regimes of high or low food. Selection was successful on both nutritional planes, but the improved performance of mice selected on high-food regimes disappeared in low-resource environments. Mice selected on low planes of nutrition carried over their improved growth performance to higher-quality resource regimes. Such results suggest that scope may be skewed, such that for a given basal level, increases are easier than decreases. Such a situation would favour risk-aversive tactics rather than maximization.

Overall, the picture that emerges is that basal metabolic rates may evolve as a compromise between maximizing productivity and minimizing risk of failure associated with variation in supplies. Once set, sustainable scope allows considerable accommodation, but beyond that range, safely tuned species would be unable to utilize resource bonanzas completely other than via numerical responses. Stress-adapted species transferred to resource-rich environments show exactly this syndrome (Hoffmann and Parsons, 1991). If correct, hypotheses based on mortality risk cannot be completely divorced from an energetic framework because internal configurations themselves may impose risk.

The principle of symmorphosis holds that physiological systems have components that are coadapted to meet the needs of maximal performance, and no more (Weibel, Taylor and Hoppeler, 1991). Dudley and Gans (1991) criticized such optimality interpretations on several grounds, notably that most subsystems are involved in multiple functions. For example, the circulatory system contributes to gas transport, thermoregulation, digestion and excretion. Maxima for one function may not apply to others. Clearly, symmorphosis might still apply if the key function could be identified.

Dudley and Gans (1991) also suggest that the existence of physiological scope may have evolved as a safety factor, in which case the degree of scope would reflect the variability of demands on the relevant systems. As Alexander (1982) showed for skeletons, the evolution of safety factors can still be interpreted in a maximizing framework, where the degree of safety reflects a cost-benefit relationship with the degree of risk and consequences of failure. In a selective framework maximizing geometric mean fitness, it seems possible that the degree of safety could be higher than that required by most organisms in their life-times (i.e. the relevant measure of risk may be applicable to lineages over multigenerational time scales). If phenotypes have been selected for risk aversion, individuals may be relatively overbuilt with respect to their actual experienced risk.

The relatively close relationship between basal, sustainable and maximal metabolic rates across species argues that metabolic features are either maximized or are subject to some universal constraints. There are good evolutionary arguments to expect that metabolic rate should be maximized for any given body size, but equally good arguments suggest that safety margins are also required. It may well be that organisms achieve the appropriate balance between power and flexibility by adjusting body size itself.

8.7 THE PHENOTYPIC BALANCE HYPOTHESIS

The relative intractability of testing life-history trade offs may also be exacerbated by the reality of coadapted genomes themselves. If phenotypes and the epigenetic machinery that deploys them function to reduce the impact of selection or stress, then the wild type may be relatively balanced in optimal environments (i.e. trade offs may be absent or minimal). Moreover, if the system is safely tuned to resource levels, then a considerable degree of accommodation may be possible before costs begin to impinge on the trade off structure. This hypothesis of phenotypic balance suggests that trade offs may be absent or difficult to discern in the wild type operating in optimal habitats.

In a system selected for balance, trade offs would become increasingly evident with decreasing resources, increasing stress (in general or with respect to single aspects such as thermoregulation), or with increasing directional selection disproportionately emphasizing any phenotypic aspect. In other words, trade offs or their signature (negative correlations) may be most strongly revealed as the phenotypic organization is pushed into a state of imbalance by either environmental or internal factors. Antagonistic pleiotropy may tend to drive the system back to equilibrium when released from stress or selection. As Roff (1993, p. 147) points out, the question is not whether costs and associated trade offs exist, but under what conditions they are expressed.

If correct, the only way to address trade offs may be via experimental manipulations, selection regimes imposing sufficient force to shift the wild type balance, or examination of phenotypes expressed under stress. What is the evidence for such a situation? A number of prominent authors have variously proposed that correlations tend to be zero or positive in optimal environments or under lack of selection, but that costs or negative correlations emerge under stress or selection (e.g. Giesel, Murphy and Manlove, 1982a; Mayer and Baker, 1984; Reznick, 1985; Bell, 1986; Bell and Koufopanou, 1986; Giesel, 1986; Mitchell-Olds and Rutledge, 1986; Newman, 1988a,b; Derting, 1989; Hillesheim and Stearns, 1991; Parsons, 1991; Spitze, Burnson and Lynch, 1991; Roff, 1993).

The case is not entirely clear cut. Bell and Koufopanou (1986) reviewed the literature and concluded that positive correlations may not only arise in rich environments, but also in novel circumstances, under inbreeding or in the presence of new mutations (e.g. Rose, 1984b; Service and Rose, 1985). Moreover, negative correlations may still occur in optimal environments and correlations reflecting rather fundamental aspects of functional organization may not change significantly with environmental or genetic perturbations (e.g. Scheiner, Caplan and Lyman, 1991).

Hoffmann and Parsons (1991, pp. 182–183) suggest how stress might reveal negative genetic correlations. Later, however, they suggest that genes that filter stress may act as an insulating envelope around resource allocation features and that genetic variation in stress resistance may show positive correlations (Hoffmann and Parsons, 1991, pp. 202–204). The generation of positive correlations obtained by comparing individuals of varying fitness was discussed earlier, and this also applies to some of the examples given by Bell and Koufopanou (1986). Such comparisons probably do not address trade offs or the balance hypothesis. Stress resistance should probably be considered as a joint adjustment of resource acquisition and the resource allocation program. When only those studies that are likely to address trade offs successfully are considered, there is strong evidence for a conversion of neutral or positive correlations to those of negative sign in response to either environmental or genetic imbalances.

An elegant experimental demonstration of the phenotypic balance hypothesis is embodied in the studies of Perrigo and Bronson (1983, 1985) and Perrigo (1987). They reasoned that varying resources are normally linked to foraging effort. Consequently, they devised an apparatus where mice had to run for specified periods on a wheel to obtain food pellets. The first lesson was that mice with *ad libitum* food spontaneously ran about 8 km/day. The allocation of energy to activity may be a feature that, although flexible, is as characteristic of the species as the allocation to growth or reproduction. Significantly, wild mice showed greater spontaneous exercise than domesticated strains (Bronson, 1984). Oster and Wilson (1978) also recognized that workers of various social insects exhibit a characteristic tempo of behaviour.

Across a broad range of foraging costs (increased wheel revolutions per food item), the amount of feeding by challenged mice remained relatively constant and there were no reductions in any measures of productivity as foraging effort compensated to meet increasing demands. In fact, mice with wheels had less fat, gained weight faster, and achieved more ovarian cycles than sedentary controls (Perrigo and Bronson, 1983). This suggests the interesting possibility that animals programmed for a given level of expenditure, may be adversely affected by having that expenditure constrained. Such results confirm positive correlations between exercise and productivity in the optimal range, a result that is inconsistent with the principle of allocation unless species operate at submaximal rates.

As foraging costs were increased, a point was reached where an impact on growth became evident, even though feeding rates were still sustained. There appeared to be an upper limit on further foraging effort, and as costs increased still further, feeding rates progressively declined despite high foraging efforts. Costs in this range impacted on most key fitness features. Body size was reduced from 24 g to 16 g and reproduction was suppressed or was controlled by infanticide. In some cases mice entered daily torpor and dropped their body temperatures by 5 °C to 10 °C below normal. These results show that within a particular range of costs, no impact may be detected in numerous key features, but once a cost threshold is exceeded, costs will impact on these features and possible trade offs may be revealed. Perrigo (1987) suggested that a critical factor shaping the global resource-allocation strategy of mice may be a configuration that ensures adequate resource returns during the most critical and costly period – lactation during brood rearing. In this context it is relevant that the negative correlations between clutch size and parental survival in great tits only occurs in years of poor food supply (Tinbergen, van Balen and van Eck, 1985). Similarly, Calow (1977) reported that differences in survival of mated or virgin *Corixa punctata* also disappeared under favourable culture conditions. Browne (1982) similarly found that significant negative correlations between longevity and reproduction in brine shrimp disappeared with improved nutrition.

Mice with extreme costs extended foraging into the daylight hours (Perrigo, 1987). We found a similar effect in cockroaches given progressively diluted diets (Kajiura and Rollo, 1994b). When cockroaches were released during daylight hours in Barbados, they became a feast for *Anolis* lizards. Temporal shifts in foraging might expose animals to a constellation of predators that are normally avoided and such risks may well limit the range of foraging flexibility in such species.

Experiments paralleling those described for mice have been performed with honey-bees. Schmid-Hempel and Wolf (1988) manipulated foragers so that the amount of time spent in expensive foraging was constrained from 0 to 8 h/day. There was no effect on survivorship among treatments, although significant correlations between activity and longevity emerged within treatments. They concluded that life-span might be reduced by increased work effort, but was unlikely to be extended by reduced activity. Wolf and Schmid-Hempel (1989) extended this study by attaching various weights to foragers. The bees showed no compensatory adjustments in frequency or duration of work, except for reduced trips by bees carrying weights close to their body mass. Bees that carried a total mass similar to that experienced during normal foraging showed little evidence of reduced survivorship. As loads were extended outside this range, however, a threshold was reached, beyond which survivorship progressively declined. The authors interpreted these results to conform to the hypothesis of maximal sustainable scope.

As will be detailed later, dietary restriction is associated with a rather widespread syndrome where productivity measures are reduced, but longevities enhanced. Here too there is evidence of balance and adaptive adjustments across a range of stress. Specifically, greater longevity is only obtained up to a point, with further dietary restriction leading to the expected reductions in longevity. Selection for enhanced longevity appears to be only effective on phenotypes that have been pushed into a stress range where the reallocation of resources into longevity assurance has been effected.

The balance hypothesis may apply even to relatively fundamental life-history features. In *Drosophila*, for example, Hiraizumi (1961) documented that high rates of development were negatively associated with fertility. However, this correlation was only obtained above a threshold in development rate, below which fertility and developmental rate were positively correlated.

The insight of Tuomi, Hakala and Haukioja (1983) that trade offs may be avoided by timing high cost activities to environmental opportunities or via reliance on storage is relevant to a balance hypothesis. Drent and Daan (1980) also recognized that the interpretation of maximal sustainable scope is complicated by species that anticipate periods of high demand by accumulating reserves. Although such features could be disturbed in ways that incur costs, organisms that are adjusted to prevailing resources or reserve sizes could avoid trade offs in optimal environments. In an early study with cockroaches I concluded that the large storage reserves of *Periplaneta americana* allowed this species to integrate across resource fluctuations without major adjustments in behavioural time budgets. Reserves acted as a buffer against short-term costs and resource failures (Rollo, 1984, 1986).

In a recent study, we reared this species on diets that were diluted with cellulose and agar to varying degrees (Kajiura and Rollo, 1994b). Even diets diluted by 75% with non-nutritive bulk had little impact on growth rates, maturation sizes or maturation rates. Compensatory feeding and respiratory adjustments allowed maintenance of targeted developmental schedules. As the compensatory ability of the system was exceeded, however, the dietary dilution then impacted negatively on all key fitness components.

Falconer (1977) suggested that conflicting fitness demands may well stabilize many phenotypic features at particular equilibria. In mice, he proposed that increased predation might limit selection for larger sizes, but smaller size may be offset by reduced fecundity. Many studies have documented that the wild type displays the maximal value for key fitness features and that deviations in any direction tend to lower fitness. For example, longevity tends to be greatest in phenotypes with intermediate growth rates, and both slower and faster growing individuals have lower life-spans (Sacher and Duffy, 1979; Economos and Lints, 1984a,b, 1985; Ingram and Reynolds, 1987).

Bell and Koufopanou (1986) concluded that most environmental correlations may display such humped distributions with decreases at either extreme. Perrin and Rubin (1990) also documented that reaction norms may be dome shaped. Temperature is a good example. Economos and Lints (1986) documented that *Drosophila* reared in temperature extremes showed decreases in body size, egg viability and longevity. Where fitness is derived at intermediate values of the phenotype, either environmental perturbations or selection pressures causing phenotypic deviations are likely to destabilize the system. Within the wild type range, however, impacts may appear relatively small or of low cost because of adaptive compensatory adjustments.

The release of genetic variation associated with inbreeding and stress has already been discussed in the context of destabilization of coadaptation. It is too early to say whether all of these phenomena are aspects of a single mechanism. For example, selection and environmental stress may shift phenotypes in organized ways via integrated systems of reaction norms and genetic correlations. Costs may not be immediately apparent, even though selectable variance becomes released. For example, it seems likely that the successful selection of enhanced longevity in flies under food restriction may involve the genetic fixation or assimilation of one part of a reaction norm that shifts allocation from productivity to longevity assurance under such conditions. Within this range, life-time metabolic output and longevity may be positively correlated (i.e. no trade off is evident).

Inbreeding may fracture the genetic organization into pockets showing various local constellations of attributes. Such release of genetic variation may actually involve disruption of coadapted complexes rather than coordinated modulation which would only break down outside umbrellas of stabilizing selection. However, it is not impossible that genetic templets could be sufficiently organized to adjust for even relatively severe changes in gene frequencies. In this case the release of genetic variation during founder events might not represent random expression of non-adaptive associations, but relative extremes of coadapted organization.

The phenotypic balance hypothesis suggests that organisms may maintain considerable flexibility embodied in epigenetic, physiological and behavioural integration, and that within certain limits, this flexibility may offset impacts associated with stress or selection. Complex adjustments are possible in the network of features outlined in Table 8.1. Bateson (1963) may have been the first explicitly to recognize such flexibility as a crucial aspect of design. He pointed out that the degree of flexibility may be related to how closely organisms are operating to their performance boundary (i.e. their degree of residual scope), and that stress in any one dimension will likely compromise the ability of the system to respond to stress in others. This may be especially true because the phenotypic adjustment of allocation under stress may be remarkably different depending on what dimension is emphasized (e.g. stress associated with temperature or with resource levels often enacts divergent constellations of adjustments). Having erected this warning flag, it is also significant that general syndromes associated with stress have been identified, probably because they all impinge on resource allocation via general reductions in metabolic rate (Hoffmann and Parsons, 1991; Parsons, 1991).

Interactions in the trade off structure may be rather complex. There may be no simple linear mapping of costs in one feature onto others. Non-linearity may also greatly complicate the detection and understanding of trade offs (Pease and Bull, 1988). A simple example detailed below concerns the fact that longevity assurance associated with antioxidant enzymes may be a threshold phenomenon rather than a continuum.

The idea that the genetic templet may be organized to filter both environmental and genetic perturbations actively to maintain relative balance, and that such balance may involve meeting numerous contingencies in different ways, suggests that evolution of coadapted complexes involves regulatory evolution at the level of subspecific lineages. Such a framework also favours the idea that speciation may involve the reorganization of how such balance is maintained.

8.8 THE PRINCIPLE OF ALLOCATION: LESSONS FROM THE SUPERMOUSE

Despite pervasive evidence that life-history theory rests fundamentally on the need to trade off various features, definitive tests are exceedingly difficult to derive. Before considering other empirical support for trade offs and the principle of allocation, I will present the results of our current research program on the Supermouse, because it provides a clear demonstration of the reality of trade offs in phenotypic evolution. It also illustrates the complexity of the linkages that may be involved. Since papers are in progress (Lachmansingh and Rollo, 1994; Kajiura and Rollo, 1994a) only a qualitative overview will be provided here.

The Supermouse was created by the insertion of numerous copies of rat growth-hormone genes into normal mice, creating the world's most famous genetically engineered mammal (Palmiter *et al.*, 1982). The genes have integrated into a single chromosome so that they segregate in a classical Mendelian manner. Mice receiving the extra genes express circulating growth hormone at levels 100–400 times normal and grow to roughly twice the size of normal mice. This system is ideal for life-history assessment because when males heterozygous for the normal and trans-genic chromosomes are mated to normal females, 50% of the mice will be transgenic and 50% from the same litters can serve as normal controls. Thus, the impact of the transgenic chromosomes can be examined against a similar genetic background. Transgenic mice are nearly identical to controls except with respect to their altered growth performance. This system was of particular interest because of the strong relationship between body size and life-history features as revealed by allometric analyses (e.g. Calder, 1984). In particular, developmentally oriented evolutionists have suggested that changes in body size might be a simple but powerful channel for telescoping life-cycles inwards or outwards to create new life-history set points (Gould, 1977; Calder, 1984; Riska, 1989).

Thus, one possible consequence of duplications in growth-hormone genes might be shifts such as those conforming to interspecific allometric equations (i.e. the mouse shifts to become rat-like). The alternative, predicted by the Phenotypic Balance Hypothesis, was that the change would simply create a very high-rate mouse that must pay for its extra growth by robbing other functions. Such costs might be reflected in longevity, reproduction, thermogenesis or behavioural activity. A key factor is whether the animals proportionally increase their resource acquisition as a result of the extra growth (Van Noordwijk and de Jong, 1986). Thus, if the larger mouse simply eats relatively more to pay for its greater growth rate, then trade offs will not necessarily follow. Otherwise, trade offs should be an unavoidable consequence.

In what turns out to be remarkably fortuitous, the Supermouse does eat more relative to its size, but on a per gram basis, it eats the same amount as normal mice (actually a little less). Thus, it has not increased its specific feeding rate to pay for the extra growth. This was true whether mice of relatively the same size or same age were compared. The consequence is that the entire trade off structure of the mouse is revealed by comparing the transgenic and normal animals.

Our approach was three-fold. Firstly, we reared large numbers of mice in a breeding program that obtained longevity and reproductive parameters for normal, heterozygous and some homozygous transgenic mice. Secondly, we examined the mass budgets of these mice to determine their basic physiological resource allocation program. Thirdly, we video-taped mice in larger cages partitioned into nesting, foraging, exercise and drinking areas to detect how their behavioural time budget was affected.

As mentioned, we found that the Supermice did not engage compensatory feeding to pay for the extra growth (Kajiura and Rollo, 1994a). One possibility was that the feeding rate of mice is already maximal so they were unable to improve resource acquisition. A study that measured the ability of normal and Supermice to elevate their feeding to offset a period of starvation showed that compensatory elevation occurred, and Supermice showed equal elevations to normal mice (DelCotto and Rollo, unpublished). The problem appears to be that the signal used to regulate feeding does not incorporate the cost of the elevated growth rate (i.e. there is no inherent constraint on increased resource acquisition). In fact, the failure to increase the specific feeding rate is probably related to a key adaptive strategy in rodents. Animals on restricted diets decrease their body size, reproductive effort and metabolic costs to match available food, thus maintaining a constant resource input per gram of tissue (e.g. Masoro, Yu and Bertrand, 1982; Holehan and Merry, 1986; Holliday, 1989b). In such a system, feeding would respond to the total body size, as well as availability and quality of food, but an artificial elevation in growth rate would not elicit increased input on a per gram basis.

If we consider that resources are cut up like a pie for various processes, it is clear that something has to give (Figure 8.1). On a life-time level, we found that the Supermouse had remarkably reduced longevity. This was obvious to earlier workers keeping these animals, but we have now quantified the results. Transgenic females rarely live longer than one year whereas normal females live two or more. Thus the two-fold increase in growth roughly halves longevity. The same result applies to males.

If the mice eat the same amount of food and digest it with relatively similar efficiencies (which they do), then the only way to pay for growth is via improved growth efficiency. Although a connection between growth rate and various measures of efficiency in converting food to biomass has often been noted, the interpretation has never been clarified (Malik, 1984). The Supermouse showed a remarkably strong relationship between growth rate and conversion of digested food into body mass (net production efficiency). Such increased growth efficiency can only be obtained by robbing other functions.

If there are extra demands from growth and maintenance, then, there will be less energy available for behavioural activity and the time budgets illustrate this clearly (Lachmansingh and Rollo, 1994). In a photoperiod of 12 h light and 12 h dark, normal mice slept 13 h out of every 24 h. The Supermouse slept a remarkable 17 h every day, leaving only 7 h for activity. Moreover, during activity Supermice patrolled their environment much less, preferring to sit around doing nothing. A running wheel was provided in one region of the cage, and normal mice ran vigorously on the wheel for nearly 2 h/day. Transgenic mice rarely used the wheel, their average spontaneous exercise being restricted to about half an hour. Supermice were also very unexcitable and lethargic, being extremely easy to capture and handle compared with normals. Thus, the extra cost of growth appears to impinge on important parameters such as reaction times, activity levels and sleep. Lethargy was also an associated characteristic in long-lived strains of the nematode, *Caenorhabditis elegans* (Friedman and Johnson, 1988b). It is easy to see how the element of risk can interface here. A habitat stocked with large, slow-moving, placid mice would be no challenge at all for sharply honed predators like hawks, owls, cats and foxes.

Small endotherms may use upwards of 80% of their available energy for thermoregulation so this is an obvious avenue for energetic savings under stress. Rodents with reduced rations do reduce their body temperatures (Forsum *et al.*, 1980; Duffy *et al.*, 1989; Nakamura *et al.*, 1989). Comparisons of rectal temperatures in Supermice indicated no significant difference between normal and transgenic males, but Supermouse females were about 1 °C hotter than normals. Thus, the increased efficiency of converting food into tissue was not obtained by reduced thermogenesis, other than possible savings that might accrue from relative reductions in surface area associated with larger bodies (McCarthy, 1980; Malik, 1984).

In terms of reproduction, Supermouse males were of normal fertility, but more than 50% of Supermice females were infertile. Those that did reproduce produced fewer litters, although the brood size was similar to normal mice. Transgenic females also ate more of their litters, consistent with a stress paradigm. Some rodents, including mice, may hormonally block pregnancy in response to stress. Ecologically, this is a sound strategy if deferred reproduction during periods of high risk or cost can increase the likelihood of parental survival and enhance later fecundity (Totter, 1985; Weindruch and Walford, 1988; Holliday, 1989b; Saitoh, 1990). The mechanism for effecting the pregnancy block is via hormonal balance, restricted rodents having lower levels of progesterone and follicle stimulating hormone (Yu, 1987). We predicted that Supermouse females are locked into a state of perpetual pregnancy block associated with the stress imposed from extra growth. The pregnancy block might be removed by injections of progesterone. We had progressed to the point of carrying out the relevant experiment when we found that Bartke *et al.* (1988) and Naar *et al.* (1991) had restored fertility via progesterone injections in other transgenic mice containing human or bovine growth-hormone genes. Finch (1990) reviewed literature showing that continued breeding is stressful for rodents which is also consistent with this interpretation, and considerable empirical support is marshalled by Holliday (1989b).

Thus, the Supermouse illustrates that various features of the life-history, physiology and behaviour of a species are tightly bound into an adaptive constellation of trade offs. Artificial alteration of a single feature, while holding available resources constant, impinges on a host of inter-connected features. In the mouse, there appear to be relatively direct connections between longevity, growth rates, reproduction, sleep and re-action time, as well as rates and durations of activity.

This experiment might be criticized because the insertion of extra copies of growth-hormone genes is quite a different process than select-ing for mice of increased size. However, modification of growth hormone has proved to be a key factor modified among strains of mice varying widely in body size (Pidduck and Falconer, 1978). In fact, artificial selec-tion for large and small size has been performed on mice numerous times. Significantly, at their selection limit, the largest strains of mice were similar in size to Supermice, and the constellation of attributes they displayed (including reduced activity, lethargy, infertility and reduced longevity) were identical to those of the Supermouse (e.g. Goodale, 1938; MacArthur, 1949; Falconer, 1953, 1981; Falconer and King, 1953; Eklund and Bradford, 1977; Goodrick, 1977). Thus, the Supermouse appears to reflect the same phenotype that is obtained by selection for large size, although some experiments have also obtained increased rates of feeding (Fowler, 1962; McCarthy, 1980; Roberts, 1981).

We were also fortunate in that we obtained some recessive phenotypes in our colony that were apparently metabolic mutants. These spinner mice were hyperactive and twice as excitable as normal mice. Such mice belong to a group of hyperactive mutants variously classified as pygmies, dwarfs, spinners or waltzers. They were extremely difficult to capture and showed a strong tendency to run in circles, or in one case, throw repeated back flips. We found that such mice grew more slowly and achieved smaller adult sizes which is consistent with a trade off between metabolic expenditure and growth. Although we obtained no data on longevity for such mice, a similar shaker mutant in *Drosophila* does have the predicted reduction in longevity associated with its abnormally hyperactive behaviour (Trout and Kaplan, 1970).

Sacher and Duffy (1979) examined metabolic rates and longevities among mouse strains and their crosses. They concluded that longevity declined with increasing metabolic rates and increased with body size. In fact their data reveal that the relationship is parabolic, longevity declining in both smaller and larger mice. In the current model, small mice might have reduced longevities associated with relatively higher metabolic rates, larger mice via reduced allocation to longevity assurance features like DNA repair or oxidative defence systems (Chapter 9). A fascinating correlate of dietary restriction in rodents is a remarkable extension of longevity associated with reductions in productivity (Holehan and Merry, 1986; Weindruch and Walford, 1988; Holliday, 1989b). The reduced longevity of the Supermouse may reflect the inverse aspect of this adaptive program – reduced somatic support under conditions of high productivity. In support of this, it seems highly significant that hypophysectomy followed by hormonal replacement can obtain responses identical to dietary restriction (reviewed by Cutler, 1984a,b; Holehan and Merry, 1986; Finch, 1990).

Interestingly, MacArthur (1949) obtained hyperactivity in mouse lineages selected for small adult sizes. Of particular interest in our study was the fact that out of more than 100 spinner mice obtained over the years, only three transgenic mice ever showed such behaviour, and then to a lesser degree. This suggests the interesting possibility that such genes may offset the lethargy accompanying increased growth rates in some way, and adaptively persist with low-penetrance in normal selection regimes. Selection for small size (or reduced growth hormone) might then unmask these genes, whose presence may be rather more widespread than suspected from examining wild type mice. Significantly, other authors observed the spontaneous appearance of such dwarfs after about nine generations of selection for small size (Falconer, 1953).

Evidence in favour of this interpretation was obtained by Havlicek, Rezek and Frieson (1976) in rats. In this case, growth-hormone secretion was inhibited using somatostatin. The rats became hyperactive and tended to run in circles. Such a situation is suggestive of how polymorphism and hidden genetic variation may be maintained in coadapted genomes occupying a particular selective equilibrium. The spinner mice appear to represent an opposite association of features to that of Supermice, high behavioural activity being associated with lower growth. Taken together, these results resoundingly support the principle of allocation and point out the multidimensional nature of the trade off structure.

A complex meshwork of integration apparently involves virtually every key physiological feature outlined in Table 8.1, confirming MacArthur's (1949) original insight that pulling on any single string impacts on numerous others via their direct and indirect linkages.

Darwin said essentially the same thing. The implications are that intra-specific selection for increased body size will not necessarily realize attributes associated with interspecific allometry, but rather may create high growth rate animals with disrupted co-adaptation. In order for a mouse to become a rat, a considerable genetic reorganization may be required, and in particular, increases in body size must be achieved in a way that does not compromise other functions.

Another interesting result is that Supermice allowed to self-select between their normal diet and sugar cubes had restored levels of behavioural activity and perhaps improvements in fertility (consistent with energy limitation). Such a dietary shift does not reduce their high growth rates, but preliminary data suggest that longevity may be improved. Before doing this experiment we predicted that Supermice could not survive in the field. However, by shifting their feeding niche to accommodate their altered physiology, they might well survive, and even avoid some competition with normal animals. The lesson for release of genetically engineered organisms is obvious.

The Supermouse cuts to the crux of the argument for genic *versus* lineage selection. A genic theorist might point out that the extra rat growth-hormone genes, because they confer lower fitness, would probably be eliminated in nature if normal and transgenic animals competed. Such a result could be interpreted as selection against these particular genes. Ehrlich and Raven (1964) pointed out the fallacy of considering one component as set against an invariant background when they invented the term coevolution. In mice, growth hormone is a key signal achieving integration and is consequently linked to numerous features (e.g. growth, production efficiencies, respiration, reproduction, behaviour and longevity).

The reduced fitness conferred by extra growth hormone in Supermice appears to result from disrupted integration. If larger size was advantageous, selection in this system could operate on growth-hormone genes, but it must simultaneously act on other aspects of coadaptation as well. For example, selection on Supermice to restore fitness might just as easily act via increasing relative feeding rates rather than reducing the number of growth-hormone genes. In fact, if we started with 100 demes of Supermice, selection for restored coadaptation would likely be achieved via numerous alternative routes (i.e. several subspecific lineages with different organizational attributes). The possible displacement of Supermice by normals in a natural setting would be better viewed as competition between two gene complexes varying in their degree of coadaptation, rather than selection against growth-hormone genes *per se*. Selection acting on growth-hormone genes acts more on the correlated responses to growth hormone rather than on the gene itself. The unit selected is the coadapted complex within which growth hormone operates.

The tools of genetic engineering could be applied to address such questions. If lineages of mice were produced with varying numbers of growth-hormone genes, and selection was then applied to restore fertility while maintaining extra growth, it is likely that a reorganization of the genome could be achieved that would allow phenotypes to transcend the current limitations of the organization. The key factor selected might well be coadaptation associated with regulatory integration, rather than the assembly of selfishly independent genes.

9

The evolution of senescence and longevity assurance

9.1 OVERVIEW OF SENESCENCE AND LONGEVITY

Despite pervasive evidence for physiological trade offs contributing to life-history strategies, it has proved difficult to extrapolate this framework to general comparative studies (Harvey, Read and Promislow, 1989). A crucial dimension is how age-specific fecundity schedules are related to mortality risks. Does extrinsic mortality exclusively shape fecundity schedules (and indirectly other life-history variables), or does survivorship also involve intrinsic factors that are constraining or open to programmed control? If so, resource allocation may be shaped within the context of a global ontogenetic allocation strategy. In this case, both intrinsic and extrinsic factors may be reflected in the evolved program. To resolve this issue requires forging longevity and senescence into the resource allocation framework. Below I review the literature on senescence and longevity and outline how longevity assurance can be viewed as a specific investment of the resource allocation program.

Senescence is maladaptive and so cannot be considered a life-history attribute. Life-span, however, is a key life-history feature. Excellent reviews on longevity and senescence are provided by Sacher (1977), Comfort (1979), Weindruch and Walford (1988), Finch (1990), Arking (1991), Rose (1991) and Bernstein and Bernstein (1991). Stearns (1992) provides a chapter particularly relevant to life-history theory. This is a huge and complex literature, but there are two basic explanatory frameworks. Firstly, the differentiated soma may deteriorate due to unavoidable physical and chemical insults that disrupt organ function, cell machinery or DNA fidelity. Variation in rates of such deterioration may occur due to differences in rates of living and variation in mechanisms preventing or repairing damage (Cutler, 1984a,b; Sohal, 1986). There are a plethora of possible proximal mechanisms, but the underlying theme is that damage accumulates due to exogenous and endogenous forces, leading to senescence and death. Evolution in this framework then involves features assuring longevity by offsetting or preventing damage (Sacher, 1978; Cutler, 1984a,b).

The other explanations are evolutionary hypotheses, which do not specify a particular proximal mechanism causally driving senescence but suggest it will be an inevitable ultimate consequence of attenuating selection acting on differentiated phenotypes (Rose, 1991). We have already discussed how mutations in early developmental genes may have greater fitness consequences than later-acting ones. The epigenetic hierarchy can be extended outwards to encompass the entire life-time, and many gerontologists consider senescence as a natural extension of the developmental program. Evolutionary hypotheses of senescence include older ideas that death may be an adaptive feature at the species level (now a generally rejected view), or that natural selection acts less forcefully on later stages of the epigenetic program. Senescence then emerges as a deleterious, unselected consequence, something akin to the dissolution of functional coherence that occurs in pseudogenes.

The evolutionary theories propose several processes that theoretically lead to a diminishing of the fitness contribution of older age classes to a population. Firstly, in an example modified from Medawar (1952), consider a restaurant that opens with an original stock of 200 glasses and a clumsy dishwasher who breaks five glasses per week. The manager replaces the breakage. Even though there is no intrinsic difference among any of the glasses, as time passes the number of original glasses will constitute a diminishing proportion of the existing stock. In the same way, a population of immortal organisms will still suffer attenuation due to extrinsic sources of accident, so older age classes constitute a progressively smaller proportion of the population. In terms of fitness, the older organisms would contribute a diminishing proportion of individuals to subsequent generations (Medawar, 1946).

Schmalhausen (1949) captured this idea as well when he pointed out that predators may impose selection on a species that requires degenerative adaptations of smaller size, higher reproductive efforts and reduced longevities. Escape from predation, such as that postulated to occur in birds and bats, might consequently allow longer, senescent-free lives simply because natural selection can now act on more advanced ages (Williams, 1957; Edney and Gill, 1968). Similarly, molluscs with highly developed shells may have resulted from evolutionary arms races with predators (Vermeij, 1987). Such molluscs show strikingly greater longevities than species with low investments in shells (Heller, 1990). Evolutionary theory holds that such reduction in extrinsic mortality allows older age classes to be selected for reduced senescence (i.e. genes with harmful late-acting effects are removed or selected against).

The second aspect pertains to the fact that early reproduction has potentially greater fitness consequences. On the one hand it decreases the chance of total failure (Sibly and Calow, 1986; Stearns, 1992), and also allows a greater life-time contribution per individual. More importantly, early reproduction has a strong effect on the intrinsic rate of increase (or Malthusian parameter). Thus, imagine an asexual organism with a reproductive rate of two offspring per generation, but mother dies. What single change would be the most effective way to increase the rate of population growth: double the brood size, become immortal, or halve the generation time? Table 9.1 illustrates the unconstrained growth of populations reflecting these various solutions. Halving the generation time is clearly just as effective as doubling the brood size and both are better than becoming immortal.

When a population is growing rapidly, decreasing maturation time may be a more effective mechanism for increasing relative fitness than increasing individual fecundity when mortality schedules are added on. This arises due to the compounding effects of genetic contributions made by off-spring produced earlier. The consequence in a Malthusian framework of fitness is that selection on older organisms attenuates as the genetic contribution of these organisms is diluted by the contributions of the kin they have already produced (a kind of inverse kin selection) (Hamilton, 1966).

The first evolutionary hypothesis (the mutation accumulation hypothesis) suggests that genes that express harmful effects early in life will be strongly selected against, whereas those that act only in older adults may accumulate because they are rarely seen (Medawar, 1952; Edney and Gill, 1968). Medawar's (1946) argument was actually slightly more sophisticated. He proposed that the timing of gene expression may be modified by selection. Activity of favourable genes would be shifted to younger age classes and deleterious ones to older age classes. This is actually a model of regulatory evolution. The mutation accumulation

Table 9.1 The exponential growth of populations showing the impact of doubling the brood size, becoming immortal or halving the generation time for a population of parthenogenic females that otherwise die after reproduction

Season	Mortals R = 2	Mortals R = 4	Immortal R = 2	Doubled generations R = 2
1	1	1	1	1
2	2	4	3	4
3	4	16	9	16
4	8	64	27	64
5	16	256	81	256
6	32	512	243	512

hypothesis is generalized from the occurrence of genetic diseases expressed only in later life. A number of these are well documented, but otherwise there is actually little direct empirical support for this hypothesis (e.g. Finch, 1990, pp. 23, 242, 463). An alternative explanation for such age-related diseases might be their association with a deterministic process of senescence that unmasks deleterious impacts only in later life. For example, where the age of onset may vary, degenerative nerve diseases like amyotrophic lateral sclerosis (ALS) have much slower rates of progress in younger individuals. The penetrance of deleterious features may increase as system performance declines.

Rose (1991) reviews the strongest arguments in favour of the mutation accumulation hypothesis. Mueller (1987) compared lines of *Drosophila* in which selection was applied for early fecundity (only young females made contributions) or in which there was no selection for such early fecundity. Flies selected for early fecundity had lower fecundity in later ages compared with the controls, suggesting that lack of selection allowed deleterious late-age impacts to accumulate. Such results could, however, also be explained by inbreeding or drift (Rose, 1990). Rose and Charlesworth (1980) and Rose (1984b) suggested that failure to detect increased additive genetic variance in the fecundity of older flies argued against this hypothesis. Kosuda (1985) did find increased variance in male virility in older flies which was interpreted as declining selection with age. Finally, Service, Hutchinson and Rose (1988) subjected long-lived flies that were originally selected for late-life fecundity to reversed selection for early-life fecundity. They found that increased desiccation resistance and alcohol tolerance originally obtained in the long-lived strains were not diminished by reversed selection to control levels of early fecundity. They proposed that this was evidence that such features were absent in the control strains of flies because of mutation accumulation. A problem with this interpretation is that the decreased performances of control flies should only be evident in later, weakly selected ages, and not across all ages which appears to be the case here.

Flight performance is similarly improved at all ages in long-lived strains (Graves, Luckinbill and Nichols, 1988; Luckinbill *et al.*, 1988). The mutation accumulation theory does not predict that strongly selected early-age fitness features should be compromised by lack of selection in later ages. Thus, the finding that long-lived strains are more resistant to desiccation and alcohol vapour at early as well as later ages does not constitute support for mutation accumulation. Similarly, the failure of these features to regress during reverse selection does not constitute a test of the mutation accumulation hypothesis. What is predicted from this theory is that organisms that usually undergo programmed death but which have their lives prolonged by castration or other short-circuiting manipulations should express immediate and rapid senescence. Instead, numerous examples show significant prolongation of life and delayed senescence into ages that selection normally rarely views. Finally, domesticated animals that have been removed from extrinsic mortality factors like predation might be expected to evolve increased longevity according to this hypothesis. Among rodents, comparisons of domesticated and feral strains show large differences in the form of survivorship curves, but no increase in maximal longevities (Sacher, 1977).

This idea of age-specific selection has been further extended to a second hypothesis (the antagonistic pleiotropy hypothesis), which considers that genes enhancing early reproductive performance may be selected even if they have negative consequences in older phenotypes (Medawar, 1952; Williams, 1957). A possible example is the *abnormal abdomen* mutation of *Drosophila*. The enhanced early reproductive success of such mutants is favoured when stress from drought reduces adult survival. Thus, *abnormal abdomen* may spread in a population, even though the mutation also reduces median adult longevity (Templeton *et al.*, 1990).

This example is also of interest with respect to regulatory evolution. The *abnormal abdomen* phenotype is expressed when a significant portion of the tandem rDNA genes contain an insertion that inactivates them. Compensatory regulation may occur, however, under-replicating such duds in polytene tissues so the wild type is still achieved. Selection appears to act both on the proportion of genes that contain inactivating sequences and also on the system modulating their recruitment in polytene tissues (Templeton *et al.*, 1990).

Support for the pleiotropy theory has been obtained in *Drosophila melanogaster* where selection for deferred reproduction successfully obtained increased longevity (Rose and Charlesworth, 1980, 1981a,b; Rose, 1983a,b, 1991; Luckinbill *et al.*, 1984; Luckinbill and Clare, 1985; Clare and Luckinbill, 1985; Arking, 1991). LeBourg (1987) and LeBourg *et al.* (1988) argue against this hypothesis, but these studies involved simple phenotypic comparisons of longevity with other fitness measures, an approach that is unlikely to detect trade offs (Chapter 8).

Another evolutionary theory is that of the disposable soma (Kirkwood, 1977, 1987, 1990; Holliday, 1989b). This model incorporates the assumption that reproduction and somatic support are unavoidably traded off. Because failure due to extrinsic causes is inevitable and contributions of older age classes decline, investments in the soma will ultimately be lost or will provide diminishing returns. Consequently, evolution will favour some compromise such that reproduction is ultimately favoured over somatic support. Reproduction will be maximized while potential reproductive losses due to somatic investment will be minimized. Thus, the disposable soma theory incorporates the idea of antagonistic pleiotropy into a framework where extrinsic mortality patterns directly contribute to life-history programs. Such programs may deliberately reduce somatic support or even re-allocate previous somatic investments into reproduction (e.g. annual plants).

From the standpoint of individual phenotypes, the evolution of senescence appears to be a paradox akin to the mystery of sex. From an individual perspective it appears that with longer functional lives, we could have more children and consequently increase our individual fitness. Thus, there should always be some directional selection for improved longevity (Williams, 1957; Edney and Gill, 1968). Furthermore, increases in size, physical skills and learning may accumulate in older age classes so older organisms are not necessarily of intrinsically lower fitness compared with youngsters (Promislow, 1991). Such attributes could improve fecundity or success in producing viable young if senescence did not intervene. The evolutionary theories hold that lack of selection causes senescence. Thus, if organisms tend to be killed by predators so they are unlikely to live to an older age, then senescence will occur when predation is removed (i.e. if you are not eaten at that age, you senesce and die anyway because your ancestors never got that old). If the external mortality is removed permanently, then senescence should be further offset and longevity enhanced via selection.

An alternative reason why older age classes may not exist, however, is because they are senescent. Predators may keenly discern weaknesses, and even very early stages of senescent decline might be selected against. Similar circularities could be applied to phenomena such as competition for resources, resistance to environmental stress or sexual prowess. The crux of this issue is whether senescence exerts fitness consequences that selection can see, or whether its occurrence is restricted to unusual 'excess' longevity. To answer this question is difficult, mainly because the contribution of early senescent declines to failures from extrinsic causes cannot be easily assessed in the field. Thus, Rose (1984c) argued that a fatal weakness for group-selectionist theories of senescence is that the process rarely occurs in natural populations but is an artefact that emerges only in artificial settings. This was also a key plank in Williams' (1957) theory, even though he also predicted that senescence should begin to accumulate immediately following the onset of maturation. Roff (1993), in an excellent review of life-history theory, dismissed senescence as an important factor because organisms die before it occurs.

Sacher (1977) rejected the idea that natural populations do not express senescence based on examination of life tables. Finch (1990) suggested that if accelerating rates of age-specific mortality or decline in fecundity are found, they likely arise from phenotypic deterioration associated with senescence. On this basis he concluded that the impact of senescence was widespread in numerous natural populations (see also Collatz, 1986). Using a similar criterion, Nesse (1988) analyzed life tables of numerous species and concluded that senescent impacts were widespread and large. In fact, the data suggested that the contribution of senescence might exceed all environmental hazards combined. Promislow and Harvey (1991) also review evidence suggesting that the impacts of senescence are widespread in natural populations. Promislow (1991) presents mammalian data documenting widespread senescence in natural populations. Senescence was most evident in short-lived species with high reproductive efforts in accordance with the disposable soma theory. In such cases individual selection might be expected to act continually to extend longevity. Over the long term, even very small selection pressures might accrue large advances, unless balanced by mutation rates, or unless senescence is in fact ultimately unavoidable.

A paradox arises because, from an individual perspective, senescence causes death either directly or indirectly by increasing the probabilities of failures in meeting extrinsic demands. This can only be offset by reproduction. The evolutionary theories instead hold that senescence results from a combination of extrinsic mortality and successful reproduction, a view that turns individual intuition on its head. The crucial conclusion of the evolutionary theories is that senescence is simply a consequence of the selective regime, there being no intrinsic reason for decline or failure (i.e. immortality is feasible). In this framework, shifts in selection pressure should easily improve longevity, there being no upper limit constraining improvements.

There are some additional nuances in the evolutionary framework that suggest caution in extrapolating conclusions. Firstly, these ideas apply only strongly where r is positive, and reverse expectations may hold if r is negative. The existence of carrying capacity and consequences of equilibrium or fluctuating density are usually ignored. Positive values of r are unsustainable for significant periods in ecological settings. In a multigenerational evolutionary framework, the integration of such dynamics may forge the relevant selective regime. Roff (1993) showed that the relative advantages of early reproduction only occur with high values of r. When r approaches zero (i.e. stable population sizes), then increasing fecundity is more advantageous. This would favour longer-lived adults. Increasing fecundity would allow individuals to transmit more of their genomic identity than could be accomplished by earlier maturation, a possible individual advantage for mechanisms of population regulation.

This raises the point that the foundation of life-history theory in general is imbedded in a framework of age-structured evolution of mortality and fecundity rates with the Malthusian parameter as the relevant measure of fitness (Charlesworth, 1980; Caswell, 1989; Dingle, 1990; Stearns, 1992; Roff, 1993). However, modifying the fitness of individuals due to their age-specific frequency or discounting an individual's fitness proportional to its compounding inclusive fitness are not strictly individual fitness measures but pertain to lineages. The Malthusian parameter is relevant either to genes in populations (the genic view) or to populations (the coadapted genetic templet). If the genic view is rejected, the next stop is the lineage. In either case, individuals are bypassed. The same arguments put forward for a lineage interpretation for sex apply equally well to organismal longevity.

Most evolutionary models of senescence assume that parents produce genetically identical offspring, or that the model applies only to selection on a particular locus with age-specific fitness consequences. Where organisms produce sexually variable young, and recombine with numerous genetically diverse mates, the discounting of individual fitness by offspring contributions might be rapidly attenuated. Recall that sexual reproduction is relevant to multi-locus combinations of genes and strictly genic models do not apply (even recombinational modifiers require something to hitchhike on). Could recombination enhance individual fitness by offsetting the age-specific dilutions in fitness that would theoretically be associated with production of clones in rapidly growing populations? Could outbreeding with unrelated and numerous mates make further contributions? Is it possible that sexual reproduction could lead to evolution of longer lived phenotypes than could evolve in parthenogenetic lineages? Would dispersal of young improve adult longevity? These seem like viable and interesting questions that spark from striking theories of senescence and sexual evolution together. Balancing such questions would be Shield's (1993) observation that increasing longevity would increase inbreeding and consequently the congruence of parental and offspring identities.

The mutation accumulation and pleiotropy theories require that genes negatively affecting only specific age classes are widespread, and that negative impacts will accumulate whenever selection on an age class is relaxed. Although evidence for age-specific gene expression is well established, and antagonistic pleiotropy among age-specific functions is also known, most evidence suggests that there is no intrinsic genetic reason for age-specific deleterious changes in expression of functional genes. Nor is there any obvious intrinsic reason why phenotypic organizations should suddenly become unravelled when selection is relaxed (Finch, 1990).

If we consider a clonal organism that reproduces once and then dies, Cole's (1954) analysis tells us that the effect of becoming immortal will be advantageous to the extent of adding a single offspring to the brood size (i.e. immortality might be relatively more advantageous for small broods and of course is essential if there are none). This simplistic analysis points out the crux of the evolutionary model. The parental organisms must be segregated as separate age-specific compartments rather than simply as another identically functioning adult as Cole (1954) assumed. But why should any adult phenotype be singled out by its age, unless there is some intrinsic force changing its basic character to begin with? Such circularity in the evolutionary theories has been criticized earlier (e.g. Kirkwood, 1977, 1987, 1990; Kirkwood and Holliday, 1979). If an iteroparous adult does not express some deleterious genes in generation 5, 15 or 25, then it is difficult to see why such effects should suddenly occur spontaneously in generation 75 or 2000. The mutation accumulation hypothesis requires that they do. Medawar's (1946) regulatory framework provides some escape from this problem, but the question that then arises is why a mechanism to suppress early expression of deleterious genes should break down once it has been put in place. It seems significant that Promislow (1991) found no evidence that senescence begins at the age of first reproduction as Williams (1957) predicted.

Evolutionary theories of senescence only postulate aging in organisms that have differentiated adult phenotypes that reproduce other than clonally or via fission (Rose, 1991); in other words, in organisms that have epigenetic hierarchies. The epigenetic program has access to excellent clocks and the timing of genetic events is definitely age-structured. However, such features largely represent specific regulatory control rather than some process amenable to incorporation of deleterious genes as selection is relaxed within a single organism's life-time (Chapters 2–4). The evolutionary theories would apply if the final expression of the adult phenotype somehow becomes dissociated and loses coherence or the developmental program continues onwards into a further genetic holon that is not strongly constrained by natural selection. Rosen (1978) proposed a model involving feedforward mechanisms (those that anticipate future states of the system) that predicts a general failure of complex systems, even though no subsystems may actually fail. More recently, chaos theory predicts that if the rate of a process exceeds the ability of feedback systems to damp oscillations, then the process can enter a trajectory that is virtually unpredictable. Chaos has been widely detected in the dynamics of biological systems and in the fractal nature of morphological patterns.

Both Rosen's (1978) theory and the chaos hypothesis are similar to Orgel's (1963) error catastrophe theory of aging (i.e. errors compound more errors), except there is no need for failure in any subsystem in order for the entire system to fail. With numerous feedback systems and cycles, it could easily follow that the nearly infinite possible trajectories of particular subprocesses could escape control and lead to a chaotic dissolution of coherence with age unless selection acted in older age classes to maintain an optimal global integration. In other words, coadaptation of the phenotype would be disrupted. If Kauffman (1993) is correct in his assertion that selection acts to achieve states that border on chaos, then loss of control in later ages or diminishing selection on functions in older age classes could well result in chaotic dissolution.

Finch (1990) concluded that senescence may arise at numerous levels of biological organization. He coined the term 'dysorganisational senescence' to refer to two sources of decline: accumulation of damage or loss of irreplaceable components; and changes in regulatory feedback and feedforward which may not be intrinsically associated with damage. He emphasized the hierarchical cascading of physiological processes as a key framework. Conceptually we might also consider a system approaching some targeted optimum from numerous epigenetic directions. If these various trajectories continue onwards after reaching the target there might be a necessary divergence, just as light passing through a pinhole diverges again on the other side. Rose (1991) rejects such 'systems breakdown' hypotheses for senescence, but if complex systems do tend to disorganization, and if extension of selection into older ages could reduce or prevent it, then this is actually an excellent framework for the evolutionary theories.

The existence of an extra holon expressing garbage (i.e. mutation accumulation), however, seems both unnecessary and unlikely. Within the epigenetic hierarchy, there is no intrinsic reason why a gene that transcribes a particular protein should spontaneously change its structure or expression pattern later in development (barring mutations or changes in chromatin structure). Similarly, transcription factors have rather precise recognition sequences so there is no reason to expect regulatory recognition to change spontaneously or break down. Once genes have engaged in autoregulatory transcription, there is no reason to expect changes in output due to intrinsic genetic features. There appears to be nothing in the epigenetic regulatory hierarchy to suggest that, once the adult phenotype is achieved, the pattern of gene expression achieving this state needs to change simply as a consequence of age.

Alternatively, it is easy to see how damage could accumulate and disrupt the phenotype, particulary where some components are non-regenerating phenotypic endpoints. Thus, a major cause of death among some mammals is tooth loss in old age (e.g. elephants). Finch's (1990, p. 64) review suggests that the number of tooth replacements may be controlled by an open-ended program requiring new epigenetic circuitry for older age classes (i.e. regeneration is not automatically a cyclical circuit but may be programmed into later age classes sequentially). In such cases, selection must see the older age classes before further tooth replacements can be incorporated into the program. If such a situation were widespread among phenotypic features associated with damage prevention, repair or replacement, then senescence could be viewed as arising from attenuation of the epigenetic hierarchy in older age classes, rather than as accumulation of deleterious age-specific genes. Cell proliferation potential could well fit this model. Lack of a subsequent tier of regulatory circuitry seems a more likely situation than one consisting of regulatory garbage. Longer-lived species, in this framework, would represent epigenetic extensions and actual organizational advances. Hall (1992), for example, lists continuously growing incisors in rodents in this category of significant organizational advance.

The initial elaboration of a particular phenotype may represent considerably different regulatory machinery than that required to repair and regenerate this optimal conformation. In particular, it may be more difficult (or expensive) to offset damage in aging features or replace those that fail than it would be to elaborate the functional phenotype in the first place. Those of us who have experienced the mounting repair bills for an old car will appreciate the possible truth of this statement. Thus, the evolutionary replacement of deteriorating spinal discs is a potentially more difficult problem than the initial deployment of healthy discs in newly developing phenotypes. This discussion brings to the fore a mixed model incorporating the likelihood of unavoidable damage within an evolutionary framework of attenuating selection. This differs from the current evolutionary hypotheses in postulating senescent decline due to a lack of extended genetic organization (and unavoidable extrinsic sources of deterioration), rather than the accumulation of a level of genetic garbage (that if removed might confer immortality).

The encyclopedic review of Finch (1990) reveals that limitations of life-span may be associated with numerous levels of phenotypic organization specific to various phylogenies or species. Thus, the ultimate limitations may be molecular. At the bottom rung may be the maintenance of integrity in DNA structure, or regulation of transcription and translation (see also Arking, 1991; Bernstein and Bernstein, 1991). Damage to other macromolecules, such as proteins or lipids, may be the next tier. Cellular integrity and the potential for cell proliferation could then limit the ability of organisms to maintain tissue and organ function. At the level of organs and specific phenotypic structures (e.g. teeth, brains, sense organs or insect wings) specific regenerative potentials may set the stage for ultimate failures. At the level of the deployed phenotypes, design features, such as lack of functional mouthparts, can strongly delimit the potential life-span of adults. Finally, the reproductive strategy may vary the support of the soma or involve reallocation of resources from somatic functions to reproductive outputs. That is, the life history may vary in terms of the disposability of the soma (Kirkwood, 1977, 1987, 1990).

Presumably, there will be strong correlations across levels, largely set by the key limiting factor. Thus, organisms with programmed early death or life-times limited by aphagy may well show lower levels of molecular defence and repair or lower cell proliferation potentials. This model suggests that selection for enhanced longevity may act downwards across the hierarchy of phenotypic organization in eukaryotes. As constraints are removed at one level, new lower-level constraints may then exert their effect. Highly advanced organizations may be largely limited at the cellular or molecular level whereas more primitive forms may be constrained more by failures at higher levels (e.g. wing wear in butterflies, or exhaustion of unsustainable reserves).

Superimposed upon this framework will be the degree of phenotypic complexity. Clearly, yeast and plants that have simple or modular phenotypes may largely avoid failures associated with deployment of specialized terminal structures (e.g. eye lenses, teeth, irreplaceable neurons). If there is some ultimate constraint on longevity, it is likely to reside at lower levels. Lineages constrained by higher-order features thus may have considerably more scope for increasing longevity than those that are currently experiencing molecular-level limitations. A hierarchy of constraining levels and the probabilistic nature of damage accumulation at any of them impart an arrow of time on the functional integrity of phenotypes. This provides a visible marker that may be discerned by natural selection, allowing the discrimination of age classes for life-history evolution.

The problem with existing evolutionary theories is that antagonistic pleiotropy is not inconsistent with a framework of unavoidable deterioration and ultimate failure due to intrinsic processes. Just as the disposable soma theory allows the degree of somatic or reproductive allocation to be varied according to the extrinsic mortality regime, it is possible that predictable patterns of senescent decline could also contribute to such programs. Both predictable increases in age-specific mortality and decreased performance of senescent phenotypes might be incorporated into the life-history trade off structure.

9.2 THE COST OF LONGEVITY: FACTORING LONGEVITY INTO THE RESOURCE ALLOCATION PARADIGM

The evidence in favour of a modified 'rate of living' hypothesis is pervasive (Chapter 10), but a unifying synthesis requires a causal linkage between metabolic rate and senescent decline (i.e. use equals abuse and abuse equals decline in metabolic rate and survival). Cutler (1984a,b) and Sohal (1986) formulated the modern form of the rate of living theory, based on the idea that senescence occurs due to imperfect physiological processes that allow damage to accumulate. Thus, longevity is inversely related to metabolic rate, but variation may occur due to differences in damage prevention and repair.

DNA damage appears to be a likely candidate for disrupting metabolic performance, but there is no need for the relevant damage to be all genetic. There are a multitude of theories pertaining to aging that invoke mechanisms such as mitochondrial deterioration, damage due to free oxygen radicals, finite cellular divisions, accumulation of metabolic junk (lipofuscin), positive feedbacks in error accumulation, denaturing of proteins via glycation and cross-linkages, or deterioration in DNA methylation patterns (Orgel, 1963; Sacher, 1977; Comfort, 1979; Hayflick, 1980; Finch, 1990; Arking, 1991; Hirsch and Witten, 1991; Rose, 1991). There is no room to review these here and all that is necessary is acceptance that there are general syndromes linking senescence with rates of living (Cutler, 1984a,b).

Of all the proposed mechanisms causing senescence, the evidence that metabolic by-products such as free oxygen radicals or other toxins increase damage is highly favoured (reviewed by Cutler, 1984a,b; Sohal, 1986, 1987, 1991; Stadtman, 1988; Weindruch and Walford, 1988; Finch, 1990; Sohal and Allen, 1990; Arking, 1991; Bernstein and Bernstein, 1991; Sestini, Carlson and Allsop, 1991). The generation of harmful oxygen species including superoxide anions (O_2-), hydrogen peroxide (H_2O_2) and hydroxyl radicals (OH), mainly by mitochondria, is well documented and correlates strongly with maximum longevity (Sohal, 1987, 1991; Sohal and Allen, 1990; Sohal, Svenson and Brunk, 1990b). Generation of such materials is directly linked to metabolic rate and Sohal (1991) argues that this may be the prime determinant of physiological age within species, and correlations between longevity and metabolism interspecifically. In fact, the density of mitochondria per gram of body mass scales interspecifically with an exponent (−0.28) virtually identical to that of specific metabolic rate (Schmidt-Nielsen, 1984). Weibel, Taylor and Hoppeler (1991) obtained excellent correspondence between mitochondrial volume and maximal metabolic performance.

Within species, increased longevity obtained via altered metabolic rate (i.e. a two-fold increase in life-span in houseflies prevented from engaging in flight) is closely associated with lower levels of H_2O_2 (Sohal, 1991). Evidence also suggests that the diverse impacts of radiation that mimic senescence in some respects are mainly associated with the generation of damaging oxygen species within cells (Sohal, 1987). Damage could well be genetic, although the main focus has been on other aspects of metabolism and cell integrity, particularly that associated with proteins, lipids and membranes.

In the light of the increasing interest in the nuclear membrane as a potential regulator of post-transcriptional RNA release (Gilbert, 1991) it is significant that the nuclear envelope shows signs of senescence consistent with accumulation of oxidative damage (Miquel and Philpott, 1986). Sohal and Allen (1990) reviewed extensive evidence for impacts of oxidative stress on age-specific genetic integrity and genetic regulatory mechanisms.

The damage caused by various agents can be offset (but not entirely prevented) either by antioxidant factors reducing damaging agents or, in the case of DNA, by improved repair (Cutler, 1984a,b; Sohal, 1986; Sohal and Allen, 1990; Finch, 1990; Arking, 1991; Bernstein and Bernstein, 1991). Significantly, *Drosophila* with a mutation preventing production of superoxide dismutase (SOD, a strong antioxidant) live only 11.8 days compared with 60 days in controls (Campbell, Hilliker and Phillips, 1986; Phillips *et al.*, 1989). *Drosophila* with null mutations for catalase, another key antioxidant, also had strongly reduced viability (MacKay and Bewley, 1989). Moreover, Arking and Dudas (1989) found that their longer-lived strains of *Drosophila* had 75% more SOD coded by the *SOD* gene on chromosome 3. Flies lacking this mechanism produced immotile sperm and imaginal discs may have deteriorated as well (Phillips *et al.*, 1989). Munkres and Furtek (1984) postulate that the AGE-1 and AGE-2 genes of *Neurospora* may derive their effects on longevity via antioxidant processes. Long-lived strains of *Caenorhabditis elegans* likewise produce elevated levels of SOD and catalase (Rusting, 1992). The extended longevity of rodents on dietary restriction regimes (Chapter 10) is also correlated with increased levels of SOD in their tissues (Sohal, 1987). Whereas such materials (e.g. SOD, catalase and metalothionein) show age-related declines in normal rodents, high levels are maintained into older ages under dietary restriction (Semsei and Richardson, 1986). Weindruch and Walford (1988) make a strong case for catalase as the key material modulated to obtain extended longevity in rodents under dietary restriction.

In a remarkably thorough review, Cutler (1984a,b) marshalled convincing evidence that antioxidant defence was a key factor linking longevity to metabolic rate. He showed that the ratio of SOD to specific metabolic rate was highly correlated with mammalian longevity. Levels of important antioxidants, like SOD and plasma urate, correlate very well with life-time metabolic scope. Serum carotenoid levels and concentrations of vitamin E also show strong positive correlations with longevity. In fact, examination of the endogenous oxidation processes in tissue homogenates from various species correlated very strongly with life-time energy potentials and longevity.

A problem with testing the 'damage prevention' hypothesis is that there are numerous possible ways of achieving antioxidant activity (or damage control in general). Examining a particular enzyme or antioxidant material will not necessarily measure the investment in such homeostasis. For example, humans have lost the ability to manufacture vitamin C, a potent antioxidant, but instead achieve antioxidant activity via high levels of urate. Plasma urate levels are correlated with longevity in primates. Interestingly, near saturation levels of plasma urate are achieved by humans via loss of the enzyme uricase that normally would convert urate to allantoin (reviewed by Weindruch and Walford, 1988;

Finch, 1990, pp. 285–286). In a regulatory sense this is possibly an example of a successful compensatory mutation (Chapter 5). Further adjustments represent what could be interpreted as a regulatory revolution. MacKay and Bewley (1989) also found evidence that the relationship between viability and catalase activity may conform more to a threshold rather than a continuous response.

Several studies have detected positive correlations between antioxidants and life-span (e.g. Cutler, 1984a,b; Sohal, Sohal and Brunk, 1990a), but some antioxidants scale negatively, or have different relationships depending on the tissue assayed. The picture emerging is that there are a battery of possible antioxidant defences (including superoxide dismutase, catalase, glutathione peroxidase, vitamin C, vitamin E, urate, ß-carotene and metal chelators), and that both the overall level of defence as well as metabolic rate itself are involved in a regulated balance. Thus, changes in either levels of antioxidants or metabolic rate may mutually influence one another and increases or decreases in one antioxidant may effect compensatory adjustments in others (see Cutler, 1984a,b; Sohal and Allen, 1990; Sohal, Sohal and Brunk, 1990). This suggests the system is under programmed control and that evolution is likely to involve regulatory adjustments. Cutler (1984a,b) points out, for example, that the SOD gene is relatively conserved in structure across phylogenies and that variation is largely limited to the timing, distribution and intensity of production.

Relevant to the phenotypic balance hypothesis, Cutler (1984a,b) shows that compensatory adjustments for dietary deficiencies in one antioxidant may be largely offset by increases in others within a certain range. Severe deficiencies in key antioxidants like vitamin E, for example, may elicit pathologies with symptoms consistent with accelerated aging. In some organisms such as rotifers, it is well established that vitamin E supplements extend longevity (Enesco, Bozovic and Anderson, 1989).

By their nature, antioxidants have negative impacts on metabolic rates at high levels (i.e. suppression or toxicity), so complete defence may be impossible (Sohal, 1987) or antagonistic to selection for high rates. Antioxidant materials also have secondary functions and interactions (i.e. ion and electrical balances) that may underly the deployment of different materials in different tissues or limit their sustainable concentrations (Sohal and Allen, 1990; Sohal, Sohal and Brunk, 1990a). Consequently, a certain proportion of harmful oxygen radicals and hydrogen peroxide necessarily escapes elimination (Sohal, 1987). Clearly this is a complex area that defies clear interpretation at this point. However, the existence of such defence systems as ubiquitous elements across diverse phylogenies, their association with longevity, their linkage to metabolic rate, and their logical linkage to damage prevention, all suggest a key role in determining life-time scope. Sohal (1991) argued that defences may remain relatively constant across mammals with diverse longevities, so that the generation of damaging agents may be more important. It should be remembered, however, that a relatively constant level of antioxidants with increasing body mass represents a proportional increase in defence relative to associated metabolic rates.

General surveys of phyla detect positive associations between DNA repair ability and longevity (Hart and Setlow, 1974). Although numerous studies have found associations between DNA repair ability and improved longevity, some others have not (Sacher, 1978; Hart, Sacher and Hoskins, 1979; Finch, 1990; Arking, 1991; Rose, 1991). The ubiquity of such repair systems suggests that they are important for organismal integrity, but linking measures of DNA damage to physiological performance may be difficult (e.g. Newton, Ducore and Sohal, 1989). There is also evidence for repair, removal or replacement of other macromolecules (e.g. proteins). The known increases in karyotypic-level abnormalities and cancers with age, coupled with other available evidence, argue that genetic-level deterioration occurs in the soma.

Molecular studies of sequence divergence among phylogenies indicate wide variation in rates of substitutions (Britten, 1986b; Li and Tanimura, 1987). The slowest rates are observed in primates and birds. Britten (1986b) suggested that wide variation in DNA repair might explain these results and this would be consistent with the unusually long lives of these taxa compared with other vertebrates (Chapter 10). Rodents have rates of nucleotide substitution four times higher than primates, and among the latter, hominids may be the slowest (Li and Tanimura, 1987). These authors emphasize differences in generation time as the key factor, rather than DNA repair. The entire question is likely to be complex. Long-lived species may need superior repair and perhaps greater recombination during reproduction. Short-lived species may derive greater contributions to repair of functional genes via more frequent sexual reproduction (offsetting the ratchet) but non-functional genes might diverge more rapidly. Regardless, repair is likely to be of crucial importance and it is likely to vary phylogenetically and with generation time. In most cases this will be closely associated with body size.

Integrating these dimensions derives a model of longevity linked to the relative balance between metabolic rate and the generation of harmful by-products on the one hand, and the joint investment in damage prevention and repair on the other. The model suggests that metabolic rate and antioxidant defence may be directly antagonistic so the balance may represent a selectable evolutionary compromise between operating rates and sustainable life-time. The probable expense of defence and repair places longevity firmly in the allocation trade off structure common to other life-history features. In addition, this linkage demarks an ultimate temporal boundary within which the principle of allocation may be applied to derive a global theory of phenotypic design. For a given metabolic stress, there will be an ultimate limit on longevity associated with repair and replacement capabilities and the maximum allowable defensive investments.

Several complexities may arise in such a framework. Firstly, there may be limitations on the efficiency or possible magnitude of defensive systems such that longevity is not linearly related to metabolism (i.e. very high rates may be less defendable). Thus a 42-fold difference in H_2O_2 levels among vertebrates was strongly correlated with longevity. When insects are included the variation spans a 300-fold range (Sohal, Svensson and Brunk, 1990b). The efficiency of a given level of defensive investment may be greater at lower metabolic rates (McArthur and Sohal, 1982). Thus, low rates could improve longevity in two ways.

Secondly, the rate of damage production may not be simply linked to global respiration rates, but may differ, depending on the relative allocation to different physiological sinks such as maintenance, growth, endothermy, storage, behaviour or reproduction. In particular, since somatic defence is one aspect of maintenance, strong allocation to any other feature that compromises longevity investments might accelerate aging. Unpublished results with the Supermouse suggest that longevity is compromised by the costs of growth. Although behavioural activity was restored by sugar supplements, longevity was marginally improved. If growth hormone regulates antioxidant defensive systems, this could explain why improved energy might not completely restore normal longevity. Rodents appear to have a regulatory conformation where high rates of growth (and growth hormone) are associated with declining longevity.

Complete understanding of life-history trade offs may require knowledge of costs specific to particular functions. Whatever factors actually drive senescence, they also impair metabolic functions both in the organism as a whole (which often has declining cell numbers) and intracellularly (deteriorating mitochondria are one obvious cause). On a life-time level, a decline in power and an ultimate termination may be probabalistic but predictable, a fact well established by successful life insurance companies. Given that decline is inevitable, coupled with the selective value of early reproduction, evolution would likely favour the channelling of resources in this direction and away from achieving the ultimate in longevity.

Selection could also act to prevent 'damaged goods' from entering the gene pool and accelerating Müller's ratchet (Chapter 11). Consequently, no existing species are likely to live to their ultimate potential life-spans, and selection for improved longevity could be universally successful. Here the modern rate of living theory based on proximal mechanisms intermeshes flawlessly to the evolutionary theory of the disposable soma. In keeping with the other evolutionary theories, organisms with extrinsically imposed mortality may reallocate investments away from somatic defence. Thus, organisms with programmed early death designed to maximize immediate reproduction probably have considerably lower life-time energetic scope than potentially available. This might occur because there is a limit to the degree that immediate output can be raised above basal levels. The greatest life-time scope might be achieved by operating at lower rates for longer times. The question of constraints on immediate output thus comes to the fore.

The pleiotropic theory of aging postulates that there is selection for features maximizing early reproductive performance at the expense of later performance because of differential contributions to fitness. However, there is little attention to what kinds of features might be involved or why there should be pervasive trade offs between early and late performance. Given a choice between a system that does not involve fitness-decreasing trade offs, and one that does, the trade off should be selected against even by very small selection pressures. A distinct possibility is that there are costs for somatic homeostasis and DNA repair that impose short-term trade offs, even though these same features provide longer-term benefits.

In a system with limited immediate scope, short-term demands for high performance in any given dimension (e.g. foraging, reproduction, growth) may limit resources available for longevity assurance (DNA repair and oxidative defence). Such a situation would lead to longevity reductions associated with maximizing early performance. Alternatively, the maximal output that can be achieved without causing immediate mortality risks is also limited by short-term scope. Where considerable potential scope is possible by living longer, the relative amount of immediate output that can be realized will be proportionally smaller. In this case, short-term constraints may favour extending the life-cycle over greater ages. The limit of sustainable scope can undoubtedly be exceeded by depleting reserves and robbing the soma for short-term gains. This could prove to be a viable strategy wherever there are sufficient fitness gains associated with intense early efforts, even if this compromises life-time scope.

These results are relevant to the phenotypic balance hypothesis with respect to longevity and life-time scope. The Supermouse supports this contention. The respiration rate of these mice is similar to normal mice. However, the doubled growth rate clearly impacts directly on longevity. It may well be that any exceptional cost, whether it be growth, reproduction, activity or thermoregulation, may impact on investments into somatic defence and repair, leading to reduced longevity. The yields from longevity assurance investments may also be non-linear. Thus, the decreases in basal metabolic rates in primates and bats (Hayssen and Lacy, 1985) are associated with greatly extended life-times. Sibly and Calow (1986) suggest that investments in somatic defence may only be reduced above particular levels of stress, other avenues being compromised first. This fits well with a hypothesis of stress impacts on metabolic rates and somatic support (Hoffmann and Parsons, 1991).

Alternatively, altered support to the soma may be part of an adaptive plastic program that could well serve in stress situations. Thus, Holliday (1989b) reviewed evidence that dietary restriction leading to reduced growth and enhanced longevity (to defer reproduction) was associated with increased levels of SOD. If growth rates and somatic defence have a negative interrelationship, then the doubling of growth rate in Supermice may decrease longevity via reductions in somatic defensive systems. Measuring such defences should provide a strong inference test of the disposable soma theory.

The existence of evolutionary strategies specifically varying somatic support via regulated shifts in defence, repair and replacement becomes a likely possibility when all evidence is considered. Cutler (1984a,b) suggests that the evolution of such a program dedicated to longevity assurance falls outside the evolutionary theories that posit aging as a consequence of age-specific expression of deleterious genes. Once again, the evolution of such a program is not one based entirely on individual fitness and consequently is unlikely to have evolved in a genic context. The lineage model, however, would allow for the evolution of programs forged by differential stress and varying resource levels experienced sequentially within sublineages and recombination of 'evolutionary wisdom' obtained by parallel sublineages experiencing different conditions. In other words, longevity and rates of senescence may be modulated by adaptive reaction norms or as parts of larger adaptive plastic trajectories. Modulation of longevity by diet in rodents (Weindruch and Walford, 1988) and differences associated with summer and overwintering longevities of honey-bees (Moritz and Southwick, 1992) may reflect such regulatory control. Canalization of longevity within optimal temperature ranges (Economos and Lints, 1986) may be another example.

Weindruch and Walford (1988) review considerable evidence that changes in the longevity of rodents are associated with the conformation of a relatively small subset of intimately connected genes. Thus, dietary restriction not only improves maximal longevity, but also immune responses (i.e. it represents an integrated response for longevity assurance). The genes for SOD and catalase are linked to the major histo-compatibility complex on chromosome 17 (Weindruch and Walford, 1988, p. 248). Gathered under this common umbrella are genes that influence DNA repair, free radical damage, immune responses, hormonal regulation of reproduction, and aspects of aerobic metabolism (Weindruch and Walford, 1988:282). Such revelations strongly suggest that the evolution of longevity and rates of senescence are firmly imbedded in a framework of regulatory evolution.

It would be most interesting to know the relationship between metabolism, somatic defence and sustainable scope in insects. Flight activity can require metabolic elevations in flies 100 times above resting levels (Sohal, 1986). The direct oxygen supply of muscles by trachea and a concentrated supply of circulating energy in the form of trehalose (which is converted to glucose in muscles) are but two features allowing such high outputs.

The very high specific metabolic rates predicted for small species such as insects might override repair ability to some extent. When coupled with expression of high unsustainable scope during adulthood, it may well be that prevention and repair systems cannot offset high rates of damage. If so, it could explain their very low life-time scopes relative to those of vertebrates (Chapter 10). Overall, the balance model fulcrums on adequate investment in somatic defence. Very high metabolism generates more damaging agents and requires higher defensive investments to offset them. Defensive costs and direct impacts of antioxidants on metabolic rates are antagonistic to maximizing short-term performance. Very high values of population mortality rates relative to individual fecundity would greatly favour maximal allocation of resources to offset mortality via early intense reproduction. Possibly somatic defence systems fail at such high rates. Alternatively, if the ratio of the Malthusian parameter to individual fecundity is large, selection may favour investing all resources into reproduction, to the detriment of longevity.

A more comprehensive model of somatic defence probably also involves cellular turnover rates and the ultimate number of cellular replications. Adult insects are post-mitotic, and consequently adult longevity is restricted by deterioration in terminally deployed cells. Weindruch and Walford (1988, p. 286) recognized that rates of cell turnover in vertebrates may be varied without changing the ultimate number of possible divisions. Thus, longevity may be a function of how well cells are defended, how long they are exposed to damage before replacement and how many replacements are possible.

This discussion posits that longevity can be directly modified as one of the slices in the resource-allocation pie. The basic idea is that for any given metabolic rate, there may be diminishing returns for somatic defence (prevention, repair, replacement), up to a point that represents the boundary for ultimate longevity. The amount of investment yielding a given return may vary with metabolic rate, lower rates perhaps requiring proportionally less investment to obtain a given return. The genetic program may adaptively vary this investment in response to predictable stress (e.g. food restriction in rodents). Stress that requires diversion of resources away from this investment will impact on rates of aging. When longevity is considered as one component of the resource-allocation system, any other single aspect may impact directly on longevity, a combination of demands may have an additive impact, or a high demand from one aspect may be offset by compensatory adjustments in others so the predicted trade off is not seen.

Although metabolic rate itself is a crucial component of this model, it is the partitioning of metabolism that is important. The relationship of longevity investments to the Malthusian parameter is also tightly linked to the overlay of extrinsic risk that is associated with this factor (i.e. high values of r are correlated with high extrinsic mortality). If extrinsic mortality risks are high, longevity assurance would be wasteful and resources are better marshalled into high early performance.

The nature of the stress or risk may also be crucial. Where it is chronic, hopeful deferment of reproduction is not useful. However, if extended growth to larger sizes or deferment of reproduction to later times can off-set or escape immediate limitations, then longevity investments may be favoured. Finally, the degree of stress will be important. At very high levels of stress, all functions will be unavoidably compromised or the system may entirely fail. There may be numerous adaptive peaks that are defined by different physiological/environmental constellations. The key factors defining the best strategy are likely to include metabolic rates, resource levels, environmental costs, the Malthusian parameter, extrinsic mortality and individual fecundity.

Relevant to the topic of storage is whether adult organisms are in fact operating with sustainable metabolisms or whether they engage in negative energy balance that eventually depletes reserves accumulated during earlier growth. Finch (1990, p. 52) provides a valuable discussion relevant to insects. He points out that many adult insects are aphagous which means that any scope at all is unsustainable. He further points out, however, that many insects with functional mouthparts may also be operating at unsustainable scopes. Thus, butterflies, moths or other insects that ingest only nectar may sustain energetic fuel while depleting other nutritional stores accumulated as larvae. In fact, depletion of the fat body with age is a marked aspect associated with decline in many species, including *Drosophila* (Miquel and Philpott, 1986). In such cases, active feeding may offset the rate of storage depletion or provide an early positive balance, but ultimately, longevity may still be determined by the size of initial reserves.

The idea that longevity of *Drosophila* may be related to accumulation of reserves or fat body size determined during the larval period is consis-tent with the increased longevity of strains with slower development (Lints and Lints, 1971) and the increased resistance of long-lived strains to starvation (Service, 1989) or the presence of mates (Luckinbill *et al.*, 1988). If nutrient reserves make an important contribution to adult longevity, then longevity could be greatly modified via selection on this aspect (Finch, 1990, p. 53). Long-lived strains of *Drosophila* may show no major changes in daily metabolic rates or efficiencies (Arking *et al.*, 1988), but the initiation of senescence is simply delayed to older ages. Once initiated, the period and rate of senescent decline is similar to controls (Arking, 1987). Such a pattern is also quite consistent with a mechanism of reserve depletion.

The very odd fact that selection for increased longevity fails when *Drosophila* larvae are reared at low densities (e.g. Lints *et al.*, 1979) might also be relevant here (reviewed by Arking, 1987; Rose, 1991). It is now generally accepted that selection for long-lived strains of *Drosophila* requires that phenotypic canalization be destabilized by competition at high larval densities (Luckinbill and Clare, 1985, 1986; Arking and Clare, 1986; Arking, 1987; Service, Hutchinson and Rose, 1988). This is exactly similar to our finding that phenotypic variance in developmental rates was greatly increased by nutritional stress in cockroaches (Kajiura and Rollo, 1994b). But why should larval crowding be relevant to selection for adult longevity? If larval competition increases the variation in individual reserve size or accumulation rates, then selection for greater longevity might act more strongly where competitive stress increases the phenotypic variance in fat body size or function that is visible to selection.

Reduction in early fecundity in lines selected for later reproduction (e.g. Clare and Luckinbill, 1985) is also consistent with reallocation of larval stores and positive reserve balances during early adulthood. Changes in reserve size are also consistent with failure to detect such allocation in lines that achieve great longevity via larger sizes (Partridge and Fowler, 1992). Long-lived strains of *Drosophila* have been shown to have higher lipid contents (Service *et al.*, 1985; Service, 1987, 1989). They also show increases of three to five times in the duration of tethered flight, probably associated with larger glycogen reserves (Luckinbill *et al.*, 1988; Graves, Luckinbill and Nichols, 1988; Graves and Rose, 1990). Taken together, there is good evidence that enhanced adult longevities are somehow linked to reserves. Even positive correlates of reproduction and protein synthesis in long-lived strains could be associated with vitellogenesis which takes place in the fat body (Arking and Wells, 1990)

It seems most significant that those factors plastically altered under high larval densities in *Drosophila* (including increased longevity, starvation resistance and fat content (Zwaan, Bijlsma and Hoekstra, 1991)) represent the same suite of characters expressed in long-lived strains obtained by selection for deferred reproduction. Furthermore, such features are not selectable under low larval densities. It is as though such selection regimes have genetically fixed and accentuated a genotype representing one portion of an adaptive plastic trajectory.

An overview suggests that environmental stress (i.e. high larval densities) or otherwise novel environments (Service and Rose, 1985) results in a reallocation of scope into storage reserves. This then brings about an extension of longevity. If selection on *Drosophila* for extended life has actually involved genetic changes consistent with the mutation accumulation or pleiotropic theories of senescence, then it is also troubling that this enhanced longevity largely disappears when the long-lived strains are reared at low densities (Luckinbill and Clare, 1986; Arking and Wells, 1990). Conflicting evidence comes from the observation that starvation resistance increases with age, and this is associated with increasing accumulations of lipid (Service *et al.*, 1985; Service, 1987). There appears to be a positive balance in reserve accumulation, especially in the first 2 weeks of adulthood. Miquel and Philpott (1986) nevertheless detected fat-body shrinkage in old flies, and 'neutral' lipid accumulation consistent with reserve exhaustion in senescence.

Luckinbill *et al.* (1989) provide an alternative explanation for these diverse results. Genetic analysis identified four differences in enzymes associated with the long-lived strain, and these enzymes are all linked to both carbohydrate utilization during activity and fat body metabolism of lipids and glycogen. They also recognized the crucial involvement of energetics and reserves, but they pointed out that the long-lived strain may well have adjusted their metabolic enzymes to utilize sucrose available in the novel laboratory environment. In this case, the energy available to the long-lived and short-lived strains would differ, allowing the long-lived flies to be generally better at everything (i.e. total energetic scope in short-lived flies is environmentally constrained). This interpretation is consistent with the improved performance of the long-lived strains across all ages, but not with loss of longevity at low densities.

This is an attractive hypothesis and, if true, illustrates a nice case of regulatory evolution in a novel environment, the achievement of a new coadapted genome and confirmation that stress (i.e. larval density) may expose regulatory variation not otherwise available for selection. The other question this poses is why the control lines do not spontaneously achieve this new state without this additional age-specific selection for longevity. Given that *Drosophila* has become an important model for evolutionary interpretations of senescence, the possibility that selection has been for use of novel food, variable reserve sizes, or timing of positive/negative reserve balances needs serious attention.

In particular, the study of Arking *et al.* (1988), that is so damaging to the rate of living theory (because there is no apparent trade off involved in achieving enhanced longevity), might require re-interpretation. Although development time (relevant to reserve accumulation) did not vary in these strains (except perhaps at 25 °C), there was secondary selection for short developmental periods that may have obscured this aspect (Arking, 1987). Other avenues of reserve modification could still have been affected. Genetic changes in reserves or for improved utilization of a novel substrate do not necessarily address either of the evolutionary theories of aging.

The disposable soma theory (Kirkwood, 1977, 1987, 1990; Kirkwood and Holliday, 1979; Holliday, 1989b) posits that somatic investments for longevity may likely be in direct conflict with allocation to reproduction. Thus, natural selection may favour diversion of such resources to reproduction, especially early in life. Kirkwood (1990) provides a model supporting this idea. Exogenous supplementation of organisms with antioxidants is largely ineffective because compensatory decreases in endogenous antioxidants occur (Sohal, 1986). This indicates that the degree of antioxidant defence is regulated, consistent with a disposable soma model. If we consider the costs of improved homeostasis and repair to a system with limited sustainable scope it is clear that short-term available resources might directly compromise longevity assurance. Unfortunately, the possible costs of mechanisms that prevent or repair DNA or other damage remain unknown, but they are very likely to be expensive. Without maintained DNA integrity, metabolic and cellular functions may degrade but such maintenance may have associated immediate costs that constrain short-term responses. Selection may thus entail a balance between the benefits of short-term and long-term performance.

The disposable soma theory in its current form assumes that immortality could be achieved by sufficient investments in somatic support, but selection to maximize population growth prevents this from being realized (Kirkwood, 1987, 1990). It might be more realistic to consider that decline is inevitable and repair can only partially offset rates of deterioration (Chapter 5). Sacher (1978) also favoured an evolutionary framework for senescence based on the idea that longevity assurance was selectable. However, if the potential metabolic returns of such investments decline in older organisms this may also be factored into the cost-benefit equation. This will be especially important if agents that damage DNA or other important cellular machinery accumulate over an organism's life-time (Hirsch and Witten, 1991). Finally, it is well established that immediate sustainable scope declines with age. In the nematode, *Caenorhabditis elegans*, for example, the rate of decline in motor performance with age was tightly correlated with maximal life-span across genotypes displaying three-fold variation in longevity (Johnson *et al.*, 1987). Such a situation would also need to be factored into models of antagonistic pleiotropy – younger ages may be able to generate more intense bursts of output and higher sustainable scopes. The overall life-history strategy may specifically incorporate senescent decline as an inevitable component of the adaptive program, with the added nuance that longevity assurance investments may defer such declines to varying degrees.

When longevity assurance is viewed as a cost, then generally high extrinsic mortality may be met by higher rates and reduced longevity assurance. This would result in small, high-rate individuals with short lives and numerous small young. Severe risk could even lead to programs that are detrimental to the soma, at least near the end point of the life cycle. Such species may have reduced need of longevity assurance, and may have much greater life-time scope than is ever realized.

Species with generally low extrinsic mortality risks may reduce variance in fitness, or obtain improved fecundity or offspring survival via longevity assurance. One aspect of this might be lowered metabolic rate, and increases in various defence and repair systems. Such tactics would result in low-rate organisms that slowly grow to large sizes, mature later and produce fewer, larger offspring.

Other tactics are possible. In species where juvenile mortality is high, but adults face relatively low risk, longevity assurance should be favoured. Relatively long-lived adults can contribute repeated bouts of reproduction and achieve eventual success, even if most cohorts fail or do poorly. Such tactics appear to be expressed by landsnails like *Cepaea* and by *Sequoia* trees. Such life histories may still contain trade offs between the need to grow rapidly through high-risk stages while engaging in strong longevity assurance. Such possibilities point out how extrinsic and intrinsic forces may jointly contribute to the global life history. As part of the allocation program intrinsic risk may constrain rates of growth, reproduction and performance. Extrinsic risks may shape how longevity assurance is deployed and may shape other features independently of allocation considerations. For example, where juvenile mortality is high, the same amount of reproductive allocation may be assigned to more numerous, smaller offspring.

Several recent studies based on mortality schedules of very large fly populations claim to demonstrate that senescence does not occur (Carey *et al.*, 1992; Curtsinger *et al.*, 1992). This conclusion was based only on expressed rates of mortality, and not on examination of performance features of the flies. Clearly, any claim that senescence has not occurred must be related to the performance features of the flies, and age-specific deterioration of *Drosophila* is undeniable.

These studies have complementary flaws. That of Curtsinger *et al.* (1992) is confounded by density. It is well established that adult longevity of flies is strongly reduced by increasing density (Chapter 10). In this study density progressively declined in groups with age such that older flies lived in lower-density environments. These authors claim that, holding age constant, mortality rates increased with decreasing density, an interpretation at odds with established facts, including the results of Carey *et al.* (1992). The data are consistent with a reduction in environmental stress with age associated with progressively declining densities.

Against the claim that senescence is absent, one must ask why there is an inexorable decline in surviving flies, and why no immortal flies were obtained. Mortality rates may have declined with age, but they did not fall to zero. Finally, the authors apply a smoothing technique that looks suspicious. If only the upper maxima of calculated mortality rates are considered, then rates do appear to increase throughout life. This is also true of the data of Carey *et al.* (1992). The smoothing technique and the method of calculating mortality rate tended to count numerous values of zero mortality in older age classes that were sparsely represented. If no fly died on a given day, then this received a zero. Higher mortality on any given day was offset by strings of zeros giving the impression that mortality was now constant.

The study of Carey *et al.* (1992) addressed the density issue by also examining flies (*Ceratitis capitata*), kept in solitary confinement. As argued above, such flies lived considerably longer than those reared in groups. The authors claimed that the mortality curve is parabolic, mortality rates declining in older age classes. Their data show that this was really only pronounced in group cages which would be subject to changing density regimes with age. Furthermore, this study used genetically hetero-geneous stock, and flies of advanced age may have been of superior fitness (Carey *et al.*, 1992). The claim that senescent decline does not occur implied by these two studies appears to be premature and should be coupled to performance evaluation of the older flies.

10

The principle of allocation: limitations, life-times and metabolic relativity

10.1 INTRODUCTION: THE RATE OF LIVING THEORY AND LIFE-TIME METABOLIC SCOPE

It would require numerous volumes to explore all the potential interconnections among key life-history traits (Stearns, 1992) and physiological features (Table 9.1). The following discussion explores the linkages of key features largely to a single axis defined by metabolic rates and longevity, because this currently represents the weakest aspect of the allocation paradigm.

The principle of allocation holds that individual phenotypes are ultimately limited both in terms of available life-time potential, as well as their maximal and sustainable performance on shorter time scales (Rollo, 1986; Sibly and Calow, 1986; Peterson, Nagy and Diamond, 1990). If we are to accept that the principle of allocation is a fundamental dimension dictating phenotypic form and function, then it is necessary to explore to what extent these assumptions of immediate and life-time scope actually hold. Without such limits, organisms could presumably increase all aspects simultaneously without incurring any costs. This chapter explores the reality of trade offs with emphasis on those pertaining to aging and physiological functions of the phenotype. The goal is to assess to what extent an allocation model consistent with the principle of allocation can unify longevity (and changes in age-specific performance) with other sinks contributing to the metabolic trade off structure.

Allocation strategies will be strongly affected by whether life-time productivity is limited. A species with potentially infinite life-time energy (i.e. immortality) might make quite different decisions about deferring reproduction under stress than one that is burning limited metabolic potential while sitting idle. The evolutionary theories require that the arrow of time needed to provide grist for the mill of age-structured evolution arises as a consequence of such selection in the first place (Rose, 1991). Here I argue that specific longevity assurance investments may be involved in determining life-spans, although most species will realize only a portion of their potential.

Deterioration associated with senescence would predictably lower the potential performance and fitness of later age classes and lead to life histories specifically incorporating such information. Life histories elaborated within such an ultimate constraint structure would have largely the same pattern as predicted by the current evolutionary theories. Thus, the evolutionary theories are not necessarily mutually exclusive to deterministic aging (see Sohal and Allen, 1990). A finite somatic integrity may create an unavoidable diminishment of intrinsic fitness in older organisms, thus allowing selection to distinguish age classes.

A truly universal theory of organism design based on the principle of allocation would be possible if there are similar and fundamental limitations on metabolic rates and longevities for all organisms. If different lineages have inherently different potentials, then it may still be possible to derive a framework with universal underpinnings: a kind of biological theory of metabolic relativity. Although considerable variation does in fact exist, this is largely compartmentalized within phylogenetic lineages (e.g. Stearns, 1983, 1984c; Saether, 1988; Harvey, Read and Promislow, 1989; Read and Harvey, 1989). Variation decreases further as the hierarchy of sublineages descends to the specific and subspecific levels.

There is good reason to believe that phenotypes represent different allocative and morphological packaging of a nearly universal fundamental biochemistry. In an epigenetic framework, the same structural gene products are organized and developmentally interpreted by a hierarchy of regulatory features. Some variation in biochemical potentials and organization may be overlaid on these fundamental aspects (particularly in symbiotic organizations) but the common descent of all life suggests that deviations may be considered within a shared biochemical context with limited variation. Within mammals, for example, a 75 million-fold variation in body size is laid over common foundations of skeletal architecture, organ systems, biochemistry and even body temperature (Lindstedt and Swain, 1988).

It is well established that the basal rate of metabolism scales interspecifically to body mass with an exponent close to a value of 0.75 (Peters, 1983; Calder, 1984, 1987; Schmidt-Nielsen, 1984; McNab, 1986a,b). Within species this is closer to 0.66 (Heusner, 1985). A rather tight coupling of metabolic rate to body size is predicted based on surface area laws (Western, 1979). In a more recent extension of Kleiber's (1961) classic derivation, Hayssen and Lacy (1985) excluded domesticated species and found that body mass explained 80% of the variation in the basal metabolic rate of eutherian mammals. They obtained an exponent of 0.70 relating organismal metabolic rate to body size, and −0.30 for that relating specific metabolism to body mass (respiration per gram of body mass). McNab (1987a) cites very similar values (0.71 and −0.29, respectively). McNab (1980, 1986a,b, 1987a,b, 1988) documented that ecological factors, such as food habits, climate, behavioural activity and predation, may also explain considerable variation, and a combination of body mass and such considerations can account for up to 98% of the variance in fitted lines (McNab, 1987a,b, 1988). Such considerations argue for a strong energetic basis coupling phenotypic organization and ecological adaptation.

A note of caution is in order, however. The implication of the above discussion is that larger organisms are simply scaled up versions of small ones and that simple extrapolation might allow prediction of most features across a size gradient of various species. The fact that intraspecific and interspecific allometries have different mass exponents (0.67 *versus* 0.75, respectively) alerts us to the fact that something else besides surface area is dictating interspecific trends. Sliding along interspecific allometric lines is not achieved by simply extrapolating intraspecific trends, but may require power shifts that would be reflected by shifts in intercepts as well. The discussion on the Supermouse suggests that complex reorganizations may be necessary to cross the size range near selection limits within a species.

To date, no satisfactory explanation for the difference between intra- and interspecific allometries has been discovered. It may well be that it has to do with balancing a complex set of functions in a coadapted way, one of which must involve respiratory constraints. Body size (and indirectly metabolic rates) is very strongly correlated with most key life-history features of organisms (Blueweiss *et al.*, 1978; Western and Ssemakula, 1982; Peters, 1983; Calder, 1984; Schmidt-Nielson, 1984; Read and Harvey, 1989). For example, Western (1979) found that 98% of variation in mammalian offspring size was accounted for by adult body size. Similar correlations occur for annual biomass production ($r = 0.96$) and litter weight ($r = 0.99$) (Read and Harvey, 1989). In terms of longevity, 85% of the variance among 85 species of mammals was accounted for by an equation including brain size, body mass, specific metabolic rate and body temperature (Sacher, 1978). Given the measurement errors in the data, such a fit suggests that there are few other factors that could improve the relationship (Sacher, 1978). Sacher's equation is:

$$L = 8B^{0.6} M^{-0.4} R^{-0.5} \cdot 10^{0.025T}$$

where L is longevity in months, B is brain size (g), M is body mass (g), R is specific metabolic rate (cal/g/h) and T is body temperature. Economos

(1980) also confirmed the strong association between body size and longevity among mammals. Prothero and Jurgens (1987), using single factor analysis, obtained mass exponents ranging from 0.116 to 0.174 (depending on the data base) which explained 40–60% of the variation in mammalian maximal longevity. A similar value was obtained for birds (0.189). They argued that ecological factors might explain the remaining variation.

The idea that all organisms may have similar finite life-time energy and that varying its rate of utilization may alter longevity is a venerable hypothesis known as the rate of living theory. This was first proposed by Rubner (1908) but was first fully articulated by Pearl (1928). Sacher (1959, 1977) provided further supporting data. Boddington (1978) reiterated this theory in the context of absolute metabolic scope. He interrelated three key components of life-history evolution – metabolic rate, longevity and body size (mass) – and concluded that they defined a balanced unitary relationship. Thus:

- Metabolic rate = $a_1.\text{Mass}^{0.75}$
- Longevity = $a_2.\text{Mass}^{0.25}$
- Metabolic rate . Longevity = $a_3.\text{Mass}^{-1}$
- Metabolic rate . $\text{Mass}^{-1} = a_3.\text{Longevity}^{-1}$
- Metabolic rate . $\text{Mass}^{-1} = a_1.\text{Mass}^{-0.25}$

where the parameters a_1 to a_3 are empirically derived constants.

In a thorough analysis based on mammals Cutler (1983, 1984a,b) also concluded that each species may have a characteristic life-time energy potential per gram of body mass, but there were differences associated with phylogeny. Thus, life-time energy for most eutherian mammals amounted to 220 kcal/g, but primates (458 kcal/g), and a few others such as humans (815 kcal/g), were exceptional. Birds also have even greater life-time energy (about 1000 kcal/g) (Western and Ssemakula, 1982; Calder, 1985). Austad and Fischer (1991) estimated that eutherian mammals had an average of 275 kcal/g per life-time, whereas bats were more like birds (595 kcal/g per life-time). They found that marsupials have an average of only 138 kcal/g per life-time. Monotremes have too little data available for an allometric analysis, but an overview suggests that they have life-time scopes similar to eutherian mammals. However, they may achieve this by operating at lower rates for longer times (Economos, 1980; Calder, 1985).

Life-time scopes may be remarkably lower among invertebrates. For several species of flies this amounted to only 25 kcal/g body mass per life-time, and for the milkweed bug, *Oncopeltus fasciatus*, it may be as low as 8 kcal/g body mass per life-time (Sohal, 1986). What is required is an expanded analysis specifically examining life-time scope within and across phylogenies. Thus, although eutherian mammals have on average a life-time scope of nearly 300 kcal/g, rats, which are relatively short lived, have values in the range of 90–130 (Masoro, Yu and Bertrand, 1982). Given that high metabolic rate and oxidative defence may be mutually antagonistic, it may well be that life-span is not linearly related to rate of living (Lints, 1989), higher rates being associated with greater proportional declines.

When the relationship between life-time scope and longevity is examined across phylogenies, it is not surprising that species with greater longevities also have greater scopes. Bats, birds and primates have life-time energy considerably in excess of other vertebrates (e.g. Prothero and Jurgens, 1987), and they also live the longest. Longer life means longer time to respire which promotes a positive relationship. This does not dismiss the possibility that longevity and metabolic rate are traded off within a lineage (which will be documented below). There is also evidence suggesting that high specific metabolisms (rates per gram) are inversely related to life-time scopes. Clearly the crucial question must be how longevity can be extended phylogenetically in the face of a possible trade off between metabolic rates and longevity.

The trends suggest that there may be evolutionary progress in life-time scope for more advanced genetic organizations. Birds tend to have higher body temperature, faster growth rates (Case, 1978) and longer lives than mammals. There is no remaining avenue for a trade off except in their delayed reproduction and perhaps reproductive effort. Various lineages have indeed achieved differences in potential life-time energy. Given the enormous diversity of life, however, what is most remarkable is that differences among phylogenies amount to such a restricted range. Stearns (1983, 1992) considered that such differences could be treated as phylogenetic constraints, which could allow a modified 'rate of living' hypothesis to be applied within but not across these evolutionary compartments. Calder (1985) also argued that the rate of living hypothesis appears viable within these constraints.

Elgar and Harvey (1987) found that, among mammals, most of the residual variation in the body mass to metabolic rate relationship was associated with the level of taxonomic orders. Within orders, body size and metabolism showed a relatively strong invariant relationship. Promislow and Harvey (1990) also found that 80% of the variation associated with life-history tactics resided from the family level upwards. Hayssen and Lacy (1985) and Saether (1988) also observed that significant variation persisted at this level. Read and Harvey (1989) estimated that 63% of variation in life histories was derived across orders.

It is clear that the relevant variation in life-history tactics usually declines rapidly with descending taxonomic levels, so it should be possible to control for higher-order phylogenetic shifts and still formulate a relatively general framework. Such efforts may be associated with the idea of different Bauplans being associated with higher taxonomic levels. It would be most interesting to know if key determinants of such variation may be associated with fixation of higher-order structure in the epigenetic hierarchy.

In general, increased longevity is associated with decreased metabolic rate. Boddington (1978) pointed out the complication that scope invested in body size might increase longevity but utilization for other purposes would always decrease available life-time. In this case, increasing size must also correlate with reduced metabolic rate, all other things being equal.

Living fast means dying young (Jones, 1990). However, a schism has developed in interpretations of what causally underlies the apparent fast–slow continuum in life-history tactics clearly identified by Stearns (1983), Calder (1984) and Read and Harvey (1989). Besides the metabolic interpretation that is developed here and related to body size, extrinsic mortality rates may also derive similar predictions (e.g. Harvey, Read and Promislow, 1989; Promislow and Harvey, 1990, 1991; Reznick, Bryga and Endler, 1990). The correct interpretation is probably that both are inextricably interwoven because risk and allocation tactics feed back on one another.

Despite the existence of a pervasive relationship among body mass, longevity and metabolic rates, the idea that longevity can be simplistically explained by a universal rate of living theory has been emphatically renounced by nearly all recent authorities (e.g. Lints, 1989; Finch, 1990; Arking, 1991: Austad and Fischer, 1991; Rose, 1991). This rejection is largely concerned with whether there is a single unifying value in life-time metabolic scope applicable across all species (which is clearly not true). The idea that metabolic limitations exist and that higher rates are associated with decreased longevity is not rejected by most authors. Unfortunately, no unifying synthesis has emerged from the field of gerontology that can clearly guide us in extensions to life history and phenotypic evolution (other than evolutionary theories that offer a framework but specify no particular physiological mechanisms). Despite this pessimistic note, there does appear to be pervasive evidence for an underlying metabolic constraint structure, such that even the existence of some serious anomalies does not warrant abandoning the rate of living hypothesis in a radically modified form.

The basic picture that emerges is that investments in longevity may be considered as an immediate cost in their own right. Thus, future phenotypic performance may rest partially on current investments that offset deterioration. Such investments are traded off against other immediate physiological costs which arise either due to unavoidable constraints or to adaptive allocation schedules forged by long-term selection regimes. The degree of such investments in longevity and their relative effectiveness will be strongly dependent on metabolic rate. Thus, the old idea of a system directly driven by 'rate of living' is combined with an evolutionary flexible allocation program that can lead to rather complex adjustments in physiological organization.

As will be discussed, such adjustments, although still resting completely within the metabolic trade off structure, may modulate longevity in unexpected ways (e.g. enhanced longevity of food-restricted animals). Evidence supporting the reality of limited metabolic scope (both in terms of immediate sustainable performance or in terms of life-time energy) comes from numerous sources (see reviews by Finch, 1990; Peterson, Nagy and Diamond, 1990; Arking, 1991; Rose, 1991). These are considered sequentially below.

10.2 ALLOMETRIC ANALYSES

Allometric analyses relating body size to other design features suggest fundamental relationships linking body size, metabolic rates, ontogenetic schedules and longevities within phylogenies. Even variation across phylogenies is relatively constrained (Peters, 1983; Stearns, 1983; Calder, 1984; Schmidt-Nielsen, 1984; Lindstedt and Swain, 1988; Reiss, 1989; Harvey, Pagel and Rees, 1991). Specifically, as the basal metabolic rate per gram of tissue increases (with smaller body sizes) longevity inversely declines. It should be noted that in all of the research selecting organisms for increased longevity, the upper limits achieved remain within the limits predicted allometrically. For example, although *Drosophila* cultures can be selected for increased longevity, they still show longevities typical of fly-sized organisms.

Nature has produced no organisms with metabolic rates like a shrew with life-times similar to elephants. The undeniable success of allometric analyses argues that either all organisms have some fundamental common mechanism(s) driving senescence, or that all selection for altered patterns of age-specific survivorship and reproduction have acted almost exclusively on the body size and metabolic rate dimension. Some caution is in order because the proposal that body size may constrain metabolic rates contains a strong asymmetry. An upper limit of metabolism for a given body size may act as a constraint, but a wide range of lower metabolic rates may be permissible within any given size (thus desert rodents have lower than predicted rates for their size).

Allometric analyses are simply statistically derived descriptions of relationships, and deviations from the average are as instructive as the fitted line itself. Thus, small species that have exceptionally long lives compared with allometric predictions, such as the spiny anteater or the pocket mouse, have compensatory features, such as reductions in metabolism, daily torpor or reduced body temperature (Sacher, 1978;

Dawson, Fanning and Bergin, 1978). Species with exceptionally short lives typically have exceptionally higher metabolisms, or bouts of suicidal metabolic activity (e.g. Bradley, McDonald and Lee, 1980). In general, exceptionally long-lived species grow more slowly or have reduced reproductive rates (Western, 1979).

An unexplained aspect revealed by allometric analyses is that life-history schedules tend to retain a constant proportionality across species of various sizes and longevities. Thus, time to maturation tends to scale directly to longevity, and both are strongly correlated to body size (Charnov, 1991). In essence, the life history of larger species appears to be a telescoping of the ontogeny of small species (Lindstedt and Calder, 1981; Calder, 1984; Lindstedt and Swain, 1988; Charnov and Berrigan, 1990). In fact, the rate of all processes from the cellular and physiological to the ecological reflect a telescoping of temporal scaling with body size. Finch (1990) regards this phenomenon as indicative of some important epigenetic mechanism that is at odds with the evolutionary theories of senescence. Charnov and Berrigan (1990) also emphasize that such proportionality requires theoretical explanation that is currently lacking. The requirements for tight integration of features like those related to rates of nutrient processing and respiration and necessary adjustments to body size seem likely candidates.

A requirement for increased time to grow to maturational size, coupled with a trade off between growth and reproduction, might be contributing constraints (Charnov, 1991). Case (1978) found that adult body size, birth weight and litter size explained 97% of the variation in postnatal growth rates. If longevity assurance requires lower rates, then longer lived organisms may need to perform more slowly to achieve longer durations. Alternatively, Harvey and Zammuto (1985) incorporated the life expectancy at birth into an analysis of body mass and age at maturity in mammals. They found that life expectancy was positively correlated to maturation time, and this even explained some residual variation after body size was controlled. But other things being equal, why should species with greater life expectancy delay maturation? These authors suggest several ideas including greater eventual yields in fecundity. An alternative interpretation is that life expectancy is reduced by earlier reproductive maturation. Such a situation could well explain why birds, bats and primates live longer (Western and Ssemakula, 1982; Read and Harvey, 1989).

In a similar vein, Wootton (1987) found that body mass was strongly related to age at first reproduction in an analysis of 547 mammalian species. A mass exponent of 0.251 was obtained for a regression that explained 56% of the observed variation. Wootton (1987) concluded that age at first reproduction was best explained by body size and phylogenetic constraints rather than ecological factors. He suggested that a targeted maturation size might impose a minimal developmental period, or alternatively, a selected maturation schedule could limit adult size. Wootton (1987) pointed out that even where longevity is better correlated with age at maturation than with body size, the additional variation accounted for is not associated with ecological factors. Intrinsic organization of the lineage seemed more important.

Charlesworth (1990) suggests instead that allometric proportionalities may be explained by a common selective framework imposed by ecological and demographic aspects relevant to body size. Williams (1957) also suggested that, since senescence should begin at the age of first reproduction, this will tend to lead to a fixed developmental schedule. Williams and Taylor (1987) pointed out that, since all populations are finite, the environment must ultimately remove individuals at a rate generally equal to their addition. In this context, rates of birth and death may have been selected as an incidental selection for body size. Even the analysis of Harvey and Zammuto (1985), however, showed that body size alone accounts for 89% of the variations in age at first reproduction.

Finch (1990) suggested that some physiological constraint, such as the ability of the circulatory system to supply tissues with metabolic resources, could forge a constraining linkage between maximal metabolic performance and body size. This is strongly reminiscent of the earlier discussion on sustainable scope for which digestive bottlenecks were equally likely constraints. A similar role might be played by the requirement to maintain a relatively targeted body temperature despite changes in body size and associated dissipative surface areas. Thus, the coadaptation of systems associated with scaling of features coupled to metabolic rates may impose specific and relatively strong constraints on metabolic rates associated with particular body sizes. For example, consider a population of elephants that is selected for increasing metabolic rates. This might arise due to a need to increase reproductive rates due to increased predation (based on the common assumption that fecundity must balance mortality).

Elephants are already of a size where heat dissipation may be a crucial problem. If thermogenesis cannot be easily disassociated from metabolism, then increasing metabolic rate may be constrained within the current body design. A simple shift to smaller sizes might allow the population to achieve higher adult metabolic rates. Such a change would also reduce maturation time and increase the Malthusian parameter. Schmalhausen (1949) outlined such a scenario earlier with respect to impacts of predators. Which is more important in this evolutionary response, generation time or body size? I think the answer is they are inseparable to some extent, regardless of any overlay of variation.

Given that ecological and demographic factors are notoriously variable, whereas biochemical and physiological features are less so, it seems that the relative explanatory power of allometric analyses must involve some intrinsic constraints that are interacting with an ecological overlay via the dimension of body size. Allometry clearly suggests that there may be an underlying unity amidst the diversity of life, regardless of how this is imposed. A key question remaining is how much of this variation relates to the intrinsic organization of species associated with their historical and functional constraints, as opposed to the shaping of species by their ecology and demographic environment.

These questions remain outstanding. Whereas McNab (1980, 1986a,b, 1987a,b, 1988) and Wootton (1987) convincingly demonstrated the reality of ecological overlays on the allometric organization across species (i.e. characteristics are influenced by climate, food, habitat type and behavioural habits), recent analyses that remove body size or phylogenetic contributions find very little ecological impacts on residual life-history variation (Harvey and Read, 1988; Elgar and Harvey, 1987; Wootton, 1987; Promislow and Harvey, 1990, 1991). McNab's framework agrees fairly well with the phenotypic balance model outlined earlier and suggests that allometry may represent a further aspect of coadaptation.

One possible resolution of this paradox is Wootton's (1987) suggestion that phylogenies with particular life-history constellations have been shuffled by selection to be disproportionately represented in habitats that suit their predisposition. That is, their particular attributes have determined the ecological niches they select, natural selection contributing little to the moulding process. I would instead prefer the perspective that ecological specialization may be reflected at higher levels in diversifying clades so that ecological aspects tend to be associated with particular phylogenies. Thus, controlling for a terrestrial, desert, seed-eating niche would largely impinge on only two specialized, but diverse phylogenies, small rodents and seed-eating ants. Invasion of a new seed niche would be more likely to arise within these existing taxa, than from another of discrepant size and evolutionary specialization. Similarly, the niche associated with eating fruit is largely occupied by only a few key clades including primates, fruit bats (which may be primate derivatives), some specialized groups of birds and particular families of insects. Stearns (1992) provides a similar interpretation. Regardless, it may be a mistake to reject or ignore the ecological contributions to allometric variation, just because such variation may be largely associated with higher taxonomic levels (i.e. higher-order lineages).

Classical allometric approaches using body size as the independent variable have been criticized lately based on the argument that they do not reflect an evolutionary framework. Specifically it is argued that if life-history features are related to extrinsic mortality regimes, then generation time or time to maturity are more appropriate key signals (e.g. Harvey and Zammuto, 1985; Partridge and Harvey, 1988; Harvey, Read and Promislow, 1989; Charnov, 1991; Promislow and Harvey, 1990, 1991).

Because development time is highly correlated with adult size, those favouring this approach usually rescale their data to control for the effects of body size. This is usually accomplished by analyzing the residual variation remaining after removing that explained by the body size relationship.

This is a worthwhile and logical extension of the allometric approach, but major problems in interpretation can emerge. If features like body size, metabolism, maturation time and longevity are causally linked together (which seems highly likely), then removing variation associated with body size will also remove most of that associated with metabolic rate (e.g. Trevelyan, Harvey and Pagel, 1990). If no relationship is then found between some life-history feature and the residual variation, it emphatically does not mean that there is no overall relationship. In the study of Trevelyan, Harvey and Pagel (1990) a significant negative correlation of metabolic rate and longevity still emerged even after removing body size. The authors argued that no relationship exists because they could achieve non-significance by removing one taxonomic group (petrels).

In terms of life-history features that are rate dependent (such as longevity) and perhaps for all features that might involve altering body size to achieve an appropriate fecundity/mortality balance and temporal scaling, specific metabolic rates (metabolism per gram of tissue) are undoubtedly most crucial. Those studies which conclude that metabolic rate is unimportant after removal of body size effects deliberately exclude this measure in favour of total respiration 'to avoid problems which can arise when mass specific BMR is correlated with body weight' (e.g. Read and Harvey, 1989). This correlation, however, is the whole point of arguments based on allometry.

It remains, however, that analysis of residuals does provide insights into life-history evolution that might not otherwise be obtained (e.g. Read and Harvey, 1989; Promislow and Harvey, 1991). Such analyses reveal the arrow of selection in fine tuning life histories for a given body size and suggest that the need to balance mortality with fecundity or reduced generation times are probably the correct framework for phenotypic evolution generally. A significant aspect is that life-history features tend to covary in very specific ways. Thus a fast–slow continuum spans suites of correlated character, such as small body size, large litter size, short lives, rapid development, small offspring and high specific metabolism (rabbits), *versus* large size, small litters, long lives, large offspring, slow development, and low specific metabolisms (elephants). Thus the fast–slow continuum persists even when the body size signal is removed (Read and Harvey, 1989).

The relative ranking of phylogenetic groups changes, however. Rabbits still fall at the fast-end, and the relatively slow rates of living for primates and bats is consistent with their unusually long lives (Read and Harvey, 1989). However, rodents and cetaceans (whales) fall close together. Lindstedt and Swain (1988) suggested that cetaceans may have relatively high rates of living for their size due to the need to offset thermal conductance in water. Rodents may have high relative rates for quite different reasons (e.g. predation), although thermogenesis is also a dominating expense. Thus, the two kinds of analyses provide different sorts of relevant insight and may be viewed as complementary rather than antagonistic. The reality of the fast–slow continuum suggests that some relatively universal theory of life history may be possible (Harvey and Nee, 1991).

Allometric analyses of residual variation show that some life-history variation may be explained independently of body size, particulary if related to development time or extrinsic mortality pressures (e.g. Harvey and Zammuto, 1985; Bennett and Harvey, 1988; Saether, 1988; Read and Harvey, 1989; Harvey, Read and Promislow, 1989; Promislow and Harvey, 1990, 1991; Harvey, 1991; Harvey, Pagel and Rees, 1991). For example, our hypothetical elephant could have responded by producing more, smaller young rather that altering body size. It must be significant, however, that very strong correlations that exist among most life-history features are greatly reduced or even disappear when the body size signal is removed (see Read and Harvey, 1989).

Regardless, the reality of significant life-history variation independent of body size has led to a schism between those recognizing an energetic linkage between body size and associated features (McNab, 1980, 1986a,b, 1987a,b; Calder, 1987; Western, 1979; Western and Ssemakula, 1982; Lindstedt and Swain, 1988; Millar and Hickling, 1990, 1992) and those rejecting any strong linkage of metabolic rate to life-history evolution (e.g. Read and Harvey, 1989; Harvey, Read and Promislow, 1989; Trevelyan *et al.*, 1990; Promislow and Harvey, 1991; Harvey, 1991; Harvey, Pagel and Rees, 1991; Speakman, 1992). Thus, Hayes, Garland and Dohm (1992) found that life-history features of mice are unrelated to metabolic rate after the body size signal is removed. Compare these conclusions to the interpretation obtained with Supermice or Derting's (1989) experimental manipulation of cotton rats.

If metabolic rates are relatively maximized for a given body mass (within the context of safe turning) then body size will explain a relatively large amount of the variation in metabolic rate. Empirical analyses show that body size alone may account for as much as 85% of the variation in metabolic rate among mammals (Hayssen and Lacy, 1985; Harvey, Pagel and Rees, 1991). Removal of the body mass signal could well remove most of the significant variation in metabolic rate as well. If some further relationship is found in this residual variation, it clearly cannot mean that body size or metabolism were unimportant, only that there is a secondary signal explaining the residual variation. Similarly, if no relationship is found, it does not imply that metabolic rate was unimportant.

As an extreme example, imagine that 99% of the variation in metabolic rate was explained by body size instead of the actual 85% (Hayssen and Lacy, 1985; Harvey, Pagel and Rees, 1991). If no significant relationship between metabolic rate and longevity was obtained after removal of the body size signal, this would clearly not reject the rate of living hypothesis. Similarly, a positive correlation of even 0.95 with some other factor using this remaining 1% of the variation would not imply that longevity was independent of body size just because this other factor contributes to the overall equation.

Using partial correlations allows one to view the relationship among particular variables after the variation in another has been removed. This is like a window into a single room in a building. For a global perspective, multiple regression analysis would provide a more comprehensive model. As the causal relationships among features become known, these key factors could be forced into the equation. This would ensure that the residual variation is only that left unexplained by the known causal factors.

In conclusion, allometric analyses, do show considerable correlations between key life-history features, and in particular, that metabolic rate and longevity are related in a fairly strong way. Those proposing that an evolutionary framework is required are absolutely correct. However, body size must be part of that framework. Deleting body size also deletes a relatively large associated variance in correlated life-history features. An either–or solution is unlikely to be possible where phenotypes represent highly coadapted suites of characters and where size has major consequences on both internal organization and ecological interfacing. Charnov (1991) comes closest to a unified approach. He assumes that growth rates are size limited, and these determine time to maturity. This in turn is a key target of selection.

The key issue fulcrums on whether body size should be simply considered as another life-history trait shaped by factors like mortality schedules, or whether it acts as a strong constraint on phenotypic design (Partridge and Harvey, 1988). The answer appears to be that both are true. Moreover, the evidence for the linked scaling of numerous key features of the respiratory system to metabolic rate and body size has been very well documented. Consequently, those arguing that size acts as a constraint are neither guilty of vague thinking nor lack an evolutionary framework as the other camp has implied.

In an integrated theory of phenotypic evolution, the evolutionary framework related to maximizing the Malthusian parameter and balancing mortality is ultimately important. It must be significant, however, that the Malthusian parameter scales with body mass ($M^{-0.27}$) with an exponent that suggests a fundamental linkage to specific metabolic rates

(Fenchel, 1974; Blueweiss *et al.*, 1978; Hennemann, 1983). Others are perfectly correct that body size is unlikely to be the sole factor varied to achieve this so residual variation within this framework is expected. As discussed earlier with respect to scope, there are tight linkages between body size and key physiological processes that impinge ubiquitously on fitness features. There are undoubtedly constraints on the range of features that can be expressed within the realm of any particular body size. Body size may also be selected directly because of ecological consequences of scaling (e.g. food sizes, foraging ranges), competitive interactions, homeostasis or risks from predation (McNab, 1980, 1986a,b, 1988). Disease could also be a factor, shorter generations providing more evolvability in Red Queen races (Chapter 5).

Unfortunately, most tests of theory are based on analysis of homeothermic lineages. It may be that metabolic rate is not as constrained by body size in ectotherms, and the analysis of Stearns (1984c) suggests that size is not as strongly related to life-history variation in reptiles as it is in mammals. Regardless, it is unlikely that a clear picture of phenotypic evolution could ever be derived by factoring out such a key dimension as size. Thus, Stearns (1983) concluded, 'size is a trait so important that its consequences retain its force regardless of lineage'. In birds, body size is still correlated with most life-history features even after controlling for metabolic rate (Trevelyan, Harvey and Pagel, 1990).

A complex interplay appears to be involved where maximizing r involves maximizing specific metabolic rates and reducing maturation time. All of these are associated with decreasing body sizes (i.e. an epigenetic framework is important). All arguments pertaining to how large animals arise are related to the assumption that improved fecundity or reproductive success offsets the advantages of earlier reproduction and its contributions to compounding reproduction. Resolution of these conflicts may lie in the realm of redefining fitness in terms of persistence and reductions in the variance of fitness of lineages associated with increases in the body size of deployed phenotypes (e.g. Boyce, 1988). To a large extent, body size may be regarded as a filter, larger species representing more advanced organizations (Chapter 1) that have improved survival, despite lower values of r.

10.3 EVIDENCE FOR ALTERATIONS IN LONGEVITY ASSOCIATED WITH METABOLIC RATES

10.3.1 Mutants, races and chemicals

Evidence that metabolic rate impacts on longevity comes from observations of mutants, comparisons among strains or races and environmental manipulations. For example, the reduced life-span of the high-metabolism

Shaker mutant in *Drosophila* supports this contention (Trout and Kaplan, 1970, 1981). Among mice, a thorough comparison of strains and their crosses (Sacher and Duffy, 1979) showed a very clear relationship: those with higher metabolisms died earlier (Figure 10.1). Among bees, species with higher worker metabolic rates and foraging efforts have correlated reductions in longevity (Dyer and Seeley, 1991).

Ethidium bromide inhibits mitochondrial function (i.e. reduces metabolic rate). Treatment of *Drosophila melanogaster* with this material increased development time by 32%, reduced body sizes by more than 20–30% and increased maximum longevity by 22–30% compared with controls. Metabolic rate was reduced as expected by the rate of living theory in the context of free radicals (Fleming, Leon and Miquel, 1981).

10.3.2 Temperature and longevity

The greatest accumulation of evidence pertaining to this aspect involves changes in longevity associated with environmental temperature, particularly in poikilotherms. Within their normal physiological range, organisms tend to live longer at cooler temperatures, a fact documented by an enormous amount of literature. Such results hold well for comparisons among organisms maintained for life at single constant temperatures within their optimal range. Most species show parabolic responses to

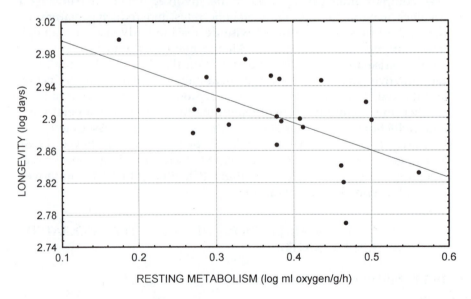

Figure 10.1 The relationship between metabolic rate and longevity among strains of mice. Redrawn from Sacher and Duffy (1979), with permission of Federation Proceedings.

temperature with declines in performance or increasing mortality in either high or low extremes.

Maynard Smith (1963) contested this wisdom, showing that the life-cycle could be viewed as an aging phase followed by a dying phase. A threshold model in which aging was relatively independent of temperature, but where the threshold for the dying phase decreased with increasing temperature, appeared to explain the mortality pattern best. These patterns were derived by rearing flies at one temperature for varying periods, and then switching them to another. Such studies will be confounded by the strong compensatory adjustments in metabolism organisms employ to acclimate to prevailing conditions, and the response of various systems to thermal stress (i.e. heat shock proteins and changes in antioxidant defence systems) (Miquel *et al.*, 1976; McArthur and Sohal, 1982).

For example, if an animal gears its metabolism upwards in the cold, a shift to a higher temperature may then result in abnormally high expenditures for some period. Organisms acclimated to heat may show abnormally low metabolisms when shifted to the cold. Unless very careful controls are implemented, tests of the rate of living theory should probably be confined to comparisons among organisms maintained at single constant temperatures. Careful analyses by Miquel *et al.* (1976) and McArthur and Sohal (1982) obtained an excellent fit of temperature to the rate of living theory in *Drosophila* and *Oncopeltus*, respectively. Sestini, Carlson and Allsopp (1991) provide more recent confirmation for *Drosophila*.

Complex and unexpected interactions complicate interpretations if organisms are switched among temperatures. Maynard Smith (1958) obtained the expected declining relationship of longevity with increasing temperatures in flies kept continuously at one temperature. However, female *Drosophila subobscura* exposed to extremes over 31°C and then reared at 20°C showed increased longevity. This turned out to be due to sterilization at the high temperature that then alleviated these flies of reproductive costs.

In a relatively comprehensive treatment Economos and Lints (1986) showed that *D. melanogaster* has an optimal temperature range centred near 23–25°C. Many physiological aspects including longevity and egg viability were adversely affected by extremes of temperature deviating from this optimum. Moreover, longevity was relatively constant across a temperature range of 17–28°C. This may have been achieved partly by acclimatization.

These authors rejected a causal relationship between growth rate and longevity because growth rate increased dramatically with increasing temperature, whereas longevity remained constant. A cost framework is still suggested by their data, however, because body sizes of flies progressively declined with increasing temperature as well (Figure 10.2).

The truth of this is evident if a new independent axis is obtained incorporating temperature, larval growth rates and duration of growth. This also yields a response function with evidence of intermediate canalization (Figure 10.2). In other words, increased rates of growth may have been offset by reduced durations. In a framework of phenotypic balance, it is not impossible to obtain regulated longevity, but the sources of compensatory adjustment may be complex.

Another interpretation of these results underscores the complexities involved in studies of longevity. Economos and Lints (1986) consider only adult longevities (otherwise animals in the cold would show longer total life-times). It must be remembered that the adult body is derived from imaginal discs that are quiescent in the larvae and consequently may be relatively unaffected by temperatures in the optimal range. Adult cells are recently proliferated from these discs.

Animals adjust their metabolic rates to compensate for temperature. Typically, animals transferred from cold or hot environments to an intermediate one show relatively high or low metabolic rates, respectively. The association of larger body size with colder temperatures in ectotherms also appears to be a relatively universal rule. I know of only a few studies reporting otherwise. It may well be that body size adjustments with temperature are a critical component of the acclimation response that has been previously overlooked. It is tantalizing that increased body size has been identified as a general trend associated with colder climates (Bergman's rule).

The advantages of maintaining higher body temperatures are difficult to factor into the equation in studies examining endotherms maintained in laboratory cage environments like those employed with rats and mice. The most relevant literature may be that associated with thermoregulation of exotherms, such as lizards. Potential benefits include improved foraging efficiency, reduced predator risk, faster growth rates, faster maturation, better competitive ability and higher reproductive rates. Benefits in such attributes in the field could well offset the decreases in longevity that might be expected from operating at higher temperatures. Thus, given appropriate environments, lizards will attempt to thermoregulate, but such strategies may be abandoned when the costs associated with finding available 'hot spots' becomes too great (Huey and Slatkin, 1976). Thus, the trade off structure of endotherms contains a rather important metabolic dimension that can be modulated and which entails significant ecological costs. Lessells (1991) reviews data suggesting that endotherms typically mature at much earlier ages relative to their potential life-span and Case (1978) shows the advantages in terms of enhanced growth rates. Coupled with enhanced rates of synthesis, relevant to reproduction, endothermy could make a major contribution to the Malthusian parameter (McNab, 1980).

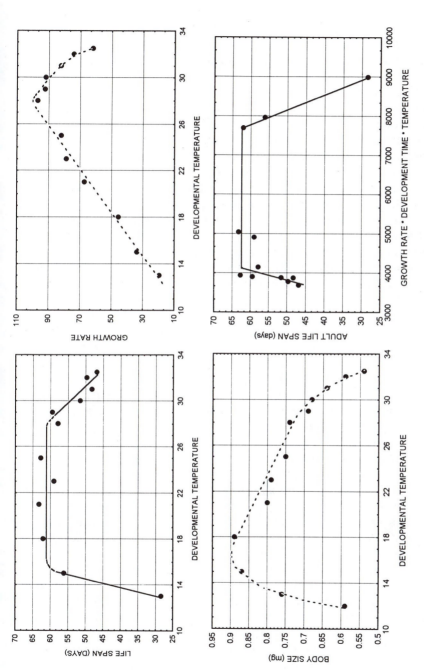

Figure 10.2 The relationship of adult longevity, larval growth rate and adult body size in *Drosophila melanogaster* in response to temperature. After Economos and Lints (1986), with permission of *Gerontology*, S. Karger A. G., Basel. Adult flies show a relatively constant longevity across their optimal temperature range (top left), despite the fact that growth rate increases with temperature across that range (top right). Decreasing body size with temperature may represent a trade off that allows such canalization of longevity (bottom left). This is supported by the fact that an x axis derived by multiplying growth rate, development time and temperature together, also yields canalized longevity within mid-range values.

For endotherms, temperatures falling below their thermal neutral zones impose metabolic costs if body temperature is to be maintained. Experimental manipulations exposing rodents to chronic and relatively severe cold (<10 °C) increases metabolic rates and decreases life-span by as much as 30%. This agrees with predictions from the rate of living theory (reviewed by Cutler, 1984a,b; Holloszy and Smith, 1986; Finch, 1990). Holloszy and Smith (1986) rejected the rate of living theory, however, because in their own experiment cold-exposed and control animals showed no differences in longevity. Problems with their study with respect to the rate of living theory arise because not all metabolic dimensions were considered. Although metabolic rate was estimated to rise about three-fold during 4 h periods of cold exposure (5 days/week), possible compensatory metabolic adjustments could have been engaged during other periods (i.e. increased sleep, reduced activity, decreased temperature during sleep).

The stress imposed was relatively weak (i.e. immersion in cold water at 23 °C) compared with the control environment (20 °C). Metabolic adjustments in other areas were also not adequately considered (e.g. reproduction or growth). In fact, the cold-exposed animals were only 78% as large as controls at 20 months of age, indicating that considerable metabolic savings from growth reduction had been engaged.

Although male rats were employed that do not have the high costs of producing litters, it could well be that sperm production in males represents a significant avenue of investment, and potential changes in such allocation appear to be virtually ignored in the literature. At first glance such results conflict with the above arguments that larger body sizes are obtained in the cold. This trend, however, refers to animals reared continuously under particular regimes. In general, mice reared continuously in the cold do tend to be larger, and mice selected by such regimes have larger body sizes (Barnett and Dickson, 1989). Selection for body size is a complex phenomenon in endotherms, however, and, as will be seen, some groups of small mammals show reversal of the general trend for body size to decrease with temperature.

A complication in studying the impact of thermal regimes is that changes that constitute a shock may induce production of heat-shock proteins and peroxidase that offset damage and possibly impact on aging. Holloszy and Smith (1986) found reduced incidence of cancer in their cold-exposed animals, supporting such a linkage. It may well be that moderate temperature stress in endotherms may engage metabolic reorganizations that might in fact increase investments in somatic support and maintain or even extend longevity. A similar interpretation of feeding restriction in rodents is elaborated below. Where temperature stress is too great to effect compensatory adjustments, animals may either increase overall metabolic costs at the price of longevity, or enter torpor and endure the stress period in an inactivated state.

Clearly the impact of temperature on longevity does not lend itself to simple interpretations. However, the reality of increased longevity of poikilotherms in colder environments is relatively robust, and in other cases, the adjustments that provide compensation appear to be congruent with an allocation program modifying the global rate of living. The fact that the impact of cold temperature on longevity of homeotherms is opposite to that for poikilotherms also argues for a central role of metabolic rate in these phenomena.

10.3.3 Hibernation, diapause, torpor and rates of living

Organisms that normally hibernate or diapause, and which are prevented from doing so, die sooner than normal. Organisms that normally engage in periods of high activity but are prevented from doing so live longer. Very significant extensions of longevity are usually obtained by castration, particularly in organisms like salmon and marsupial mice that engage in intense semelparous reproduction. In Turkish hamsters, experimental manipulation of the duration of hibernation revealed increased longevity with increased hibernation (Lyman *et al.*, 1981). Hochachka and Guppy (1987) reviewed the evidence that biological time is extended by torpor, hibernation, estivation or freezing.

In *Caenorhabditis elegans*, mutations conferring longevity act either by inducing diapausing 'dauer' larvae, or by significantly reducing reproduction (Friedman and Johnson, 1988a,b). Among insects, 10-fold differences in life-span are associated with variation in diapause among crickets and grasshoppers, and overwintering bees have life-spans 6 months longer than their summer counterparts (Finch, 1990, p. 469). Resting stages including dormant seeds, encysted protozoans and tardigrades may persist for hundreds of years (Finch, 1990, p. 462). Such capabilities are associated with highly reduced metabolic rates.

In rejecting the rate of living theory Finch (1990) placed considerable weight on evidence that degree of hibernation and torpor does not show a strong correlation with longevity when temperate zone and tropical bats are compared (e.g. Herreid, 1964). However, Calder (1985) concluded that the rate of living hypothesis fitted the data for hummingbirds very well if torpor was considered. Jurgens and Prothero (1987) found that the life-time energy of temperate zone bats was comparable to other eutherian mammals when periods of torpor and hibernation were factored in, but tropical homeothermic bats still had life-time scopes 2.5 times higher than other mammals.

Austad and Fischer (1991) confirmed that tropical bats with homeothermic temperature regulation had longevities greatly in excess of comparable-sized eutherians. The analysis of Austad and Fischer (1991) utilized extrapolations of life-time scope from estimates of basal metabolic rates. Although there is generally good agreement between basal and active rates, it is possible that bats have lower operational costs than comparably sized rodents. Thus, the field metabolism of the bat, *Macrotus californium*, at 21.5 kJ/day (measured via doubly labelled water), is considerably lower than a house mouse of comparable size (39.8 kJ/day) (Nagy, 1987).

The assumption that tropical bats do not show metabolic savings associated with reduced endothermy may also not be entirely true. In his extensive analysis, McNab (1969) pointed out that even tropical bats may reduce core temperatures in roosting sites (i.e. 20–25 °C), amounting to a 30% reduction in metabolic rate. Social clustering in roosts could yield further thermoregulatory benefits. In a careful and extensive analysis that included 35 species, Hayssen and Lacy (1985) showed that bats have generally lower metabolic rates than rodents. In fact, both long-lived primates and bats do have lower basal metabolic rates than mammals generally. Hayssen and Lacy (1985) also pointed out that long-lived groups are mainly of tropical origin. In an energetic framework this suggests that the energy subsidy of warm tropical living might also be a factor (Bozinovic and Rosenmann, 1989). Among birds, the metabolic rate of tropical species is definitely lower than that of temperate species (Bennett and Harvey, 1987).

The naked mole rat, *Heterocephalus glabes*, is another exceptionally long-lived rodent. These animals are similar in size to housemice (about 34 g), but may live up to 16 years (Jarvis and Bennett, 1991). Housemice live about 3 years. The warm (29–33 °C), stable temperatures of mole rats' underground burrows may have allowed them to become essentially poikilothermic. In fact, it may have been also adaptive to avoid over-heating and dehydration (Yahav *et al.*, 1989; Jarvis and Bennett, 1991). Poikilothermy allows a metabolic rate only 43% that of comparably sized rodents (Lovegrove, 1986). Such features may be adaptations for xeric conditions, or may allow relatively more energy to be diverted to burrowing which may be 300–3000 times more expensive than surface travel (Lovegrove, 1986; Lovegrove and Wissel, 1988). Finally, Sacher (1978) points out that a considerable portion of the increased metabolism of birds is associated with body temperatures that are generally 4.5 °C higher than the mammalian average of 37.5 °C. If this component is taken into account, the two groups are energetically more comparable.

This discussion points out that allometric analyses and considerations of aging often ignore numerous factors that could explain considerable variation. A holistic framework is rarely applied. For example, there is rarely any correction for environmental temperature let alone the integration of temperature variations across life-times and the potential price of heating bills. Many of the examples of extreme longevity in complexly differentiated animals (e.g. rockfish, lobsters, mussels, anemones) are based on species that live in frigid water, or that are relatively sessile and consequently have low behavioural costs. Herreid (1964) points out that bats have severely reduced fecundity, most species producing only one offspring once each year. For example, wild housemice may produce 45 young per year whereas the comparably sized bat *Myotis lucifugus* (and almost all other bats) produces only one (Kirkwood and Holliday, 1979). Such differences must be corrected for offspring size. Bats have very large offspring for their size, but these are probably equivalent to no more than a single litter of mice. Thus, bats show evidence of reduced reproductive effort.

In fact, when body size is controlled for, those groups of mammals with exceptional longevity for their size fall out at the extreme upper end of the slow continuum. Thus, bats and primates are the latest maturing, have the longest gestation periods, lowest fecundity and largest offspring relative to their size (Read and Harvey, 1989). Thus, they still appear to be the slowest living mammals consistent with a metabolic framework for longevity. Relatively slow rates of living may allow more effective longevity assurance features. The point is that interpretation of longevity in a metabolic trade off framework requires that all sources of cost or savings be factored into the analysis. When this is done, metabolic correlates to longevity are clearly visible.

A striking feature of Austad and Fischer's (1991) analysis is the fact that nearly all tropical bats are homeothermic and weigh at least 50 g, whereas all temperate bats are heterothermic and smaller. The marsupials showed an even clearer watershed at a value of 100 g. Regardless of whether this is real or an artefact of sampling, it must impact on the analysis. Hinds and MacMillen (1985) provide data demonstrating that, among heteromyid rodents, larger species were predominantly tropical. In both these animals and bats (McNab, 1969), larger species also had higher body temperatures, possibly facilitated by a combination of environmental heat subsidies, greater food availability and reduced surface area to volume ratios. Such exception to Bergman's rule appears to hold for rodents in general (Ingram and Reynolds, 1987). Housemice may differ because they do not engage torpor as a normal component of temperature adaptation.

Wootton (1987) found that tropical mammals (and bats in particular) tend to mature at later ages than mammals generally. These differences in body size and maturation schedules suggest that growth rates and reproductive costs should be factored into the cost equation as well. If bats have slower growth rates than rodents or other animals, it might represent another avenue of metabolic savings that could offset senescence. This analysis suggests that the appropriate comparisons for conclusions about bat longevity are probably temperate zone bats *versus* temperate mammals generally, or tropical bats *versus* tropical mammals. Although this discussion does not refute the conclusions of Austad and Fischer (1991) that bats have greater longevities and (possibly) life-time metabolic scopes, they do argue that energetics and rates of living are important aspects contributing to these allometric relationships.

Further complications pertaining to bats are revealed by an analysis by McNab (1980) that shows that the metabolic rates of insectivorous and vampire bats are indeed lower than the general mammalian curve. Those eating nectar, fruit or meat have basal metabolic rates equal to or greater than average. Recently, a controversy has erupted over whether fruit bats may be an independent phylogeny that has converged on a bat-like phenotype from primate ancestors. If so, their higher life-time scope may be associated with this phylogenetic derivation. Analyses incorporating body size, absolute metabolic scope (including reproductive costs) and life-time day-degrees experienced in the habitat might reveal some very interesting trends. Rather than highlighting the variation among phylogenies and species as a reason for dismissing a metabolic interpretation (Austad and Fischer, 1991), it seems more productive to emphasize the relatively great explanatory power of even relatively crude allometric analyses. The variation appears largely instructive rather than theoretically lethal. Bats do live at relatively slow rates, and such rates may yield non-linear benefits from longevity assurance.

10.3.4 Behavioural activity

Behavioural activity is another major dimension of metabolic costs and species may be selected for characteristic tempos of activity and emotionality (Oster and Wilson, 1978; Perrigo, 1987; Garland, 1988; Dyer and Seeley, 1991). Experimental evidence suggests that there is a rather tight linkage between levels of physical activity and longevity when costs exceed the compensatory flexibility of the system. Among insects there is good evidence that increased activity associated with flight or sexual interactions increases metabolic rate and reduces longevity (Neukirch, 1982; Sohal, 1986; Newton, Ducore and Sohal, 1989). In houseflies, behavioural activity is associated with a 2.5-fold variation in life-span, which actually exceeds the increases obtained via artificial selection regimes in *Drosophila* (Sohal and Runnels, 1986; Sohal, 1991). One characteristic of *Drosophila* selected for greater longevity via delayed reproduction was a decrease in the activity of younger flies (Service, 1987).

The *Shaker* mutants of *Drosophila* are metabolically hyperactive, consuming nearly twice as much oxygen as controls. Most of this is associated with elevated physical activity. Phenotypic correlations include slightly smaller body sizes and longevities half that of normal flies (Trout and Kaplan, 1970). The long-lived dauer larvae of *C. elegans* express low activity levels which is generally true of most hibernating or diapausing organisms. Long-lived strains of *C. elegans* are noticeably lethargic (Friedman and Johnson, 1988a,b). Beauvais and Enesco (1985) lowered the physical activity of rotifers using curare and obtained a 30% increase in longevity.

In contrast, Lints *et al.* (1984) and Le Bourg (1987) carried out a careful analysis of *Drosophila* and found no relationship between activity and life-span. Such studies demonstrate why progress in understanding trade offs has been so slow. These studies represent simple phenotypic comparisons under single environmental regimes. The earlier discussion highlighted why it is unlikely that such studies would reveal any trade off. In fact, a positive correlation might be expected due to differential fitness or quality among flies. Despite the meticulous and extensive work that these studies represent, they cannot be taken as evidence against a trade off between activity and other allocation demands, or as a basis for rejection of the rate of living hypothesis (e.g. Lints, 1989). Moreover, the costs of spontaneous (voluntary) exercise may be absorbed by a flexible allocation system in the wild type range. The trade off might only become visible as the exercise regime is pressed to greater extremes (Perrigo, 1987; Wolf and Schmid-Hempel, 1989).

At first glance the Supermouse is a counter example, living half as long despite remarkable decreases in activity. However, if activity is recognized as a metabolic sink that is directly traded off with growth, then the reduction in activity is consistent with the principle of allocation and limited metabolic scope where, in this case, both behaviour and longevity are compromised by growth. The finding that high carbo-hydrate diets improve both longevity and behaviour is consistent with this interpretation (Lachmansingh and Rollo, 1994).

Animals under resource stress typically increase activity initially, but decrease it if this proves ineffective. Reduced activity in response to lower resources commonly occurs in rodents if increasing foraging effort derives no benefit (Westerterp, 1977; Forsum *et al.*, 1981). The reduced longevity of Supermice suggests that the cost of the extra growth exceeds the savings that can be obtained via reduced activities. Given that these animals may be operating at close to their behavioural minimum, it is very likely that enforced exercise regimes would show direct and large impacts on their growth rates, longevities or both. The few transgenic mice that expressed the spinner hyperactive phenotype do show the predicted reduction in growth rate. A similar syndrome is obtained in parthogenetic strains of *Drosophila* which are strongly selected for very early and intense reproduction due to low egg viability (Templeton, 1982a). Such flies tend to have reduced longevity (17 days rather than a normal 54 days), and some strains are remarkably lethargic and inactive (Templeton, 1982a).

The best evidence for rodents suggests that voluntary exercise on running wheels may actually improve longevity compared with sedentary controls with *ad libitum* feeding (Goodrick, 1980; Goodrick *et al.*, 1983a,b; Perrigo and Bronson, 1983, 1985; Perrigo, 1987). Thus, exercise may be necessary to offset health problems associated with sedentary life-styles. Evidence that energetic trade offs are still important comes from the fact that under dietary restriction, these trends are reversed: sedentary diet-restricted animals live longer than those that exercise voluntarily (Goodrick, 1980; Goodrick *et al.*, 1982, 1983a,b; Holloszy *et al.*, 1985; Holloszy, 1988). Thus, moderate exercise may be beneficial, but costs may become apparent at exceptionally high levels or when other demands tax available energy. Even then, moderate exercise only appears to improve survival (mean longevity), but not maximal life-span (Holloszy *et al.*, 1985; Holloszy, 1988). High levels of exercise are associated with trade offs in reproduction and growth (Perrigo, 1987; Derting, 1989). Such compensation is indicative of a trade off, but they may obscure impacts on longevity.

There is relatively little appropriate literature pertaining to the role of exercise in insect performance. Given that adults are unable to increase in size, body building (plastic modification of muscular capacities) may be a less viable option for these short-lived animals. Marden (1989) documented variation in muscle mass among male dragonflies. Up to 60% of body mass constituted muscle, and those with greater musculature had the lowest gut contents and fat reserves. Marden (1989) suggested that enhanced mating success of such males is probably traded off against nutrition and longevity.

Rankin and Burchsted (1992) provide a review of the costs of migration in insects. Although the expected trade offs are seen in some cases, the only really consistent cost is delayed maturation and reproduction. In some cases, the expected costs of flight and associated structures are offset because migrants can better locate and exploit superior habitats. These authors suggest that migration may sometimes be associated with an adaptive suite that is generally superior (e.g. associated with accelerated growth or increased feeding) so that performance is accentuated rather than traded off. The idea that migrants might be intrinsically superior, however, begs the question of why residential forms should express submaximal performance.

Vertebrates have morphologies more amenable to modification, and life-times long enough to warrant phenotypic adjustments. It may well be that the regulatory system has evolved under conditions that always require a minimal level of performance, so that sedentary life-styles obtained in human or artificial environments engage maladaptive reductions in physical capacity (i.e. atrophy) as well as unhealthy levels of storage accumulation (i.e. obesity). Thus, moderate exercise may be required to achieve the long-term optimum for vertebrates (Holloszy *et al.*, 1985), but this may not hold for short-lived invertebrates with constraining exoskeletons and in which cells are post-mitotic (Collatz, 1986). Kim *et al.* (1993) recently found that antioxidant defensive systems are upregulated during exercise in rats. Thus, vertebrates may have a linkage of longevity assurance investments regulated according to their normal tempo of activity.

Turning this coin over, the Supermouse strongly suggests that sleep is a crucial dimension in the metabolic equation. Sleep may allow accumulation of reserves to allow higher levels of operation than could otherwise be obtained during activity. The lethargy and slow responses of Supermice compared with normals might arise if growth is robbing reserves that might otherwise be accumulated during sleep. Alternatively, a constant production of growth hormone in Supermice instead of the normal cyclical release may disrupt the partitioning of circadian allocation. Growth may normally occur during the day and activity at night. Although numerous authors have hypothesized that sleep is an energy-conserving adaptation, experimental demonstration has not been previously possible. Clearly, sleep could allow recovery from even strenuous activity, so any studies addressing the impact of exercise on the rate of living theory must consider both aspects. This has not been done.

Sleep may serve several masters. In the Supermouse it appears to help pay for extra growth. In small endothermic vertebrates, sleep is so deep that it constitutes a state of torpor. The occurrence of such torpor in small endotherms is widespread (McNab, 1969). Thus, sleep may be involved in conserving energy for numerous metabolic sinks such as growth, thermoregulation, exercise or reproduction. Where sustainable scope is limited, excess expenditures must be balanced by rest (Kirkwood, 1983). It would be most interesting to examine the possible circadian rhythms of repair systems to see whether they may be engaged more during sleep periods.

Behaviour is a crucial feature because it determines the impact of resources which then determines the selective regime for resource allocation. Risk enters the equation here because activity may also expose organisms to greater predation or environmental risks. Species that coexist with predators often show reduced activity (Sih, 1987), and this could impact on other features of design differentially depending on how much resource acquisition is impacted *versus* reduction in behavioural costs.

10.3.5 Reproduction

The costs of reproduction on other features like growth, survivorship and longevity are among the most well-established trade offs (e.g. Calow, 1979, 1981; Bell, 1980, 1984a,b; Sibly and Calow, 1983, 1986; Partridge and Harvey, 1985, 1988; Bell and Koufopanou, 1986; Partridge, 1989, 1992;

Lessells, 1991; Stearns, 1992; Reznick, 1992a,b; Roff, 1993). High reproductive effort early in life is causally associated with reduced longevity and reduced later productivity. Selection for delayed reproduction and increased fecundity in older age classes increases longevity, and often is associated with reductions in early performance. Thus, early and late life fitness features are often negatively associated (e.g. Wattiaux, 1968; Sokal, 1970; Mertz, 1975; Law, 1979; Rose and Charlesworth, 1980, 1981a,b; Rose, 1983b, 1984a,b; Luckinbill *et al.*, 1984; Clare and Luckinbill, 1985; Arking, 1987; Arking *et al.*, 1988; Tucic, Cvetkovic and Milanovic, 1988). Some long-lived strains of *Drosophila* showed lower levels of mean activity and depressed respiration rates early in life compared with controls, a result consistent with the rate of living theory (Service, 1987). Arking *et al.* (1988) obtained contradictory results, however, their flies being apparently superior in all fitness aspects.

In experimental manipulations altering rates of oviposition in insects, an inverse relationship between reproduction and longevity is usually obtained (reviewed by Lessells, 1991). Female *Drosophila* that are sterilized by high temperature, virgins, or flies that have no ovaries as in the *grandchildless* mutant, have greatly extended lives (Maynard Smith, 1958). Radiation may also unexpectedly extend the lives of insects if it results in sterility. Alternatively, the *dunce* mutant has highly elevated mating activity that is associated with reduced life-span in the presence of males (Bellen and Kiger, 1987).

Many studies of reproductive costs have focused on females, based on the assumption that sperm is cheap. Van Voorhies (1992), however, clearly documented a strong cost for sperm production in the nematode *C. elegans*. Mutations reducing sperm production yielded a 65% increase in mean life-span. Among butterflies, hibernation is associated with greater allocation of resources to the thorax and extended longevity whereas directly reproducing insects allocate more to the abdomen (i.e. reproduction) and die earlier (Karlsson and Wickman, 1989). In fact, in most cases where reproduction is deferred, such as under dietary restriction, longevity is enhanced (e.g. Austad, 1989). Among plants, prevention or removal of flowers often leads to greatly enhanced longevity and growth (Bazzaz *et al.*, 1987).

Unfortunately, some of the best research in this area has uncovered some inconsistencies. Both male and female flies that are exposed to mates at an early age show decreased longevity as predicted (e.g. Raglan and Sohal, 1973; Partridge and Farquhar, 1981; Partridge and Andrews, 1985; Partridge *et al.*, 1986, 1987; Service, 1989; Partridge and Fowler, 1990). Giesel, Lanciani and Anderson (1989) found that insemination, or even simple exposure to males, elevates metabolic rates in *Drosophila simulans*, which argues in favour of the rate of living theory. However, in males of *D. melanogaster*, mortality rates of flies separated from females return to levels similar to that of virgins. This suggests that immediate risks of elevated activity are contributing to mortality (Partridge and Andrews, 1985; Partridge, 1986, 1987; Partridge *et al.*, 1986, 1987; Partridge and Fowler, 1992) (i.e. sustainable scope is being exceeded).

In females, elevated mortality was associated both with immediate impacts of males as well as the cost of egg production (Partridge *et al.*, 1986, 1987). Luckinbill *et al.* (1988) also confirmed that both immediate impacts on survivorship and pleiotropic impacts of reproduction were involved in determining longevity of *Drosophila*. In their long-lived strains, however, the direct impact of the opposite sex on immediate survival was relatively small. Partridge and Fowler (1992) further confirmed the existence of both immediate and longer-term impacts, but their long-lived strain showed no evidence of age-specific fecundity trade offs.

Taken together, these results argue that energetics may determine survivorship via both immediate and deferred costs. The immediate impact of the opposite sex on survivorship is consistent with a decrease in energy available for immediate somatic support (Partridge and Andrews, 1985). The likelihood that the long-lived strains used by Luckinbill *et al.* (1988) are energetically superior (discussed earlier) supports this idea since these strains were also more refractive to immediate mortality impacts. The generally more severe impacts of reproduction on female survivorship compared with males are also consistent with a metabolic interpretation of costs.

The results of Partridge and Fowler (1992) are similar to those of Arking *et al.* (1988) in that comparison of strains revealed no life-history trade off to obtain greater longevity. If the selection regimes have yielded strains with greater energetic scope (Luckinbill *et al.*, 1988), or perhaps some change in the contribution of larval reserves, then comparisons between strains may be inappropriate for detecting the trade off structure. Varying reproduction within strains would be more appropriate. Alternatively, the compensatory dimension may have been overlooked in these studies.

10.3.6 Growth

Growth and reproduction are in general mutually antagonistic (e.g. Lawlor, 1976). Hormonal regulation of reproduction and growth is often integrated such that increases in one reduce the rate in the other (Rollo and Hawryluk, 1988). In snails, parasites often castrate their hosts, effectively prolonging their growth into giants with associated increases in longevity (food for parasites). Snails that detect such parasites respond by increasing their reproductive effort at some cost in longevity (Minchella and Loverde, 1981). Both growth and reproduction are production processes that share some resource demands and this necessitates mutual inhibition (Rollo and Hawryluk, 1988).

Complications arise because increased size often yields fecundity benefits. Where growth is rapid, sufficient size may be achieved to obtain fecundity advantages without decreasing contributions to r associated with delayed reproduction. In such situations investments in longevity may be reduced and the best strategy may be to dispose of the soma in favour of reproduction. Where growth rates are slow, fecundity benefits will be delayed and deferment of reproduction may require longevity assurance. Such a situation would lead to a negative intra-specific association between growth rate and longevity across dietary regimes. This appears to be the basis for longevity enhancement under dietary restriction that has been well documented in rodents and other animals (see below).

Growth appears to be a highly regulated (as opposed to maximized) process, with considerable compensatory capacity (Rollo and Hawryluk, 1988). An important aspect associated with growth is the accumulation of storage reserves. These may act to ameliorate short-term fluctuations in resources, or as deferred investments in reproduction (Chapin, Schulze and Mooney, 1990). Storage entails costs in terms of acquisition, mainte-nance and mobilization, but appears to be favoured in slow-growing species.

10.3.7 Defence

Defensive adaptations incur costs in growth rates or reproduction (Bazzaz *et al.*, 1987). Organisms with inducible defence are ideal material for testing this hypothesis, and there is a burgeoning literature documenting the trade offs in important aspects such as growth and reproduction associated with defence (Chapter 6). The deployment of defensive features such as secondary plant metabolites, or the maintenance of de-toxification systems in herbivores, may be associated with costs in aspects of fitness, such as growth or reproduction (Hoffmann and Parsons, 1991).

The relationship of defence to survivorship is likely to be complex. Although defence is an expense, investments are likely to assure longevity (in the case of antioxidant processes) or survivorship (with respect to parasites or predators). Successfully defended phenotypes may have intrinsically longer lives which, according to theory, may allow older age classes to be viewed by selection and defer senescence. The converse should follow for ineffective defence systems. The reality of a trade off is implied by the existence of inducible defences. Where there are low risks, constitutive defence may impose a cost that lowers fitness.

10.4 STRESS AND THE DIETARY RESTRICTION PARADIGM

Adaptations deployed in response to stress incur costs or require increasing conservation and efficiency that often results in lower metabolic rates. An emerging synthesis of stress-related syndromes has been spearheaded by Parsons (1991, 1992) and Hoffmann and Parsons (1991). The overall conclusion is that stress is generally associated with lower metabolic rates. This in turn can lead to increased longevity, particularly where reproduction is also curtailed (Hoffmann and Parsons, 1989). Lower metabolic rates conserve available resources or increase metabolic efficiency, and consequently may be favoured in harsh environments. These aspects of stress amelioration may explain why selection for increased longevity or delayed reproduction in *Drosophila* also result in increased resistance to starvation, desiccation or toxic levels of alcohol (Service, Hutchinson and Rose, 1988; Rose, Hutchinson and Graves, 1990). Selection for enhanced early reproduction also reduces resistance to starvation (Service, 1989), which is consistent with a metabolic interpretation for this factor.

Of all the evidence available pertaining to stress and longevity, the association between reduced food intake and significant life extension is well documented (reviewed by Holehan and Merry, 1986; Ingram and Reynolds, 1987; Yu, 1987; Masoro, 1988; Weindruch and Walford, 1988; Holliday, 1989b; Finch, 1990; Arking, 1991; Rose, 1991). Most of the research pertains to rodents, but the phenomenon appears to span broad phylogenies (Weindruch and Walford, 1988). Although originally touted as a support for the rate of living hypothesis via an expected reduction in metabolic rates, the story has become very muddled indeed. Although some studies found the expected decrease in metabolism as well as in body temperature (e.g. Forsum *et al.*, 1981; Nakamura *et al.*, 1989), others found no decline in metabolic rate per gram of lean body mass in diet-restricted rodents (Masoro, Yu and Bertrand, 1982; McCarter, Masoro and Yu, 1985: Masoro, 1988; Duffy *et al.*, 1989). Instead, dietary restriction appears to involve a reorganization of metabolism that may actually improve physiological performance, immunological responses and cancer resistance. In the most thorough review, however, Weindruch and Walford (1988) pointed out that the entire phenomenon ultimately hinges on available energy, and consequently must be linked to metabolic rates.

Such conflicting findings have lead many authors to reject a 'rate of living' interpretation to explain enhanced longevity under dietary restriction. Here I will argue that this rejection is due to a lack of a holistic framework. Dietary restriction may at first glance appear to enhance longevity with no metabolic consequences, but in fact, important energetic savings are effected by reductions in body size, reproduction and thermogenesis.

Most studies on dietary restriction have been carried out on endothermic rodents. Even in those studies reporting no decline in metabolic rate, core body temperatures are reduced (Finch, 1990). Energy savings of at least 15% of normal costs may result (Forsum *et al.*, 1981). Moreover, the circadian rhythm may be skewed to generate a greater amplitude in daily temperature around reduced mean values (Duffy *et al.*, 1989; Nakamura *et al.*, 1989). This points out that a 24 h assessment of metabolic rate is a minimal time frame for addressing questions pertaining to rates of living.

Reduced rations in endotherms appear to involve a reorganization of metabolism, shunting scope away from endothermy to support other functions more fully. If endothermy is relatively more expensive than some of these other demands, then increased longevity could be obtained due to lowered temperatures even though daily metabolic rates show no decrease. The other possibility is that metabolic pathways associated with endothermy make a larger contribution to aging than do those features that are more fully supported in the reorganized physiological environment. Masoro, Katz and McMahan (1989) found that food reduction was associated with lowered blood concentrations of glucose and reduced glycation of haemoglobin. Life extension might be associated with reduced temperatures or reduced accumulation of glycation damage. The fact that body temperature is relatively independent of body size in mammals has sidetracked serious consideration of metabolic consequences of thermoregulation to aging (e.g. Finch, 1990). Holding body temperatures constant across sizes means that small endotherms pay relatively enormous heating bills. Even a small change in core body temperature in such species represents a very large change in life-time energy allocation. It is perhaps significant that there are no known homeotherms that do not show age-related senescence (Finch, 1990, p. 221). A clearer picture might be obtained by examining poikilotherms and the studies available emphatically show that longevity increases with dietary restriction in numerous species (e.g. Weindruch and Walford, 1988; Austad, 1989; Kaitala, 1991; Spencer, 1990; Ernsting and Isaaks, 1991).

A further complication is that dietary restriction reduces reproduction, and this sink must be factored into the global analysis. In fact, moderate dietary restriction in rodents is strongly associated with reductions or curtailment of reproductive competence to later ages (Ingram and Reynolds, 1987; Yu, 1987; Holliday, 1989b) which amounts to a reduction in immediate performance rates. That is, reproductive effort is stretched out over longer time frames to adjust for thinner resource supplies. Thus, the dietary restriction results are consistent with a hypothesis of adaptive deferral of reproduction (Totter, 1985; Weindruch and Walford, 1988; Holliday, 1989b). Among invertebrates (or plants), some situations may also favour reducing metabolism and riding out periods of low resource availability. In others, a relative increase in metabolism might be expected (e.g. where the soma is cannibalized in big-bang reproduction). Comparisons among species, life cycle stages or environments using poikilotherms appear to be the kind of crucial test needed to interrelate resource levels, metabolic rates and longevities. Significantly, Hillesheim and Stearns (1992) obtained results very similar to those found for rodents in selecting for body size. Large flies had increased early reproduction but decreased longevity. Small flies deferred reproduction more to later ages and lived longer.

A major dimension that must be carefully considered with respect to food restriction is the fact that animals compensate strongly to maintain intakes as diets are diluted or restricted (reviews in Rollo, 1984; Rollo and Harwyluk, 1988; Rollo and Shibata, 1991). Animals offered diluted diets or restricted opportunities to feed can compensate via increased ingestion rates. Most gerontological studies circumvent this problem by providing measured portions of food or every-other-day feeding regimes. Where dietary restriction reduces feeding rates, body size is adjusted in rodents to maintain relatively constant per gram resource levels. Thus, rats on moderately restricted diets do not reduce their intakes per gram of lean body mass (Masoro, Yu and Bertrand, 1982, 1989; Masoro, 1988; Duffy *et al.*, 1989), but body sizes are instead scaled down. In some cases, in fact, the restricted rodents may show slightly higher per gram feeding rates than *ad libitum* controls (e.g. Masoro, Yu and Bertrand, 1982). Our results with cockroaches suggest that moderate dietary dilutions may be met by increased activity, respiration and feeding that allows maintenance of normal body sizes. Further dilutions, however, will yield the reductions in metabolism expected to be associated with restricted food availability (Kajiura and Rollo, 1994b). The crucial factors appear to be whether dietary restriction (i.e. providing food only intermittently) or dietary dilution overrides the ability of the system to compensate, and whether the species attempts to maintain a constant growth trajectory.

In some recent experiments on cockroaches (Geissler and Rollo, in preparation), we offered two food choices: a diet of pure sucrose (highly attractive to cockroaches) and one of high protein and vitamin quality. The animals struck a balance between these two diets, preferring to ingest more of the carbohydrate.

A very striking result was obtained when the carbohydrate diet was progressively diluted with indigestible agar. With moderate dilutions the animals increased their carbohydrate consumption while maintaining protein intake. With severe carbohydrate dilution, the insects compensated by rejecting the carbohydrate diet and eating more of the protein. The result was a significant increase in reproduction. Yu, Masoro and McMahan (1985) documented the reverse case in rodents. Diets with low protein, but high carbohydrate led to greater life-time caloric intake, probably as a compensatory response to obtain protein. In the case of the cockroaches, their free choice did not obtain maximal fitness as measured by reproductive performance. Better yields (at least in the short term) were obtained by carbohydrate restriction. This lends support to the idea that *ad libitum* feeding may be an unnatural state for rodents (Arking, 1991, p. 263), and also suggests that restriction or dilution may sometimes improve nutrition via compensatory feeding responses. Experiments restricting food are likely to derive different results than those employing *ad libitum* diluted diets.

Clearly, any manipulations of metabolic rate via dietary alterations may have independent and perhaps conflicting impacts on nutrition and health. The physiology of the species may be adapted to specific food qualities as well as quantities. Diets sufficiently diluted to reduce metabolic rates drastically may ultimately lead to shortened lives due to malnutrition, even though reduced metabolism might be expected to extend longevity (Rollo and Shibata, 1991). Evidence in this direction comes from Forsum *et al.* (1981) who showed that on severely restricted diets, rats do reduce their metabolic rates, even when corrected for body size. Beauchene *et al.* (1986) review evidence that severe dietary restriction of rodents reduces rather than extends longevity. Such results support the balance hypothesis outlined earlier.

Our Supermouse results illustrate that growth must be considered to compete for metabolic resources just like any other energy sink. Studies that claim to reject the rate of living theory often do not factor this dimension into the cost analysis. Diet-restricted rodents generally adjust their lean body mass on restricted diets to maintain a constant caloric intake per gram of body mass (i.e. they grow more slowly and/or reduce their size) (see Masoro, Yu and Bertrand, 1982; McCarter, Masoro and Yu, 1985; Weindruch *et al.*, 1986; Ingram and Reynolds, 1987). Thus, although these studies found no reduction in specific metabolic rates the lean body mass of the restricted rodents was much smaller than those fed *ad libitum*. Evidently a major energetic saving has been effected via reduced growth rate. Dietary restriction is also effective in mature animals but improvements in longevity decline precipitously as growth slows. Even full grown animals have protein turnover rates that represent a continuing cost that may be adjusted by dietary restriction, so weak extensions of longevity in such animals cannot be taken to show that growth is not a dimension in trade offs.

A fundamental strategy to maintain relatively constant rates of energy flux per gram of body tissue neatly explains why the Supermouse does not increase its per gram feeding rate in response to growth enhancement. The mice do regulate whole-animal food intake to account for larger body size (i.e. big mice do eat more), but specific rates of feeding and respiration do not vary much. Thus, the only avenue available for modulation is the relative efficiency of growth, and improvements can only be obtained by robbing other functions. Thus, whereas reduced growth under dietary restriction obtains an energetic saving for normal rodents, the forced growth of the Supermouse superimposed on this adaptive trajectory represents the imposition of a significant additional cost. The marked decreases in longevity in Supermice then agrees with that predicted as an extension of the trajectory revealed by dietary restriction studies (Ingram and Reynolds, 1987). Rapid growth is usually coupled to maximal reproduction and shorter lives (see also Hillisheim and Stearns, 1992).

Holliday (1989b) reviewed earlier literature on rats which showed that dietary restriction that reduced adult size by 50% obtained proportional increases in life-span of 36–42%. Masoro, Yu and Bertrand (1982) and McCarter, Masoro and Yu (1985) clearly show that dietary restriction is directly linked to body size reductions in rats, a qualitative reduction in diet to 60% of controls yielding body sizes 59–66% of controls. Associated with this adjustment was a prolongation of life such that controls had survivorship only 68% of the restricted rodents. Weindruch *et al.* (1986) obtained similar results with mice. A 50% dietary restriction yielded mice half their normal size and a nearly proportional doubling of longevity.

Yu (1987) claims that growth reductions associated with dietary restriction are simply a coincidental feature since growth reduction is not necessarily associated with restriction. However, one of the papers cited to support this conclusion (Yu, Masoro and McMahan, 1985) showed the growth reduction very clearly (i.e. maximal sizes of *ad libitum* fed animals was 500–600 g compared with less than 300 g for those on 60% diets). This study concluded, in fact, that there was reduction in lean body mass proportional to food intake. In a very thorough review, Ingram and Reynolds (1987) explored the relationship between body size and longevity in rodents. They showed that dietary restriction almost inevitably impacted negatively on body size, although the impact on longevity varied widely among strains. They rejected any causal relationship between body size and longevity, however, largely based on the confusion resulting from comparing correlations across treatments and strains as well as those within treatments and strains. For example, while experimental manipulations clearly documented the linkage between dietary restriction, reduced body size and enhanced longevity across treatments (the appropriate framework for addressing trade offs), comparisons within each treatment or across strains yielded no absolute relationship between body size and longevity.

As discussed earlier, it is quite possible for correlations between experimental groups to be opposite in direction to those within treatments.

Trends among animals in a single regime may merely represent variation in fitness for example. Thus, dietary restriction reduces body size and enhances longevity, but within treatments, individuals with superior metabolisms might both grow larger and live longer in either treatment. Similar caution applies to comparisons within and among strains (e.g. Goodrick, 1977; Goodrick *et al.*, 1982, 1983a,b; Beauchene *et al.*, 1986; Weindruch *et al.*, 1986; Ingram and Reynolds, 1987). Consequently, positive correlations within treatments do not reject the experimentally demonstrated trade off involving growth, metabolism and longevity.

Such results have generally been considered to reject the necessity for any trade off between growth and longevity. The Supermouse suggests that high growth rates may be associated with reduced longevity. In comparisons among mouse strains that achieve high body weights, these generally also express reduced longevity (Goodrick, 1977). If only the results of experimental manipulation are considered, there appears to be a negative correlation of longevity with growth rates and a positive correlation with extending reduced growth rates over longer durations. However, these may only be apparent under stress, consistent with the balance hypothesis.

Significantly, the impact of high densities on *Drosophila* (which is equivalent to restricted diet treatments in rodents) yields the same phenotypic impacts: adult body size is reduced in conjunction with enhanced longevity (Zwaan, Bijlsma and Hoekstra, 1991). Baur and Baur (1992) reported similar results with snails. In ectotherms, dietary restriction is more commonly associated with real reductions in metabolic rates. Ernsting and Isaaks (1991) concluded that the dietary restriction paradigm is consistent with the rate of living theory, and appears to have its major linkage to deferral of reproduction which entails high metabolic expenditures. This echoes the conclusions of Holliday (1989b) for endotherms.

This points out that the cost of growth must centre not simply on ultimate body size, but also on the dimensions of the rate and duration of growth. Low metabolism animals could still obtain large sizes, albeit at slower rates. Thus, there is a general trend for poikilothermic species to express larger body sizes and longer lives at colder temperatures – the opposite association of body size and longevity to that obtained by dietary restriction (e.g. Economos and Lints, 1986). In this case, resources are not necessarily limiting and the adaptive strategy has an entirely different cost-benefit structure.

The resource allocation strategy may vary depending on what feature represents a target for canalization. Thus, molluscs and most rodents appear to maintain relatively constant rates of specific energy supply and metabolism, and vary body sizes as one dimension to achieve this. In the cockroach, *Periplaneta americana*, however, body size itself appears to be the target of canalization, relatively large dilutions of the diet having almost no impact on rates of growth or final body mass (Kajiura and Rollo, 1994b). Possibly, different results might be obtained by dietary restriction which would prevent the massive compensatory adjustments in feeding employed by these insects on diluted diets. Among rodents, the A/J strain of mice also appears to be exceptional in that body size is not strongly adjusted in response to dietary restriction, and, as might be expected, associated improvements in longevity are also reduced compared with other strains (Ingram and Reynolds, 1987).

A striking conclusion of Yu, Masoro and McMahan (1985) was that among a diverse range of dietary regimes producing divergent longevities, the total caloric intake on a whole rat basis was relatively constant at about 36 000 kcal per life-time. The Supermouse appears to represent a maladaptive extension of a continuum of response associated with maintaining a constant per gram food intake while body size and growth rates vary (i.e. a plastic response to resource availability). In this case, the doubled rate of growth halves longevity. The entire response continuum conforms very well to an adaptive plastic trajectory in the sense of Stearns and Crandall (1984), and thus to a framework of phenotypic plasticity (Chapter 6). In essence, rodents obtaining *ad libitum* food at minimal costs may reallocate resources away from supporting longevity in order to maximize reproduction. The two may be mutually exclusive because reproduction has high metabolic costs (e.g. Ernsting and Isaaks, 1991).

Growth rate may be a primary signal used to modulate other factors in the resource allocation strategy. For example, Bronson (1984) examined the constellation of attributes associated with wild strains of mice compared with those adapted to laboratory conditions of stable temperatures, *ad libitum* food and sedentary lives. The domesticated mice grew to larger sizes (which enhances litter size) and also had earlier maturation and higher reproduction rates. They also displayed lower voluntary exercise. Unfortunately, longevity was not measured, but the suite of features is consistent with differences seen in restricted *versus* non-restricted laboratory animals and in the Supermouse. The reproductive failure of the Supermouse follows as growth is pushed beyond the normal range of allocative balances, and actively reduces resources available for reproduction. As Bronson (1984) points out, maturation and the commencement of reproduction in rodents is believed to hinge on a 'permissive' signal denoting that a suitable size or body composition has been achieved (reviewed by Glass and Swerdloff, 1980). Because of the high cost of growth, this signal is likely deferred indefinitely in the Supermouse.

These results also relate to the contention of several authors that rodents under dietary restriction are closer to the normal operational mode of wild rodents and *ad libitum*, sedentary animals do not represent suitable controls. They postulate that dietary restriction does not represent a mechanism for extending longevity, but simply offsets an unhealthy life-style associated with overeating and lack of exercise (e.g. Sohal, 1987). The current proposal is instead that laboratory animals represent an adaptive suite geared to capitalize on a rich environment by maximizing early and high reproduction and reducing longevity assurance. It is perhaps signifi-cant that the two major research initiatives addressing longevity (dietary restriction in mice and selection for longevity via deferred reproduction in high-density *Drosophila* cultures) are both related to the expression of similar plastic programs apparently adapted to deal with food stress. Figure 10.3 provides a general conceptual model. The evolution of such a complex adaptive trajectory is clearly relevant to lineage selection.

Ingram and Reynolds (1987) derived a conceptual model relating the intra-specific relationship of body size to longevity for rodents. Their model converged remarkably upon that of Sacher and Duffy (1979) for metabolic rate. Maximum longevity occurs at intermediate sizes and

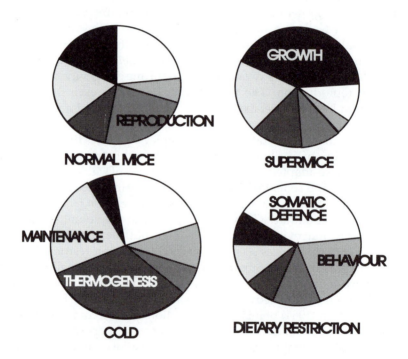

Figure 10.3 A conceptual model outlining the possible adjustments associated with various stresses such as dietary restriction, cold temperatures and enforced growth (Supermice). In each case the reorganization of allocative investments may reflect the presence of reaction norms or genetic correlation structure forged by evolutionary experience with variation in temperature and growth perfor-mance. The size of the pies is indicative of specific feeding rates.

decreases both in large obese animals and those that are exceptionally small. Such a model is also consistent with derivations by Economos and Lints (1984a,b, 1985) for *Drosophila*. Such models help to explain why the allometric trends across species (i.e. larger species live longer) are generally opposite to those within species (larger animals die sooner), and suggest that reorganizations of the resource allocation strategy may be involved in interspecific differences. Thus, other things being equal, selection for large mice reduces their longevity, a phenomenon that is also reflected in the reduced longevity of larger breeds of dogs (Sohal, 1987).

In summary, the best evidence available from the dietary restriction studies and the Supermouse suggests that there is a complex readjustment of resource allocation in response to environmental resources that conforms to an adaptive trajectory. Both environmental impacts and selection regimes tend to reflect the same underlying trajectory. The dimensions of endothermy, reproduction, behavioural activity, growth and longevity appear to be mutually linked, and the relative allocation to each appears to be adaptively modulated along a response continuum. Under conditions unfavourable to productivity, savings may be effected from other resource sinks and invested in longevity. Under highly favourable conditions, resources are marshalled into growth and reproduction, and somatic support is reduced. A similarly complex but different response occurs with variation in temperature.

10.5 CONCLUSION

Taken in isolation, nearly every dimension relevant to a metabolic interpretation of longevity has been vigorously attacked in the literature, resulting in rejection of the rate of living theory. Many of these criticisms can be answered directly. Many others stem from considering only one or two aspects in what is clearly a complex multifunctional integration of physiological dimensions. A holistic treatment in the context of the phenotypic balance framework is required. Numerous studies use comparisons that are not only inadequate for revealing trade offs, but they are likely to be misleading. This problem is so pervasive that the entire literature relevant to life-history evolution needs careful re-interpretation.

When all of the evidence is marshalled, and emphasis placed on experiments that can adequately address trade offs, there appears to be very strong support for the rate of living theory. The key trade off appears to be immediate rates *versus* life-time durations. This probably comes about because high rates are less defendable, and greater returns in longevity assurance can be overlaid on organisms that are operating at lower rates. The empirical evidence shows that slow-rate organisms may actually achieve more in their total life-time than an otherwise similar, faster-rate organism. The slower-rate organism may expend less on the short-term, but accumulate equal or more per life-time.

This framework can be linked to evolutionary theories based on fitness returns via relative contributions of different strategies to the Malthusian parameter, improved fecundity with longer durations, or reduced mortality via longevity assurance adaptations. If contributions to the Malthusian parameter dominate, faster short-term rates and reduced durations are expected. High gains in fecundity or improved offspring survival may select for longer durations at slower rates if they offset this benefit. This may be associated with stable populations where contributions to individual fecundity may be of greater value than early reproduction (Roff, 1993). Longer durations may also be selected if longevity assurance reduces mortality and increases the reproductive success of organisms with reduced rates. In other words, reduced durations may increase fitness in terms of persistence and extended variance in fitness. This brings the metabolic and evolutionary frameworks into juxtaposition.

11

Life-times and limitations: speculations in a lineage paradigm

Taken together, there is clearly enormous evidence for ultimate short-term and life-time metabolic limitations on performance and these diverse sources of support weave a mutually consistent tapestry. The one clear consensus emerging from gerontology, however, is that there is no single and simple process or explanation for senescence, so any generalizations must be cautious. Given the immense complexities associated with all of these intercorrelated and interacting features (Harvey and Nee, 1991), it is rather remarkable that metabolic rate is so highly correlated with other key features. It is premature to dismiss a metabolic basis for life-history evolution and senescence. On the contrary, the ubiquitous evidence for trade offs discussed in the last few chapters suggests that a comprehensive theory of life-history evolution fulcrums on understanding how a balance of competing functions is best accommodated within limits imposed by both immediate and life-time metabolic scope.

Organizational selection incorporates the idea that subcomponents may be sacrificed for higher-order function. Thus, programmed cell death and tissue necrosis are essential epigenetic processes (e.g. to remove webs between digits), and individual sacrifice of sterile workers in social insects is well known. Lineage selection consequently incorporates the idea that individual death could be a programmed feature for both parental and lineage benefit. Such ideas, originally attributed to A. Weismann and A.R. Wallace, are generally regarded as group selectionist, and consequently, unworthy of serious consideration (e.g. Kirkwood and Holliday, 1979). A lineage framework forces a re-evaluation of two important aspects: can death or reduced longevity assurance be selectable via lineage-level benefits, and associated with this, is there evidence for regulatory evolution targeted to death or longevity assurance?

Longevity is intimately linked to numerous key features including metabolic rates, body sizes, brain size, body temperatures and extrinsic mortality pressures. There are numerous genes contributing to patterns of age specific mortality and senescence. The evolution of a polygenic, integrated organizational feature likely involves regulatory evolution relevant to a multigenerational framework. Thus, Nevo and Shkolnik (1974) observed that the metabolic rates of mole rats (*Spalax*) were highly correlated with their diploid chromosome number, suggesting chromosomal-level rearrangements as relevant evolutionary events. Both metabolic rates and karyotypes showed clear correlations with an ecological cline in increasing aridity.

In *Neurospora crassa*, Munkres and Furtek (1984) traced numerous mutations that reduced longevity to a complex of linked genes (the *AGE-1* complex). All of these genes were related to the functioning of enzymes involved in anti-oxidant defence strongly implicated in deferring damage associated with aging (Chapter 9). In this context it may be significant that the genes associated with carbohydrate metabolism that have been altered in long-lived strains, as well as important oxidative defence enzymes such as superoxide dismutase and catalase, all are located on the third chromosome in *Drosophila* (Campbell, Hilliker and Phillips, 1986; Luckinbill *et al.*, 1989; MacKay and Bewley, 1989; Arking and Wells, 1990). Such findings suggest a diversified regulatory ensemble dedicated specifically to somatic defence. The linkage relationships could reflect evolution via gene duplication, regulatory integration, or both.

In studies of cellular senescence, the hypothesis that senescence is genetically programmed is rapidly gaining support. Sugawara *et al.* (1990) demonstrated that one arm of the human chromosome I contained genetic elements that dominantly bestowed senescence upon immortalized hamster cells. Vaux, Weissman and Kim (1992) found that the human *bcl-2* gene prevents normal programmed cell death of unwanted cells in *Caenorhabditis elegans*. Taken together the accumulating evidence suggests that regulatory evolution involving numerous dispersed genes and some linked complexes may be involved in determining longevity. Not only may the forces forging the mortality rates act in a transgenerational framework, the relevant genetic changes may not conform to a strictly genic framework.

It is remarkable that the two lineage-level forces that appear to be strongly related to the evolution of sex (i.e. evolvability and damage control) appear equally paramount to the evolution of senescence. Bell (1988b), for example, directly adopted Orgel's (1963) error catastrophe theory of somatic aging (i.e. that errors in repair of regulatory systems can lead to an avalanche of new errors) to the action of Müller's ratchet in lineages.

Early explanations of senescence suggested that death may be a programmed attribute to remove outdated or damaged adults and make room for the new models. Thus, death becomes part of the evolvability function of lineages. With the dismissal of an evolvability function for sex, consideration of death in this light has also been ignored. From the discussion outlining the possible selective advantages of sex, it follows, however, that populations dominated by potentially immortal adults (in established competitive positions) would be stalled in an evolutionary sense. With no wind in its sails, such a lineage would be literally dead in the water in terms of environmental tracking or coevolutionary arms races. Such a lineage would represent a living corpse for rapidly evolving diseases.

Sexual reproduction would offset such problems to some extent, but the persistence of immortal adults could still greatly slow evolutionary responses of lineages. Such 'good of lineage' ideas, because they merge ultimately with the upper-level species holon, admittedly have the appearance of group selection in its worst sense. It remains, however, that adaptations for evolvability necessarily transcend individual adaptiveness and may be better understood in a lineage framework. The current model suggests that such selection could occur on relatively short time scales. If selection can act on multigenerational time frames, then death, like sex, may be encompassed within a selectable transgenerational framework. The real individual that is being selected is the common genetic templet itself, via the indirect and global success of the phenotypes it elaborates. Where lineage persistence is favoured by rapid elaboration of highly variable phenotypes, death may be a necessary complement to sex to keep the road forward clear of obstruction and reduce genetic inertia.

Whereas sexual reproduction could enhance individual longevity via offsetting the dilution effects of successful reproduction, lineage selection for evolvability could favour both sexual reproduction and truncation of individual longevity. Thus, sexual organisms could pay for lineage membership with both loss of their transmitted genetic identities and programmed termination of their phenotype. Of course, this cost may be relatively small if organisms are destined to die anyway, and the question then becomes how much potential life-span, if any, might be traded off to achieve improved evolvability. These are difficult questions, but it is clear that if an evolvability function of sex is accepted, then an evolvability function for death must also be seriously reconsidered.

Numerous cases are already known documenting the cost in parental longevity associated with reproduction and raising offspring. In many cases, programmed parental death ensues as organisms engage in high rate activities that are unsustainable, or they may actively cannibalize somatic resources and re-invest them in reproduction (e.g. Watkinson, 1992). Thus, the idea that parents might sometimes improve the fitness of their lineage by dying is well supported by abundant evidence. Rose (1991) attributed this idea to A.R. Wallace, the co-founder of the theory of evolution via natural selection and Finch (1990) reprints Wallace's original essay on the subject. Rose (1991) points out that in this framework (parental investment in offspring), the idea of selectable death is not group selectionist but represents a viable neo-Darwinian theory. The critical criteria that must accompany parental sacrifice are that the donated resources are bequeathed directly to the parental line, or that evolvability is extremely valuable.

The lineage paradigm is congruent with these ideas, but allows that all members of local demes are directly part of a genetic organization, and are also linked to other sublineages in a more global organization. As such, the definition of what constitutes kin is simply extended. In an organizational sense, it is not even necessary that senescence be the cause of death. Sacrifices by males could be facilitated by nasty females that ensure that somatic resources are directed to eggs by eating their mates. It would be most instructive, for example, to know whether male praying mantises or spiders that allow females to devour their bodies without sufficient complaint, in fact had much functional time left. Even then, non-compliance could be overridden by organizational selection. Removal of subcomponents for organizational benefit is well established at the sub-organismal level (e.g. Schwartz *et al.*, 1990) and commonly occurs in social insect colonies (e.g. removal of excess soldiers by termites when they are no longer needed, or removal of infertile queens or surplus males by worker honey-bees). Wynne-Edwards's (1962, 1986) ideas on group selection were based on the requirement that local lineages reside with their resources over long periods of time. In such circumstances parental death could indirectly yield long-term evolutionary and ecological benefits to descendants. Selection among sublineages could favour parental death if this improves their ability to bequeath resources directly to descendants, or facilitates the invasion of resources sequestered by other lineages.

Finch (1990, p. 631) specifically recognized the importance of trans-generational evolution with respect to patterns of mortality, particularly where parental bodies may provide food for offspring, or non-feeding parents prevent depletion of potential offspring resources. The study of Wilson (1978) also drew such a conclusion regarding the evolution of prudent predation in the tiger beetle *Cicindella repanda*. Adult beetles appear to have shifted their feeding strategies to avoid impacts on offspring food supplies. The key factor here is that adults vacate the niche occupied by their offspring. This is a temporary solution, however, as offspring eventually enter the adult niche. Wilson (1980) elaborated a viable evolutionary framework for such features.

Parental death or sacrifice will not always be a worthwhile strategy.

There are numerous examples of inverse features. Parents that are subject to resource failure may block pregnancy, resorb developing oocytes or embryos, abort fetuses or cannibalize newly born offspring. During their early formative stages, queen ants that do not obtain sufficient yields from their initially small worker complement will eat both their brood and their workers (personal observation). Housemice also adjust brood sizes via cannibalism (Perrigo, 1987). In cockroaches, lineages may persist on diets that are inadequate for any single individual, by cannibalizing their slower-growing relatives which serve as biological amplifiers. This also represents boot-strapping selection for performance on the diet since the best performers eat the poorer (Gordon, 1959). Among fish, shifting from a niche dominated by small food items to a richer larger-sized food niche is possibly facilitated by cannibalism by larger individuals on smaller size classes of the same species (Nellen, 1986).

Although cannibalism can be interpreted as a selfish trait driven by individual selection, the idea that it may be forged at higher lineage levels is suggested by the phenomenon of cannibalistic salamanders (Collins and Cheek, 1983). In this case, the specialized cannibalistic morph is expressed at high densities. However, the program represents induction of a specific phenotype in direct response to the environment rather than a change in gene frequencies (Hall, 1992, p. 127). In this case, the expression of cannibalism does not reflect the selfish evolution of a particular program, unless it is hitchhiking along with the remainder of the successful genotype. The fact that only a single cannibal arose in each tank of animals smacks of a regulated expression for this program (Collins and Cheek, 1983). Such examples point out that death can be programmed into a lineage in numerous ways, senescence not being the only aspect that needs to be addressed in an evolutionary framework.

Death might be selectable in a lineage framework whenever the removal of certain kinds of individuals (e.g. cell lineages, age classes, males, insect castes) yields the greatest long-term persistence of the lineage. A more global framework might consider that selection varies longevity via regulatory changes in longevity assurance systems, and death can be considered as emerging as a reflection of such selection. However, longevity assurance may not only be abandoned under certain conditions, it may be worthwhile to reallocate the resources constituting essential somatic support, or even the entire soma of the individual to off-spring, or other individuals in the lineage. Selection for such features may act to reduce the variance in fitness of the lineage, or minimize failure under harsh conditions. As discussed earlier, if selection acts only on the Malthusian parameter, it is difficult to reconcile why the most highly developed biological organizations (e.g. mammals *versus* prokaryotes) show a consistent trend for reduced values of r. Certainly the largest eukaryotes are not the least fit, and yet they have the lowest intrinsic rates of increase, the slowest development, the lowest fecundities and the longest lives.

There are only a few criteria that might favour the adaptive evolution of death: where the soma of the individual can derive higher lineage benefit via its conversion to eggs or support of relatives; where the disappearance of the organism donates a resource supply to offspring or relatives; where death contributes to evolvability; or where death reduces risk to offspring or relatives. Vacating resources could just as well be met by dispersal or niche shifts, so selectable death for this reason does not seem like the most viable alternative, unless this aspect were forged at a very high level in the lineage hierarchy. Dispersal or niche shifts would not be alternatives in an evolvability context, because the genetic contributions of immortal parents would still ultimately choke evolvability, regardless of how they shuffled about spatially or ecologically.

Parental death could contribute to lineage persistence by simultaneously freeing resources to offspring and by promoting evolvability. Evolvability would be particularly enhanced where the resources donated by the exit of a single large parent support numerous small offspring to be sifted by natural selection for improved fitness or damage-free status. Current evolutionary theory would argue that senescence is largely a result of lack of selection in older ages (e.g. Rose, 1991). King (1982) recognized that death might be advantageous if parental exit reduced competition. He concluded that even with parental care, death might be advantageous to grandchildren, leading to the prediction that life-span may be limited at most to twice the generation time.

In a lineage framework, however, selection for evolvability could also be a contributing factor to the ubiquity of metazoan death, and it may be no accident that sex and death, which both seem paradoxical to individual fitness, are nearly universal characteristics of multicellular phenotypes. Such selection could be rather subtle, requiring only that natural processes of deterioration not be offset by repair or prevention. The critical question is to what extent long-term benefits such as evolvability could contribute to the apparent ubiquity of death.

The idea that senescence occurs due to the accumulation of late-acting deleterious genes that act epigenetically or the potential deterioration of DNA and dependent functions (both within and across generations) are different, but related phenomenon. Mechanistic theories may focus on intrinsic damage as the causal mechanism. Evolutionary theories propose deleterious expression that may be unrelated to damage. They imply that organisms might otherwise be immortal.

There is convincing evidence that natural selection is necessary to purge lineages of accumulating mutations and DNA damage and maintain the fittest genotypes re-created by sex (Chapter 5). These observations suggest that at least some classes of DNA damage (such as double-stranded breaks) either cannot be recognized or cannot be repaired effectively by somatic cellular machinery. Although the idea that accumulations of genetic damage disrupt DNA functional integrity as organisms age is still controversial (Kondrashov, 1988; Finch, 1990; Arking, 1991; Bernstein and Bernstein, 1991), there is certainly abundant evidence of karyotypic abnormalities, increased cancers and decreased fertility as a function of age (Kirkwood, 1987; Finch, 1990). Recently, increasing attention has been directed to possible age-related changes in DNA methylation patterns and their potential impact on regulatory integration. Kondrashov (1988) provides figures that suggest high rates of somatic mutations that must be offset by repair.

If even lineages have finite persistence in the absence of sex and natural selection, then there must be an even shorter ultimate longevity for a single soma. Among those organisms known to have no senescence (e.g. prokaryotes), individuals still disappear during fission events, and it is still only the lineage that retains true immortality.

Finch (1990) generally concluded that senescence is avoidable, even somatic cell lineages being potentially immortal. This view conflicts strongly with theory developed relevant to sexual reproduction that posits that vegetative reproduction is ultimately subject to failure unless large populations can be exposed to selection to offset Müller's ratchet (Bell, 1988a,b). It remains true that in complex differentiated phenotypes, even those showing negligible senescence, longevities are restricted to maximum durations of perhaps 100–200 years. A quantum jump occurs in plants with modular phenotypes and in clonally reproducing forms that may have longevities of at least 1000 or 10 000 years, respectively.

Modular organisms, such as long-lived coniferous trees, generate parts from meristems with unrestricted cell proliferation potential. Gametes are also derived from such cell lineages within each module. It may well be that this theoretical conflict can be resolved if there is selection occurring within and among these meristematic cell lineages. Selection could possibly eliminate deficient cell lineages if there is competition to form successive modules or to differentiate into gametes. Modules could also compete with one another on the same plant. Such a process could allow selection to act within the organism to at least partially offset Müller's ratchet. In organisms that differentiate a final phenotype of fixed size, cellular replacement (if any) would be too slow, and cell populations too small to offset the ratchet intra-phenotypically and accumulations of somatic mutations would be unavoidable.

Klekowski and Godfrey (1989) compared the mutation rate leading to albinism in long-lived mangroves (generation times of at least 20 years) to that in shorter-lived annual plants. They concluded that the much greater mutation rate observed in mangroves (5.9–7.4×10^{-3} mutations per haploid genome per generation) was probably due to the greater number of cellular divisions occurring before gametes were formed compared with annuals (mutation rates of about 3.2×10^{-4}). This amounts to a 25-fold difference in mutation rates.

Bosch (1971) found that the seed viability of coastal redwoods that live possibly 2200 years is at best about 35% even at relatively young ages. When seedling failure is considered, 160 000 seeds yield only 26 new trees. For old trees (over 800 years old) viability is reduced to only 8% and 160 000 seeds produce only six new trees. These results document the decline in viability with age in this long-lived species, and suggest that the remarkably small seed size of these trees may be a mechanism for providing more numerous genotypes for selection. Such a hypothesis for factors affecting seed size or number is a new one. Generally, however, seed size increases with increasing plant size and longevity, so other factors such as competitive ability of seeds must dominate. Fire-selected trees, however, may have reduced competition during their germination period. It would be most interesting to know the rates of recombination taking place in such long-lived plants and whether they vary with age. Theory predicts they should be high and could be greater in older age classes.

Another interesting aspect is that a cut redwood produces hundreds of sprouts from the remaining stump (Bosch, 1971). Given that channelling all resources into one competitive shoot would be the best ecological strategy, it seems possible that such a response might be a tactic to allow natural selection to act on a potentially age-damaged soma. In this case, the most viable shoot wins. Although tentative, such evidence suggests that even very long-lived plants cannot completely offset genetic decline. It would be most instructive to examine this problem taking into account the size of meristematic cell populations, branching tendencies and rates of cell turnover across species with various rates of aging.

Clonal organisms may reach even greater group ages, but in this case, numerous individuals are exposed to selection. This allows selection for competence, both in the vegetative cell lineages competing to form the new individual, and also among individuals that may compete for local resources. Such processes could well offset the ratchet to some extent, allowing longevities considerably in excess of individuals with terminal differentiation and finite cell proliferation potentials. Theoretically, relatively large population sizes (perhaps 10^5 for eukaryotes) are required to offset Müller's ratchet in the absence of sex (Bell, 1988a,b). Although long-lived clonal species may appear to have too few individuals to effect such selective repair, if the process is traced downwards to particular cell lineages contributing to the successive organisms, the cellular population under selection may be sufficiently large to meet theoretical expectations. Orgel (1963) explicitly recognized that selection at the cellular level may be important in eliminating cell lineages that have accumulated high error loads.

The immortality of cancer cells in tissue culture may similarly be interpreted in such a context. There is evidence that cells in such cultures randomly convert to finite potential and are lost. Rosenberger, Gounaris and Kolletas (1991) suggest that such conversion may reflect a regulatory mechanism that normally protects organisms from cancer or impaired cellular lineages (see also Kirkwood, 1987). They suggest that cells may recognize their intrinsic level of damage, and that a growth inhibitor is released when a certain threshold of deterioration is exceeded. Cancer cells may have largely escaped such control, a hypothesis that is consistent with evidence for irregular karyotypes and widespread accumulation of damage in such immortalized lineages (Rosenberger, Gounaris and Kolettas, 1991). Kirkwood and Cremer (1982) suggest that a culture size of perhaps 10^{10} cells may be necessary to offset this process and achieve true immortality. Significantly, this is twice the estimate Bell (1988b) derived for offsetting Müller's ratchet in clonal species. Thus, the apparent immortality of cancer cell cultures may not be in conflict with a theory of ultimate damage accumulation.

Genome size is also a factor. Highly advanced genotypes with larger functional genomes are subject to greater impacts from the ratchet. This brings up the important point that organisms with great longevities tend to be more primitive or have indefinite cell proliferation potentials. Such an attribute is largely restricted to single-celled organisms or those with morphologies compatible with unrestricted growth (such as modular plants). In organisms with complex phenotypes of targeted size and structure, or those with growth constrained by inherent design (e.g. respiratory and strength restrictions in insects), cell population sizes and proliferation potentials may be epigenetically programmed to particular end points. In insects, for example, somatic cell proliferation is absent in adults. Such organisms have no potential for offsetting accumulation of somatic mutations among cell lineages via internal selection (Orgel, 1963). Thus, species with terminal phenotypes may differ fundamentally from those with indeterminate final forms.

The germ line could largely escape from the deterioration occurring in the active soma if cells are held as a metabolically quiescent reservoir of gametes or as a bank of generative stem cells. Sequestered somatic cells may also be buffered from senescence. Thus, the adult features of *Drosophila* are retained in imaginal discs until hormonally engaged during pupation. Imaginal discs that are held in culture without access to their initiating hormonal signal will deploy adult features even after 10 years (Arking, 1991), even though the life-span of a fly is a fraction of this time.

In addition to low rates of damage accumulation, the germ line may have access to both on-board repair systems and the intervention of recombination and natural selection to restore and maintain competent genotypes. In organisms with a fixed complement of oocytes, gradual losses may represent elimination of damaged gametes. Selection may act very early on. At the embryonic stage, 30% of human abortions have obvious genetic damage. Thus, damage tends to be eliminated even during very early formative stages and may represent significant editing by natural selection that is not even visible at the organismal level. In such ways the germ line may obtain immortality whereas the soma must ultimately deteriorate (Kirkwood, 1987, 1990).

Once again we return to the idea that the sexual lineage is an organizational transition necessary to accomplish multicellular phenotypic evolution. Bell (1988b) provides models that suggest that the ratchet would strongly resist a transition to multicellularity without sex. Multicellular organisms may well be dead ends with respect to immortality so reproduction is necessary to offset loss. Furthermore, damage accumulates in such organisms and may require recombination to facilitate repair and offer corrected variants to the judgement of natural selection.

If we consider a potentially immortal organism, we must also consider that it is faced with difficult problems of DNA parasites (viruses) and selection for selfish DNA that can divert resources to their own ends or cause irreparable damage to the code for organismal function. The well established increased incidence of cancer in older age classes (Finch, 1990) is highly relevant here. Even if no other form of senescence were acting, the organismal level might be impossible to maintain against such forces. Sexual lineages, however, by using organisms as temporary stepping stones, can achieve both potential immortality and ecological fitness by deploying transient packages of disposable soma. Where particular organisms are disrupted by cheating selfish DNA, their organization will fail and be eliminated from the lineage.

Within the lineage paradigm, negative impacts of damaged parents upon offspring could also lead to active selection for programmed death. Thus, if older adults accumulated disease or parasitic loads that infect younger age classes, there could be active selection for mechanisms of removal. Relevant to this is the demonstration by McAllister, Roitberg and Weldon (1990) that parasitized aphids are more likely to choose escape behaviours that are suicidal than younger or unparasitized animals. They argued that inclusive fitness theory predicts such behaviour if the advantages of extended individual fitness are offset by contagion of kin. This idea can be extended to disease generally, and suggests that if virus loads or residual pathogens increase with age, fitness may not simply attenuate with age, it may actually shift strongly to the negative from a lineage-level perspective.

A similar argument might be applied to genetic contagion if even small amounts of DNA damage accumulate in the germ line of older individuals. In this case, sexual selection based on visible markers of somatic senescence would also be engaged, actively curtailing the ability of older age classes to contribute reproductively. A fascinating complexity might come into play as well, since age is some measure of success and continued survival smacks of the handicap principle. Shields (1993) adds that increasing longevity would accentuate inbreeding in sexual lineages, and this in turn might accentuate the expression of damaged goods.

The suggestion here is that selection to offset the ratchet might partially work by a lineage-level adaptation to remove individuals predictably contributing greater genetic damage (i.e. older age classes). There has in fact been considerable controversy in attempting to demonstrate the reality of a phenomenon known as the Lansing effect. The supporting evidence suggests that lineages that are successively derived from older age classes die out or have decreased viability. The idea has fallen into disfavour due to numerous conflicting results (Lints and Hoste, 1974; Comfort, 1979; Rose, 1991). A really crucial problem is that any possible expression of such a phenomenon is confounded by selection for reproduction at later ages and for longer life. However, the paradigm outlined here suggests that such a phenomenon may well be real, even if very careful experiments might be needed for its demonstration.

For example, studies with *Drosophila* have failed to demonstrate the Lansing effect. Longevity in these flies, however, may not be due to genetic deterioration, but to other limitations associated with fat body reserves or lack of cellular replication in adults. If these ideas are correct, genetic programs might be expected that inflict somatic failure on organisms before appreciable genetic damage can be transmitted. Longevity would thus represent a balance between individual selection for greater longevity *versus* lineage selection to eliminate such age classes if they accumulate excessive deleterious mutations. Progress in this case would require both improvements in damage prevention, and removal of previously selected programmed termination. Remember that the idea of programmed death may simply entail decline or termination of somatic defensive systems in older ages. Significantly, in rodents with *ad libitum* diets, levels of antioxidant enzymes (and protection) decline markedly in advanced ages. Rodents with extended longevities associated with dietary restriction do not show such declines (Semsei and Richardson, 1986).

In the marsupial mouse, *Antechinus stuartii*, it seems significant that stress associated with marshalling enormous reproductive effort is associated with relaxed immunological defence so that death is often associated with invasions of parasites and disease (Bradley *et al.*, 1980). This underscores the interesting fact that modulation of longevity via dietary restriction impacts in parallel on both oxidative defence systems and immunological systems (Weindruch and Walford, 1988). Relaxation of such investments to achieve superior reproduction is essentially programmed death by default.

There is no reason to expect a linkage between these systems unless we are seeing a global program dedicated to the modulation of longevity assurance.

Finch (1990) suggests that some species do not show senescent decline. Many of these are modular or clonal as already discussed. Watkinson (1992) argues that even long-lived clonal plants do show senescence. In sea anemones reproduction is via binary fission (Rose and Graves, 1989). Many others appear restricted to cold water, often accompanied by indeterminate growth. In most cases that have been carefully examined, even where age-specific mortality rates are relatively constant, there is at least some diminishment in performance late in life (e.g. birds). Regardless of the degree of senescent decline, when plants and clones are excluded, realized longevities are restricted to less than 200 years. Perhaps it is possible for selection for early competence to offset senescence well beyond the maximum longevity allowed by extrinsic mortality rates. Given that good examples of non-senescing phenotypes are sparse and poorly studied, it appears that a theory of organism design based on inevitable senescence and predictable ultimate failure is not inconsistent with available evidence.

Where programmed death occurs, its explanation is usually near at hand (i.e. immediate increases in offspring number, offspring quality or increased survival of relatives) (Watkinson, 1992). A lineage paradigm suggests that the long-term benefits of individual death are worth considering, and may constitute an important extra dimension intersecting sexuality and life-history evolution. If natural selection can operate with respect to lineage persistence rather than simply on organismal fecundity or the Malthusian parameter (that must approach zero at carrying capacity), then sex and death could jointly contribute to reduced risk at the lineage level, even though the costs to individuals may be considerable (i.e. both genetic and phenotypic transience). As King (1982) pointed out, the current evolutionary theories do not regard longevity as a primary life-history feature, but rather as a secondary consequence of the selection regime. The idea that death may be a regulated aspect of the epigenetic program (in fact the dominant paradigm in plant research) brings longevity assurance firmly into the fold of primary life-history trade offs.

12

The ecological triumvirate: phenotypic correlations of stress, disturbance and biotically diverse habitats

12.1 OVERVIEW

Earlier discussion underscored that most major hypotheses related to the evolution of recombination are equally relevant to discussions of phenotypic plasticity, to the extent that failure to consider both leads to confounded interpretations of data and theory. Moreover, discussions of phenotypic plasticity appear to be interwoven with questions of asexuality, general-purpose genotypes and polyploidy (Glesener and Tilman, 1978; Bell, 1982, 1985; Lynch, 1984; Bierzychudek, 1985, 1989; Michaels and Bazzaz, 1989). However, plasticity and polyploidy are possibly forged, at least to some degree, by different selective regimes. Thus, phenotypic plasticity is thought to be favoured in disturbed habitats, and yet Bierzychudek (1985) showed that in plants, the expected association of early successional or disturbed habitats with asexuality was absent.

Although there is insufficient evidence to draw any general conclusions, it appears that polyploids have reduced plasticity (MacDonald, Chinnappa and Reid, 1988). Schlichting (1986) argued that there is no intrinsic reason why sexually reproducing organisms cannot also be phenotypically plastic but other studies have documented strong associations between asexuality and high plasticity (e.g. Jain, 1979; Levin, 1988; Harvell, 1990). Several authors have suggested that weedy species tend to have intermittent sexual reproduction, self fertilization and reduced chromosome numbers (that would decrease recombination) (Stebbins, 1950; Ghiselin, 1974; Selander and Kaufman, 1973; Selander and Hudson, 1976).

Plasticity, by uncoupling the genotype from the phenotype, might be expected to reduce the value of outcrossing and sexual reproduction (Levin, 1988). Taken together, the evidence suggests that outcrossing, plasticity and polyploidy (especially when associated with adversity) may be mutually incompatible tactics or may be forged by different selection regimes.

There have been numerous attempts to classify environments into categories characterized by their main selective features. These attempts have largely converged on the dominant regimes associated with disturbance, stress or high biological diversity (Grime, 1977, 1988; Greenslade, 1983; Sibly and Calow, 1986; Southwood, 1977, 1988). Figure 12.1 outlines a conceptual model that holistically integrates the evolution of sex, plasticity and stress-related specialization. The key to resolving these issues may be the recognition that stress and disturbance are fundamentally different selective regimes. In addition, the degree of uncertainty of the environment must be considered. It is likely that no environment conforms entirely to a stress, disturbance or Red Queen classification. Thus, absolute trends in organismal strategies are not expected. However,

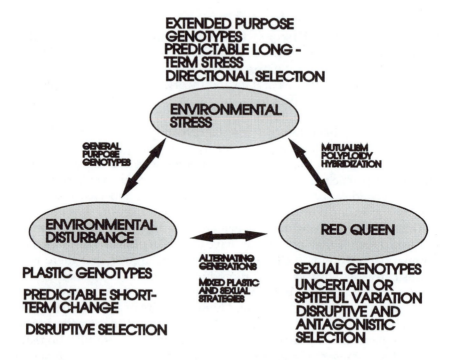

Figure 12.1 The ecological triumvirate. Selection may be categorized along axes linking regimes of stress, disturbance or high biotic diversity (Red Queen).

painting a picture of pure extremes helps to clarify the kinds of forces that may be acting in any particular instance. Previous discussions sometimes failed to discriminate between stress and disturbance, probably because the two concepts are intimately related (in a single habitat one organism's stress may be another's disturbance).

12.2 THE STRESS PARADIGM

Stress environments as defined here are not unpredictable or unstable relative to the generation time of the individual organism. The stress represents constant, reliable information associated with unrelenting dominant directional selection on time scales approaching or exceeding the generation time (e.g. metal or salt contamination of soil for plants or drought in deserts). Most current examinations of the stress paradigm fail to differentiate stress and disturbance relevant to the appropriate scale (i.e. the generation time of the organism). Kolasa and Rollo (1991) considered this aspect in the context of spatial and temporal heterogeneity and identified the appropriate scale to be relative to the organism itself rather than any absolute defined by the observer.

Parsons (1991) suggested that stress periods are 'unpredictable in occurrence and tend to be of short duration'. In the current discussion, such dynamics would be classified as strong disturbance rather than stress. Clearly the concept of stress has different meanings to different authors. Stress environments in the sense applied here may have intrinsically lower carrying capacities or they may require resource allocation investments in particular features that restricts flexibility in others. Thus, several authors have recognized that stressful habitats favour species with low metabolic rates, slow growth and low fertility (Greenslade, 1983; Vermeij, 1987; Hoffmann and Parsons, 1989, 1991; Parsons, 1991, 1992). Hoffmann and Parsons (1991) provide a thorough review, so there is no need to elaborate a detailed discussion here.

The suite of features associated with stress in animals is exactly mirrored in those of stress-adapted plants. This includes low rates of growth, reproduction and photosynthesis, as well as strong allocation to storage and roots (Coley, Bryant and Chapin, 1985; Chapin, 1991). In a thorough review, Levin (1983) concluded that polyploidy results in slower growth over longer time scales. Essentially this represents an upward shift along the r–K continuum. He recognized that such a circumstance may make polyploidy a pre-adaptive feature for stress environments.

Vermeij (1987) suggested that low energy supplies and metabolic rates will limit the scope for adaptation (e.g. for Red Queen arms races), because a higher proportion of available resources are strongly committed to essential functions. Thus, investment in heavy armour requires metabolic resources that may be unavailable in a system stressed in other dimensions. Although this is logical in a resource allocation framework, organisms with highly constrained fecundity may actually respond by investing more in adult survivorship. Thus, the induction of defensive features in *Daphnia* is stronger under conditions of low resources and low fecundity (Parejko and Dodson, 1991), and in plants, stress habitats are characterized by exceptionally high defensive investments (Coley, Bryant and Chapin, 1985; Grime, 1988; Chapin, 1991).

Coley, Bryant and Chapin (1985) provide a model for plants that suggests that relative defensive investments should increase as growth rates decline. A given amount of herbivory has a relatively greater impact on a slower-growing plant. The level of defensive investment then depends upon achieving a balance between the improvement in growth achieved via reduced herbivory and the physiological costs of defence. The patterns in plant growth and correlations in defensive investments strongly support the model. The reader may be struck by the similarity of this model to the dietary restriction paradigm in animals: where growth is slower, greater somatic defences are engaged.

In animals, predators are more likely to kill their prey rather than simply nibble off a bit. The model of Coley, Bryant and Chapin (1985) might still apply to parasite resistance, but anti-predator strategies in animals may require investment tactics that reflect the likelihood of total failure or serious incapacity if defences fail. The trade off here is not with growth rate (and eventual reproductive yields) but with mortality. As Vermeij (1987) suggests, animals may be forced to increase defensive investments in resource-rich habitats where predation risks are greater. More data are needed to clarify the trends in animals, in contrast to the well-established pattern recognized in plants. This discussion points out why understanding organismal design requires integration of both aspects pertaining to resource allocation and to mortality risk.

Stress environments tend to have lower biological diversity and low population densities (Glesener and Tilman, 1978; Bell, 1982; Greenslade, 1983; Moore, 1983; Bierzychudek, 1985) so competition and Red Queen coevolution may be reduced. Stress environments should favour species with strong environmental competence; we might call them extended purpose genotypes. This appellation is meant to capture the idea that the phenotype may be accentuated in particular dimensions, which, according to the phenotypic balance hypothesis, may consequently accentuate trade offs in other dimensions.

Previous discussions have emphasized the concept of general purpose genotypes. This, however, implies that the organisms may be better at everything. Most such organisms are in fact better considered as extended purpose genotypes. Responses to stress include reduced niche specialization and temporally expanded life cycles (Greenslade, 1983) which may give the appearance of general purpose. However, such species may not be competitive with their genetic derivatives living in environmentally more favourable habitats (Levin, 1983; Moore, 1983; Southwood, 1988; Parsons, 1992), and they generally have lower metabolic rates.

In attempting to clarify the relationship between competition and stress, Grace (1991) recognized that stress associated with declining resources and stress associated with environmental constraints, like temperature, pH, heavy metals or salt, might be profitably separated. In the case of declining resources, competition might still be expected to be a major consideration. If stress syndromes involve realignment of allocation strategies as a response to slower growth, however, then it may not matter if this arises due to resource restriction or the necessity to pay extra costs due to external contingencies. If growth rate is a component of competition (which is undoubtedly true for plants), then it remains that stress-adapted species may be inferior competitors compared with those adapted to less constraining habitats.

Extended purpose genotypes would arise via hard directional selection eliminating all but those genotypes capable of handling the stress (specialized monomorphic genotypes in a genic sense). Although metabolic rates may be reduced, and flexibility constrained in the short term, stress environments should select for features that lend greater phenotypic power, efficiency and flexibility in the long term. Gene duplications may constitute one avenue. Single genomes may adaptively deploy duplicated genes during stress (e.g. Cullis, 1986), and this is also a common route taken by bacteria to enhance enzyme synthesis.

Asexual plant species show significant correlations in abundance with recently glaciated areas, with higher latitudes, and with higher altitudes. This might also extend to xeric habitats. Such environments, although not lacking in variability, are better classified as stress rather than disturbance regimes. Generally, asexual lineages have wider geographic and ecological ranges than their sexual counterparts (Bierzychudek, 1985). In many cases, these asexual lineages are polyploids. The ultimate in gene duplication is via polyploidy, and there is considerable evidence that polyploids are often more resistant to environmental extremes than their diploid counterparts (reviewed by Bierzychudek, 1985). At least in some cases, such polyploids have increased enzymatic power (Levin, 1983).

Polyploidy is often achieved via hybridization, particularly in plants. Such constructs may have greater ecological adaptability than either parent (Bierzychudek, 1989), due to greater allelic diversity (Garcia *et al.*, 1991). Symbiotic species associations or fusions may also be involved in achieving extended purpose genotypes (e.g. lichens) (Grime, 1977; Odum, 1985). Lynch (1984) and Hebert (1987) proposed that asexual reproduction may be associated with lineages with altered genetic architecture, such as polyploidy, in order to stabilize the genome against disruption arising from backcrossing to diploid ancestors.

There is a general trend for reduced levels of genetic variation and increased homozygosity in populations subjected to long-term stress (Hoffmann and Parsons, 1991). Although recombination might be valuable in initially assembling the best phenotypic constellation for meeting a particular stress, the value of recombination might be locally diminished as genetic variation becomes reduced. Crosstalk could still remain a valuable aspect, however. The diversity and propagation of parasites is expected to be reduced in stress environments that have lower species diversity and lower population densities. Thus, there are a number of reasons to believe that asexuality might be associated with extended purpose genotypes.

If persistence in a particularly stressful environment requires a particular genetic constitution, then gene flow from other habitats might disrupt local adaptation. Given time, mate recognition systems might offset this, but if there is a sufficient immediate cost to outbreeding, then extended purpose genotypes might switch to inbreeding, self fertilization or asexuality. In the light of the fact that stress is a relatively constant environmental attribute, the value of recombination in a stress-dominated habitat may be reduced once effective extended purpose genotypes have consolidated sufficient competence.

Species occupying stressful environments frequently express non-dispersing phenotypes (e.g. wingless insects; Harrison, 1980) indicative of reduced outcrossing. Disturbance regimes would not favour such tactics, but rather plastic polymorphism in dispersal phenotypes. Sexuality is usually associated with greater degrees of dispersal and outcrossing than occurs in stressful environments. In species with intermittent sexuality, the dispersing forms usually reproduce sexually.

12.3 THE RED QUEEN PARADIGM

Red Queen environments are dominated by high biotic diversity, strong competition and coevolution. In many cases, the ecological and evolutionary adjustments of associated species are mutually spiteful. Thus, adaptations in one direction may be advantageous in one or a few generations, but later may require opposite or novel adjustments due to adaptations of competitors, predators or parasites to thwart features deployed against them. Moreover, the complexity of biological dynamics and unpredictability of evolution make such environments highly uncertain.

High uncertainty and spiteful interactions favour polymorphic populations with high recombination and evolvability (Chapter 5). Persistent asexuality may be impossible in such environments, except perhaps for lineages with considerable organizational advancement (e.g. polyploids) that can out-perform their diploid competitors within particular regions of the relevant niche and which obtain considerable genetic hetero-zygosity via their usual hybrid origin. Even then, such species might be selected to restore sexuality in the long term.

12.4 THE DISTURBANCE PARADIGM

Chapter 6 suggests that phenotypic plasticity will be favoured under the following conditions: clear discrimination of a signal from a predictable environmental factor; environments that fluctuate with periodicities close to or somewhat longer than the generation time; a regime of hard selection in a heterogenous environment (mainly temporal); a low cost of plasticity or a high cost of constitutive expression; or where a coadapted complex is difficult to achieve via recombination (i.e. the best man is unlikely to be derived fast enough by microevolution). This last factor suggests that plasticity would be particularly favoured where complex adjustments are required, where populations are small or where recom-bination is reduced.

Plasticity may be somewhat exclusive of stress adaptation. It may not be capable of evolving across really large spatial and temporal scales. Parsons (1991) adds that if plasticity entails a cost, it may be constrained by low metabolic rates under strong stress. Gause (1942) also obtained results strongly suggesting that extreme phenotypes may have reduced plasticity, at least in the direction of the selected feature. Thus, there may be a direct trade off between high evolutionary adaptation in any particular direction and the degree of plasticity that can be achieved. At best, plastic responses may be limited to asymmetrical responses downwards in such circumstances.

Environments characterized by disturbance involve fluctuations on various scales and with varying levels of predictability but such fluctua-tions are oblivious to the adaptation of the organism, unlike spiteful Red Queen dynamics. This favours the evolution of specific adaptive features deployed with respect to predictable cues, rather than random recombination of features in each generation. Short-term disturbance may be accommodated by homeostasis, behavioural adjustments or simply integration across the variation such as can be achieved using storage reserves. Mrosovsky (1990) has emphasized that many features of organisms are adaptively dynamic.

Variation on levels close to or larger than a single generation may be met via phenotypic plasticity. Thus, amphibious plants subjected to seasonal variation in flooding may produce adaptive varieties of leaves for both aquatic or terrestrial function in different patches of a heterogenous environment or even on the same plant. We can expect, wherever reliable information is associated with environmental variation, that lineages may adapt to use it. In a species with sufficient plasticity, there would be reduced need for genetic polymorphism or recombination with respect to the environmental variability. A single plastic phenotype would become invisible to selection. Where variation is unpredictable, or plasticity in any given genotype is insufficient to span the entire range experienced by the species, genetic polymorphism would be retained. There are some very thorny questions here related to the evolution and maintenance of sex.

A number of prominent evolutionists have noted that life tends to evolve in such a way as to reduce the impact of the environment and natural selection itself (Wright, 1931; Lewontin, 1957). Thus, progressive evolution tends towards organizations that absorb predictable disturbances (and some unpredictable ones as a bonus) via features associated with plasticity, resilience, robustness and resistance. Even the spectre of the Red Queen is met in large part by a plastic immune system of enormous effectiveness (Wills, 1989). Plastic responses to predators, and competitors, are also well documented (Harvell, 1990). The question is, could organizations that have reached a very high level of phenotypic adaptability be selectively favoured to abandon sexuality? Such a scenario could apply to some bacteria, which seem to represent consolidated organizations of high evolutionary stability. Once a globally optimal phenotype has been achieved, the value of recombination will be negative or considerably reduced, except on very long time scales (especially where genetic information can still be exchanged as in bacterial plasmids). Thus, the association of plasticity with asexuality may come about because plastic species can forego sex without severe penalty (as long as the ratchet can be avoided).

The above argument is exactly opposite to the usual interpretation of why plastic species may be more often asexual. The usual question is, does recombination hinder or prevent the evolution of plasticity? This is a bigger question than it appears, because the same problem arises when we ask, do outcrossing and recombination hinder or interfere with the local diversification and adaptation of a species? The answer is clearly relevant to theories of sympatric *versus* allopatric speciation, and hinges on the nexus of whether recombination is necessarily disruptive in organizational evolution.

Possibly the reason that this problem has not been adequately resolved is that the answer may be both yes and no. The model of allopatric speciation, best exemplified by Mayr (1963), is based on the idea that sexual panmixia in single populations will prevent phenotypic divergence. There is considerable evidence suggesting that local consolidation of specialized phenotypes can be disrupted by gene flow, and so there is a strong empirical basis for such models. Theoretically, we also know that, if we have a locus with two possible alleles, fixation of two alternative homozygotes in a single panmictic sexual population will be impossible because heterozygotes will continue to be formed. For multiple loci, the fixation of alternative extremes representing different epistatic optima is equally difficult.

Theory aside, the reality of populations with limited dispersal ability and a high degree of inbreeding does allow local diversification to proceed to a greater extent than was previously imagined. For example, Endler (1973, 1977) elegantly demonstrated the local diversification of *Drosophila* along artificial clines of selection, despite even very high levels of gene flow. Presumably mate recognition systems, dispersal characteristics or habitat choice mechanisms could be subsequently modified to reduce disruptive gene flow and full genetic diversification could follow (see the thorough review of Shields, 1982). Empirical support for such diversification includes the elaboration of local ecotypes, along selective clines, despite gene flow, and the genetic diversification of races to the extent that opposite ends of the continuum may be genetically incompatible.

Of greater importance is the fact that previous ideas concerning diversification are couched in genic models that assume the only important factor is the competition of alleles for residence at a fixed locus. White (1978) argued that changes in genetic structure may be even more important. If we imagine that the evolution of regulatory variants in two sub-niches occurs via competition of alleles at particular loci, and that this genetic architecture is fixed, and parsimonious, then the genic interpretation of necessary disruption must hold. However, if genetic architecture varies, or has highly redundant components, then the diversification does not necessarily require that different alleles must compete. Thus, West-Eberhard (1989) has resoundingly shown that a single genome may achieve a diversity of phenotypes, either by the mechanism of environmental or allelic switches. The important concept is that the same epigenetic structure can diverge in alternative directions at once. The nature of the switch (i.e. allelic or environmental) is largely irrelevant.

Related to these ideas is the fact that most genes are front-ended by regulatory regions that allow complex handshaking among genes. Changes in such elements constitute changes in the connectance of genomic circuitry. Moreover, large complex circuits involving many genes may become selectable units in their own right, subordinating all genes that contribute to them to invariance via stabilizing selection. Where alternative circuits map on to alternative adaptive peaks, lower-level genes could develop regulatory regions, such that the same genes could function differently in each circuit. This appears to be the case for deriving segment-specific appendages in insects, which is wonderfully illustrated by occasional errors that convert antennae to legs, for example. In this way rather large diversification in regulatory functions could be achieved across numerous genes, and recombination would have little disruptive impact on most lower-level elements. If such processes are involved in specialization of parts within an organism, then why not also among polymorphic phenotypes?

These ideas bring us full circle. Whether recombination and sex are disruptive will likely depend on the extent to which the alternative phenotypes depend on common loci. The genetic architecture, and in particular the structure of potential regulatory hierarchies, determines the ease of diversification or the strength of disruption. Clearly, gene duplications will allow increases in organizational complexity. Thus, the evolution of phenotypic plasticity might be strongly facilitated by duplications, particularly of higher-order switches.

The evolution of branching regulatory structure seems critical, and a rich environment for such potential evolution may be provided by polyploid events, particularly when associated with hybridization. Although sex and plasticity are not necessarily disassociated, a high proportion of plastic species are asexual (Lynch, 1984; Harvell, 1990). It seems unlikely that this could be due to a direct conflict of recombination in the evolution of plasticity since plasticity necessarily involves an expanded regulatory structure, which might be assembled more quickly in a sexual organization. Significantly, many plastic species that are asexual are also sessile, or have weak mechanisms for habitat choice. In these cases, inbreeding depression may be reduced or absent and combined with high phenotypic plasticity this might facilitate the loss of outcrossing or recombination in some organisms. Alternatively, adopting asexual reproduction or an inbreeding system (philopatry) might also enhance local adaptation (Shields, 1982).

In many species deploying plastic responses to predators and competitors, environmental cycles are the major driving factor for both the predator and host populations (e.g cyclomorphosis in cladocerans). This may reduce the uncertainty of Red Queen interactions, allowing specific plasticity effectively to substitute for recombination. For example, if key predators are constrained by a strong environmental cycle they then become environmentally predictable themselves (Harvell, 1990). In other cases, plasticity may be associated with genetic reorganizations that are also selected to prevent destabilization by adopting parthenogenesis (Lynch, 1984).

Finally, the relationship of disturbance to the generation time of the organism is crucial. Clearly, disturbances occurring on scales much shorter than the generation time must be met by behavioural or physiological adjustments. Disturbances on scales of one or several generations are probably ineffectively tracked via circulating alleles, and phenotypic plasticity should be favoured. Where environmental information is not detected, or where the time frame is very long, circulating alleles may be favoured. Annual cycles are our strongest obvious environmental fluctuation, and annual or multivoltine species might be expected to show more plasticity than perennial species. Where perennial species do show adjustments (i.e. loss of leaves in winter, changes in coat thickness and colour in mammals), these must be reversible, and begin to fall into the category of acclimation and homeostatic adjustments.

There is clearly a considerable complexity in forces shaping the empirical trends in plasticity, recombination and stress resistance detected in nature. Only a holistic view that incorporates all these (and perhaps some other) factors can hope to separate their confounding overlaps and reveal the causal underpinnings of various life-history strategies.

12.5 GENERAL CONCLUSION

This concludes a rather tortuous journey from molecular organization through epigenetics and into the realm of ecology. My intent, and I hope that I have been successful, was to show that it is the integration of biological systems that is most crucial to evolutionary success and that the aspect most relevant to this is the interaction among subcomponents. If nothing else, an emphasis on holism and organizational evolution generates interesting ideas that cannot be derived from a genic, reductionist view.

Lineage selection provides a key to convert a genic perspective into a hierarchical framework of selection and organizational evolution. This applies both to the functional units associated with hierarchies of regulatory control in the genome, as well as to the hierarchy of evolutionary entities associated with lineages, nested sublineages and individuals. Such a view increases the scope of genomic features and dynamics relevant to evolutionary processes. Lineage selection is open to direct investigation and testing using new molecular techniques, particularly DNA hybridization and sequencing that, incidentally, lead automatically to a lineage perspective.

Lineage selection, as illustrated by the discussion of phenotypic plasticity, is largely concerned with evolution of genetic architecture and/or polygenic systems where some interactions are regulatory. Such phenomena have only recently been emphasized in the modern synthesis, perhaps because the genic perspective cannot address them (Stanley, 1979; Raff and Kaufman, 1983). Rendel (1967) elegantly made this point with reference to the evolution of canalized development

The lesson is that the traditional view of multiple alleles jockeying for position at a single locus must be expanded to consider multiple loci that are mutually interdependent or regulated by switches keyed to environmental, developmental or physiological signals. If the traditional view leans toward a 'bean bag' perspective, lineage selection evokes an integrated genetic templet with coadapted loci and evolutionarily determined homeostatic or homeorhetic targets (Alexander, 1975; Stearns, 1980, 1984a,b; Brooks, 1983; Wake, Roth and Wake, 1983; Calow, 1984; Dingle, 1984; Kingsolver and Wiernasz, 1987). In this framework evolution among loci is equally, if not more, relevant to the evolution of multicellular phenotypes than allele frequency changes at single loci.

Although Maynard Smith (1989) suggested that lineage selection is a form of group selection where groups belong to different generations, other authors recognize that it is decidedly different than group selection since it can act within single populations (Williams, 1975; Jain, 1979; Brooks, 1988). Maynard Smith's objection breaks down in species with overlapping generations, so globally at least, lineage selection should be considered as selection of a higher-order organization rather than a process of group selection (Hull, 1980). There is a wide assortment of evolutionary phenomena that various authors have found difficult to interpret within the genic paradigm. This failure may have been due to not identifying relevant organizations as distinct from mere groups or categories (Hull, 1980).

Some of these difficult features include sexual reproduction and recombination rates, phenotypic plasticity, linkage and chromosome structure, norms of reaction, epistasis, risky dispersal, regulated mutation rates, homeostasis, submaximal rates, species stasis, senescence, iteroparity, canalization and dominance (see also Rendel, 1967; Williams, 1975; Haukioja, 1982; Reid, 1985; Trivers, 1985; Eldredge, 1985; Stearns, 1986, 1992; Keller, 1987). Genic models for the evolution and persistence of many of these phenomena, if successful at all, make evolution seem more like a contortionist than a tinkerer. Lineage selection offers an alternative, testable paradigm, that has high explanatory power for many features previously ascribed to group selection, or poorly described with genic mechanisms. For lineages, all of the features listed above are important to internal cohesion and long-term persistence, and consequently, may be adaptive elements of a higher-order genetic design.

References

Abbott, R.J. (1992) Plant invasions, interspecific hybridization and the evolution of new plant taxa. *TREE*, **7**, 401–5.

Addicott, J.F. (1986) On the population consequences of mutualism, in *Community Ecology* (eds J. Diamond and T.J. Case), Harper and Row, NY, pp. 425–36.

Adler, F.R. and Harvell, C.D. (1990) Inducible defenses, phenotypic variability and biotic environments. *TREE*, **5**, 407–10.

Akam, M. (1987) The molecular basis for metameric pattern in the *Drosophila* embryo. *Development*, **101**, 1–22.

Akam, M. (1989) Hox and HOM, homologous gene clusters in insects and vertebrates. *Cell*, **57**, 347–9.

Akam, M., Dawson, I. and Tear, G. (1988) Homeotic genes and the control of segment diversity. *Development*, (Suppl.) **104**, 123–33.

Akimoto, S. (1992) Shift in life-history strategy from reproduction to defense with colony age in the galling aphid, *Hemipodaphis persimilis* producing defensive first-instar larvae. *Res. Popul. Ecol.*, **34**, 359–72.

Alberch, P. (1982) Developmental constraints in evolutionary processes, in *Evolution and Development* (ed J.T. Bonner), Springer Verlag, Berlin, pp. 313–32.

Alberch, P. (1991) From genes to phenotype, dynamical systems and evolvability. *Genetica*, **84**, 5–11.

Alberch, P., Gould, S.J., Oster, G.F. and Wake, D.B. (1979) Size and shape in ontogeny and phylogeny. *Paleobiology*, **5**, 296–317.

Alexander, R.D. (1991) Some unanswered questions about naked mole-rats, in *The Biology of the Naked Mole-rat* (eds P.W. Sherman, J.U.M. Jarvis and R.D. Alexander), Princeton University Press, Princeton, NJ, pp. 446–65.

Alexander, R.M. (1975) Evolution of integrated design. *Am. Zool.*, **15**, 419–25.

Alexander, R.M. (1982) *Optima for Animals*, Edward Arnold, London.

Allen, T.F.H. and Starr, T.B. (1982) *Hierarchy: Perspectives for Ecological Complexity*, University of Chicago Press, Chicago.

Ambros, V. (1989) A hierarchy of regulatory genes controls a larva-to-adult developmental switch in *C. elegans*. *Cell*, **57**, 49–57.

Anderson, W.W. *et al.* (1991) Four decades of inversion polymorphism in *Drosophila pseudoobscura*. *Proc. Natl. Acad. Sci. USA*, **88**, 10367–71.

Andersson, M. (1984) The evolution of eusociality. *Ann. Rev. Ecol. Syst.*, **15**, 165–89.

Antonovics, J. (1976) The nature of limits to natural selection. *Ann. Missouri Bot. Gard.*, **63**, 224–47.

Antonovics, J. and Ellstrand, N.C. (1984) Experimental studies of the evolutionary significance of sexual reproduction. I. A test of the frequency-dependent selection hypothesis. *Evolution*, **38**, 103–15.

Arking, R. (1987) Successful selection for increased longevity in *Drosophila*: analysis of the survival data and presentation of a hypothesis on the genetic regulation of longevity. *Expt. Geront.*, **22**, 199–220.

Arking, R. (1991) *Biology of Aging: Observations and Principles*, Prentice Hall, Englewood Cliffs, NJ.

Arking, R., Buck, S., Wells, R.A. and Pretzlaff, R. (1988) Metabolic rates in genetically based long lived strains of *Drosophila*. *Expt. Geront.*, **23**, 59–76.

Arking, R. and Clare. M. (1986) Genetics of aging: effective selection for increased longevity in *Drosophila*, in *Insect Aging* (eds K.G. Collatz and R.S. Sohal), Springer Verlag, Berlin, pp. 217–36.

Arking, R. and Dudas, S.P. (1989) Review of genetic investigations into the aging process in *Drosophila*. *J. Am. Geriatr. Soc.*, **37**, 757–73.

Arking, R. and Wells, R.A. (1990) Genetic alteration of normal aging processes is responsible for extended longevity in *Drosophila*. *Devel. Gen.*, **11**, 141–8.

Arnold, A.J. and Fristrup, K. (1982) The theory of evolution by natural selection: a hierarchical expansion. *Paleobiology*, **8**, 113–29.

Arnold, M.L., Buckner, C.M. and Robinson, J.J. (1991) Pollen-mediated introgression and hybrid speciation in Louisiana irises. *Proc. Natl. Acad. Sci. USA*, **88**, 1398–402.

Arnold, S.J., Alberch, P., Casanyi, V. *et al.* (1989) How do complex organisms evolve? in *Complex Organismal Functions* (eds D.B. Wake and G. Roth), J. Wiley and Sons, NY, pp. 403–33.

Arthur, W. (1984) *Mechanisms of Morphological Evolution*, J. Wiley and Sons, NY.

Arthur, W. (1988) *A Theory of the Evolution of Development*, J. Wiley and Sons, NY.

Atchley, W.R. and Hall, B.K. (1991) A model for development and evolution of complex morphological structures. *Biol. Rev.*, **66**, 101–57.

Atchley, W.R., Riska, B., Kohn, L.A.P. *et al.* (1984) A quantitative genetic analysis of brain and body size associations, their origin and ontogeny: data from mice. *Evolution*, **38**, 1165–79.

Atlan, H. (1974) On a formal definition of organization. *J. Theor. Biol.*, **45**, 295–304.

Austad, S.N. (1989) Life extension by dietary restriction in the bowl and doily spider, *Frontinella pyramitela*. *Expt. Geront.*, **24**, 83–92.

Austad, S.N. and Fischer, K.E. (1991) Mammalian aging, metabolism, and ecology: evidence from the bats and marsupials. *J. Geront.*, **46**, B47–B53.

Avise, J.C. (1989) Gene trees and organismal histories: a phylogenetic approach to population biology. *Evolution*, **43**, 1192–208.

Avise, J.C. (1990) Flocks of African fishes. *Nature*, **347**, 512–13.

Axelrod, R. (1984) *The Evolution of Cooperation*, Basic Books, NY.

Axelrod, R. and Dion, D. (1988) The further evolution of cooperation. *Science*, **242**, 1385–90.

Axelrod, R. and Hamilton, W.D. (1981) The evolution of cooperation. *Science*, **211**, 1390–6.

Ayala, F.J. (1988) Can 'progress' be defined as a biological concept? in *Evolutionary Progress* (ed M.H. Nitecki), University of Chicago Press, Chicago, IL, pp. 75–96.

Azuma, A. and Yasuda, K. (1989) Flight performance of rotary seeds. *J. Theor. Biol.*, **138**, 23–53.

Baerlocher, F. (1990) The Gaia hypothesis: a fruitful fallacy? *Experientia*, **46**, 232–8.

Baird, D.J., Linton, L.R. and Davies, W. (1987) Life history flexibility as a strategy for survival in a variable environment. *Funct. Ecol.*, **1**, 45–8.

Bak, P. and Chen, K. (1991) Self-organized criticality. *Sci. Amer.*, **264**, 46–53.

Baker, B.S. (1989) Sex in flies: the splice of life. *Nature*, **340**, 521–4.

Baldauf, S.L. and Palmer, J.D. (1990) Evolutionary transfer of the chloroplast tufA gene to the nucleus. *Nature*, **334**, 262–5.

Baldwin, I.T. and Schultz, J.C. (1983) Rapid changes in tree leaf chemistry induced by damage: evidence for communication between plants. *Science*, **221**, 277–9.

Bandziulis, R.J., Swanson, M.S. and Dreyfuss, G. (1989) RNA-binding proteins as developmental regulators. *Genes Develop.*, **3**, 431–7.

Banerjee, S., Sibbald, P.R. and Maze, J. (1990) Quantifying the dynamics of order and organization in biological systems. *J. Theor. Biol.*, **143**, 91–111.

Barbosa, P., Cranshaw, W. and Greenblatt, J.A. (1981) Influence of food quantity and quality on polymorphic dispersal behaviours in the gypsy moth, *Lymantria dispar*. *Can. J. Zool.*, **59**, 293–6.

Bargmann, C. and Horvitz, H.R. (1991) Control of larval development by chemosensory neurons in *Caenorhabditis elegans*. *Science*, **251**, 1243–6.

Barlow, C. (ed) (1991) *From Gaia to Selfish Genes*, MIT Press, Cambridge, MA.

Barnett, S.A. and Dickson, R.G. (1989) Wild mice in the cold: some findings on adaptation. *Biol. Rev.*, **64**, 317–40.

Bartke, A., Steger, R.W., Hodges, S.L. *et al.* (1988) Infertility in transgenic female mice with human growth hormone expression: evidence for luteal failure. *J. Expt. Zool.*, **248**, 121–4.

Barton, N.H. (1989) Founder effect speciation, in *Speciation and its Consequences* (eds D. Otte and J.A. Endler), Sinauer Associates, Sunderland, MA, pp. 229–56.

Barton, N.H. and Charlesworth, B. (1984) Genetic revolutions, founder effects and speciation. *Ann. Rev. Ecol. Syst.*, **15**, 133–64.

Barton, N. and Clark, A. (1990) Population structure and processes in evolution, in *Population Biology* (eds K. Wohrmann and S. Jain), Springer Verlag, Berlin, pp. 115–73.

Barton, N.H. and Turelli, M. (1989) Evolutionary quantitative genetics: how little do we know? *Ann. Rev. Genet.*, **23**, 337–70.

Bateson, G. (1963) The role of somatic change in evolution. *Evolution*, **17**, 529–39.

Baumgartner, S. and Noll, M. (1991) Network of interactions among pair-rule genes regulating paired expression during primordial segmentation of *Drosophila*. *Mech. Dev.*, **33**, 1–18.

Baur, A. and Baur, B. (1992) Responses in growth, reproduction and life span to reduced competition pressure in the land snail *Balea perversa*. *Oikos*, **63**, 298–304.

Bazzaz, F.A., Chiariello, N.R., Coley, P.D. and Pitelka, L.F. (1987) Allocating resources to reproduction and defense. *BioScience*, **37**, 58–67.

Beardsley, T. (1991) Smart genes. *Sci. Amer.*, **265**, 86–95.

Beauchamp, G.K., Yamazaki, K. and Boyse, E.A. (1985) The chemosensory recognition of genetic individuality. *Sci. Amer.*, **253**, 86–92.

Beauchene, R.E., Bales, C.W., Bragg, C.S. *et al.* (1986) Effect of age of initiation of feed restriction on growth, body composition, and longevity of rats. *J. Geront.*, **41**, 13–19.

Beaumont, M.A. (1988) Stabilizing selection and metabolism. *Heredity*, **61**, 437–38.

Beauvais, J.E. and Enesco, H.E. (1985) Life span and age-related changes in activity levels of the rotifer *Asplanchna brightwelli*: influence of curare. *Expt. Geront.*, **20**, 359–66.

Beeman, R.W., Stuart, J.J., Haas, M.S. and Denell, R.E. (1989) Genetic analysis of the homeotic gene complex (HOM–C) in the beetle *Tribolium castaneum*. *Devel. Biol.*, **133**, 196–209.

Bell, G. (1980) The costs of reproduction and their consequences. *Amer. Nat.*, **116**, 45–76.

Bell, G. (1982) *The Masterpiece of Nature, the Evolution and Genetics of Sexuality*, University of California Press, Berkeley, CA.

Bell, G. (1984a) Measuring the cost of reproduction. I. The correlation structure of the life table of a plankton rotifer. *Evolution*, **38**, 300–13.

Bell, G. (1984b) Measuring the cost of reproduction. II. The correlation structure of life tables of five freshwater invertebrates. *Evolution*, **38**, 314–26.

Bell, G. (1985) Two theories of sex and variation. *Experientia*, **41**, 1235–45.

Bell, G. (1986) Reply to Reznick *et al*. *Evolution*, **40**, 1344–6.

Bell, G. (1988a) *Sex and Death in Protozoa: The History of an Obsession*, Cambridge University Press, Cambridge.

Bell, G. (1988b) Recombination and the immortality of the germ line. *J. Evol. Biol.*, **1**, 67–82.

Bell, G. (1988c) Uniformity and diversity in the evolution of sex, in *The Evolution of Sex* (eds R.E. Michod and B.R. Levin), Sinauer Associates, Sunderland, MA, pp. 126–38.

Bell, G. and Koufopanou, V. (1986) The cost of reproduction. *Oxford Surv. Evol. Biol.*, **3**, 83–131.

Bell, G. and Lechowicz, M.J. (1991) The ecology and genetics of fitness in forest plants. I. Environmental heterogeneity measured by explant trials. *J. Ecol.*, **79**, 663–85.

Bell, G., Lechowicz, M.J. and Schoen, D.J. (1991) The ecology and genetics of fitness in forest plants. III. Environmental variance in natural populations of *Impatiens pallida*. *J. Ecol.*, **79**, 697–713.

Bell, G. and Maynard Smith, J. (1987) Short-term selection for recombination among mutually antagonistic species. *Nature*, **328**, 66–8.

Bellen, H.J. and Kiger Jr., J.A. (1987) Sexual hyperactivity and reduced longevity of dunce females of *Drosophila melanogaster*. *Genetics*, **119**, 153–60.

Bellig, R. and Stevens, G. (eds) (1988) *The Evolution of Sex*, Nobel Conference XXIII, Harper and Row, San Francisco, CA.

Bennett, P.M. and Harvey, P.H. (1987) Active and resting metabolism in birds: allometry, phylogeny and ecology. *J. Zool.*, **213**, 327–63.

Bennett, P.M. and Harvey, P.H. (1988) How fecundity balances mortality in birds. *Nature*, **333**, 216.

Benton, M.J. (1987) Progress and competition in macroevolution. *Biol. Rev.*, **62**, 305–38.

Berg, R.L. (1960) The ecological significance of correlation pleiades. *Evolution*, **14**, 171–80.

Bergman, A. and Feldman, M.W. (1990) More on selection for and against recombination. *Theor. Popul. Biol.*, **38**, 68–92.

Bergner, A.D. (1928) The effect of prolongation of each stage of the life-cycle on crossing over in the second and third chromosomes of *Drosophila melanogaster*. *J. Expt. Zool.*, **50**, 107–63.

Bernardo, J. (1991) Manipulating egg size to study maternal effects on offspring traits. *TREE*, **6**, 1–2.

Bernays, E.A. (1986) Diet-induced head allometry among foliage-chewing insects and its importance for graminivores. *Science*, **231**, 495–7.

Bernstein, C. and Bernstein, H. (1991) *Aging, sex and DNA repair*. Academic Press, San Diego.

Bernstein, H., Byerly, H., Hopf, F. and Michod, R.E. (1985a) DNA repair and complementation: the major factors in the origin and maintenance of sex, in *The Origin and Evolution of Sex* (eds H.O. Halvorson and A. Monroy), Alan R. Liss, NY, pp. 29–45.

Bernstein, H., Byerly, H.C., Hopf, F.A. and Michod, R.E. (1985b) Genetic damage, mutation and the evolution of sex. *Science*, **229**, 1277–81.

Bernstein, H., Hopf, F.A. and Michod, R.E. (1988) Is meiotic recombination an adaption for repairing DNA, producing genetic variation, or both? in *The Evolution of Sex* (eds R.E. Michod and B.R. Levin), Sinauer Associates,

Sunderland, MA, pp. 139–60.

Berryman, A.A. (1981) *Population Systems*, Plenum Press, NY.

Bierzychudek, P. (1985) Patterns in plant parthenogenesis. *Experientia*, **41**, 1255–64.

Bierzychudek, P. (1989) Environmental sensitivity of sexual and apomictic Antennaria: do apomicts have general-purpose genotypes? *Evolution*, **43**, 1456–66.

Black, A.R. and Dodson, S.I. (1990) Demographic costs of *Chaoborus*-induced phenotypic plasticity in *Daphnia pulex*. *Oecologia*, **83**, 117–22.

Blackburn, D.G. (1984) From whale toes to snake eyes: comments on the reversibility of evolution. *Syst. Zool.*, **33**, 241–5.

Blau, H.M. (1988) Hierarchies of regulatory genes may specify mammalian development. *Cell*, **53**, 673–4.

Blueweiss, L., Fox, H., Kudzma, V. *et al.* (1978) Relationships between body size and some life history parameters. *Oecologia*, **37**, 257–72.

Boddington, M.J. (1978) An absolute metabolic scope for activity. *J. Theor. Biol.*, **75**, 443–9.

Bodnar, J.W., Jones, G.S., Ellis Jr., C.H. (1989) The domain model for eukaryotic DNA organization 2: a molecular basis for constraints on development and evolution. *J. Theor. Biol.*, **137**, 281–320.

Boggs, C.L. (1992) Resource allocation: exploring connections between foraging and life history. *Funct. Ecol.*, **6**, 508–18.

Bonner, J.T. (1973) Hierarchical control programs in biological development, in *Hierarchy Theory* (ed H.H. Pattee), George Braziller, NY, pp. 49–70.

Bonner, J.T. (1974) *On Development: the Biology of Form*, Harvard University Press, Cambridge, MA.

Bonner, J.T. (1984) The evolution of chemical signal-receptor systems (from slime moulds to man). *Oxford Surv. Evol. Biol.*, **1**, 1–15.

Bonner, J.T. (1988) *The Evolution of Complexity*, Princeton University Press, Princeton, NJ.

Bookstaber, R. and Langsam, J. (1985) On the optimality of coarse behaviour rules. *J. Theor. Biol.*, **116**, 161–93.

Bosch, C.A. (1971) Redwoods: a population model. *Science*, **172**, 345–9.

Boyce, M.S. (1988) Evolution of life histories, theory and patterns from mammals, in *Evolution of Life Histories of Mammals* (ed M.S. Boyce), Yale University Press, New Haven, CT, pp. 3–30.

Boyce, M.S. and Perrins, C.M. (1987) Optimizing great tit clutch size in a fluctuating environment. *Ecology*, **68**, 142–53.

Bozinovic, F. and Rosenmann, M. (1989) Maximum metabolic rate of rodents: physiological and ecological consequences on distributional limits. *Funct. Ecol.*, **3**, 173–81.

Brace, C.L. (1963) Structural reduction in evolution. *Amer. Nat.*, **97**, 39–49.

Bradley, A.J., McDonald, I.R. and Lee, A.K. (1980) Stress and mortality in a small marsupial (*Antechinus stuartii*, Macleay). *Gen. Comp. Endocrinol.*, **40**, 188–200.

Bradshaw, A.D. (1965) Evolutionary significance of phenotypic plasticity in plants. *Adv. Genet.*, **13**, 115–55.

Bradshaw, A.D. and Hardwick, K. (1989) Evolution and stress – genotypic and phenotypic components, in *Evolution, Ecology and Environmental Stress* (eds P. Calow and R.J. Berry), Academic Press, NY, pp. 137–55.

Brandon, R.N. and Burian, R.M. (eds) (1984) *Genes, Organisms, Populations: Controversies over the Units of Selection*, Bradford Books, MIT Press, Cambridge, MA.

Bremermann, H.J. (1987) The adaptive significance of sexuality, in *The Evolution of Sex and its Consequences* (ed S. Stearns), Birkhauser, Basel, Switzerland, pp. 135–61.

Brett, M.T. (1992) *Chaoborus* and fish-mediated influences on *Daphnia longispina* population structure, dynamics and life history strategies. *Oecologia*, **89**, 69–77.

Britten, R.J. (1986a) Intraspecies genomic variation, in *Genetics, Development and Evolution* (eds J.P. Gustafson, G.L. Stebbins and F.J. Ayala), Plenum Press, London, pp. 289–306.

Britten, R.J. (1986b) Rates of DNA sequence evolution differ between taxonomic groups. *Science*, **231**, 1393–8.

Britten, R.J. and Davidson, E.H. (1969) Gene regulation for higher cells: a theory. *Science*, **165**, 349–57.

Britten, R.J. and Davidson, E.H. (1971) Repetitive and non-repetitive DNA sequences and a speculation on the origins of evolutionary novelty. *Quart. Rev. Biol.*, **46**, 111–33.

Brody, E.J., Marie, J., Goux-Pelleton, M.G. and d'Orval, B.C. (1988) Alternative splicing to tissue specific splicing – an evolutionary pathway? in *Evolutionary Tinkering in Gene Expression* (eds M. Grunberg-Manago, B.F.C. Clark and H.G. Zachau), Plenum Press, London, pp. 203–13.

Brody, M.S. and Lawlor, L.R. (1984) Adaptive variation in offspring size in the terrestrial isopod, *Armadillidium vulgare. Oecologia*, **61**, 55–9.

Bronmark, C. and Miner, J.G. (1992) Predator-induced phenotypic change in body morphology in crucian carp. *Science*, **258**, 1348–50.

Bronson, F.H. (1984) Energy allocation and reproductive development in wild and domestic house mice. *Biol. Repr.*, **31**, 83–8.

Brooks, D.R. (1983) What's going on in evolution? A brief guide to some new ideas in evolutionary theory. *Can. J. Zool.*, **61**, 2637–45.

Brooks, L.D. (1988) The evolution of recombination rates, in *The Evolution of Sex* (eds R.E. Michod and B.R. Levin), Sinauer Associates, Sunderland, MA, pp. 87–105.

Brown, J.H. and Maurer, B.A. (1986) Body size, ecological dominance and Cope's rule. *Nature*, **324**, 248–50.

Browne, R.A. (1982) The costs of reproduction in brine shrimp. *Ecology*, **63**, 43–7.

Bryant, E.H. (1989) Multivariate morphometrics of bottlenecked populations, in *Evolutionary Biology of Transient Unstable Populations* (ed A. Fontdevila), Springer Verlag, Berlin, pp. 19–31.

Bryant, E.H., Combs, L.M. and McCommas, S.A. (1986a) Morphometric differentiation among experimental lines of the housefly in relation to a bottleneck. *Genetics*, **114**, 1213–23.

Bryant, E.H., McCommas, S.A. and Combs, L.M. (1986b) The effect of an experimental bottleneck upon quantitative genetic variation in the housefly. *Genetics*, **114**, 1191–211.

Bryant, E.H. and Meffert, L.M. (1988) Effect of an experimental bottleneck on morphological integration in the housefly. *Evolution*, **42**, 698–707.

Bryant, E.H. and Meffert, L.M. (1990) Multivariate phenotypic differentiation among bottleneck lines of the housefly. *Evolution*, **44**, 660–8.

Budd, P.S. and Jackson, I.J. (1991) What do the regulators regulate? First glimpses downstream. *Trends Genet.*, **7**, 74–6.

Bull, J.J. (1983) *Evolution of Sex Determining Mechanisms*, Benjamin Cummings, Menlo Park, CA.

Bull, J.J. (1987) Evolution of phenotypic variance. *Evolution*, **41**, 303–15.

Bull, J.J. and Harvey, P.H. (1989) A new reason for having sex. *Nature*, **339**, 260–1.

Bull, J.J. and Rice, W.R. (1991) Distinguishing mechanisms for the evolution of co-operation. *J. Theor. Biol.*, **149**, 63–74.

Bull, J.J., Molineux, I.J. and Rice, W.R. (1991) Selection of benevolence in a host–parasite system. *Evolution*, **45**, 875–82.

Burt, A. and Bell, G. (1987) Mammalian chiasma frequencies as a test of two theories of recombination. *Nature*, **326**, 803–5.

Bush, G.L., Case, S.M., Wilson, A.C. and Patton, J.L. (1977) Rapid speciation and chromosomal evolution in mammals. *Proc. Natl. Acad. Sci. USA*, **74**, 3924–46.

Buss, L.W. (1987) *The Evolution of Individuality*, Princeton University Press, Princeton, NJ.

Buss, L.W. and Grosberg, R.K. (1990) Morphogenetic basis for phenotypic differences in hydroid competitive behaviour. *Nature*, **343**, 63–6.

Cain, A.J. (1983) Ecology and ecogenetics of terrestrial molluscan populations, in *The Mollusca: Ecology* (ed W.D. Russell-Hunter), Academic Press, **6**, 597–647.

Cairns, J., Overbaugh, J. and Millar, S. (1988) The origin of mutants. *Nature*, **335**, 142–5.

Calder, W.A. III. (1984) *Size, Function and Life History*, Harvard University Press, Cambridge, MA.

Calder, W.A. III. (1985) The comparative biology of longevity and lifetime energetics. *Expt. Geront.*, **20**, 161–70.

Calder, W.A. III. (1987) Scaling energetics of homeothermic vertebrates: an operational allometry. *Ann. Rev. Physiol.*, **49**, 107–20.

Calow, P. (1975a) The feeding strategies of two freshwater gastropods, *Ancylus fluviatilis* Müll. and *Planorbis contortus* Linn. (Pulmonata), in terms of ingestion rates and absorption efficiencies. *Oecologia*, **20**, 33–49.

Calow, P. (1975b) Defaecation strategies of two freshwater gastropods, *Ancylus fluviatilis* Müll. and *Planorbis contortus* Linn. (Pulmontata) with a comparison of field and laboratory estimates of food absorption rate. *Oecologia*, **20**, 51–63.

Calow, P. (1977) Ecology, evolution and energetics: a study in metabolic adaptation. *Adv. Ecol. Res.*, **10**, 1–62.

Calow, P. (1979) The cost of reproduction – a physiological approach. *Biol. Rev.*, **54**, 23–40.

Calow, P. (1981) Resource utilization and reproduction, in *Physiological Ecology: an Evolutionary Approach to Resource Use* (eds P. Calow and C.R. Townsend), Sinauer Associates, Sunderland, MA, pp. 245–70.

Calow, P. (1984) Economics of ontogeny – adaptational aspects, in *Evolutionary ecology* (ed B. Shorrocks), Blackwell Scientific, Oxford, pp. 81–104.

Campbell, J.H. (1985) An organizational interpretation of evolution, in *Evolution at a Crossroads: the New Biology and the New Philosophy of Science* (eds D.J. Depew and B.H. Weber), MIT Press, Cambridge, MA, pp. 133–67.

Campbell, J.H. (1987) The new gene and its evolution, in *Rates of Evolution* (eds K.S.W. Campbell and M.F. Day), Allen and Unwin, London, pp. 283–309.

Campbell, S.D., Hilliker, A.J. and Phillips, J.P. (1986) Cytogenetic analysis of the cSOD microregion in *Drosophila melanogaster*. *Genetics*, **112**, 205–15.

Capinera, J.L. (1979) Qualitative variation in plants and insects: effect of propagule size on ecological plasticity. *Amer. Nat.*, **114**, 350–61.

Carey, J.R., Liedo, P., Orozco, D. and Vaupel, J.W. (1992) Slowing of mortality rates at older ages in large medfly cohorts. *Science*, **258**, 457–60.

Carpenter, A.T.C. (1987) Gene conversion, recombination nodules, and the initiation of meiotic synapsis. *BioEssays*, **6**, 232–6.

Carson, H.L. (1975) The genetics of speciation at the diploid level. *Amer. Nat.*, **109**, 73–92.

Carson, H.L. (1982) Speciation as a major reorganization of polygenic balances, in *Mechanisms of Speciation* (ed C. Barigozzi), Alan R. Liss, NY, pp. 411–33.

Carson, H.L. (1987a) The genetic system, the deme, and the origin of species. *Ann. Rev. Genet.*, **21**, 405–23.

Carson, H.L. (1987b) High fitness of heterokaryotypic individuals segregating naturally within a long-standing laboratory population of *Drosophila silvestris*. *Genetics*, **116**, 415–22.

Carson, H.L. (1989) Genetic imbalance, realigned selection, and the origin of species, in *Genetics, Speciation and the Founder Principle* (eds L.V. Giddings, K.Y. Kaneshiro and W.W. Anderson), Oxford University Press, NY, pp. 345–66.

Carson, H.L. (1990a) Increased genetic variance after a population bottleneck. *TREE*, **5**, 228–30.

Carson, H.L. (1990b) Evolutionary process as studied in population genetics: clues from phylogeny. *Oxford Surv. Evol. Biol.*, **7**, 129–56.

Carson, H.L. (1991) Episodic evolutionary change in local populations. *NATO ASI Series*, **H57**, 217–32.

Carson, H.L. and Templeton, A.R. (1984) Genetic revolutions in relation to speciation phenomena: the founding of new populations. *Ann. Rev. Ecol. Syst.*, **15**, 97–131.

Carson, H.L. and Wisotzkey, R.C. (1989) Increase in genetic variance following a population bottleneck. *Amer. Nat.*, **134**, 668–73.

Case, T.J. (1978) On the evolution and adaptive significance of postnatal growth rates in the terrestrial vertebrates. *Quart. Rev. Biol.*, **53**, 243–82.

Case, T.J. and Taper, M.L. (1986) On the coexistence and coevolution of asexual and sexual competitors. *Evolution*, **40**, 366–87.

Caswell, H. (1983) Phenotypic plasticity in life-history traits: demographic effects and evolutionary consequences. *Amer. Zool.*, **23**, 35–46.

Caswell, H. (1989) Life-history strategies. *Symp. Brit. Ecol. Soc.*, **29**, 285–307.

Chandra, H.S. and Nanjundiah, V. (1990) The evolution of genomic imprinting. *Development*, **1990** (Suppl.), 47–53.

Chao, L. (1990) Fitness of RNA virus decreased by Müller's ratchet. *Nature*, **348**, 454–5.

Chao, L. (1991) Levels of selection, evolution of sex in RNA viruses and the origin of life. *J. Theor. Biol.*, **153**, 229–46.

Chao, L. (1992) Evolution of sex in RNA viruses. *TREE*, **7**, 147–50.

Chao, L., Tran, T. and Matthews, C. (1992) Müller's ratchet and the advantage of sex in the RNA virus $\phi6$. *Evolution*, **46**, 289–99.

Chapin, F.S. (1991) Integrated responses of plants to stress. *BioScience*, **41**, 29–36.

Chapin, F.S., Schulze, E.D. and Mooney, H.A. (1990) The ecology and economics of storage in plants. *Ann. Rev. Ecol. Syst.*, **21**, 423–47.

Charlesworth, B. (1980) *Evolution in Age-structured Populations*, Cambridge University Press, Cambridge.

Charlesworth, B. (1988) Selection for longer-lived rodents. *Growth Develop. Aging*, **52**, 211.

Charlesworth, B. (1990) Natural selection and life history patterns, in *Genetic Effects on Aging II* (ed D.E. Harrison), Telford Press, Caldwell, NJ, pp. 21–40.

Charlesworth, B. (1991) When to be diploid. *Nature*, **351**, 273–4.

Charlesworth, B. and Rouhani, S. (1988) The probability of peak shifts in a founder population. II. An additive polygenic trait. *Evolution*, **42**, 1129–45.

Charnov, E.L. (1982) *The Theory of Sex Allocation*, Princeton Monogr. Popul. Biol. 18, Princeton University Press.

Charnov, E.L. (1991) Evolution of life history variation among female mammals. *Proc. Natl. Acad. Sci. USA*, **88**, 1134–7.

Charnov, E.L. and Berrigan, D. (1990) Dimensionless numbers and life history evolution: age of maturity versus the adult lifespan. *Evol. Ecol.*, **4**, 273–5.

Cheplick, G.P. (1991) A conceptual framework for the analysis of phenotypic plasticity and genetic constraints in plants. *Oikos*, **62**, 283–91.

Cheverud, J.M. (1988) A comparison of genetic and phenotypic correlations. *Evolution*, **42**, 958–68.

Clare, M.J. and Luckinbill, L.S. (1985) The effects of gene–environment interaction on the expression of longevity. *Heredity*, **55**, 19–29.

Clark, A.G. (1987a) Senescence and the genetic-correlation hang-up. *Amer. Nat.*, **129**, 932–40.

Clark, A.G. (1987b) Genetic correlations: the quantitative genetics of evolutionary constraints, in *Genetic Constraints on Adaptive Evolution* (ed V. Loeschcke), Springer Verlag, Berlin, pp. 25–45.

Clarke, P.H. (1983) Experimental evolution, in *Evolution from Molecules to Men* (ed D.S. Bendall), Cambridge University Press, Cambridge, p. 235–52.

Cockburn, A. (1991) *An Introduction to Evolutionary Ecology*, Blackwell Scientific, Oxford.

Cody, M.L. (1966) A general theory of clutch size. *Evolution*, **20**, 174–84.

Cohen, S. and Jurgens, G. (1991) *Drosophila* headlines. *Trends Genet.*, **7**, 267–72.

Cohen, J.E. and Newman, C.M. (1985) When will a large complex system be stable? *J. Theor. Biol.*, **113**, 153–6.

Cohn, J.P. (1992) Naked mole-rats. *BioScience*, **42**, 86–9.

Cole, L.C. (1954) The population consequences of life history phenomena. *Quart. Rev. Biol.*, **29**, 103–37.

Cole, R.K. (1967) *Ametapodia*, a dominant mutation in the fowl. *J. Heredity*, **58**, 141–6.

Coley, P.D., Bryant, J.P. and Chapin III, F.S. (1985) Resource availability and plant antiherbivore defense. *Science*, **230**, 895–9.

Collatz, K.G. (1986) Towards a comparative biology of aging, in *Insect Aging: Strategies and Mechanisms* (eds K.G. Collatz and R.S. Sohal), Springer Verlag, Berlin, pp. 1–8.

Collatz, K.G. and Sohal, R.S. (eds) (1986) *Insect Aging: Strategies and Mechanisms*, Springer-Verlag, Berlin.

Collins, J.P. and Cheek, J.E. (1983) Effect of food and density on development of typical and cannibalistic salamander larvae in *Ambystoma tigrinum nebulosum*. *Amer. Zool.*, **23**, 77–84.

Comfort, A. (1979) *The Biology of Senescence*, 3rd edn, Churchill Livingstone, Edinburgh.

Connell, J.H. (1978) Diversity in tropical rain forest and coral reefs. *Science*, **199**, 1302–10.

Conrad, M. (1979) Bootstrapping on the adaptive landscape. *BioSystems*, **11**, 167–82.

Conrad, M. (1983) *Adaptability: the Significance of Variability from Molecule to Ecosystem*, Plenum Press, NY.

Cook, S.A. and Johnson, M.P. (1968) Adaptation to heterogeneous environments I. Variation in heterophylly in *Ranunculus flammula* L. *Evolution*, **22**, 496–516.

Corces, V.G. and Geyer, P.K. (1991) Interactions of retrotransposons with the host genome: the case of the gypsy element in *Drosophila*. *Trends Genet.*, **7**, 86–90.

Covello, P.S. and Gray, M.W. (1992) Silent mitochondrial and active nuclear genes for subunit 2 of cytochrome oxidase (*cox* 2) in soybean: evidence for RNA-mediated gene transfer. *EMBO J.*, **11**, 3815–20.

Cowley, D.E. and Atchley, W.R. (1992) Quantitative genetic models for development, epigenetic selection, and phenotypic evolution. *Evolution*, **46**, 495–518.

Coyne, J.A. and Prout, T. (1984) Restoration of mutationally suppressed characters in *Drosophila melanogaster*. *J. Heredity*, **75**, 308–10.

Crick, F.H.C. (1968) The origin of the genetic code. *J. Mol. Biol.*, **38**, 367–79.

Crow, J.F. (1988) The importance of recombination, in *The Evolution of Sex*, (eds R.E. Michod and B.R. Levin), Sinauer Associates, Sunderland, MA, pp. 56–73.

Crow, J.F., Engels, W.R. and Denniston, C. (1990) Phase three of Wright's shifting-balance theory. *Evolution*, **44**, 233–47.

Crow, J.F. and Kimura, M. (1965) Evolution in sexual and asexual populations. *Amer Nat.*, **99**, 439–50.

Crow, J.F. and Kimura, M. (1969) Evolution in sexual and asexual populations: a reply. *Amer. Nat.*, **103**, 89–90.

Crowl, T.A. and Covich, A.P. (1990) Predator-induced life-history shifts in a freshwater snail. *Science*, **247**, 949–51.

Csink, A.K. and McDonald, J.F. (1990) *Copia* expression is variable among natural populations of *Drosophila*. *Genetics*, **126**, 375–85.

Cullis, C.A. (1986) Plant DNA variation and stress, in *Genetics, Development and Evolution* (eds J.P. Gustafson, G.L. Stebbins and F.J. Ayala), Plenum Press, NY, pp. 143–55.

Curio, E. (1973) Towards a methodology of teleonomy. *Experientia*, **29**, 1045–58.

Curtsinger, J.W., Fukui, H.H., Townsend, D.R. and Vaupel, T.W. (1992) Demography of genotypes: failure of the limited life-span paradigm in *Drosophila melanogaster*. *Science*, **258**, 461–3.

Cutler, R.G. (1983) Species probes, longevity and aging, in *Intervention in the Aging Process* (eds W. Regelson and F.M. Sinex), Alan R. Liss, NY, pp. 69–144.

Cutler, R.G. (1984a) Antioxidants, aging and longevity, in *Free radicals in Biology*, Vol. VI (ed W.A. Pryor), Academic Press, NY, pp. 371–428.

Cutler, R.G. (1984b) Evolutionary biology of aging and longevity in mammalian species, in *Aging and Cell Function* (ed J.E. Johnson), Plenum Press, NY, pp. 1–147.

Damuth, J. (1987) Interspecific allometry of population density in mammals and other animals, the independence of body mass and population energy-use. *Biol. J. Linn. Soc.*, **31**, 193–246.

Damuth, J. and Heisler, I.L. (1988) Alternative formulations of multilevel selection. *Biol. Philos.*, **3**, 407–30.

Dangerfield, J.M. and Hassall, M. (1992) Phenotypic variation in the breeding phenology of the woodlouse *Armadillidium vulgare*. *Oecologia*, **89**, 140–6.

Darlington, C.D. (1958) *Evolution of Genetic Systems*, Oliver and Boyd, Edinburgh.

Darwin, C. (1859) *The Origin of Species*, Mentor Books, NY.

Davidson, E.H. (1982) Evolutionary change in genomic regulatory organization: speculations on the origins of novel biological structure, in *Evolution and Development* (ed J.T. Bonner), Springer Verlag, Berlin, pp. 65–84.

Davidson, E.H. (1986) *Gene Activity in Early Development*, Academic Press, NY.

Davidson, E.H. and Britten, R.J. (1973) Organization, transcription, and regulation in the animal genome. *Quart. Rev. Biol.*, **48**, 565–613.

Davidson, E.H. and Britten, R.J. (1979) Regulation of gene expression: possible role of repetitive sequences. *Science*, **204**, 1052–9.

Davidson, E.H., Jacobs, H.T. and Britten, R.J. (1983) Very short repeats and coordinate induction of genes. *Nature*, **301**, 468–71.

Dawkins, R. (1976 and 1989a) *The Selfish Gene*, 1st edn, Paladin Press, London, 2nd edn, Oxford University Press, Oxford.

Dawkins, R. (1979) Twelve misunderstandings of kin selection. *Zeitschrift Tierpsychol.*, **51**, 184–200.

Dawkins, R. (1982) *The extended Phenotype: the Gene as the Unit of Selection*, Oxford University Press, Oxford.

Dawkins, R. (1986) *The Blind Watchmaker*, Longman Scientific and Technical, Essex, England.

Dawkins, R. (1989b) The evolution of evolvability, in *Artificial Life* (ed C. Langton), Addison-Wesley, NY, pp. 201–20.

Dawson, T.J., Fanning, D. and Bergin, T.J. (1978) Metabolism and temperature regulation in the New Guinea monotreme *Zaglossus bruijni*. *Austr. J. Zool.*, **20**, 99–103.

Dearolf, C.R., Topal, J. and Parker, C.S. (1989) The caudal gene is a direct activator of *fushi tarazu* transcription during *Drosophila* embryogenesis. *Nature*, **341**, 340–2.

De Jong, G. (1989) Phenotypically plastic characters in isolated populations, in *Evolutionary Biology of Transient Unstable Populations* (ed A. Fontdevila), Springer-Verlag, Berlin, pp. 3–18.

De Jong, G. (1990a) Genotype-by-environment interaction and the genetic covariance between environments: multilocus genetics. *Genetica*, **81**, 171–7.

De Jong, G. (1990b) Quantitative genetics of reaction norms. *J. Evol. Biol.*, **3**, 447–68.

De Jong, G. and Van Noordwijk, A.J. (1992) Acquisition and allocation of resources: genetic (co)variances, selection and life histories. *Amer. Nat.*, **139**, 749–70.

Denno, R.F., Olmstead, K.L. and McCloud, E.S. (1989) Reproductive cost of flight capacity: a comparison of life history traits in wing dimorphic planthoppers. *Ecol. Entomol.*, **14**, 31–44.

De Pomerai, D. (1990) *From Gene to Animal: an Introduction to the Molecular Biology of Animal Development*, Cambridge University Press, Cambridge.

Depew, D.J. and Weber, B.H. (1985) Innovation and tradition in evolutionary theory, an interpretive afterword, in *Evolution at a Crossroads: the New Biology and the New Philosophy of Science* (eds D.J. Depew and B.H. Weber), MIT Press, Cambridge, MA, pp. 227–60.

Derting, T.L. (1989) Metabolism and food availability as regulators of production in juvenile cotton rats. *Ecology*, **70**, 587–95.

DeVries, P.J. (1990) Enhancement of symbioses between butterfly caterpillars and ants by vibrational communication. *Science*, **248**, 1104–6.

Diamond, J.M. (1986) Why do disused proteins become genetically lost or repressed? *Nature*, **321**, 565–6.

Dickerson, G.E. (1955) Genetic slippage in response to selection for multiple objectives. *Cold Spring Harbor Symp. Quant. Biol.*, **20**, 213–24.

Dickinson, W.J. (1989) Gene regulation and evolution, in *Genetics, Speciation and the Founder Principle* (eds L.V. Giddings, K.Y. Kaneshiro and W.W. Anderson), Oxford University Press, Oxford, pp. 181–202.

Dickinson, W.J. (1991) The evolution of regulatory genes and patterns in *Drosophila*. *Evol. Biol.*, **25**, 127–73.

Dingle, H. (1984) Behavior, genes, and life histories: complex adaptations in uncertain environments, in *A New Ecology: Novel Approaches to Interactive Systems* (eds P.W. Price, C.N. Slobodchikoff and W.S. Gaud), Wiley, NY, pp. 169–94.

Dingle, H. (1990) The evolution of life histories, in *Population Biology* (eds K. Wohrmann and S. Jain), Springer Verlag, Berlin, pp. 267–89.

Dingle, H. and Evans, K.E. (1987) Responses in flight to selection on wing length in non-migratory milkweed bugs, *Oncopeltus fasciatus*. *Entomol. Expt. Appl.*, **45**, 289–96.

Dingle, H., Evans, K.E. and Palmer, J.O. (1988) Responses to selection among life-history traits in a nonmigratory population of milkweed bugs (*Oncopeltus fasciatus*). *Evolution*, **42**, 79–92.

Dixon, W.N. and Payne, T.L. (1980) Attraction of entomophagous and associate insects of Southern pine beetle to beetle- and host tree-produced volatiles. *J. Georgia Entomol. Soc.*, **15**, 378.

Dobzhansky, T. (1937) *Genetics and the Origin of Species*, Columbia University Press, NY.

Dobzhansky, T. (1970) *Genetics of the Evolutionary Process*, Columbia University Press, NY.

Dodds, W.K. (1988) Community structure and selection for positive or negative species interactions. *Oikos*, **53**, 387–90.

Dodson, S. (1989) Predator-induced reaction norms. *BioScience*, **39**, 447–52.

Doolittle, W.F. (1987) The origin and function of intervening sequences: a review. *Am. Nat.*, **130**, 155–85.

Doolittle, W.F. and Sapienza, C. (1980) Selfish genes, the phenotype paradigm and genome evolution. *Nature*, **284**, 601–3.

Dorit, R.L. (1990) The correlates of high diversity in Lake Victoria Haplochromine cichlids: a neontological perspective, in *Causes of Evolution* (eds R.M. Ross and W.D. Allmon), University of Chicago Press, Chicago, IL, pp. 322–52.

Dover, G.A. (1982) Molecular drive: a cohesive mode of species evolution. *Nature*, **299**, 111–17.

Dover, G. (1986) Molecular drive in multigene families: how biological novelties arise, spread and are assimilated. *Trends Genet.*, **2**, 159–65.

Drent, R.H. and Daan, S. (1980) The prudent parent: energetic adjustments in avian breeding. *Ardea*, **68**, 225–52.

Duboule, D. and Dolle, P. (1989) The structural and functional organization of the murine HOX gene family resembles that of *Drosophila* homeotic genes. *EMBO J.*, **8**, 1497–505.

Dudley, R. and Gans, C. (1991) A critique of symmorphosis and optimality models in physiology. *Physiol. Zool.*, **64**, 627–37.

Duffy, P.H., Feuers, R.J., Leakey, J.A. *et al.* (1989) Effect of chronic caloric restriction on physiological variables related to energy metabolism in the male Fischer 344 rat. *Mech. Age. Devel.*, **48**, 117–33.

Dyer, F.C. and Seeley, T.D. (1991) Nesting behavior and the evolution of worker tempo in four honey bee species. *Ecology*, **72**, 156–70.

Eberhard, W.G. (1980) Horned beetles. *Sci. Amer.*, **242**, 166–82.

Eberhard, W.G. (1982) Beetle horn dimorphism: making the best of a bad lot. *Amer. Nat.*, **111**, 420–6.

Eberhard, W.G. and Gutiérrez, E.E. (1991) Male dimorphisms in beetles and earwigs and the question of developmental constraints. *Evolution*, **45**, 18–28.

Echols, H. (1981) SOS functions, cancer and inducible evolution. *Cell*, **25**, 1–2.

Economos, A.C. (1980) Taxonomic differences in the mammalian life span–body weight relationship and the problem of brain weight. *Gerontology*, **26**, 90–8.

Economos, A.C. and Lints, F.A. (1984a) Growth rate and life span in *Drosophila* III. Effect of body size and developmental temperature on the biphasic relationship between growth rate and life span. *Mech. Age. Devel.*, **27**, 153–60.

Economos, A.C. and Lints, F.A. (1984b) Growth rate and life span in *Drosophila*. II. A biphasic relationship between growth rate and life span. *Mech. Ageing Devel.*, **27**, 143–51.

Economos, A.C. and Lints, F.A. (1985) Growth rate and life span in *Drosophila*. IV. Role of cell size and cell number in the biphasic relationship between life span and growth rate. *Mech. Age. Devel.*, **32**, 193–204.

Economos, A.C. and Lints, F.A. (1986) Developmental temperature and life span in *Drosophila melanogaster*. *Gerontology*, **32**, 18–27.

Edelman, G.M. (1989) Topobiology. *Sci. Amer.*, **260**, 76–88.

Edney, E.B. and Gill, R.W. (1968) Evolution of senescence and specific longevity. *Nature*, **220**, 281–2.

Edson, M.M., Foin, T.C. and Knapp, C.M. (1981) 'Emergent properties' and ecological research. *Amer. Nat.*, **118**, 593–6.

Ehrlich, P.R. and Raven, P.H. (1964) Butterflies and plants: a study in coevolution. *Evolution*, **18**, 586–608.

Ehrlich, P.R. and Raven, P.H. (1969) Differentiation of populations. *Science*, **165**, 1228–32.

Eigen, M. and Schuster, P. (1979) *The Hypercycle*, Springer Verlag, Berlin.

Eklund, J. and Bradford, G.E. (1977) Longevity and lifetime body weight in mice selected for rapid growth. *Nature*, **265**, 48–9.

Eldredge, N. (1985) *Unfinished Synthesis: Biological Hierarchies and Modern Evolutionary Thought*, Oxford University Press, Oxford.

Eldredge, N. (1986) Information, economics and evolution. *Ann. Rev. Ecol. Syst.*, **17**, 351–69.

Eldredge, N. and Gould, S.J. (1972) Punctuated equilibria: an alternative to phyletic gradualism, in *Models in Paleobiology* (ed T.J.M. Schopf), Freeman, Cooper and Co., San Francisco, CA, pp. 82–115.

Eldredge, N. and Salthe, S.N. (1984) Hierarchy and evolution. *Oxford Surv. Evol. Biol.*, **1**, 184–208.

Elgar, M.A. and Harvey, P.H. (1987) Basal metabolic rates in mammals: allometry, phylogeny and ecology. *Funct. Ecol.*, **1**, 25–36.

Ellstrand, N.C. and Antonovics, J. (1985) Experimental studies of the evolutionary significance of sexual reproduction II. A test of the density-dependent selection hypothesis. *Evolution*, **39**, 657–66.

Ellstrand, N.C. and Hoffman, C.A. (1990) Hybridization as an avenue of escape for engineered genes. *BioScience*, **40**, 438–42.

Endler, J.A. (1973) Gene flow and population differentiation. *Science*, **179**, 243–50.

Endler, J.A. (1977) *Geographic Variation, Speciation, and Clines*, Princeton University Press, Princeton, NJ.

Enesco, H.E., Bozovic, V. and Anderson, P.D. (1989) The relationship between lifespan and reproduction in the rotifer *Asplanchna brightwelli*. *Mech. Age. Devel.*, **48**, 281–9.

Engebrecht, J., Hirsch, J. and Roeder, G.S. (1990) Meiotic gene conversion and crossing over: their relationship to each other and to chromosome synapsis and segregation. *Cell*, **62**, 927–37.

Ereshefsky, M. (1992) *The Units of Evolution: Essays on the Nature of the Species*, MIT Press, Cambridge, MA.

Erickson, J.W. and Cline, T.W. (1991) Molecular nature of the *Drosophila* sex determination signal and its link to neurogenesis. *Science*, **251**, 1071–4.

Ernsting, G. and Isaaks, J.A. (1991) Accelerated ageing: a cost of reproduction in the carabid beetle *Notiophilus biguttatus* F. *Funct. Ecol.*, **5**, 299–303.

Erwin, D.H. and Valentine, J.W. (1984) 'Hopeful monsters', transposons and Metazoan radiation. *Proc. Natl. Acad. Sci. USA*, **81**, 5482–3.

Eshel, I. and Feldman, M.W. (1970) On the evolutionary effect of recombination. *Theor. Popl. Biol.*, **1**, 88–100.

Fagen, R. (1987) Phenotypic plasticity and social environment. *Evol. Ecol.*, **1**, 263–71.

Fairbairn, D.J. and Roff, D.A. (1990) Genetic correlations among traits determining migratory tendency in the sand cricket, *Gryllus firmus*. *Evolution*, **44**, 1787–95.

Falconer, D.S. (1953) Selection for large and small size in mice. *J. Genet.*, **51**, 470–501.

Falconer, D.S. (1960) Selection of mice for growth on high and low planes of nutrition. *Genet. Res.*, **1**, 91–113.

Falconer, D.S. (1977) Why are mice the size they are? in *International Conference on Quantitative Genetics* (eds E. Pollack, O. Kempthorne and E.J. Bailey), Iowa State University Press, Ames, IA, pp. 19–21.

Falconer, D.S. (1981) *Introduction to Quantitative Genetics*, 2nd edn, Longman, London.

Falconer, D.S. and King, J.W.B. (1953) A study of selection limits in the mouse. *J. Genet.*, **51**, 561–81.

Falconer, D.S. and Latyszewski, M. (1952) The environment in relation to selection for size in mice. *J. Genet.*, **51**, 67–80.

Feibleman, J.K. (1955) Theory of integrative levels. *Brit. J. Phil. Soc.*, **5**, 59–66.

Felsenfeld, G. (1978) Chromatin. *Nature*, **271**, 115–22.

Felsenstein, J. (1974) The evolutionary advantage of recombination. *Genetics*, **78**, 737–56.

Felsenstein, J. (1981) Skepticism towards Sainta Rosalia, or why are there so few kinds of animals? *Evolution*, **35**, 124–38.

Felsenstein, J. (1988) Sex and the evolution of recombination, in *The Evolution of Sex* (eds R.E. Michod and B.R. Levin), Sinauer Associates, Sunderland, MA, pp. 74–86.

Felsenstein, J. and Yokoyama, S. (1976) The evolutionary advantage of recombination II. Individual selection for recombination. *Genetics*, **83**, 845–59.

Fenchel, T. (1974) Intrinsic rate of natural increase: the relationship with body size. *Oecologia*, **14**, 317–26.

Finch, C.E. (1990) *Longevity, Senescence, and the Genome*, University of Chicago Press, Chicago, IL.

Findlay, S. and Rowe, G. (1990) Computer experiments on the evolution of sex: the haploid case. *J. Theor. Biol.*, **146**, 379–93.

Finnegan, D.J. (1989) Eukaryotic transposable elements and genome evolution. *Trends. Genet.*, **5**, 103–7.

Fisher, R.A. (1930) *The Genetical Theory of Natural Selection*, Clarendon Press, Oxford.

Fleming, J.E., Leon, H.A. and Miquel, J. (1981) Effects of ethidium bromide on development and aging of *Drosophila*: implications for the free radical theory of aging. *Expt. Geront.*, **16**, 287–93.

Fontdevila, A. (1992) Genetic instability and rapid speciation: are they coupled? *Genetica*, **86**, 247–58.

Ford, N.B. and Seigel, R.A. (1989) Phenotypic plasticity in reproductive traits: evidence from a viviparous snake. *Ecology*, **70**, 1768–74.

Forsum, E., Hillman, P.E. and Nesheim, M.C. (1981) Effect of energy restriction on total heat production, basal metabolic rate, and specific dynamic action of food in rats. *J. Nutr.*, **111**, 1691–7.

Fowler, R.E. (1962) The efficiency of food utilization, digestibility of foodstuffs and energy expenditure of mice selected for large or small body size. *Genet. Res.*, **3**, 51–68.

Frank, S.A. and Slatkin, M. (1990) Evolution in a variable environment. *Amer. Nat.*, **136**, 244–60.

Franklin, I. and Lewontin, R.C. (1970) Is the gene the unit of selection? *Genetics*, **65**, 707–34.

Friedman, D.B. and Johnson, T.E. (1988a) Three mutants that extend both mean and maximum life span of the nematode, *Caenorhabditis elegans*, define the age-1 gene. *J. Geront.*, **43**, B102–B109.

Friedman, D.B. and Johnson, T.E. (1988b) A mutation in the age-1 gene in

Caenorhabditus elegans lengthens life and reduces hermaphrodite fertility. *Genetics*, **118**, 75–86.

Futuyma, D.J. and Moreno, G. (1988) The evolution of ecological specialization. *Ann. Rev. Ecol. Syst.*, **19**, 207–33.

Gabriel, W. and Lynch, M. (1992) The selective advantage of reaction norms for environmental tolerance. *J. Evol. Biol.*, **5**, 41–59.

Gadgil, M. and Bossert, W.H. (1970) Life historical consequences of natural selection. *Amer. Nat.*, **104**, 1–24.

Galiana, A., Ayala, F.J. and Moya, A. (1989) Flush-crash experiments in *Drosophila*, in *Evolutionary Biology of Transient Unstable Populations* (ed A. Fontdevila), Springer Verlag, Berlin, pp. 58–73.

Gallistel, C.R. (1980) From muscles to motivation. *Amer. Sci.*, **68**, 398–409.

Garcia, P., Morris, M.I., Saenz-de-Miera, L.E. *et al.* (1991) Genetic diversity and adaptedness in tetraploid *Avena barbata* and its diploid ancestors *Avena hirtula* and *Avena wiestii*. *Proc. Natl. Acad. Sci. USA*, **88**, 1207–11.

Garland, T. Jr. (1988) Genetic basis of activity metabolism. I. Inheritance of speed, stamina, and antipredator displays in the garter snake *Thamnophis sirtalis*. *Evolution*, **42**, 335–50.

Gause, G.F. (1942) The relation of adaptability to adaptation. *Q. Rev. Biol.*, **17**, 99–114.

Gebhardt, M.D. and Stearns, S.C. (1988) Reaction norms for developmental time and weight at eclosion in *Drosophila mercatorum*. *J. Evol. Biol.*, **1**, 335–54.

Geissler, T.G. and Rollo, C.D. (1987) The influence of nutritional history on the response to novel food by the cockroach, *Periplaneta americana*. (L.). *Anim. Behav.*, **35**, 1908–16.

Ghiselin, M.T. (1974) *The Economy of Nature and the Evolution of Sex*, University California Press, Berkeley, CA.

Ghiselin, M.T. (1975) A radical solution to the species problem. *Syst. Zool.*, **23**, 536–44.

Ghiselin, M.T. (1987) Species concepts, individuality and objectivity. *Biol. Phil.*, **2**, 127–43.

Ghiselin, M.T. (1988) The evolution of sex: a history of competing points of view, in *The Evolution of Sex* (eds R.E. Michod and B.R. Levin), Sinauer Associates, Sunderland, MA, pp. 7–23.

Giesel, J.T. (1979) Genetic co-variation of survivorship and other fitness indices in *Drosophila melanogaster*. *Exp. Geront.*, **14**, 323–8.

Giesel, J.T. (1986) Genetic correlation structure of life history variables in outbred, wild *Drosophila melanogaster*: effects of photoperiod regimen. *Amer. Nat.*, **128**, 593–603.

Giesel, J.T., Lanciani, C.A. and Anderson, J.F. (1989) Effects of parental photoperiod on metabolic rate in *Drosophila melanogaster*. *Florida Entomol.*, **72**, 499–503.

Giesel, J.T., Murphy, P. and Manlove, M. (1982a) An investigation of the effects of temperature on the genetic organization of life history indices in three populations of *Drosophila melanogaster*, in *Evolution and Genetics of Life Histories* (eds H. Dingle and J.P. Hegmann), Springer Verlag, Berlin, pp. 189–207.

Giesel, J.T., Murphy, P.A. and Manlove, M.N. (1982b) The influence of temperature on genetic interrelationships of life history traits in a population of *Drosophila melanogaster*: what tangled data sets we weave. *Amer. Nat.*, **119**, 464–79.

Giesel, J.T. and Zettler, E.E. (1980) Genetic correlations of life historical parameters and certain fitness indices in *Drosophila melanogaster*: r_m, r_s, diet breadth. *Oecologica*, **47**, 299–302.

Gilbert, J.J. (1966) Rotifer ecology and embryological induction. *Science*, **151**, 1234–7.

Gilbert, J.J. (1980) Further observations on developmental polymorphism and its evolution in the rotifer *Brachionus calyciflorus*. *Freshwater Biol.*, **10**, 281–94.

Gilbert, N. (1986) Control of fecundity in *Pieris rapae* IV. Patterns of variation and their ecological consequences. *J. Anim. Ecol.*, **55**, 317–29.

Gilbert, S.F. (1991) *Developmental Biology*, 3rd edn, Sinauer Associates, Sunderland, MA.

Gilbert, W. (1978) Why genes in pieces? *Nature*, **271**, 501.

Gilbert, W. (1985) Genes in pieces revisited. *Science*, **228**, 823–4.

Gillespie, J.H. (1974) Natural selection for within-generation variance in offspring number. *Genetics*, **76**, 601–6.

Gillespie, J.H. (1977) Natural selection for variance in offspring numbers: a new evolutionary principle. *Amer. Nat.*, **111**, 1010–14.

Gillespie, J.H. and Turelli, M. (1989) Genotype–environment interactions and the maintenance of polygenic variation. *Genetics*, **121**, 129–38.

Gillis, A.M. (1991) Can organisms direct their evolution? *BioScience*, **41**, 202–5.

Gilpin, M.E. (1975) *Group Selection in Predator–prey Communities*, Princeton University Press, Princeton, NJ.

Gingerich, P.D., Smith, B.H. and Simons, E.L. (1990) Hind limbs of Eocene *Basilosaurus*: evidence of feet in whales. *Science*, **249**, 154–7.

Glass, A.R. and Swerdloff, R.S. (1980) Nutritional influences on sexual maturation in the rat. *Fed. Proc.*, **39**, 2360–4.

Gleick, J. (1987) *Chaos: making a new science*, Penguin Books, NY.

Glesener, R.R. and Tilman, D. (1978) Sexuality and the components of environmental uncertainty: clues from geographic parthenogenesis in terrestrial animals. *Amer. Nat.*, **112**, 659–73.

Goff, L.J. (1991) Symbiosis, interspecific gene transfer, and the evolution of new species: a case study in the parasitic red algae, in *Symbiosis as a Source of Evolutionary Innovation* (eds L. Margulis and R. Fester), MIT Press, Cambridge, MA, pp. 341–63.

Goldschmidt, R. (1940) *The Material Basis of Evolution*, Yale University Press, New Haven, CT.

Gomulkiewicz, R. and Kirkpatrick, M. (1992) Quantitative genetics and the evolution of reaction norms. *Evolution*, **46**, 390–411.

Goodale, H.D. (1938) A study of the inheritance of body weight in the albino mouse by selection. *J. Hered.*, **29**, 101–12.

Goodman, D. (1979) Regulating reproductive effort in a changing environment. *Amer. Nat.*, **113**, 735–48.

Goodnight, C.J. (1987) On the effect of founder events on epistatic genetic variance. *Evolution*, **41**, 80–91.

Goodnight, C.J. (1988) Epistasis and the effect of founder events on the additive genetic variance. *Evolution*, **42**, 441–54.

Goodnight, C.J. (1990a) Experimental studies of community evolution I: The response to selection at the community level. *Evolution*, **44**, 1614–24.

Goodnight, C.J. (1990b) Experimental studies of community evolution II: The ecological basis of the response to community selection. *Evolution*, **44**, 1625–36.

Goodnight, C.J., Schwartz, J.M. and Stevens, L. (1992) Contextual analysis of models of group selection, soft selection, hard selection, and the evolution of altruism. *Amer. Nat.*, **140**, 743–61.

Goodrick, C.L. (1977) Body weight change over the life span and longevity for C57BL/6J mice and mutations which differ in maximal body weight. *Gerontology*, **23**, 405–13.

Goodrick, C.L. (1980) Effects of long-term voluntary wheel exercise on male and female Wistar rats. I. Longevity, body weight, and metabolic rate. *Gerontology*, **26**, 22–33.

Goodrick, C.L., Ingram, D.K., Reynolds, M.A. *et al.* (1982) Effects of intermittent feeding upon growth and life span in rats. *Gerontology*, **28**, 233–41.

Goodrick, C.L., Ingram, D.K., Reynolds, M.A. *et al.* (1983a) Differential effects of intermittent feeding and voluntary exercise on body weight and lifespan in adult rats. *J. Geront.*, **38**, 36–45.

Goodrick, C.L., Ingram, D.K., Reynolds, M.A. *et al.* (1983b) Effects of intermittent feeding upon growth, activity, and lifespan in rats allowed voluntary exercise. *Expt. Aging Res.*, **9**, 203–9.

Gordon, H. (1959) Minimal nutritional requirements of the German roach, *Blattella germanica* L. *Ann. N.Y. Acad. Sci.*, **77**, 290–351.

Gordon, H. (1972) Interpretations of insect quantitative nutrition, in *Insect and Mite Nutrition* (ed J.G. Rodriguez), North Holland, Amsterdam, pp. 73–105.

Gordon, K.R. (1989) Adaptive nature of skeletal design. *BioScience*, **39**, 784–90.

Gottlieb, G. (1992) *Individual Development and Evolution: the Genesis of Novel Behaviour*, Oxford University Press, NY.

Gould, A.P., Brookman, J.J., Strutt, D.I. and White, R.A.H. (1990) Targets of homeotic gene control in *Drosophila*. *Nature*, **348**, 308–12.

Gould, S.J. (1977) *Ontogeny and Phylogeny*, Belknap Press, Harvard University Press, Cambridge, MA.

Gould, S.J. (1980) Is a new and general theory of evolution emerging? *Paleobiology*, **6**, 119–30.

Gould, S.J. (1982) Changes in developmental timing as a mechanism of macro-evolution, in *Evolution and Development* (ed J.T. Bonner), Springer Verlag, Berlin, pp. 333–46.

Gould, S.J. (1989a) *Wonderful Life: the Burgess Shale and the Nature of History*, W.W. Norton and Company, NY.

Gould, S.J. (1989b) A developmental constraint in *Cerion*, with comments on the definition and interpretation of constraint in evolution. *Evolution*, **43**, 516–39.

Gould, S.J. and Eldredge, N. (1977) Punctuated equilibria: the tempo and mode of evolution reconsidered. *Paleobiology*, **3**, 115–51.

Gould, S.J. and Eldredge, N. (1986) Punctuated equilibrium at the third stage. *Syst. Zool.*, **35**, 143–8.

Gould, S.J. and Lewontin, R.C. (1979) The spandrels of San Marco and the Panglossian paradigm: a critique of the adaptationist programme. *Proc. R. Soc. Lond. B.*, **205**, 581–98.

Gould, S.J. and Vrba, E.S. (1982) Exaptation – a missing term in the science of form. *Paleobiology*, **8**, 4–15.

Gouyon, P.H., Gliddon, C.J. and Couvet, D. (1988) The evolution of reproductive systems: a hierarchy of causes, in *Plant Population Ecology* (eds A.J. Davy, M.J. Hutchings and A.R. Watkinson), Blackwell Scientific, Oxford, pp. 23–33.

Grace, J.B. (1991) A clarification of the debate between Grime and Tilman. *Funct. Ecol.*, **5**, 583–7.

Graham, A., Papalopulu, N. and Krumlauf, R. (1989) The mouse and *Drosophila* homeobox gene clusters have common features of organization and expression. *Cell*, **56**, 367–78.

Graham-Smith, W. (1978) Organization in natural systems. *Ecol. Quart.*, **2**, 114–21.

Grant, B. and Wiseman, L.L. (1982) Fossil genes: scarce as hen's teeth? *Science*, **215**, 698–9.

Graves, J.L., Luckinbill, L.S. and Nichols, A. (1988) Flight duration and wing beat frequency in long- and short-lived *Drosophila melanogaster*. *J. Insect Physiol.*, **34**, 1021–6.

Graves, J.L. and Rose, M.R. (1990) Flight duration in *Drosophila melanogaster* selected for postponed senescence, in *Genetic Effects of Aging II* (ed D.E. Harrison), Telford Press, Caldwell, NJ, pp. 57–63.

Gray, M.W. (1989) The evolutionary origins of organelles. *Trends Genet.*, **5**, 294–9.

Greene, E. (1989) A diet-induced developmental polymorphism in a caterpillar. *Science*, **243**, 643–6.

Greenslade, P.J.M. (1983) Adversity selection and the habitat templet. *Amer. Nat.*, **122**, 352–65.

Grene, M. (1987) Hierarchies in biology. *Amer. Sci.*, **75**, 504–10.

Grime, J.P. (1977) Evidence for the existence of three primary strategies in plants and its relevance to ecological and evolutionary theory. *Amer. Nat.*, **111**, 1169–94.

Grime, J.P. (1988) The C-S-R model of primary plant strategies – origins, implications and tests, in *Plant Evolutionary Biology* (eds L.D. Gottleib and K.S. Jain), Chapman & Hall, London, pp. 371–93.

Groeters, F.R. and Dingle, H. (1988) Genetic and maternal influences on life history plasticity in milkweed bugs (*Oncopeltus*): response to temperature. *J. Evol. Biol.*, **1**, 317–33.

Gross, M.R. (1985) Disruptive selection for alternative life histories in salmon. *Nature*, **313**, 47–8.

Grunstein, M. (1990) Nucleosomes: regulators of transcription. *Trends Genet.*, **6**, 395–400.

Grunstein, M. (1992) Histones as regulators of genes. *Sci. Amer.*, **267**, 68–74B.

Guerrero, R. (1991) Predation as prerequisite to organelle origin: *Daptobacter* as example, in *Symbiosis as a Source of Evolutionary Innovation* (eds L. Margulis and R. Fester), MIT Press, Cambridge, MA, pp. 106–17.

Gupta, A.P. and Lewontin, R.C. (1982) A study of reaction norms in natural populations of *Drosophila pseudoobscura*. *Evolution*, **36**, 934–48.

Haig, D. and Grafen, A. (1991) Genetic scrambling as a defence against meiotic drive. *J. Theor. Biol.*, **153**, 531–58.

Haldane, J.B.S. (1932) *The Causes of Evolution*, Longmans, Green and Co., London.

Haldane, J.B.S. (1964) A defense of beanbag genetics. *Perspt. Biol. Med.*, **7**, 343–59.

Hall, B.G. (1982) Evolution of a regulated operon in the laboratory. *Genetics*, **101**, 335–44.

Hall, B.G. (1991) Adaptive evolution that requires multiple spontaneous mutations: mutations involving base substitutions. *Proc. Natl. Acad. Sci. USA*, **88**, 5882–6.

Hall, B.G., Yokoyama, S. and Calhoun, D.H. (1983) Role of cryptic genes in microbial evolution. *Mol. Biol. Evol.*, **1**, 109–24.

Hall, B.K. (1984) Developmental mechanisms underlying the formation of atavisms. *Biol. Rev.*, **59**, 89–124.

Hall, B.K. (1992) *Evolutionary Developmental Biology*, Chapman & Hall, London.

Hamilton, W.D. (1964a,b) The genetic evolution of social behaviour I and II. *J. Theor. Biol.*, **7**, 1–52.

Hamilton, W.D. (1966) The moulding of senescence by natural selection. *J. Theor. Biol.*, **12**, 12–45.

Hamilton, W.D. (1980) Sex versus non-sex versus parasite. *Oikos*, **35**, 282–90.

Hamilton, W.D. (1988) Sex and disease, in *The Evolution of Sex* (eds R. Bellig and G. Stevens), Harper and Row, San Francisco, CA, pp. 65–99.

Hamilton, W.D. (1993) Inbreeding in Egypt and in this book: a childish perspective, in *The Natural History of Inbreeding and Outbreeding* (ed N.W. Thornhill),

University of Chicago Press, Chicago, IL, pp. 429–50.

Hamilton, W.D., Axelrod, R. and Tanese, R. (1990) Sexual reproduction as an adaptation to resist parasites. *Proc. Natl. Acad. Sci. USA*, **87**, 3566–73.

Hamilton, W.D., Henderson, P.A. and Moran, N.A. (1981) Fluctuation of environment and coevolved antagonistic polymorphisms as factors in the maintenance of sex, in *Natural Selection and Social Behavior* (eds R.D. Alexander and D.W. Tinkle), Chirnon Press, NY, pp. 363–81.

Hanski, I. (1991) Single-species metapopulation dynamics: concepts, models and observations. *Biol. J. Linn. Soc.*, 42, 17–38.

Hanski, I. and Gilpin, M. (1991) Metapopulation dynamics: brief history and conceptual domains. *Biol. J. Linn. Soc.*, **42**, 3–16.

Hansson, L. (1991) Dispersal and connectivity in metapopulations. *Biol. J. Linn. Soc.*, **42**, 89–103.

Hardie, J. and Lees, A.D. (1985) Endocrine control of polymorphism and polyphenism, in *Comprehensive Insect Physiology, Biochemistry and Pharmacology. Vol. 8, Endocrinology II* (eds G.A. Kerkut and L.I. Gilbert), Pergamon Press, Toronto, pp. 441–90.

Harris, G.P. (1986) *Phytoplankton Ecology: Structure, Function and Fluctuation*, Chapman & Hall, London.

Harrison, R.G. (1980) Dispersal polymorphisms in insects. *Ann. Rev. Ecol. Syst.*, 11, 95–118.

Harrison, R.G. (1991) Molecular changes at speciation. *Ann. Rev. Ecol. Syst.*, 22, 281–308.

Harrison, S.C. (1991) A structural taxonomy of DNA-binding domains. *Nature*, 353, 715–19.

Hart, R.W., Sacher, G.A. and Hoskins, T.L. (1979) DNA repair in a short- and a long-lived rodent species. *J. Geront.*, 34, 808–17.

Hart, R.W. and Setlow, R.B. (1974) Correlation between deoxyribonucleic acid excision repair and life-span in a number of mammalian species. *Proc. Natl. Acad. Sci. USA*, 71, 2169–73.

Hartung, J. (1981) Genome parliaments and sex with the Red Queen, in *Natural Selection and Social Behaviour* (eds R.D. Alexander and D.W. Tinkle), Chiron Press, NY, pp. 382–402.

Harvell, C.D. (1984) Predator-induced defense in a marine bryozoan. *Science*, 224, 1357–9.

Harvell, C.D. (1990) The ecology and evolution of inducible defences. *Quart. Rev. Biol.*, 65, 323–40.

Harvell, C.D. and Padilla, D.K. (1990) Inducible morphology, heterochrony, and size hierarchies in a colonial invertebrate monoculture. *Proc. Natl. Acad. Sci. USA*, 87, 508–12.

Harvey, P.H. (1985) Intrademic group selection and the sex ratio, in *Behavioural Ecology: Ecological Consequences of Adaptive Behaviour* (eds R.M. Sibly and R.H. Smith), Blackwell Scientific, Oxford, pp. 59–73.

Harvey, P.H. (1991) Comparing life histories, in *Evolution of Life* (eds S. Osawa and T. Honjo), Springer Verlag, Tokyo, pp. 215–28.

Harvey, P.H. and Nee, S. (1991) How to live like a mammal. *Nature*, 350, 23–4.

Harvey, P.H., Nee, S., Mooers, A.O. and Partridge, L. (1992) These hierarchical views of life: phylogenies and metapopulations. In press.

Harvey, P.H., Pagel, M.D. and Rees, J.A. (1991) Mammalian metabolism and life histories. *Amer. Nat.*, 137, 556–66.

Harvey, P.H. and Read, A.F. (1988) How and why do mammalian life histories vary? in *Evolution of Life Histories: Pattern and Theory from Mammals* (ed M.S. Boyce), Yale University Press, New Haven, CT, pp. 213–32.

Harvey, P.H., Read, A.F. and Promislow, D.E.L. (1989) Life history variation in placental mammals: unifying the data with theory. *Oxford Surv. Evol. Biol.*, **6**, 13–31.

Harvey, P.H. and Zammuto, R.M. (1985) Patterns of mortality and age at first reproduction in natural populations of mammals. *Nature*, **315**, 319–20.

Hassell, M.P., Comins, H.N. and May, R.M. (1991) Spatial structure and chaos in insect population dynamics. *Nature*, **353**, 255–8.

Haukioja, E. (1982) Are individuals really subordinated to genes? A theory of living entities. *J. Theor. Biol.*, **99**, 357–75.

Haukioja, E. (1990) Induction of defenses in trees. *Ann. Rev. Entomol.*, **36**, 25–42.

Havel, J.E. (1987) Predator-induced defenses: a review, in *Predation: Direct and Indirect Impacts on Aquatic Communities* (eds W.C. Kerfoot and A. Sih), New England Press, Hanover, NH, pp. 263–78.

Havel, J.E. and Dodson, S.I. (1987) Reproductive costs of *Chaoborus*-induced polymorphism in *Daphnia pulex*. *Hydrobiologia*, **150**, 273–82.

Havlicek, V., Rezek, M. and Friesen, H. (1976) Somatostatin and thyrotropin releasing hormone: cental effect on sleep and the motor system. *Pharmacol. Biochem. Behav.*, **4**, 455–9.

Hayes, J.P., Garland Jr., T. and Dohm, M.R. (1992) Individual variation in metabolism and reproduction of *Mus*: Are energetics and life histories linked? *Funct. Ecol.*, **6**, 5–14.

Hayflick, L. (1980) The cell biology of human aging. *Sci. Amer.*, **242**, 58–65.

Hayssen, V. and Lacy, R.C. (1985) Basal metabolic rates in mammals: taxonomic differences in the allometry of BMR and body mass. *Comp. Biochem. Physiol.*, **81A**, 741–54.

Hebert, P.D.N. (1987) Genotypic characteristics of cyclic parthenogens and their obligately asexual derivatives, in *The Evolution of Sex and its Consequences* (ed S.C. Stearns), Birkhäuser, Basel, Switzerland, pp. 175–95.

Hebert, P.D.N. and Grewe, P.M. (1985) *Chaoborus* induced shifts in the morphology of *Daphnia ambigua*. *Limnol. Oceanogr.*, **30**, 1291–7.

Hedrick, P.W. (1986) Genetic polymorphism in heterogenous environments: a decade later. *Ann. Rev. Ecol. Syst.*, **17**, 535–66.

Hedrick, P.W., Ginevan, M.E. and Ewing, E.P. (1976) Genetic polymorphism in heterogeneous environments. *Ann. Rev. Ecol. Syst.*, **7**, 1–32.

Hedrick, P.W. and McDonald, J.F. (1980) Regulatory gene adaptation: an evolutionary model. *Heredity*, **45**, 83–97.

Hedrick, P.W. and Whittam, T.S. (1989) Sex in diploids. *Nature*, **342**, 231.

Hegmann, J.P. and Dingle, H. (1982) Phenotypic and genetic covariance structure in milkweed bug life history traits, in *Evolution and Genetics of Life Histories* (eds H. Dingle and J.P. Hegmann), Springer Verlag, NY, pp. 177–85.

Heinrich, B. (1979) *Bumblebee Economics*, Harvard University Press, Cambridge, MA.

Heisler, I.L. and Damuth, J. (1987) A method for analysing selection in hierarchically structured populations. *Amer. Nat.*, **130**, 582–602.

Heller, J. (1990) Longevity in molluscs. *Malacologia*, **31**, 259–95.

Hendriks, W., Leunissen, J., Nevo, E. *et al.* (1987) The lens protein αA-crystallin of the blind mole rat, *Spalax ehrenbergi*: evolutionary change and functional constraints. *Proc. Natl. Acad. Sci. USA*, **84**, 5320–4.

Hennemann, W.W. (1983) Relationship among body mass, metabolic rate and the intrinsic rate of natural increase in mammals. *Oecologia*, **56**, 104–8.

Herreid, C.F. II. (1964) Bat longevity and metabolic rate. *Expt. Geront.*, **1**, 1–9.

Heslop-Harrison, J.S. (1990) Gene expression and parental dominance in hybrid plants. *Development*, **1990** (Suppl.), 21–8.

Heusner, A.A. (1985) Body size and energy metabolism. *Ann. Rev. Nutr.*, **5**, 267–93.

Hickey, D.A., Bally-Cuif, L., Abukashawa, S. *et al.* (1991) Concerted evolution of duplicated protein-coding genes in *Drosophila*. *Proc. Natl. Acad. Sci. USA* **88**, 1611–15.

Hickey, D.A. and Rose, M.R. (1988) The role of gene transfer in the evolution of eukaryotic sex, in *The Evolution of Sex* (eds R.E. Michod and B.R. Levin), Sinauer Publ., Sunderland, MA, pp. 161–75.

Hickman, J.C. (1975) Environmental unpredictability and plastic energy allocation strategies in the annual *Polygonun cascadense* (Polygonaceae). *J. Ecol.*, **63**, 689–701.

Hill, W.G. and Robertson, A. (1966) The effect of linkage on limits to artificial selection. *Genet. Res.* **8**, 269–94.

Hillesheim, E. and Stearns, S.C. (1991) The responses of *Drosophila melanogaster* to artificial selection on body weight and its phenotypic plasticity in two larval food environments. *Evolution*, **45**, 1909–23.

Hillesheim, E. and Stearns, S.C. (1992) Correlated responses in life-history traits to artificial selection for body weight in *Drosophila melanogaster*. *Evolution*, **46**, 745–52.

Hillis, D.M. and Bull, J.J. (1991) Of genes and genomes. *Science*, **254**, 528.

Hillis, D.M., Dixon, M.T. and Jones, A.L. (1991) Minimal genetic variation in a morphologically diverse species (Florida tree snail, *Liguus fasciatus*). *J. Heredity*, **82**, 282–6.

Hinds, D.S. and MacMillan, R.E. (1985) Scaling energy metabolism and evaporative water loss in heteromyid rodents. *Physiol. Zool.*, **58**, 282–98.

Hiraizumi, Y. (1961) Negative correlation between rate of development and female fertility in *Drosophila melanogaster*. *Genetics*, **46**, 615–24.

Hiromi, Y. and Gehring, W.J. (1987) Regulation and function of the *Drosophila* segmentation gene *fushi tarazu*. *Cell*, **50**, 963–74.

Hirsch, H.R. and Witten, M. (1991) The waste-product theory of aging: simulation of metabolic waste production. *Expt. Geront.*, **26**, 549–67.

Ho, M.W., Tucker, C., Keeley, D. *et al.* (1983) Effects of successive generations of ether treatment on penetrance and expression of the *Bithorax* phenocopy in *Drosophila melanogaster*. *J. Expt. Zool.*, **225**, 357–68.

Hochachka, P.W. and Guppy, M. (1987) *Metabolic Arrest and the Control of Biological Time*, Harvard University Press, Cambridge.

Hoffman, M. (1991a) How parents make their mark on genes. *Science*, **252**, 1250–1.

Hoffman, M. (1991b) An RNA first: its part of the gene-copying machinery. *Science*, **252**, 506–7.

Hoffman, R.J. (1978) Environmental uncertainty and evolution of physiological adaptation in *Colias* butterflies. *Amer. Nat.*, **112**, 999–1015.

Hoffmann, A.A. and Parsons, P.A. (1989) An integrated approach to environmental stress tolerance and life-history variation. Desiccation tolerance in *Drosophila*. *Biol. J. Linn. Soc.*, **37**, 117–36.

Hoffmann, A.A. and Parsons, P.A. (1991) *Evolutionary Genetics and Environmental Stress*, Oxford University Press, Oxford.

Hoffmann, F.M. (1991) *Drosophila abl* and genetic redundancy in signal transduction. *Trends. Genet.*, **7**, 351–5.

Holehan, A.M. and Merry, B.J. (1986) The experimental manipulation of ageing by diet. *Biol Rev.*, **61**, 329–68.

Holland, P.W.H. (1990) Homeobox genes and segmentations: co-option, co-evolution, and convergence. *Devel. Biol.*, **1**, 135–45.

Hölldobler, B. and Wilson, E.O. (1990) *The Ants*, Belknap Press, Harvard University Press, Cambridge, MA.

Holliday, R. (1984) The biological significance of meiosis, in *Controlling Events in*

Meiosis (eds C.E. Evans and H.G. Dickinson), Cambridge University Press, Cambridge, pp. 381–94.

Holliday, R. (1988) A possible role for meiotic recombination in germ line reprogramming and maintenance, in *The Evolution of Sex* (eds R.E. Michod and B.R. Levin), Sinauer Associates, Sunderland, MA, pp. 45–55.

Holliday, R. (1989a) A different kind of inheritance. *Sci. Amer.*, **260**, 60–73.

Holliday, R. (1989b) Food, reproduction and longevity: is the extended lifespan of calorie-restricted animals an evolutionary adaptation? *BioEssays*, **10**, 125–7.

Holliday, R. (1990) Mechanisms for the control of gene activity during development. *Biol. Rev.*, **65**, 431–71.

Hollis, G.F., Hieter, P.A., McBride, O.W. *et al.* (1982) Processed genes: a dispersed human immunoglobin gene bearing evidence of RNA-type processing. *Nature*, **296**, 321–5.

Holloszy, J.O. (1988) Exercise and longevity: studies on rats. *J. Geront.*, **43**, B149–B151.

Holloszy, J.O. and Smith, E.K. (1986) Longevity of cold-exposed rats: a re-evaluation of the 'rate-of-living-theory'. *J. Appl. Physiol.*, **61**, 1656–60.

Holloszy, J.O., Smith, E.K., Vining, M. and Adams, S. (1985) Effect of voluntary exercise on longevity of rats. *J. Appl. Physiol.*, **59**, 826–31.

Holtzman, E. (1992) Intracellular targeting and sorting: how are macromolecules delivered to specific locations. *BioScience*, **42**, 608–20.

Hoshijima, K., Inoue, K., Higuchi, I. *et al.* (1991) Control of doublesex alternative splicing by transformer and transformer-2 in *Drosophila*. *Science*, **252**, 833–6.

Houck, M.A., Clark, J.B., Peterson, K.R. *et al.* (1991) Possible horizontal transfer of *Drosophila* genes by the mite *Proctolaelaps regalis*. *Science*, **253**, 1125–9.

Houle, D. (1991) Genetic covariance of fitness correlates: what genetic correlations are made of and why it matters. *Evolution*, **45**, 630–48.

Howard, K., Ingham, P. and Rushlow, C. (1988) Region-specific alleles of the *Drosophila* segmentation gene *hairy*. *Genes Dev.*, **2**, 1037–46.

Huey, R.B. and Slatkin, M. (1976) Cost and benefits of lizard thermoregulation. *Quart. Rev. Biol.*, **51**, 363–84.

Hull, D.L. (1976) Are species really individuals? *Syst. Zool.*, **25**, 174–91.

Hull, D.L. (1978) A matter of individuality. *Philos. Sci.*, **45**, 335–60.

Hull, D.L. (1980) Individuality and selection. *Ann. Rev. Ecol. Syst.*, **11**, 311–32.

Hull, D.L. (1981) Units of evolution: a metaphysical essay, in *The Philosophy of Evolution* (eds U.L. Jensen and R. Harré), Harvester Press, Brighton, pp. 23–44.

Hull, D.L. (1988) Progress in ideas of progress, in *Evolutionary Progress* (ed M.H. Nitecki), Chicago University Press, Chicago, IL, pp. 27–48.

Hunding, A., Kauffman, S.A. and Goodwin, B.C. (1990) *Drosophila* segmentation: supercomputer simulation of pre-pattern hierarchy. *J. Theor. Biol.*, **145**, 369–84.

Hunkapiller, T., Huang, H., Hood, L. *et al.* (1982) The impact of modern genetics on evolutionary theory, in *Perspectives on Evolution* (ed R. Milkman), Sinauer Assoc., Sunderland, MA, pp. 164–89.

Hunt, R.H. (1980) Toad sanctuary in a tarantula burrow. *Nat. Hist.*, **89**, 48–53.

Hurst, L.D., Hamilton, W.D. and Ladle, R.J. (1992) Covert sex. *TREE*, **7**, 144–5.

Hurst, L.D. and Nurse, P. (1991) A note on the evolution of meiosis. *J. Theor. Biol.*, **150**, 561–3.

Huston, M., DeAngelis, D. and Post, W. (1988) New computer models unify ecological theory. *BioScience*, **38**, 682–91.

Hutchinson, G.E. (1959) Homage to Santa Rosalia, or why are there so many kinds of animals? *Amer. Nat.*, **43**, 146–59.

Huxley, J.S. (1932) *Problems of Relative Growth*, Dover, NY.

Huxley, J.S. (1942) *Evolution: The Modern Synthesis*, George Allen and Unwin, London.

Immerglück, K., Lawrence, P.A. and Bienz, M. (1990) Induction across germ layers in *Drosophila* mediated by a genetic cascade. *Cell*, **62**, 261–8.

Ingham, P.W. (1988) The molecular genetics of embryonic pattern formation in *Drosophila*. *Nature*, **335**, 25–34.

Ingham, P.W. and Gergen, P. (1988) Interactions between the pair-rule genes *runt*, *hairy, even-skipped* and *fushi tarazu* and the establishment of periodic pattern in the *Drosophila* embryo. *Development*, **104** (Suppl.), 51–60.

Ingham, P.W., Baker, N.E. and Martinez-Arias, A. (1988) Regulation of segment polarity genes in the *Drosophila* blastoderm by *fushi tarazu* and *even-skipped*. *Nature*, **333**, 73–75.

Ingram, D.K. and Reynolds, M.A. (1987) The relationship of body weight to longevity within laboratory rodent species, in *Evolution of Longevity in Animals* (eds A.D. Woodhead and K.H. Thompson), Plenum Press, NY, pp. 247–82.

Jablonka, E. and Lamb, M.J. (1989) The inheritance of acquired epigenetic variations. *J. Theor. Biol.*, **139**, 69–83.

Jacob, F. and Monad, J. (1961a) Genetic regulatory mechanisms in the synthesis of proteins. *J. Mol. Biol.* **3**, 318–56.

Jacob, F. and Monad, J. (1961b) On the regulation of gene activity. *Cold Spring. Harb. Symp. Quant. Biol.*, **26**, 193–210.

Jaenike, J. (1978) An hypothesis to account for the maintenance of sex within populations. *Evol. Theor.* **3**, 191–4.

Jain, S. (1979) Adaptive strategies: polymorphism, plasticity, and homeostasis, in *Topics in Plant Population Biology* (eds O.T. Solbrig, S. Jain, G.B. Johnson and P.H. Raven), Columbia University Press, pp. 160–87.

Janzen, D.H. (1977) What are dandelions and aphids? *Amer. Nat.*, **111**, 586–9.

Janzen, D.H. (1985) On ecological fitting. *Oikos*, **45**, 308–10.

Jarvis, J.U.M. and Bennett, N.C. (1991) Ecology and behavior of the family Bathyergidae, in *The Biology of the Naked Mole-rat* (eds P.W. Sherman, J.U.M. Jarvis and R.D. Alexander), Princeton University Press, Princeton, pp. 66–96.

Jeffreys, A.J. (1982) Evolution of globin genes, in *Genome Evolution* (eds G.A. Dover and R.B. Flavell), Academic Press, NY, pp. 157–76.

Jeffreys, A.J., Harris, S., Barrie, P.A. *et al.* (1983) Evolution of gene families: the globin genes, in *Evolution from Molecules to Men* (ed D.S. Bendall), Cambridge University Press, Cambridge, pp. 175–95.

John, B. and Miklos, G.L.G. (1988) *The Eukaryote Genome in Development and Evolution*, Allen and Unwin, London.

Johnson, T.E., Friedman, D.B., Fitzpatrick, P.A. and Conley, W.L. (1987) Mutant genes that extend life span, in *Evolution of Longevity in Animals* (eds A.D. Woodhead and K.H. Thompson), Plenum Press, NY, pp. 91–100.

Johnston, T.D. and Gottlieb, G. (1990) Neophenogenesis: a developmental theory of phenotypic evolution. *J. Theor. Biol.*, **147**, 471–95.

Jones, J.S. (1990) Living fast and dying young. *Nature*, **348**, 288–9.

Jones, J.S., Leith, B.H. and Rawlings, P. (1977) Polymorphism in *Cepaea*: a problem with too many solutions? *Ann. Rev. Ecol. Syst.*, **8**, 109–43.

Jurgens, K.D. and Prothero, J. (1987) Scaling of maximal lifespan in bats. *Comp. Biochem. Physiol.*, **88A**, 361–7.

Kafatos, F.C., Mitsialis, S.A., Spoerel, N. *et al.* (1985) Studies on the developmentally regulated expression and amplification of insect chorion genes. *Cold Spring Harbor Symp. Quant. Biol.*, **50**, 537–47.

Kaitala, A. (1987) Dynamic life history strategy of the waterstrider *Gerris thoracicus* as an adaptation to food and habitat variation. *Oikos*, **48**, 125–31.

Kaitala, A. (1991) Phenotypic plasticity in reproductive behaviour of water-striders: trade-offs between reproduction and longevity during food stress. *Funct. Ecol.*, **5**, 12–18.

Kajiura, L. and Rollo, C.D. (1994a) A mass budget for Supermice engineered with extra growth hormone genes: evidence for energetic limitations. *Can. J. Zool.* In press.

Kajiura, L. and Rollo, C.D. (1994b) Canalization of development under dietary stress in the American cockroach, *Periplaneta americana*, and its evolutionary implications. In preparation.

Kaplan, R.H. and Cooper, W.S. (1984) The evolution of developmental plasticity in reproductive characteristics: an application of the 'adaptive coin-flipping' principle. *Amer. Nat.*, **123**, 393–410.

Karasov, W.H. (1986) Energetics, physiology and vertebrate ecology. *TREE*, **1**, 101–4.

Karasov, W.H., Phan, D., Diamond, J. and Carpenter, F.L. (1986) Food passage and intestinal nutrient absorption in hummingbirds. *Auk*, **103**, 453–64.

Karban, R. (1989) Fine-scale adaptation of herbivorous thrips to individual host plants. *Nature*, **340**, 60–1.

Karban, R. and Myers, J.H. (1989) Induced plant responses to herbivory. *Ann. Rev. Ecol. Syst.*, **20**, 331–48.

Karlsson, B. and Wickman, P.O. (1989) The cost of prolonged life: an experiment on a nymphalid butterfly. *Funct. Ecol.*, **3**, 399–405.

Katz, M.J. (1982) Ontogenetic mechanisms: the middle ground of evolution, in *Evolution and Development* (ed J.T. Bonner), Springer Verlag, Berlin, pp. 207–12.

Kauffman, S.A. (1983) Developmental constraints: internal factors in evolution, in *Development and Evolution* (eds B.C. Goodwin, N. Holder and C.C. Wylie), Cambridge University Press, pp. 195–225.

Kauffman, S.A. (1985) New questions in genetics and evolution. *Cladistics* **1**, 247–65.

Kauffman, S.A. (1993) *The Origins of Order*, Oxford University Press, Oxford.

Kauffman, S.A. and Goodwin, B.C. (1990) Spatial harmonics and pattern specification in early *Drosophila* development. Part II. The four colour wheels model. *J. Theor. Biol.*, **144**, 321–45.

Keller, E.F. (1987) Reproduction and the central project of evolutionary theory. *Biol. Phil.*, **2**, 383–96.

Kelly, S.E. (1989a) Experimental studies of the evolutionary significance of sexual reproduction V. A field test of the sib-competition lottery hypothesis. *Evolution*, **43**, 1054–65.

Kelly, S.E. (1989b) Experimental studies of the evolutionary significance of sexual reproduction. VI. A greenhouse test of the sib-competition hypothesis. *Evolution*, **43**, 1066–74.

Kelly, S.E., Antonovics, J. and Schmitt, T. (1988) A test of the short-term advantage of sexual reproduction. *Nature*, **331**, 714–16.

Kendeigh, S.C. (1972) Energy control of size limits in birds. *Amer. Nat.*, **106**, 79–88.

Kendrick, B. (1991) Fungal symbioses and evolutionary innovations, in *Symbiosis as a Source of Evolutionary Innovation* (eds L. Margulis and R. Fester), MIT Press, Cambridge, MA, pp. 249–61.

Kennison, J.A. (1993) Transcriptional activation of *Drosophila* homeotic genes from distant regulatory sites. *Trends Genet.*, **9**, 75–9.

Kenyon, C. and Wang, B. (1990) A cluster of *Antennapedia*-class homeobox genes in a non-segmented animal. *Science*, **253**, 516–17.

Kerszberg, M. (1989) Developmental canalization can enhance species survival. *J. Theor. Biol.*, **139**, 287–309.

Ketterson, E.D. and Nolan Jr., V. (1992) Hormones and life histories: an integrative approach. *Amer. Nat.*, **140** (Suppl.), S33–S62.

Kim, J.D., McCarter, R. and Yu, B.P. (1993) Effects of exercise, dietary restriction and aging on oxygen free radical formation and membrane fluidity. FASEB, **7**, A81.

Kimura, M. (1983) *The Neutral Theory of Molecular Evolution*, Cambridge University Press, Cambridge.

Kimura, M. (1991) Recent development of the neutral theory viewed from the Wrightian tradition of theoretical population genetics. *Proc. Natl. Acad. Sci. USA*, **88**, 5969–73.

King, C.E. (1982) The evolution of life span, in *Evolution and Genetics of Life Histories* (eds H. Dingle and J.P. Hegmann), Springer Verlag, Berlin, pp. 121–38.

King, M.–C. and Wilson, A.C. 1975. Evolution at two levels in humans and chimpanzees. *Science*, **188**, 107–16.

Kingsolver, J.G. (1985) Butterfly engineering. *Sci. Amer.*, **253**, 106–13.

Kingsolver, J.G. and Wiernasz, D.C. (1987) Dissecting correlated characters: adaptive aspects of phenotypic covariation in melanization pattern of *Pieris* butterflies. *Evolution*, **41**, 491–503.

Kingsolver, J.G. and Wiernasz, D.C. (1991) Development, function, and the quantitative genetics of wing melanin pattern in *Pieris* butterflies. *Evolution*, **45**, 1480–92.

Kirkpatrick, M. and Jenkins, C.D. (1989) Genetic segregation and the maintenance of sexual reproduction. *Nature*, **339**, 300–1.

Kirkpatrick, M. and Lande, R. (1989) The evolution of maternal characters. *Evolution*, **43**, 485–503.

Kirkwood, J.K. (1983) A limit to metabolisable energy intake in mammals and birds. *Comp. Biochem. Physiol.*, **75A**, 1–3.

Kirkwood, T.B.L. (1977) Evolution of ageing. *Nature*, **270**, 301–4.

Kirkwood, T.B.L. (1987) Immortality of the germ-line versus disposability of the soma, in *Evolution of Longevity in Animals* (eds A.D. Woodhead and K.H. Thompson), Plenum Press, NY, pp. 209–18.

Kirkwood, T.B.L. (1990) The disposable soma theory of aging, in *Genetic Effects on Aging II* (ed D.E. Harrison), Telford Press, Caldwell, NJ, pp. 9–19.

Kirkwood, T.B.L. and Cremer, T. (1982) Cytogerontology since 1881: a reappraisal of August Weismann and a review of modern progress. *Hum. Genet.*, **60**, 101–21.

Kirkwood, T.B.L. and Holliday, F.R.S. (1979) The evolution of ageing and longevity. *Proc. R. Soc. Lond.*, **B205**, 531–46.

Klar, A.J.S. (1990) Regulation of fission yeast mating-type interconversion by chromosome imprinting. *Development*, Suppl. 1990, 3–8.

Kleiber, M. (1961) *The Fire of Life*, Wiley, NY.

Klekowski Jr., E.J. and Godfrey, P.J. (1989) Ageing and mutation in plants. *Nature*, **340**, 389–91.

Koestler, A. (1967) *The Ghost in the Machine*, Macmillan, NY.

Koestler, A. (1978) *Janus: a Summing Up*, Vintage Books, NY.

Kolasa, J. and Pickett, S.T.A. (1989) Ecological systems and the concept of biological organization. *Proc. Natl. Acad. Sci. USA*, **86**, 8837–41.

Kolasa, J. and Rollo, C.D. (1991) Introduction: the heterogeneity of heterogeneity, a glossary, in *Ecological Heterogeneity* (eds J. Kolasa and S.T.A. Pickett), Springer Verlag, Berlin, pp. 1–23.

Kollar, E.J. and Fisher, C. (1980) Tooth induction in chick epithelium: expression of quiescent genes for enamel synthesis. *Science*, **207**, 993–5.

Kondrashov, A.S. (1988) Deleterious mutations and the evolution of sexual reproduction. *Nature*, **336**, 435–40.

Kondrashov, A.S. and Crow, J.F. (1991) Haploidy or diploidy: which is better? *Nature*, **351**, 314–15.

Koopman, P., Gubbay, J., Vivian, N. *et al.* (1991) Male development of chromosomally female mice transgenic for *Sry. Nature*, **351**, 117–21.

Kosuda, K. (1985) The aging effect on male mating activity in *Drosophila melanogaster. Behav. Genet.*, **15**, 297–303.

Krebs, J.R. and Houston, A.I. (1989) Optimization in ecology, in (ed J.M. Cherrett), *Ecological Concepts*, Blackwell Scientific, Oxford, pp. 309–38.

Kubli, E. (1986) Molecular mehcanisms of suppression in *Drosophila. Trends Genet.*, **2**, 204–8.

Kuo, C.J., Conley, P.B., Chen, L. *et al.* (1992) A transcriptional hierarchy involved in mammalian cell-type specification. *Nature*, **355**, 457–61.

Kurten, B. (1963) Return of a lost structure in the evolution of felid dentition. *Soc. Sci. Fenn.*, **26**, 4–12.

Lacey, E.P., Real, L., Antonovics, J. and Heckel, D.G. (1983) Variance models in the study of life histories. *Amer. Nat.* **122**, 114–31.

Lachmansingh, E. and Rollo, C.D. (1994) Evidence for a trade off between growth and behavioural activity in giant 'Supermice' genetically engineered with extra growth hormone genes. *Can. J. Zool.* In press.

Ladle, R.T. (1992) Parasites and sex, catching the red queen. *TREE*, **7**, 405–8.

Lanciani, C.A., Giesel, J.T., Anderson, J.F. and Emerson, S.S. (1990) Photoperiod-induced changes in metabolic response to temperature in *Drosophila melanogaster* Meigen. *Funct. Ecol.*, **4**, 41–5.

Lande, R. (1978) Evolutionary mechanisms of limb loss in tetrapods. *Evolution*, **32**, 73–92.

Lande, R. (1979) Quantitative genetic analysis of multivariate evolution, applied to brain body size allometry. *Evolution*, **33**, 402–16.

Lande, R. (1980) The genetic covariance between characters maintained by pleiotropic mutations. *Genetics*, **94**, 203–15.

Lande, R. (1981) The minimum number of genes contributing to quantitative variation between and within populations. *Genetics*, **99**, 541–53.

Lande, R. (1982) A quantitative genetic theory of life history evolution. *Ecology*, **63**, 607–15.

Lande, R. (1986) The dynamics of peak shifts and the pattern of morphological evolution. *Paleobiology*, **12**, 343–54.

Lande, R. and Arnold, S.T. (1983) The measurement of selection on correlated characters. *Evolution*, **37**, 1210–26.

Latchman, D. (1990) *Gene Regulation: a Eukaryotic Perspective*, Allen and Unwin, London.

Lauder, G.V. and Liem, K.F. (1989) The role of historical factors in the evolution of complex organismal functions, in *Complex Organismal Functions: Integration and Evolution in Vertebrates* (eds D.B. Wake and G. Roth), J. Wiley and Sons, pp. 63–78.

Law, R. (1979) The cost of reproduction in annual meadow grass. *Amer. Nat.*, **113**, 3–16.

Law, R. (1988) Some ecological properties of intimate mutualisms involving plants, in *Plant Population Ecology* (eds A.J. Davy, M.J. Hutchings and A.R. Watkinson), Blackwell Scientific, Oxford, pp. 315–41.

Law, R. and Lewis, D.H. (1983) Biotic environments and the maintenance of sex – some evidence from mutualistic symbioses. *Biol. J. Linn. Soc.* **20**, 249–76.

Lawlor, L.R. (1976) Molting, growth and reproductive strategies in the terrestrial isopod, *Armadillidium vulgare. Ecology*, **57**, 1179–94.

Lawrence, P.A. (1992) *The Making of a Fly: the Genetics of Animal Design*, Blackwell Scientific, London.

Lawton, J.H. (1992) Feeble links in food webs. *Nature*, **355**, 19–20.

LeBourg, E. (1987) The rate of living theory. Spontaneous locomotor activity, aging and longevity in *Drosophila melanogaster*. *Exp. Geront.*, **22**, 359–69.

LeBourg, E., Lints, F.A., Delince, J. and Lints, C.V. (1988) Reproductive fitness and longevity in *Drosophila melanogaster*. *Expt. Geront.*, **23**, 491–500.

Lechowicz, M.J. and Bell, G. (1991) The ecology and genetics of fitness in forest plants. II. Microspatial heterogeneity of the edaphic environment. *J. Ecol.*, **79**, 687–96.

Leigh Jr., E.G. (1977) How does selection reconcile individual advantage with the good of the group? *Proc. Natl. Acad. Sci. USA*, **74**, 4542–6.

Lerner, I.M. (1954) *Genetic Homeostasis*, Oliver and Boyd, London.

Lerner, S.A., Wu, T.T. and Lin, E.C.C. (1964) Evolution of a catabolic pathway in bacteria. *Science*, **146**, 1313–15.

Lessells, C.M. (1991) The evolution of life histories, in *Behavioural Ecology: an Evolutionary Approach* (eds J.R. Krebs and N.B. Davies), Blackwell Scientific, London, pp. 32–65.

Letourneau, D.K. (1990) Code of ant–plant mutualism broken by parasite. *Science*, **248**, 215–17.

Levin, D.A. (1970) Developmental instability and evolution in peripheral isolates. *Amer. Nat.*, **104**, 343–53.

Levin, D.A. (1975) Pest pressure and recombination systems in plants. *Amer. Nat.*, **109**, 437–51.

Levin, D.A. (1983) Polyploidy and novelty in flowering plants. *Amer. Nat.*, **122**, 1–25.

Levin, D.A. (1988) Plasticity, canalization and evolutionary stasis in plants, in *Plant Population Ecology* (eds A.J. Davey, M.J. Hutchings and A.R. Watkinson), Blackwell, Oxford, pp. 35–45.

Levins, R. (1963) Theory of fitness in a heterogeneous environment. Developmental flexibility and niche selection. *Amer. Nat.* **97**, 75–90.

Levins, R. (1968) *Evolution in Changing Environments*, Princeton University Press, NJ.

Levinton, J.S. (1988) *Genetics, Paleontology and Macroevolution*, Cambridge University Press, Cambridge.

Lewin, R. (1987) The surprising genetics of bottlenecked flies. *Science*, **235**, 1325–7.

Lewis, D.H. (1991) Mutualistic symbioses in the origin and evolution of land plants, in *Symbiosis as a source of evolutionary innovation* (eds L. Margulis and R. Fester), MIT Press, Cambridge, MA, pp. 288–300.

Lewontin, R.C. (1957) The adaptations of populations to varying environments. *Cold Spring Harbor Symp. Quant. Biol.*, **22**, 395–408.

Lewontin, R.C. (1970) The units of selection. *Ann. Rev. Ecol. Syst.*, **1**, 1–18.

Lewontin, R.C. (1974) *The Genetic Basis of Evolutionary Change*, Columbia University Press.

Li, W.-H. (1984) Retention of cryptic genes in microbial populations. *Mol. Biol. Evol.*, **1**, 213–19.

Li, W.-H. and Tanimura, M. (1987) The molecular clock runs more slowly in man than in apes and monkeys. *Nature*, **326**, 93–6.

Liem, K.F. (1990) Key evolutionary innovations, differential diversity, and symecomorphosis, in *Evolutionary Innovations* (ed M.H. Nitecki), University of Chicago Press, Chicago, IL, pp. 147–70.

Liem, K.F. and Kaufman, L.S. (1984) Intraspecific macroevolution: functional biology of the polymorphic cichlid species *Cichlasoma minckleyi*, in *Evolution of Fish Species Flocks* (eds A.A. Echelle and I. Kornfield), University Maine, Orono, ME, pp. 203–15.

Lindahl, K.F. (1991) His and hers recombinational hotspots. *Trends Genet.*, **7**, 273–6.

Lindstedt, S.L. and Calder III, W.A. (1981) Body size, physiological time, and longevity of homeothermic animals. *Quart. Rev. Biol.*, **56**, 1–16.

Lindstedt, S.L. and Swain, S.D. (1988) Body size as a constraint of design and function, in *Evolution of Life Histories: Pattern and Theory from Mammals* (ed M.S. Boyce), Yale University Press, New Haven, CT, pp. 93–106.

Lints, F.A. (1989) The rate of living theory revisited. *Gerontology*, **35**, 36–57.

Lints, F.A. and Gruwez, G. (1972) What determines the duration of development in *Drosophila melanogaster*? *Mech. Age. Devel.*, **1**, 285–97.

Lints, F.A. and Hoste, C. (1974) The Lansing effect revisited – I. Life-span. *Expt. Geront.* **9**, 51–69.

Lints, F.A., LeBourg, E. and Lints, C.V. (1984) Spontaneous locomotor activity and life span. A test of the rate of living theory in *Drosophila melanogaster*. *Gerontology*, **30**, 376–87.

Lints, F.A. and Lints, C.V. (1971) Influence of preimaginal environment on fecundity and ageing in *Drosophila melanogaster* hybrids – III. Developmental speed and life span. *Expt. Geront.*, **6**, 427–45.

Lints, F.A., Stoll, J., Gruwez, G. and Lints, C.V. (1979) An attempt to select for increased longevity in *Drosophila melanogaster*. *Gerontology*, **25**, 192–204.

Lively, C.M. (1986a) Predator-induced shell dimorphism in the acorn barnacle *Chthamalus anisopoma*. *Evolution*, **40**, 232–42.

Lively, C.M. (1986b) Canalization versus developmental conversion in a spatially variable environment. *Amer. Nat.*, **128**, 561–72.

Lively, C.M. (1992) Parthenogenesis in a freshwater snail, reproductive assurance versus parasitic release. *Evolution*, **46**, 907–13.

Lively, C.M., Craddock, C. and Vrijenhoek, R.C. (1990) Red queen hypothesis supported by parasitism in sexual and clonal fish. *Nature*, **344**, 864–6.

Lloyd, D.G. (1984) Variation strategies of plants in heterogeneous environments. *Biol. J. Linn. Soc.*, **21**, 357–85.

Loomis, W.F. (1988) *Four Billion Years: an Essay on the Evolution of Genes and Organisms*, Sinauer Associates, Sunderland, MA.

Lovegrove, B.G. (1986) The metabolism of social subterranean rodents: adaptation to aridity. *Oecologia*, **69**, 551–5.

Lovegrove, B.G. and Wissel, C. (1988) Sociality in mole-rats: metabolic scaling and the role of risk sensitivity. *Oecologia*, **74**, 600–6.

Lovelock, J.E. (1979) *Gaia: a New Look at Life on Earth*, Oxford University Press, Oxford.

Lovelock, J.E. (1988) *The Ages of Gaia*, W. Norton, NY.

Lovtrup, S. (1974) *Epigenetics: a Treatise on Theoretical Biology*, J. Wiley and Sons, London.

Lovtrup, S. (1987) *Darwinism: the Refutation of a Myth*, Croom Helm, London.

Luckinbill, L.S., Arking, R., Clare, M.J. *et al.* (1984) Selection for delayed senescence in *Drosophila melanogaster*. *Evolution*, **38**, 996–1003.

Luckinbill, L.S. and Clare, M.J. (1985) Selection for life span in *Drosophila melanogaster*. *Heredity*, **55**, 9–18.

Luckinbill, L.S. and Clare, M.J. (1986) A density threshold for the expression of longevity in *Drosophila melanogaster*. *Heredity*, **56**, 329–35.

Luckinbill, L.S., Graves, J.L., Tomkiw, A. and Sowirka, O. (1988) A qualitative analysis of some life-history correlates of longevity in *Drosophila melanogaster*. *Evol. Ecol.*, **2**, 85–94.

Luckinbill, L.S., Grudzien, T.A., Rhine, S. and Weisman, G. (1989) The genetic basis of adaptation to selection for longevity in *Drosophila melanogaster*. *Evol. Ecol.*, **3**, 31–9.

Lyman, C.P., O'Brien, R.C., Greene, G.C. and Papafrangos, E.D. (1981) Hibernation and longevity in the Turkish hamster *Mesocricetus brandti*. *Science*, **212**, 668–70.

Lynch, M. (1984) Destabilizing hybridization, general-purpose genotypes and geographic parthenogenesis. *Quart. Rev. Biol.*, **59**, 257–90.

Lynch, M. and Gabriel, W. (1983) Phenotypic evolution and parthenogenesis. *Amer. Nat.*, **122**, 745–64.

Lynch, M. and Gabriel, W. (1990) Mutation load and the survival of small populations. *Evolution*, **44**, 1725–37.

Lynch, M., Spitze, K. and Crease, T. (1989) The distribution of life-history variation in the *Daphnia pulex* complex. *Evolution*, **43**, 1724–36.

MacArthur, J.W. (1949) Selection for small and large body size in the house mouse. *Genetics*, **34**, 194–209.

MacBeth, N. (1980) Reflections on irreversibility. *Syst. Zool.*, **29**, 402–4.

MacDonald, S.E. and Chinnappa, C.C. (1989) Population differentiation for phenotypic plasticity in the *Stellaria longipes* complex. *Amer. J. Bot.*, **76**, 1627–37.

MacDonald, S.E., Chinnappa, C.C. and Reid, D.M. (1988) Evolution of phenotypic plasticity in the *Stellaria longipes* complex: comparisons among cytotypes and habitats. *Evolution*, **42**, 1036–46.

MacKay, W.J. and Bewley, G.C. (1989) The genetics of catalase in *Drosophila melanogaster*: isolation and characterization of acatalasemic mutants. *Genetics*, **122**, 643–652.

MacMahon, J.A., Phillips, D.L., Robinson, J.V. and Schimpf, D.J. (1978) Levels of biological organization: an organism-centered approach. *BioScience* **28**, 700–4.

Malik, R.C. (1984) Genetic and physiological aspects of growth, body composition and feed efficiency in mice: a review. *J. Anim. Sci.*, **58**, 577–90.

Manning, J.T. (1976) Is sex maintained to facilitate or minimise mutational advance? *Heredity*, **36**, 351–7.

Marden, J.H. (1989) Bodybuilding dragonflies: costs and benefits of maximizing flight muscle. *Physiol. Zool.*, **62**, 505–21.

Margulis, L. (1971) Symbiosis and evolution. *Sci. Amer.*, **225**, 48–57.

Margulis, L. (1981) *Symbiosis in Cell Evolution*, W.H. Freeman and Company, San Francisco, CA.

Margulis, L. and Sagan, D. (1984) Evolutionary origins of sex. *Oxford Surv. Evol. Biol.*, **1**, 16–47.

Margulis, L. and Sagan, D. (1986) *Origins of Sex: Three Billion Years of Genetic Recombination*, Yale University Press.

Margulis, L. and Sagan, D. (1988) Sex: the cannibalistic legacy of primordial androgynes, in *The Evolution of Sex* (eds R. Bellig and G. Stevens), Harper and Row, London, pp. 23–40.

Marshall, D.L., Levin, D.A. and Fowler, N.L. (1986) Plasticity of yield components in response to stress in *Sesbania macrocarpa* and *Sesbania vesicaria* (Leguminosae). *Amer. Nat.*, **127**, 508–21.

Martinez-Arias, A., Baker, N.E. and Ingham, P.W. (1988) Role of segment polarity genes in the definition and maintenance of cell states in the *Drosophila* embryo. *Development*, **103**, 157–70.

Masaki, S. (1980) Summer diapause. *Ann. Rev. Entomol.*, **25**, 1–25.

Masman, D., Dijkstra, C., Daan, S. and Bult, A. (1989) Energetic limitation of avian parental effort: field experiments in the kestrel (*Falco tinnunculus*). *J. Evol. Biol.*, **2**, 435–56.

Masoro, E.J. (1988) Food restriction in rodents: an evaluation of its role in the study of aging. *J. Gerontol.*, **43**, B59–B64.

Masoro, E.J., Katz, M.S. and McMahan, C.A. (1989) Evidence for the glycation hypothesis of aging from the food-restricted rodent model. *J. Gerontol.*, **44**, B20–B22.

Masoro, E.J., Yu, B.P. and Bertrand, H.A. (1982) Action of food restriction in delaying the aging process. *Proc. Natl. Acad. Sci. USA*, **79**, 4239–41.

Matsuda, R. (1987) *Animal Evolution in Changing Environments with Special Reference to Abnormal Metamorphosis*, Wiley InterScience, NY.

May, R.M. (1974) *Stability and Complexity in Model Ecosystems*, Princeton University Press, Princeton, NJ.

May, R.M. (1989) Levels of organization in ecology, in *Ecological Concepts* (ed J.M. Cherrett), Blackwell Scientific, Oxford, pp. 339–63.

Mayer, P.J. and Baker III, G.T. (1984) Developmental time and adult longevity in two strains of *Drosophila melanogaster* in a constant low-stress environment. *Mech. Age. Dev.*, **26**, 283–98.

Maynard Smith, J. (1958) The effects of temperature and of egg-laying on the longevity of *Drosophila subobscura*. *J. Expt. Biol.*, **35**, 832–42.

Maynard Smith, J. (1963) Temperature and the rate of ageing in poikilotherms. *Nature*, **199**, 400–2.

Maynard Smith, J. (1968) Evolution in sexual and asexual populations. *Amer. Nat.* **102**, 469–73.

Maynard Smith, J. (1971a) What use is sex? *J. Theor. Biol.*, **30**, 319–35.

Maynard Smith, J. (1971b) The origin and maintenance of sex, in *Group Selection* (ed G.C. Williams), Aldine, Atherton, Chicago, IL, pp. 163–75.

Maynard Smith, J. (1974) Recombination and the rate of evolution. *Genetics*, **78**, 299–304.

Maynard Smith, J. (1976) A short-term advantage for sex and recombination through sib-competition. *J. Theor. Biol.* **63**, 245–58.

Maynard Smith, J. (1978a) *The Evolution of Sex*, Cambridge University Press, London.

Maynard Smith, J. (1978b) Optimization theory in evolution. *Ann. Rev. Ecol. Syst.*, **9**, 31–56.

Maynard Smith, J. (1979) The effects of normalizing and disruptive selection on genes for recombination. *Genet. Res.*, **33**, 121–8.

Maynard Smith, J. (1980) Selection for recombination in a polygenic model. *Genet. Res.*, **35**, 269–77.

Maynard Smith, J. (1983) The genetics of stasis and punctuation. *Ann. Rev. Genet.*, **17**, 11–25.

Maynard Smith, J. (1984) The ecology of sex, in *Behavioural Ecology* (eds J.R. Krebs and N.B. Davies), Blackwell Scientific, Oxford, pp. 201–21.

Maynard Smith, J. (1988a) The evolution of recombination, in *The Evolution of Sex* (eds R.E. Michod and B.R. Levin), Sinauer Associates, Sunderland, MA, pp. 106–25.

Maynard Smith, J. (1988b) The evolution of sex, in *The Evolution of Sex* (eds R. Bellig and G. Stevens), Harper and Row, San Franscisco, CA, pp. 3–19.

Maynard Smith, J. (1988c) Selection for recombination in a polygenic model – the mechanism. *Genet. Res.*, **51**, 59–63.

Maynard Smith, J. (1989) *Evolutionary Genetics*, Oxford University Press, Oxford.

Maynard Smith, J. (1990) The evolution of prokaryotes: does sex matter? *Ann. Rev. Ecol. Syst.*, **21**, 1–12.

Maynard Smith, J. (1991) A Darwinian view of symbiosis, in *Symbiosis as a Source of Evolutionary Innovation* (eds L. Margulis and R. Fester), MIT Press, Cambridge, MA, pp. 26–39.

Maynard Smith, J., Burian, R., Kauffman, S. *et al.* (1985) Developmental constraints and evolution. *Quart. Rev. Biol.*, **60**, 265–87.

Maynard Smith, J., Dowson, C.G. and Spratt, B.G. (1991) Localized sex in bacteria. *Nature*, **349**, 29–31.

Maynard Smith, J. and Haigh, J. (1974) The hitch-hiking effect of a favourable gene. *Genet. Res.*, **23**, 23–35.

Mayr, E. (1942) *Systematics and the Origin of Species*, Columbia University Press, NY.

Mayr, E. (1954) Change of genetic environment and evolution, in *Evolution as a Process* (eds J. Huxley, A.C. Hardy and E.B. Ford), Allen and Unwin, London, pp. 157–80.

Mayr, E. (1955) Integration of genotypes: synthesis. *Cold Spring Harb. Symp. Quant. Biol.*, **20**, 327–33.

Mayr, E. (1961) Cause and effect in biology. *Science*, **134**, 1501–6.

Mayr, E. (1963) *Animal Species and Evolution*, Harvard University Press, Cambridge, MA.

Mayr, E. (1975) The unity of the genotype. *Biol. Zentralbatt*, **94**, 377–88.

Mayr. E. (1978 Evolution. *Sci. Amer.*, **239**, 46–55.

Mayr, E. (1982) Processes of speciation in animals, in *Mechanisms of Speciation* (ed C. Barigozzi), Allan R. Liss, NY, pp. 1–19.

Mayr, E. (1983) How to carry out the adaptationist program? *Amer. Nat.*, **121**, 324–34.

Mayr, E. (1992) Controversies in retrospect. *Oxford Surv. Evol. Biol.*, **8**, 1–34.

McAllister, M.K., Roitberg, B.D. and Weldon, K.L. (1990) Adaptive suicide in pea aphids: decisions are cost sensitive. *Anim. Behav.*, **40**, 167–75.

McArthur, M.C. and Sohal, R.S. (1982) Relationship between metabolic rate, aging, lipid peroxidation, and fluorescent age pigment in milkweed bug, *Oncopeltus fasciatus* (Hemiptera). *J. Geront.*, **37**, 268–74.

McCall, C., Mitchell-Olds, T. and Waller, D.M. (1989) Fitness consequences of outcrossing in *Impatiens capensis*: tests of the frequency-dependent and sib-competition models. *Evolution*, **43**, 1075–84.

McCarter, R., Masoro, E.J. and Yu, B.P. (1985) Does food restriction retard aging by reducing the metabolic rate? *Amer. J. Physiol.*, **248**, E488–E490.

McCarthy, J.C. (1980) Morphological and physiological effects of selection for growth rate in mice, in *Selection Experiments in Laboratory and Domestic Animals* (ed A. Robertson), Commonwealth Agricultural Bureaux, Wallingford, pp. 100–9.

McClintock, B. (1984) The significance of responses of the genome to challenge. *Science*, **226**, 792–801.

McDonald, J.F. (1990) Macroevolution and retroviral elements. *BioScience*, **40**, 183–191.

McDonald, J.F., Strand, D.J., Lambert, M.E. and Weinstein, I.B. (1987) The responsive genome: evidence and evolutionary implications, in *Development as an Evolutionary Process* (eds R.A. Raff and E.C. Raff), Allan R. Liss, pp. 239–63.

McFarland, D.J. (1977) Decision making in animals. *Nature*, **269**, 15–21.

McKinney, F.K., Broadhead, T.W. and Gibson, M.A. (1990) Coral–bryozoan mutualism: structural innovation and greater resource exploitation. *Science*, **248**, 466–98.

McKinney, M.L. and McNamara, K.J. (1991) *Heterochrony: the Evolution of Ontogeny*, Plenum Press, NY.

McNab, B.K. (1969) The economics of temperature regulation in neotropical bats. *Comp. Biochem. Physiol.*, **31**, 227–68.

McNab, B.K. (1980) Food habits, energetics, and the population biology of mammals. *Amer. Nat.*, **116**, 106–24.

McNab, B.K. (1986a) The influence of food habits on the energetics of eutherian mammals. *Ecol. Monogr.*, **56**, 1–19.

McNab, B.K. (1986b) Food habits, energetics, and the reproduction of marsupials. *J. Zool. Lond.*, **208A**, 595–614.

McNab, B.K. (1987a) Basal rate and phylogeny. *Funct. Ecol.*, **1**, 159–67.

McNab, B.K. (1987b) The evolution of mammalian energetics, in *Evolutionary Physiological Ecology* (ed P. Calow), Cambridge University Press, Cambridge, pp. 219–36.

McNab, B.K. (1988) Complications inherent in scaling the basal rate of metabolism in mammals. *Quart. Rev. Biol.*, **63**, 25–54.

Medawar, P.B. (1946) Old age and natural death. *Modern Quart.*, **2**, 30–49.

Medawar, P.B. (1952) *An Unsolved Problem of Biology*, H.K. Lewis, London.

Meinhardt, H. (1986) Hierarchical inductions of cell states: a model for segmentation in *Drosophila*. *J. Cell Sci.*, **4**(Suppl.), 357–81.

Mertz, D.B. (1975) Senescent decline in flour beetle strains selected for early adult fitness. *Physiol. Zool.*, **48**, 1–23.

Meyer, A. (1987) Phenotypic plasticity and heterochrony in *Cichlasoma managuense* (Pisces: Cichlidae) and their implications for speciation in cichlid fishes. *Evolution*, **41**, 1357–69.

Meyer, A. (1990) Ecological and evolutionary consequences of the trophic polymorphism in *Cichlasoma citrinellum* (Pisces: Cichlidae). *Biol. J. Linn. Soc.*, **39**, 279–99.

Meyer, A., Kocher, T.D., Basasibwaki, P. and Wilson, A.C. (1990) Monophyletic origin of Lake Victoria cichlid fishes suggested by mitochondrial DNA sequences. *Nature*, **347**, 550–3.

Michaels, H.J. and Bazzaz, F.A. (1989) Individual and population responses of sexual and apomitic plants to environmental gradients. *Amer. Nat.*, **134**, 190–207.

Michod, R.E. and Gayley, T.W. (1992) Masking of mutations and the evolution of sex. *Amer. Nat.*, **139**, 706–34.

Michod, R.E. and Levin, B.R. (eds) (1988) *The Evolution of Sex: an Examination of Current Ideas*, Sinauer Associates, Sunderland, MA.

Millar, J.S. and Hickling, G.J. (1990) Fasting endurance and the evolution of mammalian body size. *Funct. Ecol.*, **4**, 5–12.

Millar, J.S. and Hickling, G.J. (1992) The fasting endurance hypothesis revisited. *Funct. Ecol.*, **6**, 496–8.

Miller, J.H. and Reznikoff, W.S. (eds) (1980) *The Operon*, Cold Spring Harbor Laboratory Press.

Minchella, D.J. and Loverde, P.Y. (1981) A cost of increased early reproductive effort in the snail *Biomphalaria glabrata*. *Amer. Nat.*, **118**, 876–81.

Miquel, J., Lundgren, P.R., Bensch, K.G. and Atlan, H. (1976) Effects of temperature on the life span, vitality and fine structure of *Drosophila melanogaster*. *Mech. Ageing Dev.*, **5**, 347–70.

Miquel, J. and Philpott, D.E. (1986) Structural correlates of aging in *Drosophila*: Relevance to the cell differentiation, rate-of-living and free radical theories of aging, in *Insect Aging: Strategies and Mechanisms* (eds K.-G. Collatz and R.S. Sohal), Springer Verlag, Berlin, pp. 117–29.

Mitchell, P.J. and Tjian, R. (1989) Transcriptional regulation in mammalian cells by sequence-specific DNA binding proteins. *Science*, **245**, 371–8.

Mitchell-Olds, T. (1992) Does environmental variation maintain genetic variation? A question of scale. *TREE*, **7**, 397–8.

Mitchell-Olds, T. and Rutledge, J. (1986) Quantitative genetics in natural plant populations: a review of the theory. *Amer. Nat.*, **127**, 379–402.

Mitsialis, S.A. and Kafatos, F.C. (1985) Regulatory elements controlling chorion gene expression are conserved between flies and moths. *Nature*, **317**, 453–6.

Mitton, J.B. (1993) Theory and data pertinent to the relationship between hetero-zygosity and fitness, in *The Natural History of Inbreeding and Outbreeding* (ed N.W. Thornhill), University of Chicago Press, Chicago, pp. 17–41.

Moller, H., Smith, R.H. and Sibly, R.M. (1989a) Evolutionary demography of a bruchid beetle. I. Quantitative genetical analysis of the female life history. *Funct. Ecol.*, **3**, 673–81.

Moller, H., Smith, R.H. and Sibly, R.M. (1989b) Evolutionary demography of a bruchid beetle. II. Physiological manipulations. *Funct. Ecol.*, **3**, 683–91.

Moller, H., Smith, R.H. and Sibly, R.M. (1990) Evolutionary demography of a bruchid beetle. III. Correlated responses to selection and phenotypic plasticity. *Funct. Ecol.*, **4**, 489–93.

Monk, M. (1990) Variation in epigenetic inheritance. *Trends Genet.*, **6**, 110–14.

Moody, M.E. and Basten, C.J. (1990) The evolution of latent genes in subdivided populations. *Genetics*, **124**, 187–97.

Moore, J.C. and Hunt, H.W. (1988) Resource compartmentation and the stability of real ecosystems. *Nature* **333**, 261–3.

Moore, P.D. (1983) Ecological diversity and stress. *Nature*, **306**, 17.

Moran, N.A. (1991) Phenotype fixation and genotypic diversity in the complex life cycle of the aphid *Pemphigus betae*. *Evolution*, **45**, 957–70.

Moran, M.A. (1992a) The evolution of aphid life cycles. *Ann. Rev. Entomol.*, **37**, 321–48.

Moran, N.A. (1992b) The evolutionary maintenance of alternative phenotypes. *Amer. Nat.*, **139**, 971–89.

Morata, G. and Struhl, G. (1990) Fly fishing downstream. *Nature*, **348**, 587–8.

Mori, K. and Nakasuji, F. (1990) Genetic analysis of the wing-form determination of the small brown planthopper, *Laodelphax striatellus* (Hemiptera, Delphacidae). *Res. Popul. Ecol.*, **32**, 279–87.

Moritz, C., McCallum, H., Dannellan, S. and Roberts, J.D. (1991) Parasite loads in pathenogenetic and sexual lizards (*Heteronotia binoei*): support for the Red Queen hypothesis. *Proc. Roy. Soc. Lond.*, **244**, 145–9.

Moritz, R.F.A. and Southwick, E.E. (1992) *Bees as Superorganisms: an Evolutionary Reality*, Springer Verlag, Berlin.

Mousseau, T.A. and Dingle, H. (1991) Maternal effects in insect life histories. *Ann. Rev. Ecol. System.*, **36**, 511–34.

Mrosovsky, N. (1990) *Rheostasis: the Physiology of Change*, Oxford University Press, Oxford.

Mueller, L.D. (1987) Evolution of accelerated senescence in laboratory popula-tions of *Drosophila*. *Proc. Nat. Acad. Sci. USA*, **84**, 1974–7.

Mueller, L.D., Guo, P. and Ayala, F.J. (1991) Density-dependent natural selection and trade-offs in life history traits. *Science*, **253**, 433–5.

Müller, H.J. (1932) Some genetic aspects of sex. *Amer. Nat.*, **66**, 118–38.

Müller, H.J. (1964) The relation of recombination to mutational advance. *Mut. Res.*, **1**, 2–9.

Munkres, K.D. and Furtek, C. (1984) Linkage of conidial longevity determinant genes in *Neurospora crassa*. *Mech. Ageing Develop.*, **25**, 63–77.

Murray, T.D. (1988) How the leopard gets its spots. *Sci. Amer.*, **258**, 80–7.

Naar, E.M., Bartke, A., Majumdar, S.S. *et al.* (1991) Fertility of transgenic female mice expressing bovine growth hormone or human growth hormone variant genes. *Biol. Reprod.*, **45**, 178–87.

Nagy, K.A. (1987) Field metabolic rate and food requirement scaling in mammals and birds. *Ecol. Monogr.*, **57**, 111–128.

Nakamura, K.D., Duffy, P.H., Lu, M.H. *et al.* (1989) The effect of dietary restriction on MYC protooncogene expression in mice: a preliminary study. *Mech. Age. Develop.*, **48**, 199–205.

Nardelli, J., Gibson, T.J., Vesque, C. and Charnay, P. (1991) Base sequence discrimination by zinc-finger DNA-binding domains. *Nature*, **349**, 175–8.

Nee, S., Harvey, P.H. and Cotgreave, P. (1992) Population persistence and the natural relationships between body size and abundance, in *Conservation of Biodiversity for Sustainable Development* (eds O.T. Sandlund, K. Hindar and A.H.D. Brown), Scandinavian University Press, Oslo, pp. 124–36.

Nellen, W. (1986) A hypothesis on the fecundity of bony fish. *Meereiforschung*, **31**, 75–89.

Nesse, R.M. (1988) Life table tests of evolutionary theories of senescence. *Expt. Geront.*, **23**, 445–53.

Neukirch, A. (1982) Dependence of life span of the honeybee (*Apis mellifica*) upon flight performance and energy consumption. *J. Comp. Physiol.*, **146**, 35–40.

Nevo, E. and Shkolnik, A. (1974) Adaptive metabolic variation of chromosome forms in mole rats *Spalax*. *Experientia*, **30**, 724–6.

Newman, R.A. (1988a) Adaptive plasticity in development of *Scaphiopus couchii* tadpoles in desert ponds. *Evolution*, **42**, 774–83.

Newman, R.A. (1988b) Genetic variation for larval anuran (*Scaphiopus couchii*) development time in an uncertain environment. *Evolution*, **42**, 763–73.

Newman, R.A. (1989) Developmental plasticity of *Scaphiopus couchii* tadpoles in an unpredictable environment. *Ecology*, **70**, 1775–87.

Newman, R.A. (1992) Adaptive plasticity in amphibian metamorphosis. *BioScience*, **42**, 671–8.

Newton, I. (1989) Synthesis, in *Lifetime Reproduction in Birds* (ed I. Newton), Academic Press, London, pp. 441–69.

Newton, R.K., Ducore, J.M. and Sohal, R.S. (1989) Relationship between life expectancy and endogenous DNA single-strand breakage, strand break induction and DNA repair capacity in the adult housefly, *Musca domestica*. *Mech. Age. Develop.*, **49**, 259–70.

Nicolini, C. (1988) Nuclear architecture and three-dimensional organization of chromatin: their role in the control of cell function, in *Heterochromatin, Molecular and Structural Aspects* (ed R.S. Verma), Cambridge University Press, Cambridge, pp. 228–49.

Nijhout, H.F. and Wheeler, D.E. (1982) Juvenile hormone and the physiological basis of insect polymorphisms. *Quart. Rev. Biol.*, **57**, 109–33.

Norris, D.M. (1990) Repellents, in *CRC Handbook of Natural Pesticides VI: Insect Attractants and Repellants* (eds E.D. Morgan and N.B. Mandava), CRC Press, Boca Raton, FL, pp. 135–49.

North, G. (1984) Multiple levels of gene control in eukaryote cells. *Nature*, **312**, 308–9.

Nunney, L. (1989) The maintenance of sex by group selection. *Evolution*, **43**, 245–57.

Nur, U. (1990) Heterochromatization and euchromatization of whole genomes in scale insects (Coccoidea, Homoptera). *Development* **1990** (Suppl.), 29–34.

Odum, E.P. (1985) Trends in stressed ecosystems. *BioScience* **35**, 419–22.

Ohno, S. (1970) *Evolution by Gene Duplication*, George Allen and Unwin, London.

Ohno, S. (1973) Ancient linkage groups and frozen accidents. *Nature*, **244**, 259–62.

Ohta, T. (1988) Multigene and supergene families. *Oxford Surv. Evol. Biol.*, **5**, 41–65.

Olivieri, I., Couvet, D. and Gouyon, P.H. (1990) The genetics of transient populations: research at the metapopulation level. *TREE*, **5**, 207–10.

O'Neill, R.V. (1989) Perspectives in hierarchy and scale, in *Perspectives in Ecological Theory* (eds J. Roughgarden, R.M. May and S.A. Levin), Princeton University Press, Princeton, NJ, pp. 140–56.

Orgel, L.E. (1963) The maintenance of the accuracy of protein synthesis and its relevance to aging. *Proc. Natl. Acad. Sci.* **49**, 517–21.

Orgel, L.E. and Crick, F.H.C. (1980) Selfish DNA: the ultimate parasite. *Nature*, **284**, 604–7.

Orians, G.H. (1981) Foraging behavior and the evolution of discriminatory abilities, in *Foraging Behaviour: Ecological, Ethological and Physiological Approaches* (eds A.C. Kamil and T.D. Sargent), Garland Publ., NJ, pp. 389–405.

Oster, G.F. and Alberch, P. (1982) Evolution and bifurcation of developmental programs. *Evolution*, **36**, 444–59.

Oster, G.F., Shubin, N., Murray, J.D. and Alberch, P. (1988) Evolution and morphogenetic rules: the shape of the vertebrate limb in ontogeny and phylogeny. *Evolution*, **42**, 862–84.

Oster, G.F. and Wilson, E.O. (1978) *Caste and Ecology in the Social Insects*, Princeton University Press, Princeton, NJ.

Pahl-Wostl, C. (1990) Temporal organization: a new perspective on the ecological network. *Oikos*, **58**, 293–305.

Paine, R.T. (1992) Food-web analysis through field measurements of per capita interaction strength. *Nature*, **355**, 73–5.

Paley, W. (1828) *Natural Theology*, J. Vincent, Oxford.

Palmer, J.O. (1985) Ecological genetics of wing length, flight propensity, and early fecundity in a migratory insect, in *Migration: Mechanisms and Adaptive Significance* (ed M.A. Rankin), *Contrib. Marine Sci. Suppl.*, **27**, 663–73.

Palmer, J.O. and Dingle, H. (1986) Direct and correlated responses to selection among life-history traits in milkweed bugs (*Oncopeltus fasciatus*). *Evolution*, **40**, 767–77.

Palmer, J.O. and Dingle, H. (1989) Responses to selection on flight behavior in a migratory population of milkweed bug (*Oncopeltus fasciatus*). *Evolution*, **43**, 1805–8.

Palmiter, R.D., Brinster, R.L., Hammer, R.E. *et al.* (1982) Dramatic growth of mice that develop from eggs microinjected with metallothionein-growth hormone fusion genes. *Nature*, **300**, 611–15.

Pankratz, M.J. and Jackle, H. (1990) Making stripes in the *Drosophila* embryo. *Trends Genet.*, **6**, 287–92.

Parejko, K. and Dodson, S.I. (1991) The evolutionary ecology of an antipredator reaction norm: *Daphnia pulex* and *Chaoborus americanus*. *Evolution*, **45**, 1665–74.

Parker, G.A. and Maynard Smith, J. (1990) Optimality theory in evolutionary biology. *Nature*, **348**, 27–33.

Parsons, P.A. (1988) Evolutionary rates: effects of stress upon recombination. *Biol. J. Linn. Soc.*, **35**, 49–68.

Parsons, P.A. (1991) Evolutionary rates: stress and species boundaries. *Ann. Rev. Ecol. Syst.* **22**, 1–18.

Parsons, P.A. (1992) Evolutionary adaptation and stress: the fitness gradient. *Evol. Biol.*, **26**, 191–223.

Partridge, L. (1986) Sexual activity and life span, in *Insect Aging, Strategies and Mechanisms* (eds K.G. Collatz and R.S. Sohal), Springer Verlag, Berlin, pp. 45–54.

Partridge, L. (1987) Is accelerated senescence a cost of reproduction? *Funct. Ecol.*, **1**, 317–20.

Partridge, L. (1989) Lifetime reproductive success and life-history evolution, in *Lifetime Reproduction in Birds* (ed I. Newton), Academic Press, NY, pp. 421–40.

Partridge, L. (1992) Measuring reproductive costs. *TREE*, **7**, 99–100.

Partridge, L. and Andrews, R. (1985) The effect of reproductive activity on the longevity of male *Drosophila melanogaster* is not caused by acceleration of ageing. *J. Insect Physiol.*, **31**, 393–5.

Partridge, L. and Farquhar, M. (1981) Sexual activity reduces longevity of male fruitflies. *Nature*, **294**, 580–2.

Partridge, L. and Fowler, K. (1990) Nonmating costs of exposure to males in female *Drosophila melanogaster*. *J. Insect Physiol.*, **36**, 419–25.

Partridge, L. and Fowler, K. (1992) Direct and correlated responses to selection on age at reproduction in *Drosophila melanogaster*. *Evolution*, **46**, 76–91.

Partridge, L., Fowler, K., Trevitt, S. and Sharp, W. (1986) An examination of the effects of males on the survival and egg-production rates of female *Drosophila melanogaster*. *J. Insect Physiol.*, **32**, 925–9.

Partridge, L., Green, A. and Fowler, K. (1987) Effects of egg-production and of exposure to males on female survival in *Drosophila melanogaster*. *J. Insect Physiol.*, **33**, 745–9.

Partridge, L. and Harvey, P.H. (1985) Costs of reproduction. *Nature*, **316**, 20.

Partridge, L. and Harvey, P.H. (1988) The ecological context of life history evolution. *Science*, **241**, 1449–55.

Partridge, L. and Sibly, R. (1991) Constraints in the evolution of life histories. *Phil. Trans. R. Soc. Lond.*, **332**, 3–13.

Patel, N.H., Ball, E.E. and Goodman, C.S. (1992) Changing role of even-skipped during the evolution of insect pattern formation. *Nature*, **357**, 339–42.

Pattee, H.H. (1973) *Hierarchy Theory*, Braziller, New York.

Pays, E. (1989) Pseudogenes, chimaeric genes and the timing of antigen variation in African trypanosomes. *Trends Genet.*, **5**, 389–91.

Pearl, R. (1928) *The Rate of Living*, University of London Press, London.

Pease, C.M. and Bull, J.J. (1988) A critique of methods for measuring life history trade-offs. *J. Evol. Biol.*, **1**, 293–303.

Pernin, P., Ataya, A. and Cariou, M.L. (1992) Genetic structure of natural populations of the free-living amoeba, *Naegleria lovaniensis*. Evidence for sexual reproduction. *Heredity*, **68**, 173–81.

Perrigo, G. (1987) Breeding and feeding strategies in deer mice and house mice when females are challenged to work for their food. *Anim. Behav.*, **35**, 1298–316.

Perrigo, G. and Bronson, F.H. (1983) Foraging effort, food intake, fat deposition and puberty in female mice. *Biol. Repr.*, **29**, 455–63.

Perrigo, G. and Bronson, F.H. (1985) Sex differences in the energy allocation strategies of house mice. *Behav. Ecol. Sociobiol.*, **17**, 297–302.

Perrin, N. and Rubin, J.F. (1990) On dome-shaped norms of reaction for size-to-age at maturity in fishes. *Funct. Ecol.*, **4**, 53–7.

Perrot, V., Richerd, S. and Valero, M. (1991) Transition from haploidy to diploidy. *Nature*, **351**, 315–17.

Peters, R.H. (1983) *The Ecological Implications of Body Size*, Cambridge University Press, Cambridge.

Peterson, C.C., Nagy, K.A. and Diamond, J. (1990) Sustained metabolic scope. *Proc. Natl. Acad. Sci. USA*, **87**, 2324–8.

Peterson, K. and Sapienza, C. (1993) Imprinting the genome: imprinted genes, imprinting genes, and a hypothesis for their interaction. *Ann. Rev. Genet.*, **27**, 7–31.

Pettifor, R.A., Perrins, C.M. and McCleery, R.H. (1988) Individual optimization of clutch size in great tits. *Nature*, **336**, 160–2.

Philippi, T. and Seger, J. (1989) Hedging one's evolutionary bets, revisited. *TREE*, **4**, 41–4.

Phillips, D.C., Sternberg, M.J.E. and Sutton, B.J. (1983) Intimations of evolution from the three-dimensional structures of proteins, in *Evolution from Molecules to Men* (ed D.S. Bendall), Cambridge University Press, Cambridge, pp. 145–73.

Phillips, J.P., Campbell, S.D., Michaud, D. *et al.* (1989) Null mutation of copper/zinc superoxide dismutase in *Drosophila* confers hypersensitivity to paraquat and reduced longevity. *Proc. Natl. Acad. Sci. USA*, **86**, 2761–5.

Pianka, E.R. (1978) *Evolutionary Ecology*, Harper and Row, New York.

Pidduck, H.G. and Falconer, D.S. (1978) Growth hormone function in strains of mice. *Genet. Res.*, **32**, 195–206.

Pimentel, D. (1968) Population regulation and genetic feedback. *Science*, **159**, 1432–7.

Prakash, S. and Lewontin, R.C. (1968) A molecular approach to the study of genic heterozygosity in natural populations. III. Direct evidence of coadaptation in gene arrangements of *Drosophila*. *Proc. Natl. Acad. Sci. USA*, **59**, 398–405.

Price, P.W. (1991) The web of life: development over 3.8 billion years of trophic relationships, in *Symbiosis as a Source of Evolutionary Innovation* (eds L. Margulis and R. Fester), MIT Press, Cambridge, pp. 262–72.

Price, T. and Langen, T. (1992) Evolution of correlated characters. *TREE*, **7**, 307–10.

Promislow, D.E.L. (1991) Senescence in natural populations of mammals: a comparative study. *Evolution*, **45**, 1869–87.

Promislow, D.E.L. and Harvey, P.H. (1990) Living fast and dying young: a comparative analysis of life-history variation among mammals. *J. Zool. Lond.*, **220**, 417–37.

Promislow, D.E.L. and Harvey, P.H. (1991) Mortality rates and the evolution of mammal life histories. *Acta Oecologica*, **12**, 119–37.

Prothero, J. and Jurgens, K.D. (1987) Scaling of maximal lifespan in mammals: a review, in *Evolution of Longevity in Animals* (eds A.D. Woolhead and K.H. Thompson), Plenum Press, NY, pp. 49–71.

Provine, W.B. (1989) Founder effects and genetic revolutions in microevolution and speciation: a historical perspective, in *Genetics, Speciation and the Founder Principle* (eds L.V. Giddings, K.Y. Kaneshiro and W.W. Anderson), Oxford University Press, Oxford, pp. 43–76.

Pyke, G.H. (1984) Optimal foraging theory: a critical review. *Ann. Rev. Ecol. Syst.*, **15**, pp. 523–75.

Raff, R.A. and Kaufman, T.C. (1983) *Embryos, Genes, and Evolution*, MacMillan, NY.

Raff, E.C., Diaz, H.B., Hoyle, H.D. *et al.* (1987) Origin of multiple gene families: are there both functional and regulatory constraints? in *Development as an Evolutionary Process* (eds R.A. Raff and E.C. Raff), Alan R. Liss, NY, pp. 203–38.

Raglan, S.S. and Sohal, R.S. (1973) Mating behavior, physical activity and aging in the housefly, *Musca domestica*. *Expt. Geront.*, **8**, 135–45.

Rahel, F.J. (1990) The hierarchical nature of community persistence: a problem of scale. *Amer. Nat.*, **136**, 328–44.

Rankin, M.A. and Burchsted, J.C.A. (1992) The cost of migration in insects. *Ann. Rev. Entomol.*, **37**, 533–59.

Rasmussen, N. (1987) A new model of developmental constraints as applied to the *Drosophila* system. *J. Theor. Biol.*, **127**, 271–99.

Ratner, V.A. and Vasilyeva, L.A. (1989) Mobile genetic elements and quantitative characters in *Drosophila*: fast heritable changes under temperature treatment, in *Evolutionary Biology of Transient Unstable Populations* (ed A. Fontdevila), Springer Verlag, NY, pp. 165–89.

Rausher, M.D. and Englander, R. (1987) The evolution of habitat preference II. Evolutionary genetic stability under soft selection. *Theor. Popul. Biol.*, **31**, 116–39.

Read, A.F. and Harvey, P.H. (1989) Life-history differences among the eutherian radiations. *J. Zool. Lond.*, **219**, 329–53.

Real, L.A. (1980a) Fitness, uncertainty, and the role of diversification in evolution and behavior. *Am. Nat.*, **115**, 623–38.

Real, L.A. (1980b) On uncertainty and the law of diminishing returns in evolution and behavior, in *Limits to Action* (ed J.E.R. Staddon), Academic Press, pp. 37–64.

Real, L.A. (1991) Animal choice behavior and the evolution of cognitive architecture. *Science*, **253**, 980–6.

Real, L.A. and Caraco, T. (1986) Risk and foraging in stochastic environments. *Ann. Rev. Ecol. Syst.*, **17**, 371–90.

Real, L.A. and Ellner, S. (1992) Life history evolution in stochastic environments: a graphical mean-variance approach. *Ecology*, **73**, 1227–36.

Reid, L. (1990) From gradients to axes, from morphogenesis to differentiation. *Cell*, **63**, 875–82.

Reid, R.G.B. (1985) *Evolutionary Theory: the Unfinished Synthesis*, Croom Helm, London.

Reik, N., Howlett, S.K. and Surani, M.A. (1990) Imprinting by DNA methylation: from transgenes to endogenous gene sequences. *Develop.*, **1990** (Suppl.), 99–106.

Reiss, M.J. (1989) *The Allometry of Growth and Reproduction*, Cambridge University Press, Cambridge.

Rendel, J.M. (1967) *Canalisation and Gene Control*, Academic Press, London.

Rendel, J.M. (1979) Canalisation and selection, in *Quantitative Genetic Variation* (eds J.N. Thomson and J.M. Thoday), Academic Press, NY, pp. 139–56.

Renkawitz, R. (1990) Transcriptional repression in eukaryotes. *Trends Genet.*, **6**, 192–7.

Rennie, J. (1992) Living together. *Sci. Amer.*, **266**, 122–33.

Rensch, B. (1960) *Evolution Above the Species Level*, Columbia University Press, NY.

Reznick, D. (1981) 'Grandfather effects': the genetics of interpopulation difference in offspring size in the mosquito fish. *Evolution*, **35**, 941–53.

Reznick, D. (1985) Costs of reproduction: an evaluation of the empirical evidence. *Oikos*, **44**, 257–67.

Reznick, D. (1992a) Measuring the costs of reproduction. *TREE*, **7**, 42–5.

Reznick, D. (1992b) Measuring reproductive costs: response to Partridge. *TREE*, **7**, 134.

Reznick, D., Bryga, H. and Endler, J.A. (1990) Experimentally induced life-history evolution in a natural population. *Nature*, **346**, 357–9.

Reznick, D., Perry, E. and Travis, J. (1986) Measuring the cost of reproduction: a comment on papers by Bell. *Evolution*, **40**, 1338–44.

Rice, S.H. (1990) A geometric model for the evolution of development. *J. Theor. Biol.*, **143**, 319–42.

Rice, W.R. (1983) Parent–offspring pathogen transmission: a selective agent promoting sexual reproduction. *Amer. Nat.*, **121**, 187–203.

Riessen, H.P. and W.G. Sprules. (1990) Demographic costs of antipredator defenses in *Daphnia pulex*. *Ecology*, **71**, 1536–46.

Rio, D.C. (1991) Regulation of *Drosophila* P element transposition. *Trends Genet.*, **7**, 282–7.

Riska, B. (1989) Composite traits, selection response, and evolution. *Evolution*, **43**, 1172–91.

Riska, B., Atchley, W.R. and Rutledge, J.J. (1984) A genetic analysis of targeted growth in mice. *Genetics*, **107**, 79–101.

Roach, D.A. and Wulff, R.D. (1987) Maternal effects in plants. *Ann. Rev. Ecol. Syst.*, **18**, 209–35.

Roberts, R.C. (1981) The growth of mice selected for large and small size in relation to food intake and the efficiency of conversion. *Genet. Res.*, **38**, 9–24.

Robertson, A. (1952) The effect of inbreeding on the variation due to recessive genes. *Genetics*, **37**, 187–207.

Robinson, G.E. (1992) Regulation of division of labor in insect societies. *Ann. Rev. Ecol. Syst.*, **37**, 637–65.

Roff, D.A. (1986) The evolution of wing dimorphism in insects. *Evolution*, **40**, 1009–20.

Roff, D.A. (1993) *The Evolution of Life Histories*, Chapman & Hall, NY.

Rollo, C.D. (1984) Resource allocation and time budgeting in adults of the cockroach, *Periplaneta americana*: the interaction of behaviour and metabolic reserves. *Res. Popul. Ecol.*, **26**, 150–87.

Rollo, C.D. (1986) A test of the principle of allocation using two sympatric species of cockroaches. *Ecology*, **67**, 616–28.

Rollo, C.D. and Hawryluk, M.D. (1988) Compensatory scope and resource allocation in two species of aquatic snails. *Ecology*, **69**, 146–56.

Rollo, C.D., MacFarlane, J.D. and Smith, B.S. (1984) Electrophoretic and allometric variation in burdock (*Arctium* spp.): hybridization and its ecological implications. *Can J. Bot.*, **63**, 1255–61.

Rollo, C.D. and Shibata, D.M. (1991) Resilience, robustness and plasticity in a terrestrial slug, with particular reference to food quality. *Can. J. Zool*, **69**, 978–87.

Rollo, C.D., Vertinisky, I.B., Wellington, W.G. *et al.* (1983) Description and testing of a comprehensive simulation model of the ecology of terrestrial gastropods in unstable environments. *Res. Popul. Ecol.*, **25**, 150–79.

Root, T. (1988) Energy constraints on avian distributions and abundances. *Ecology*, **69**, 330–9.

Rose, M.R. (1982) Antagonistic pleiotropy, dominance, and genetic variation. *Heredity*, **48**, 63–78.

Rose, M.R. (1983a) Theories of life-history evolution. *Amer. Zool.*, **23**, 15–23.

Rose, M.R. (1983b) Further models of selection with antagonistic pleiotropy, in *Population Biology* (eds H.I. Freedman and C. Strobeck), Springer Verlag, Berlin, pp. 47–53.

Rose, M.R. (1984a) Genetic covariation in *Drosophila* life history: untangling the data. *Amer. Nat.*, **123**, 565–9.

Rose, M.R. (1984b) Laboratory evolution of postponed senescence in *Drosophila melanogaster*. *Evolution*, **38**, 1004–10.

Rose, M.R. (1984c) The evolution of animal senescence. *Can. J. Zool.*, **62**, 1661–7.

Rose, M.R. (1990) Evolutionary genetics of aging in *Drosophila*, in *Genetic Effects on Aging II* (ed D.E. Harrison), Telford Press, Caldwell, NJ, pp. 41–55.

Rose, M.R. (1991) *Evolutionary Biology of Aging*, Oxford University Press, Oxford.

Rose, M.R. and Charlesworth, B. (1980) A test of evolutionary theories of senescence. *Nature*, **287**, 141–2.

Rose, M.R. and Charlesworth, B. (1981a) Genetics of life history in *Drosophila melanogaster* I. Sib analysis of adult females. *Genetics*, **97**, 173–86.

Rose, M.R. and Charlesworth, B. (1981b) Genetics of life history in *Drosophila melanogaster* II. Exploratory selection experiments. *Genetics*, **97**, 187–96.

Rose, M.R., Dorey, M.L., Coyle, A.M. and Service, P.M. (1984) The morphology of postponed senescence in *Drosophila melanogaster*. *Can. J. Zool.*, **62**, 1576–80.

Rose, M.R. and Graves Jr., J.L. (1989) What evolutionary biology can do for gerontology. *J. Geront.*, **44**, B27–B29.

Rose, M.R., Hutchinson, E.W. and Graves, J.L. (1990) Genetics of longer-lived *Drosophila*, in *Molecular Biology of Aging* (eds C.E. Finch and T.E. Johnson), Wiley–Liss, NY, pp. 19–30.

Rose, M.R., Service, P.M. and Hutchinson, E.W. (1987) Three approaches to trade-offs in life-history evolution, in *Genetic Constraints on Adaptive Evolution* (ed V. Loeschcke), Springer Verlag, Berlin, pp. 91–105.

Rosen, R. (1978) Feedforwards and global system failure: a general mechanism for senescence. *J. Theor. Biol.*, **74**, 579–90.

Rosenberger, R.F., Gounaris, E. and Kolettas, E. (1991) Mechanisms responsible for the limited lifespan and immortal phenotypes in cultured mammalian cells. *J. Theor. Biol.*, **148**, 383–92.

Rossant, J. and Joyner, A.L. (1989) Towards a molecular-genetic analysis of mammalian development. *Trends Genet.*, **5**, 277–83.

Rossiter, M.C. (1991) Maternal effects generate variation in life history: consequences of egg weight plasticity in the gypsy moth. *Funct. Ecol.*, **5**, 386–93.

Roth, V.L. (1991) Homology and hierarchies: problems solved and unresolved. *J. Evol. Biol.*, **4**, 167–94.

Rothschild, M. and Ford, B. (1964) Maturation and egglaying of the rabbit flea (*Spilopsyllus cuniculi* Dale) induced by external application of hydrocortisone. *Nature*, **203**, 210–11.

Roughgarden, J. (1979) *Theory of Population Genetics and Evolutionary Ecology: an Introduction*, Collier MacMillan, Toronto.

Roughgarden, J. (1989) The structure and assembly of communities, in *Perspectives in Ecological Theory* (eds J. Roughgarden, R.M. May and S. Levin), Princeton University Press, Princeton, NJ, pp. 203–26.

Rowe, J.S. (1961) The level-of-integration concept and ecology. *Ecology*, **42**, 420–7.

Rubner, M. (1908) *Das Problem der Lebensdauer und seine Beziehungen zum Wachstrum und Ernahrung*, Oldenbourg, Munich.

Ruse, M. (1988) Molecules to men: evolutionary biology and thoughts of progress, in *Evolutionary Progress* (ed M.H. Nitecki), University of Chicago Press, Chicago, IL, pp. 97–126.

Ruse, M. (1993) Evolution and progress. *TREE*, **8**, 55–9.

Russell-Hunter, W.D. (1978) Ecology of freshwater pulmonates, in *Pulmonates: Systematics, Evolution and Ecology. Vol. 2A* (eds V. Fretter and J. Peake), Academic Press, NY, pp. 335–83.

Russo, V.E.A., Brody, S., Cove, D. and Ottolenghi, S. (eds) (1992) *Development: the Molecular Genetic Approach*, Springer Verlag, Berlin.

Rusting, R.L. (1992) Why do we age? *Sci. Amer.*, **267**, 131–41.

Rutledge, J.J., Eisen, E.J. and Legates, J.E. (1974) Correlated response in skeletal traits and replicate variation in selected lines of mice. *Theor. Appl. Genet.*, **45**, 26–31.

Saarikko, J. and Hanski, I. (1990) Timing of rest and sleep in foraging shrews. *Anim. Behav.*, **40**, 861–9.

Sacher, G.A. (1959) Relationship of lifespan to brain weight and body weight in mammals, in *CIBA Foundation Symposium on the Lifespan of Animals* (eds G.E.W. Wolstenholme and M. O'Conner), Little and Brown, Boston, MA, pp. 115–53.

Sacher, G.A. (1977) Life table modification and life prolongation, in *Handbook of the Biology of Aging* (eds C. E. Finch and L. Hayflick), Van Nostrand-Reinhold, NY, pp. 582–638.

Sacher, G.A. (1978) Longevity and aging in vertebrate evolution. *BioScience*, **28**, 497–501.

Sacher, G.A. and Duffy, P.H. (1979) Genetic relation of life-span to metabolic rate for inbred mouse strains and their hybrids. *Fed. Proc.*, **38**, 184–8.

Sachs, T. (1988) Ontogeny and phylogeny: phytohormones as indicators of labile changes, in *Plant Evolutionary Biology* (eds L.D. Gottlieb and S.H. Jain), Chapman & Hall, London, pp. 157–76.

Saether, B.-E. (1988) Pattern of covariation between life-history traits of European birds. *Nature*, **331**, 616–17.

Saitoh, T. (1990) Lifetime reproductive success in reproductively suppressed female voles. *Res. Popul. Ecol.*, **32**, 391–406.

Salt, G.W. (1979) A comment on the use of the term emergent properties. *Amer. Nat.*, **113**, 145–61.

Salthe, S.N. (1975) Problems of macroevolution (molecular evolution, phenotype definition, and canalization) as seen from a hierarchical viewpoint. *Amer. Zool.*, **15**, 295–314.

Salthe, S.N. (1985) *Evolving Hierarchical Systems*, Columbia University Press, NY.

Sander, K. (1983) The evolution of patterning mechanisms: gleanings from insect embryogenesis and spermatogenesis, in *Development and Evolution* (eds B.C. Goodwin, N. Holder and C.C. Wylie), Cambridge University Press, pp. 137–59.

Santos, M., Ruiz, A., Quezada-Diaz, J.E. *et al.* (1992) The evolutionary history of *Drosophila buzzatii*. XX. Positive phenotypic covariance between field adult fitness components and body size. *J. Ecol. Biol.*, **5**, 403–22.

Sapienza, C. (1990) Parental imprinting of genes. *Sci. Amer.*, **263**, 52–60.

Sarkar, S. (1992) Sex, disease and evolution – variations on a theme from J.B.S. Haldane. *BioScience*, **42**, 448–53.

Saunders, P.T. (1990) The epigenetic landscape and evolution. *Biol. J. Linn. Soc.*, **39**, 125–34.

Schaffner, W. (1989) How do different transcription factors binding the same DNA sequence sort out their jobs? *Trends Genet.*, **5**, 37–9.

Scharloo, W. (1991) Canalization: genetic and developmental aspects. *Ann. Rev. Ecol. Syst.*, **22**, 65–93.

Scheiner, S.M. (1993) Genetics and evolution of phenotypic plasticity. *Ann. Rev. Ecol. Syst.*, **24**, 35–68.

Scheiner, S.M., Caplan, R.L. and Lyman, R.F. (1991) The genetics of phenotypic plasticity. III. Genetic correlations and fluctuating asymmetries. *J. Evol. Biol.*, **4**, 51–68.

Scheiner, S.M. and Goodnight, C.J. (1984) The comparison of phenotypic plasticity and genetic variation in populations of the grass *Danthonia spicata*. *Evolution*, **38**, 845–55.

Scheiner, S.M. and Istock, C.A. (1991) Correlational selection on life history traits in the pitcher-plant mosquito. *Genetica*, **84**, 123–8.

Scheiner, S.M. and Lyman, R.F. (1991) The genetics of phenotypic plasticity. II. Response to selection. *J. Evol. Biol.*, **4**, 23–50.

Schemske, D.W. (1982) Limits to specialization and coevolution in plant–animal mutualisms, in *Coevolution* (ed M.H. Nitecki), Chicago University Press, Chicago, IL, pp. 67–109.

Schlichting, C.D. (1986) The evolution of phenotypic plasticity in plants. *Ann. Rev. Ecol. Syst.*, **17**, 667–93.

Schlichting, C.D. (1989a) Phenotypic integration and environmental change. *Bioscience*, **39**, 460–4.

Schlichting, C.D. (1989b) Phenotypic plasticity in *Phlox* II. Plasticity of character correlations. *Oecologia*, **78**, 496–501.

Schlichting, C.D. and Levin, D.A. (1990) Phenotypic plasticity in *Phlox*. III. Variation among natural populations of *P. drummondii*. *J. Evol. Biol.*, **3**, 411–28.

Schluter, D. and Smith, J.N.M. (1986) Genetic and phenotypic correlations in a natural population of song sparrows. *Biol. J. Linn. Soc.*, **29**, 23–36.

Schmalhausen, I.I. (1949) *Factors of Evolution: the Theory of Stabilizing Selection*, Blakeston, Philadelphia.

Schmid-Hempel, P. (1990) In search of optima: equilibrium models of phenotypic evolution in *Population Biology* (eds K. Wohrmann and S. Jain), Springer Verlag, Berlin, pp. 321–47.

Schmid-Hempel, P. and Wolf, T. (1988) Foraging effort and life span of workers in a social insect. *J. Anim. Ecol.*, **57**, 500–21.

Schmidt, H. (1909) *Das biogenetische Grundgesetz Ernst Haeckels und seine Gegner,* Neuer Frankfurter Verlag, Frankfurt.

Schmidt-Nielsen, K. (1984) *Scaling: Why is Animal Size so Important?* Cambridge University Press, Cambridge.

Schmitt, J. and Antonovics, J. (1986) Experimental studies of the evolutionary significance of sexual reproduction. IV. Effect of neighbor relatedness and aphid infestation on seedling performance. *Evolution*, **40**, 830–6.

Schoen, D.J. and Brown, A.H.D. (1991) Intraspecific variation in population gene diversity and effective population size correlates with mating system in plants. *Proc. Natl. Acad. Sci. USA*, **88**, 4494–7.

Schoonhoven, L.M. (1990) Insects in a chemical world, in *CRC handbook of natural pesticides, Vol. VI. Insect attractants and repellants* (eds E.D. Morgan and N.B. Mandava), CRC Press, Boca Raton, FL, pp. 1–21.

Schultz, J.C. (1988) Plant responses induced by herbivores. *TREE*, **3**, 45–50.

Schwartz, L.M., Kosz, L. and Kay, B.K. (1990) Gene activation is required for developmentally programmed cell death. *Proc. Natl. Acad. Sci. USA*, **87**, 6594–8.

Schwemmler, W. (1991) Symbiogenesis in insects as a model for cell differentiation, morphogenesis, and speciation, in *Symbiosis as a Source of Evolutionary Innovation* (eds L. Margulis and R. Fester), MIT Press, Cambridge, MA, pp. 178–204.

Scott, M.P. and Carroll, S.B. (1987) The segmentation and homeotic gene network in early *Drosophila* development. *Cell*, **51**, 689–98.

Seeger, M.A. and Kaufman, T.C. (1987) Homeotic genes of the *Antennapedia* complex (ANT–C) and their molecular variability in the phylogeny of the Drosophilididae, in *Development as an Evolutionary Process* (eds R.A. Raff and E.C. Raff), Alan R. Liss, pp. 179–202.

Seger, J. and Brockmann, H.J. (1987) What is bet-hedging? *Oxford Surv. Evol. Biol.*, **4**, 182–211.

Seger, J. and Hamilton, W.D. (1988) Parasites and sex, in *The Evolution of Sex* (eds R.E. Michod and B.R. Levin), Sinauer, Sunderland, MA, pp. 176–93.

Seidel, H.M., Pompliano, D.L. and Knowles, J.R. (1992) Exons as microgenes. *Science*, **257**, 1489–90.

Selander, R.K. and Hudson, R.O. (1976) Animal population structure under close inbreeding: the land snail *Rumina* in southern France. *Amer. Nat.*, **110**, 695–718.

Selander, R.K. and Kaufman, D.W. (1973) Self-fertilization and genetic population structure in a colonizing land snail. *Proc. Natl. Acad. Sci. USA*, **70**, 1186–90.

Selander, R.K., Yang, S.Y., Lewontin, R.C. and Johnson, W.E. (1970) Genetic variation in the horseshoe crab (*Limulus polyphemus*), a phylogenetic 'relic'. *Evolution*, **24**, 402–14.

Semsei, I. and Richardson, A. (1986) Effect of age on the expression of genes involved in free radical protection. *Fed. Proc.*, **45**, 217.

Service, P.M. (1987) Physiological mechanisms of increased stress resistance in *Drosophila melanogaster* selected for postponed senescence. *Physiol. Zool.*, **60**, 321–6.

Service, P.M. (1989) The effect of mating status on lifespan, egg laying and starvation resistance in *Drosophila melanogaster* in relation to selection on longevity. *J. Insect Physiol.*, **35**, 447–52.

Service, P.M., Hutchinson, E.W., MacKinley, M.D. and Rose, M.R. (1985) Resistance to environmental stress in *Drosophila melanogaster* selected for postponed senescence. *Physiol. Zool.*, **58**, 380–9.

Service, P.M., Hutchinson, E.W. and Rose, M.R. (1988) Multiple genetic mechanisms for the evolution of senescence in *Drosophila melanogaster*. *Evolution*, **42**, 708–16.

Service, P.M. and Rose, M.R. (1985) Genetic covariation among life-history components: the effect of novel environments. *Evolution*, **39**, 943–5.

Sestini, E.A., Carlson, J.C. and Allsopp, R. (1991) The effects of ambient temperature on life span, lipid peroxidation, superoxide dismutase and phospholipase A_2 activity in *Drosophila melanogaster*. *Expt. Geront.*, **26**, 385–95.

Shapiro, A.M. (1976) Seasonal polyphenism. *Evol. Biol.*, **9**, 259–333.

Shapiro, A.M. (1984) The genetics of seasonal polyphenism and the evolution of 'general purpose genotypes' in butterflies, in *Population Biology and Evolution* (eds K. Wöhrmann and V. Loeschcke), Springer Verlag, Berlin, pp. 16–30.

Shapiro, J.A. (1991) Genomes as smart systems. *Genetica*, **84**, 3–4.

Shapiro, J.A. (1992) Natural genetic engineering in evolution. *Genetica*, **86**, 99–111.

Shapiro, D.Y. (1980) Serial female sex changes after simultaneous removal of males from social groups of a coral reef fish. *Science*, **209**, 1136–7.

Shemer, R., Kafri, T., O'Connell, A. *et al.* (1991) Methylation changes in the apolipoprotein AI gene during embryonic development of the mouse. *Proc. Natl. Acad. Sci. USA*, **88**, 11300–4.

Sherman, P.W., Jarvis, J.U.M. and Alexander, R.D. (eds) (1991) *The Biology of the Naked Mole-rat*, Princeton University Press, Princeton, NJ.

Shibata, D.M. and Rollo, C.D. (1988) Intraspecific variation in the growth rate of gastropods: five hypotheses. *Mem. Ent. Soc. Can.*, **146**, 199–213.

Shields, W.M. (1982) *Philopatry, Inbreeding, and the Evolution of Sex*, State University of New York Press, Albany, NY.

Shields, W.M. (1988) Sex and adaptation, in *The Evolution of Sex* (eds R.E. Michod and B.R. Levin), Sinauer Associates, Sunderland, MA, pp. 253–69.

Shields, W.M. (1993) The natural and unnatural history of inbreeding and outbreeding, in *The Natural History of Inbreeding and Outbreeding* (ed N.W. Thornhill), University of Chicago Press, Chicago, IL, pp. 143–69.

Shuster, S.M. and Wade, M.J. (1991) Equal mating success among male reproductive strategies in a marine isopod. *Nature*, **350**, 608–10.

Sibly, R.M. (1981) Strategies of digestion and defecation, in *Physiological Ecology: an Evolutionary Approach* (eds C.R. Townsend and P. Calow), Blackwell Scientific, Oxford, pp. 109–39.

Sibly, R.M. (1989) What evolution maximizes. *Funct. Ecol.*, **3**, 129–35.

Sibly, R. and Calow, P. (1983) An integrated approach to life-cycle evolution using selective landscapes. *J. Theor. Biol.*, **102**, 527–47.

Sibly, R.M. and Calow, P. (1986) *Physiological Ecology of Animals: an Evolutionary Approach*, Blackwell, Oxford.

Sih, A. (1987) Predators and prey lifestyles: an evolutionary and ecological overview, in *Predation: Direct and Indirect Impacts on Aquatic Communities* (eds C. Kerfoot and A. Sih), University Press of Hanover, London, pp. 203–24.

Simpson, G.G. (1944) *Tempo and Mode in Evolution*, Columbia University Press, NY.

Simpson, G.G. (1953a) *The Major Features of Evolution*, Columbia University Press, NY.

Simpson, G.G. (1953b) The Baldwin effect. *Evolution*, **7**, 110–17.

Simon, H.A. (1962) The architecture of complexity. *Proc. Am. Phil. Soc.*, **106**, 462–82.

Simon, H.A. (1973) The organization of complex systems, in *Hierarchy Theory* (ed H.H. Pattee), Braziller, NY, pp. 1–27.

Sinervo, B. (1990) The evolution of maternal investment in lizards: an experimental and comparative analysis of egg size and its effects on offspring performance. *Evolution*, **44**, 279–94.

Sinervo, B. (1993) The effect of offspring size on physiology and life history.

BioScience, **43**, 210–18.

Sinervo, B., Doughty, P., Huey, R.B. and Zamudio, Z. (1992) Allometric engineering: a causal analysis of natural selection on offspring size. *Science*, **258**, 1927–30.

Sinervo, B. and Huey, R.B. (1990) Allometric engineering: an experimental test of the causes of interpopulational differences in performance. *Science*, **248**, 1106–9.

Skelly, D.K. and Werner, E.E. (1990) Behavioural and life-historical response of larval American toads to an odonate predator. *Ecology*, **71**, 2313–22.

Slack, J.M.W. (1991) *From Egg to Embryo: Regional Specification in Early Development*, Cambridge University Press, Cambridge.

Slatkin, M. (1974) Hedging one's evolutionary bets. *Nature*, **250**, 704–5.

Slatkin, M. (1987) Gene flow and the geographic structure of natural populations. *Science*, **236**, 787–92.

Smith, C.W.J., Patton, J.G. and Nadal-Ginard, B. (1989) Alternative splicing in the control of gene expression. *Ann. Rev. Genet.*, **23**, 527–77.

Smith, T.B. (1990a) Resource use by bill morphs of an African finch: evidence for intraspecific competition. *Ecology*, **71**, 1246–57.

Smith, T.B. (1990b) Patterns of morphological and geographic variation in trophic bill morphs of the African finch *Pyrenestes*. *Biol. J. Linn. Soc.*, **41**, 381–414.

Smith-Gill, S.J. (1983) Developmental plasticity: developmental conversion versus phenotypic modulation. *Amer. Zool.*, **23**, 47–55.

Sohal, R.S. (1986) The rate of living theory: a contemporary interpretation, in *Insect Aging* (eds K.G. Collatz and R.S. Sohal), Springer Verlag, Berlin, pp. 23–44.

Sohal, R.S. (1987) The free radical theory of aging: a critique. *Rev. Biol. Res. Aging*, **3**, 431–49.

Sohal, R.S. (1991) Hydrogen peroxide production by mitochondria may be a biomarker of aging. *Mech. Age. Develop.*, **60**, 189–98.

Sohal, R.S. and Allen, R.G. (1990) Oxidative stress as a causal factor in differentiation and aging: a unifying hypothesis. *Expt. Geront.*, **25**, 499–522.

Sohal, R.S. and Runnels, J.H. (1986) Effect of experimentally-prolonged life span on flight performance of houseflies. *Expt. Geront.*, **21**, 509–14.

Sohal, R.S., Sohal, B.H. and Brunk, U.T. (1990a) Relationship between antioxidant defenses and longevity in different mammalian species. *Mech. Age. Develop.*, **53**, 217–27.

Sohal, R.S., Svensson, I. and Brunk, U.T. (1990b) Hydrogen peroxide production by liver mitochondria in different species. *Mech. Age. Develop.*, **53**, 209–15.

Sokal, R.R. (1970) Senescence and genetic load: evidence from *Tribolium*. *Science*, **167**, 1733–4.

Solter, D. (1988) Differential imprinting and expression of maternal and paternal genomes. *Ann. Rev. Genet.*, **22**, 127–46.

Southwood, T.R.E. (1977) Habitat, the templet for ecological strategies? *J. Anim. Ecol.*, **46**, 337–65.

Southwood, T.R.E. (1988) Tactics, strategies and templets. *Oikos*, **52**, 3–18.

Speakman, J.R. (1992) Evolution of animal body size: a cautionary note on assessments of the role of energetics. *Funct. Ecol.*, **6**, 495–6.

Spencer, R.P. (1990) Relationship of reproductive success and median longevity to food intake, in the captive female spider *Frontinella pryamitela*. *Mech. Age. Develop.*, **55**, 9–13.

Spitze, K. (1992) Predator-mediated plasticity of prey life history and morphology: *Chaoborus americanus* predation on *Daphnia pulex*. *Amer. Nat.*, **139**, 229–47.

Spitze, K., Burnson, J, and Lynch, M. (1991) The covariance structure of life-history characters in *Daphnia pulex. Evolution*, **45**, 1081–90.

Stadtman, E.R. (1988) Protein modification in aging. *J. Geront.*, **43**, B112–B120.

Stanley, S.M. (1979) *Macroevolution: Pattern and Process*, W.H. Freeman and Company, San Francisco, CA.

Stanojevic, D., Hoey, T. and Levine, M. (1989) Sequence-specific DNA-binding activities of the gap proteins encoded by hunchback and krüppel in *Drosophila. Nature*, **341**, 331–5.

Stanojevic, D., Small, S. and Levine, M. (1991) Regulation of a segmentation stripe by overlapping activators and repressors in the *Drosophila* embryo. *Science*, **254**, 1385–7.

Stavenhagen, J.B. and Robins, D.M. (1988) An ancient provirus has imposed androgen regulation on the adjacent mouse sex-limited protein gene. *Cell*, **55**, 247–54.

Stearns, S.C. (1976) Life-history tactics: a review of the ideas. *Quart. Rev. Biol.*, **51**, 3–47.

Stearns, S.C. (1980) A new view of life-history evolution. *Oikos*, **35**, 266–81.

Stearns, S.C. (1983) The influence of size and phylogeny on patterns of covariation among life-history traits in the mammals. *Oikos*, **41**, 173–87.

Stearns, S.C. (1984a) How much of the phenotype is necessary to understand evolution at the level of the gene? in *Population Biology and Evolution* (eds K. Wöhrmann and V. Loeschcke), Springer Verlag, Berlin, pp. 31–45.

Stearns, S.C. (1984b) The tension between adaptation and constraint in the evolution of reproductive patterns. *Adv. Invert. Reprod.*, **3**, 387–98.

Stearns, S.C. (1984c) The effects of size and phylogeny on patterns of covariation in the life history traits of lizards and snakes. *Amer. Nat.*, **123**, 56–72.

Stearns, S.C. (1986) Natural selection and fitness, adaptation and constraint, in *Patterns and Processes in the History of Life* (eds D.M. Raup and D. Jablonski), Springer Verlag, Berlin, pp. 23–44.

Stearns, S.C. (1987a) Why sex evolved and the differences it makes, in *The Evolution of Sex and its Consequences* (ed S.C Stearns), Birkhäuser, Basel, Switzerland, pp. 15–31.

Stearns, S.C. (1987b) The selection-arena hypothesis, in *The Evolution of Sex and its Consequences* (ed S.C. Stearns), Birkhäuser, Basel, Switzerland, pp. 337–49.

Stearns, S.C. (ed) (1987c) *The Evolution of Sex and its Consequences*, Birkhäuser, Basel, Switzerland.

Stearns, S.C. (1989) The evolutionary significance of phenotypic plasticity. *BioScience*, **39**, 436–45.

Stearns, S.C. (1992) *The Evolution of Life Histories*, Oxford University Press, Oxford.

Stearns, S.C. and Crandall, R.E. (1984) Plasticity for age and size at sexual maturity: a life-history response to unavoidable stress, in *Fish Reproduction* (ed R.C. Wooton), Academic Press, NY, pp. 13–33.

Stearns, S., de Jong, G. and Newman, B. (1991) The effects of phenotypic plasticity on genetic correlations. *TREE*, **6**, 122–6.

Stearns, S.C. and Koella, J.C. (1986) The evolution of phenotypic plasticity in life-history traits: predictions of reaction norms for age and size at maturity. *Evolution*, **40**, 893–913.

Stebbins, G.L. (1950) *Variation and Evolution in Plants*, Columbia University Press.

Stemberger, R.S. and Gilbert, J.J. (1987) Multiple-species induction of morphological defenses in the rotifer *Keratella testudo. Ecology*, **68**, 370–8.

Stenseth, N.C. and Maynard Smith, J. (1984) Coevolution in ecosystems: red queen or stasis? *Evolution*, **38**, 870–80.

Stouthamer, R., Luck, R.F. and Hamilton, W.D. (1990) Antibiotics cause parthenogenetic *Trichogramma* (Hymenoptera/Trichogrammatidae) to revert to sex. *Proc. Natl. Acad. Sci. USA*, **87**, 2424–7.

Strobeck, C., Maynard Smith, J. and Charlesworth, B. (1976) The effects of hitch-hiking on a gene for recombination. *Genetics*, **82**, 547–58.

Struhl, G. (1981) A homoeotic mutation transforming leg to antenna in *Drosophila*. *Nature*, **292**, 635–8.

Stuart, J.J., Brown, S.J., Beeman, R.W. and Denell, R.E. (1991) A deficiency of the homeotic complex of the beetle *Tribolium*. *Nature*, **350**, 72–4.

Suarez, R.K., Lighton, J.R.B., Moyes, C.D. *et al.* (1990) Fuel selection in rufous hummingbirds: ecological implications of metabolic biochemistry. *Proc. Natl. Acad. Sci. USA*, **87**, 9207–10.

Sugawara, O., Oshimura, M., Koi, M. *et al.* (1990) Induction of cellular senescence in immortalized cells by human chromosome I. *Science*, **247**, 707–10.

Sultan, S.E. (1987) Evolutionary implications of phenotypic plasticity in plants. *Evol. Biol.*, **20**, 127–78.

Surani, M.A., Kothary, R., Allen, N.D. *et al.* (1990) Genome imprinting and development in the mouse. *Development*, **1990** (Suppl.), 89–98.

Surbey, M.K. and Rollo, C.D. (1991) Physiological and behavioural compensation for food quality and quantity in the slug *Lehmannia marginata*. *Malacologia*, **33**, 193–8.

Syvanen, M. (1984) The evolutionary implications of mobile genetic elements. *Ann. Rev. Genet.*, **18**, 271–93.

Tantawy, A.O. and El-Helw, M.R. (1966) Studies on natural populations of *Drosophila*. V. Correlated response to selection in *Drosophila melanogaster*. *Genetics*, **53**, 97–110.

Tartof, K.D. and Bremer, M. (1990) Mechanisms for the construction and developmental control of heterochromatin formation and imprinted chromosome domains. *Development*, **1990** (Suppl.), 35–45.

Tautz, D. (1992) Genetic and molecular analysis of early pattern formation in *Drosophila*, in *Development: the Molecular Genetic Approach* (eds V.E.A. Russo, S. Brody, D. Cove and S. Ottolenghi), Springer Verlag, Berlin, pp. 308–27.

Taylor, C.R. (1987) Structural and functional limits to oxidative metabolism: insights from scaling. *Ann. Rev. Physiol.*, **49**, 135–46.

Templeton, A.R. (1979) The unit of selection in *Drosophila mercatorum*. II. Genetic revolution and the origin of coadapted genomes in parthenogenetic strains. *Genetics*, **92**, 1265–82.

Templeton, A.R. (1980) The theory of speciation via the founder principle. *Genetics*, **94**, 1011–38.

Templeton, A.R. (1981) Mechanisms of speciation – a population genetic approach. *Ann. Rev. Ecol. Syst.*, **12**, 23–48.

Templeton, A.R. (1982a) The prophecies of parthenogenesis, in *Evolution and Genetics of Life Histories* (eds H. Dingle and J.P. Hegmann), Springer Verlag, NY, pp. 75–101.

Templeton, A.R. (1982b) Genetic architectures of speciation, in *Mechanisms of Speciation* (ed C. Baragozzi), Allan R. Liss, NY, pp. 105–21.

Templeton, A.R. (1989) The meaning of species and speciation: a genetic perspective, in *Speciation and its Consequences* (eds D. Otte and J.A. Endler), Sinauer Associates, Sunderland, MA, pp. 3–27.

Templeton, A.R., Hollocher, H., Lawler, S. and Johnson, J.S. (1990) The ecological genetics of abnormal abdomen in *Drosophila mercatorum*, in *Ecological and evolutionary genetics of Drosophila* (eds J.S. Barker, W.T. Starmer and R.J. MacIntyre), Plenum Press, NY, pp. 17–35.

Thoday, J.M. (1953) Components of fitness. *Symp. Soc. Expt. Biol.*, **7**, 96–113.

Thompson, D'A. W. (1942) *On Growth and Form*, Cambridge University Press, Cambridge.

Thompson, D.B. (1992) Consumption rates and the evolution of diet-induced plasticity in the head morphology of *Melanoplus femurrubrum* (Orthoptera, Acrididae). *Oecologia*, **89**, 204–13.

Thompson, J.D. (1991) Phenotypic plasticity as a component of evolutionary change. *TREE*, **6**, 246–9.

Thompson, J.N. (1982) *Interaction and Coevolution*, John Wiley and Sons, NY.

Thompson, V. (1976) Does sex accelerate evolution? *Evol. Theor.*, **1**, 131–56.

Thomson, J.A. (1987) Evolution of gene structure in relation to function, in *Rates of Evolution* (eds K.S.W. Campbell and M.F. Day), Allen and Unwin, London, pp. 189–208.

Thornhill, N.W. (ed) (1993) *The Natural History of Inbreeding and Outbreeding*, University of Chicago Press, Chicago.

Thornhill, R. and Alcock, J. (1983) *The Evolution of Insect Mating Systems*, Harvard University Press, Cambridge, MA.

Tiivel, T. (1991) Cell symbiosis, adaptation, and evolution: insect-bacteria examples, in *Symbiosis as a source of evolutionary innovation* (eds L. Margulis and R. Fester), MIT Press, Cambridge, MA, pp. 170–7.

Timm, R.M. (1982) Fahrenholz's rule and resource tracking: a study of host–parasite coevolution, in *Coevolution* (ed M.H. Nitecki), University Chicago Press, Chicago, IL, pp. 225–65.

Tinbergen, J.M., Van Balen, J.H. and Van Eck, H.M. (1985) Density-dependent survival in an isolated great tit population. Kluyvers data reanalysed. *Ardea*, **73**, 38–48.

Tittiger, C., Whyard, S. and Walker, V.K. (1993) A novel intron site in the triosephosphate isomerase gene from the mosquito *Culex tarsalis*. *Nature*, **361**, 470–2.

Tomlinson, J. (1966) The advantages of hermaphrodism and parthenogenesis. *J. Theor. Biol.*, **11**, 54–8.

Totter, J.R. (1985) Food restriction, ionizing radiation, and natural selection. *Mech. Ageing Develop.*, **30**, 261–71.

Travers, A. (1985) Sigma factors in multitude. *Nature*, **313**, 15–16.

Travis, J. (1992) Possible evolutionary role explored for 'jumping genes' *Science*, **257**, 884–5.

Trevelyan, R., Harvey, P.H. and Pagel, M.D. (1990) Metabolic rates and life histories in birds. *Funct. Ecol.*, **4**, 135–41.

Trewavas, A. (1986) Resource allocation under poor growth conditions. A major role for growth substances in developmental plasticity. *Symp. Soc. Expt. Biol.*, **40**, 31–76.

Trivers, R.L. (1971) The evolution of reciprocal altruism. *Quart. Rev. Biol.*, **46**, 35–57.

Trivers, R. (1985) *Social Evolution*, Benjamin/Cummings Publ. Co., Menlo Park, CA.

Trivers, R. (1988) Sex differences in rates of recombination and sexual selection, in *The Evolution of Sex* (eds R.E. Michod and B.K. Levin), Sinauer Associates, Sunderland, MA, pp. 270–85.

Trout, W.E. and Kaplan, W.D. (1970) A relation between longevity, metabolic rate, and activity in shaker mutants of *Drosophila melanogaster*. *Expt. Geront.*, **5**, 83–92.

Trout, W.E. and Kaplan, W.D. (1981) Mosaic mapping of foci associated with longevity in the neurological mutants Hk and Sh of *Drosophila melanogaster*. *Expt. Geront.*, **16**, 461–74.

Tucic, N., Cvetkovic, D. and Milanovic, D. (1988) The genetic variation and covariation among fitness components in *Drosophila melanogaster* females and males. *Heredity*, **60**, 55–60.

Tuomi, J., Hakala, T. and Haukioja, E. (1983) Alternative concepts of reproductive effort, costs of reproduction, and selection in life-history evolution. *Amer. Zool.*, **23**, 25–34.

Turing, A.M. (1952) The chemical basis of morphogenesis. *Phil. Trans. R. Soc. Lond. Ser. B*, **237**, 37–72.

Turlings, T.C.J., Tumlinson, J.H. and Lewis, W.J. (1990) Exploitation of herbivore-induced plant odors by host-seeking parasitic wasps. *Science*, **250**, 1251–3.

Ulanowicz, R.E. (1980) An hypothesis on the development of natural communities. *J. Theor. Biol.*, **85**, 223–45.

Underwood, A.J. (1986) What is a community? in *Patterns and Processes in the History of Life* (eds D.M. Raup and D. Jablonski), Springer Verlag, Berlin, pp. 351–67.

Val, F.C. (1977) Genetic analysis of the morphological differences between two species of interfertile species of Hawaiian *Drosophila*. *Evolution*, **31**, 611–29.

Van Noordwijk, A.J. (1989) Reaction norms in genetical ecology. *BioScience*, **39**, 453–8.

Van Noordwijk, A.J. (1990) The methods of genetical ecology applied to the study of evolutionary change, in *Population Biology* (eds K. Wohrmann and S. Jain), Springer Verlag, Berlin, pp. 291–319.

Van Noordwijk, A.J. and de Jong, G. (1986) Acquisition and allocation of resources: their influence on variation in life history tactics. *Amer. Nat.*, **128**, 137–42.

Van Valen, L. (1973) A new evolutionary law. *Evol. Theor.*, **1**, 1–30.

Van Valen, L. (1976) Ecological species, multispecies and oaks. *Taxon*, **25**, 233–9.

Van Voorhies, W.A. (1992) Production of sperm reduces nematode lifespan. *Nature*, **360**, 456–8.

Vaux, D.L., Weissman, I.L. and Kim, S.K. (1992) Prevention of programmed cell death in *Caenorhabditis elegans* by human bcl–2. *Science*, **258**, 1955–7.

Vepsalainen, K. and Patama, T. (1983) Allocation of reproductive energy in relation to the pattern of environment in five *Gerris* species, in *Diapause and Life Cycle Strategies in Insects* (eds V.K. Brown and I. Hodek), Dr. W. Junk, The Hague, pp. 189–207.

Vermeij, G.J. (1987) *Evolution and Escalation: an Ecological History of Life*, Princeton University Press, Princeton, NJ.

Via, S. (1987) Genetic constraints on the evolution of phenotypic plasticity, in *Genetics Constraints on Adaptive Evolution* (ed V. Loeschcke), Springer Verlag, Berlin, pp. 47–71.

Via, S. (1992) Models of the evolution of phenotypic plasticity. *TREE*, **7**, 63.

Via, S. and Lande, R. (1985) Genotype–environment interaction and the evolution of phenotypic plasticity. *Evolution*, **39**, 505–22.

Via, S. and Lande, R. (1987) Evolution of genetic variability in a spatially heterogeneous environment: effects of genotype–environment interaction. *Genet. Res.*, **49**, 147–56.

Vail, S.G. (1992) Selection for overcompensatory plant responses to herbivory: a mechanism for the evolution of plant–herbivore mutualism. *Amer. Nat.*, **139**, 1–8.

Von Frisch, K. (1974) *Animal Architecture*, Harcourt Brace Janovich, NY.

Von Bertalanffy, L. (1952) *Problems of Life*, Wiley.

Vrba, E.S. (1989a) Levels of selection and sorting with special reference to the species level. *Oxford Surv. Evol. Biol.*, **6**, 111–68.

Vrba, E.S. (1989b) What are the biotic hierarchies of integration and linkage? in (eds D.B. Wake and G. Roth), *Complex Organismal Functions: Integration and Evolution in Vertebrates*, J. Wiley and Sons, pp. 379–401.

Vrijenhoek, R.C. (1984) Ecological differentiation among clones: the frozen niche variation model, in *Population Biology and Evolution* (eds K. Wohrmann and V. Loeschcke), Springer Verlag, Berlin, pp. 217–31.

Vrijenhoek, R.C. (1990) Genetic diversity and the ecology of asexual populations, in *Population Biology* (eds K. Wohrmann and S. Jain), Springer Verlag, Berlin, pp. 175–97.

Waddington, C.H. (1940) *Organizers and Genes*, Cambridge University Press, Cambridge.

Waddington, C.H. (1956) Genetic assimilation of the bithorax phenotype. *Evolution*, **10**, 1–13.

Waddington, C.H. (1957) *The Strategy of the Genes*, George Allen and Unwin, London.

Waddington, C.H. (1959) Canalization of development and genetic assimilation of acquired characters. *Nature*, **183**, 1654–5.

Waddington, C.H. (1975) *The Evolution of an Evolutionist*, Edinburgh University Press, Edinburgh.

Wade, M. (1977) An experimental study of group selection. *Evolution*, **31**, 134–53.

Wade, M.J. (1980) Kin selection: its components. *Science*, **210**, 665–7.

Wade, M.J. (1985) Soft selection, hard selection, kin selection and group selection. *Amer. Nat.*, **125**, 61–73.

Wade, M.J. (1990) Genotype–environment interaction for climate and competition in a natural population of flour beetles, *Tribolium castaneum*. *Evolution*, **44**, 2004–11.

Wade, M.J. (1992) Sewall Wright: gene interaction and the shifting balance theory. *Oxford. Surv. Evol. Biol.*, **8**, 35–62.

Wade, M.J. and Goodnight, C.J. (1991) Wright's shifting balance theory: an experimental study. *Science*, **253**, 1015–18.

Wade, M.J. and McCauley, D.E. (1980) Group selection: the phenotypic and genotypic differentiation of small populations. *Evolution*, **34**, 799–812.

Wagner, G.P. and Gabriel, W. (1990) Quantitative variation in finite parthenogenetic populations: what stops Müller's ratchet in the absence of recombination? *Evolution*, **44**, 715–31.

Wainwright, P.C., Osenberg, C.W. and Mittelbach, G.G. (1991) Trophic polymorphism in the pumpkinseed sunfish (*Lepomis gibbosus* Linnaeus): effects of environment on ontogeny. *Funct. Ecol.*, **5**, 40–55.

Wake, D.B. and Larson, A. (1987) Multidimensional analysis of an evolving lineage. *Science*, **238**, 42–8.

Wake, D.B., Roth, G. and Wake, M.H. (1983) On the problem of stasis in organismal evolution. *J. Theor. Biol.*, **101**, 211–24.

Wake, D.B. and Roth, G. (1989) The linkage between ontogeny and phylogeny in the evolution of complex systems, in *Complex Organismal Functions: Integration and Evolution Vertebrates* (eds D.B. Wake and G. Roth), John Wiley and Sons, pp. 361–77.

Wallace, B. (1991) Coadaptation revisited. *J. Heredity*, **82**, 89–95.

Walldorf, U., Fleig, R. and Gehring, W.J. (1989) Comparison of homeobox-containing genes of the honeybee and *Drosophila*. *Proc. Natl. Acad. Sci. USA*, **86**, 9971–5.

Warner, R.R., Robertson, D.R. and Leigh Jr., E.G. (1975) Sex change and sexual selection. *Science*, **190**, 633–8.

Washburn, J.O., Gross, M.E., Mercer, D.R. and Anderson, J.R. (1988) Predator-induced trophic shift of a free-living ciliate: parasitism of mosquito larvae by their prey. *Science*, **240**, 1193–5.

Washburn, J.O., Mercer, D.R. and Anderson, J.C. (1991) Regulatory role of parasites: impact on host population shifts with resource availability. *Science*, **253**, 185–8.

Wassarman, D.A. and Steitz, J.A. (1991) Alive with dead proteins. *Nature*, **349**, 463–4.

Watkinson, A. (1992) Plant senescence. *TREE*, **7**, 417–20.

Watson, J.D., Tooze, J. and Kurtz, D.T. (1983) *Recombinant DNA*, W.H. Freeman, NY.

Watt, W.B. (1985) Bioenergetics and evolutionary genetics: opportunities for new synthesis. *Amer. Nat.*, **125**, 118–43.

Wattiaux, J.M. (1968) Cumulative parental age effects in *Drosophila subobscura*. *Evolution*, **22**, 406–21.

Weibel, E.R. (1987) Scaling of structural and functional variables in the respiratory system. *Ann. Rev. Physiol.*, **49**, 147–59.

Weibel, E.R., Taylor, C.R. and Hoppeler, H. (1991) The concept of symmorphosis: a testable hypothesis of structure–function relationship. *Proc. Natl. Acad. Sci. USA*, **88**, 10357–61.

Weider, L.J. and Pijanowska, J. (1993) Plasticity of *Daphnia* life histories in response to chemical cues from predators. *Oikos*, **67**, 385–92.

Weinberg, R.A. (1985) The molecules of life. *Sci. Amer.*, **253**, 48–57.

Weindruch, R. and Walford, R.L. (1988) *The Retardation of Aging and Disease by Dietary Restriction*, Charles Thomas, Springfield, IL.

Weindruch, R., Walford, R.L., Fligiel, S. and Guthrie, D. (1986) The retardation of aging in mice by dietary restriction: longevity, cancer, immunity and life time energy intake. *J. Nutrit.*, **116**, 641–54.

Weiner, J. (1992) Physiological limits to sustainable energy budgets in birds and mammals: ecological implications. *TREE*, **7**, 384–8.

Wellington, W.G. (1965) Some maternal influences on progeny quality in the western tent caterpillar, *Malacosoma pluviale* (Dyar). *Can. Entomol.*, **97**, 1–14.

Wellington, W.G. (1980) Dispersal and population change, in *Dispersal of Forest Insects: Evolution, Theory and Management Implications* (eds A.M. Berryman and L. Safranyik), Cooperative Extension Service, Washington State University, Pullman, pp. 11–24.

Welty, C. (1955) Birds as flying machines. *Sci. Amer.*, **192**, 88–97.

West-Eberhard, M.J. (1989) Phenotypic plasticity and the origins of diversity. *Ann. Rev. Ecol. Syst.*, **20**, 249–78.

Western, D. (1979) Size, life history and ecology in mammals. *Afr. J. Ecol.*, **17**, 185–204.

Western, D. and Ssemakula, J. (1982) Life history patterns in birds and mammals and their evolutionary interpretation. *Oecologica*, **54**, 281–90.

Westerterp, K. (1977) How rats economize – energy loss in starvation. *Physiol. Zool.*, **50**, 331–62.

Westoby, M. (1981) How diversified seed germination behavior is selected. *Amer. Nat.*, **118**, 882–5.

Wheeler, D.E. (1991) The developmental basis of worker caste polymorphism in ants. *Amer. Nat.*, **138**, 1218–38.

Wheeler, D.E. and Nijhout, H.F. (1981) Soldier determination in ants: new role for juvenile hormone. *Science*, **213**, 361–3.

White, M.J.D. (1978) *Modes of Speciation*, W.H. Freeman, San Francisco, CA.

Whitham, T.G. (1989) Plant hybrid zones as sinks for pests. *Science*, **244**, 1490–3.

Whittemore, A.T. and Schaal, B.A. (1991) Interspecific gene flow in sympatric oaks. *Proc. Natl. Acad. Sci. USA*, **88**, 2540–4.

Wiener, P., Feldman, M.W. and Otto, S.P. (1992) On genetic segregation and the evolution of sex. *Evolution*, **46**, 775–82.

Wiens, J.A. (1977) On competition and variable environments. *Am. Sci.*, **65**, 590–7.

Wieser, W. (1991) Limitations of energy acquisition and energy use in small poikilotherms: evolutionary implications. *Funct. Ecol.*, **5**, 234–40.

Wilbur, H.M. and Fauth, J.E. (1990) Experimental aquatic food webs: interactions between predators and their prey. *Amer Nat.*, **135**, 176–204.

Wilkinson, G.S. (1988) Reciprocal altruism in bats and other mammals. *Ethol. Sociobiol.*, **9**, 85–100.

Wilkinson, G.S., Fowler, K. and Partridge, L. (1990) Resistance of genetic correlation structure to directional selection in *Drosophila melanogaster. Evolution*, **44**, 1990–2003.

Williams, G.C. (1957) Pleiotropy, natural selection, and the evolution of senescence. *Evolution*, **11**, 398–411.

Williams, G.C. (1966a) *Adaptation and Natural Selection*, Princeton University Press, Princeton.

Williams, G.C. (1966b) Natural selection, the costs of reproduction, and a refinement of Lack's principle. *Amer. Nat.*, **100**, 687–90.

Williams, G.C. (1975) *Sex and Evolution*, Princeton University Press, Princeton.

Williams, G.C. (1985) A defense of reductionism in evolutionary biology. *Oxford Surv. Evol. Biol.*, **2**, 1–27.

Williams, G.C. (1988) Retrospect on sex and kindred topics, in *The Evolution of Sex* (eds R.E. Michod and B.R. Levin), Sinauer Associates, Sunderland, MA, pp. 287–98.

Williams, G.C. (1992) *Natural Selection: Domains, Levels and Challenges*, Oxford University Press, Oxford.

Williams, G.C. and Mitton, J.B. (1973) Why reproduce sexually? *J. Theor. Biol.*, **39**, 545–54.

Williams, G.C. and Taylor, P.D. (1987) Demographic consequences of natural selection, in *Evolution of Longevity in Animals* (eds A.D. Woodhead and K.H. Thompson), Plenum Press, NY, pp. 235–42.

Wills, C. (1989) *The Wisdom of the Genes*, Basic Books, NY.

Wills, C. (1991) *Exons, Introns and Talking Genes*, Basic Books, NY.

Wilson, A.C. (1976) Gene regulation in evolution, in *Molecular Evolution* (ed F.J. Ayala), Sinauer Associates, Sunderland, MA, pp. 225–34.

Wilson, A.C. (1985) The molecular basis of evolution. *Sci. Amer.*, **253**, 164–73.

Wilson, A.C., Maxson, L.R. and Sarich, V.M. (1974a) Two types of molecular evolution. Evidence from studies of interspecific hybridization. *Proc. Natl. Acad. Sci. USA*, **71**, 2843–7.

Wilson, A.C., Sarich, V.M. and Maxson, L.R. (1974b) The importance of gene rearrangement in evolution: evidence from studies on rates of chromosomal, protein and anatomical evolution. *Proc. Natl. Acad. Sci. USA*, **71**, 3028–30.

Wilson, D.S. (1976) Evolution on the level of communities. *Science*, **192**, 1358–60.

Wilson, D.S. (1977) Structured demes and the evolution of group-advantageous traits. *Amer. Nat.*, **111**, 157–85.

Wilson, D.S. (1978) Prudent predation: a field study involving three species of tiger beetles. *Oikos*, **31**, 128–36.

Wilson, D.S. (1980) *The Natural Selection of Populations and Communities*, Benjamin Cummings, Menlo Park, CA.

Wilson, D.S. (1983) The group selection controversy: history and current status. *Ann. Rev. Ecol. Syst.*, **14**, 159–87.

Wilson, D.S. (1988) Holism and reductionism in evolutionary ecology. *Oikos*, **53**, 269–73.

Wilson, D.S. (1989) The diversification of single gene pools by density- and frequency-dependent selection, in *Speciation and its Consequences* (eds D. Otte and J.A. Endler), Sinauer Associates, Sunderland, MA, pp. 366–85.

Wilson, D.S. (1990) Weak altruism, strong group selection. *Oikos*, **59**, 135–40.

Wilson, D.S. and Sober, E. (1989) Reviving the superorganism. *J. Theor. Biol.*, **136**, 337–56.

Wilson, E.O. (1975) *Sociobiology: the New Synthesis*, Belknap, Harvard University

Press, Cambridge, MA.

Wilson, E.O. (1985) The sociogenesis of insect colonies. *Science*, **228**, 1489–95.

Wimberger, P.H. (1991) Plasticity of jaw and skull morphology in the neotropical cichlids *Geophagus brasiliensis* and *G. steindachneri*. *Evolution*, **45**, 1545–63.

Woese, C.R. (1983) The primary lines of descent and the universal ancestor in *Evolution from Molecules to Men* (ed D.S. Bendall), Cambridge University Press, Cambridge, pp. 207–33.

Wolf, T.J. and Schmid-Hempel, P. (1989) Extra loads and foraging life span in honeybee workers. *J. Anim. Ecol.*, **58**, 943–54.

Wolpert, L. (1983) Constancy and change in the development and evolution of pattern, in *Development and evolution* (eds B.C. Goodwin, N. Holder and C.C. Wylie), Cambridge University Press, Cambridge, pp. 47–57.

Wood, T.K., Olmstead, K.L. and Guttman, S.F. (1990) Insect phenology mediated by host-plant water relations. *Evolution*, **44**, 629–36.

Wootton, J.T. (1987) The effects of body mass, phylogeny, habitat, and trophic level on mammalian age at first reproduction. *Evolution*, **41**, 732–49.

Wright, S. (1931) Evolution in Mendelian populations. *Genetics*, **16**, 97–159.

Wright, S. (1932) The roles of mutation, inbreeding, crossbreeding and selection in evolution. *Proc 6th Int. Congr. Genetics*, **1**, 356–66.

Wright, S. (1964) Biology and the philosophy of science, in *Process and Divinity: the Hartshorne Festschrift* (eds W.R. Reese and E. Freeman), La Salle, IL, Open Court (also the *Monist*, **48**, 265–90).

Wright, S. (1980) Genic and organismic selection. *Evolution*, **34**, 825–43.

Wright, S. (1982a) Character change, speciation and the higher taxa. *Evolution*, **36**, 427–43.

Wright, S. (1982b) The shifting balance theory and macroevolution. *Ann. Rev. Genetics*, **16**, 1–19.

Wright, S. (1988) Surfaces of selective value revisited. *Amer. Nat.*, **131**, 115–23.

Wynne-Edwards, V.C. (1962) *Animal Dispersion in Relation to Social Behaviour*, Oliver and Boyd, Edinburgh.

Wynne-Edwards, V.C. (1986) *Evolution through Group Selection*, Blackwell, Oxford.

Wyss, A. (1990) Clues to the origin of whales. *Nature*, **347**, 428–9.

Yahav, S., Buffenstein, R., Jarvis, J.U.M. and Mitchell, D. (1989) Thermoregulation and evaporative water loss in the naked mole rat, *Heterocephalus glaber*. *S. Afr. J. Sci.*, **85**, 340.

Yeakley, J.A. and Cale, W.G. (1991) Organizational levels analysis: a key to understanding processes in natural systems. *J. Theor. Biol.*, **149**, 203–16.

Yodzis, P. (1980) The connectance of real ecosystems. *Nature*, **284**, 544–5.

Yokouchi, Y., Sasaki, H. and Kuroiwa, A. (1991) Homeobox gene expression correlated with the bifurcation process of limb cartilage development. *Nature*, **353**, 443–5.

Young, J.P.W. (1981) Sib competition can favour sex in two ways. *J. Theor. Biol.*, **88**, 755–6.

Young, L.S., Dunstan, H.M., Witte, P.R. *et al.* (1991) A class III transcription factor composed of RNA. *Science*, **252**, 542–6.

Yu, B.P. (1987) Update on food restriction and aging. *Rev. Biol. Res. Aging*, **3**, 495–505.

Yu, B.P., Masoro, E.J. and McMahan, C.A. (1985) Nutritional influences on aging in Fischer 344 rats, I. Physical, metabolic and longevity characteristics. *J. Geront.*, **40**, 657–70.

Zeng, Z.-B. (1988) Long-term correlated response, interpopulation covariation, and interspecific allometry. *Evolution*, **42**, 363–74.

Zhuchenko, A.A., Korol, A.B. and Kovtyukh, L.P. (1985) Change of the crossing-over frequency in *Drosophila* during selection for resistance to temperature fluctations. *Genetics*, **67**, 73–8.

Zwaan, B.T., Bijlsma, R. and Hoekstra, R.F. (1991) On the developmental theory of ageing. I. Starvation resistance and longevity in *Drosophila melanogaster* in relation to pre-adult breeding conditions. *Heredity*, **66**, 29–39.

Index

Page numbers appearing in **bold** refer to figures and page numbers appearing in *italic* refer to tables.

Acclimation 351–2
Adaptability, range of 239
Adaptive radiation 194–5
Aging
  error catastrophe theory 318, 376
  mechanisms of 321–2
  see also Longevity
Allocation 279, 282, 303–9, 336–74
Allometric analyses 342–9
Allomones 55
Altruism 38–40, 116, 119–20
  reciprocal 38–9
Amplification 36
Antagonism 112
Antagonistic pleiotropy 281, 314, 317, 321
Antioxidant factors, longevity and 323–4, 326, 329, 376
Aphids 169
Artificial selection 262
Asexuality and stress environments 391–2
Autoregulation 85

Bacteria, substrate utilization by 63–4
  see also Prokaryotes
Balance hypothesis 253
Baldwin effect 223
Basal metabolic rate
  and body mass 337–8, 341
  evolution of 296
Bats, longevity 355–8
Bauplan 26, 253, 277, 341
Beetles, scarab **206**

Behaviour
  and longevity 358–61
    in *Drosophila* 358–9
    exercise 359–61
    migration 360
    sleep 361
    of Supermouse 359
  and morphology 277–8
  social cooperative, evolution of 116
Behavioural time budget **274**, 274–5
Body
  plans, *see* Bauplan; Gastropods
  reserves, size of 300–1, 330–2 size
    adjustments with temperature 352–3
    correlation with life–history features 338–9, 344, 346, 348–9
    and longevity 338–9, 369–73
    and metabolic rate 337–8, 341–2, 346–8
    and sustainable scope 291–3
Boolean logic 24
Brain size, evolution of 12
Burdock 181, 277

Cambrian, explosion of diversity 44, 275–6
Canalization 61, 198–9, **214**, 218–22, 243, 370–1
Cancer 383, 384
  cells, immortality of 383
Cannibalism 379
Cascade, epigenetic 96
Centre court advantage, *see* Hypothesis

Chaos 24–5
  theory 318
Chloroplasts 27–8, 41, 43
Chromatin 64, 67, 70–1
  structure, regulation of 64–71
Chromosome
  inversions, polymorphisms for 132
  number, intraspecific variation 132
  structure
    eukaryotic 64, 67
    intraspecific variation 132
Cichlid fish, evolution of 93–4,
    227–8
Clade selection 145
Cladogenesis 143, 227
Coadaptation 121, 150–1, 252
  and genetic revolutions 258–67
Coevolution
  in ecological communities 48, 50–1
  of parasites and host 192–3
  of pathogen and host 191–2
  of plants and herbivores 227
  of symbiotic organisms 46
Coevolutionary disequilibria 189
Communication, sexual 111, 138–44
Community, ecological 48–50, 55
  coevolution in 48, 50–1
  stability of 52–4
Competition 276
  and stress 390–1
Computer technology 3
  evolution of 44
Connectance 53–5
Connectedness 28
Convergence 18, **19**, **30**, 278
Cooperation 32
Cope's rule 27
Crosstalk, among subspecific lineages
    174–82, 192
Cytosine, methylation of 70

Darwin, Charles 1–2, 17
Dawkins, Richard 2, 4, 6
Death
  criteria favouring adaptive
    evolution of 380
  importance of 376–80, 386
  programmed 384–6
Defence
  inducible 202–4, 236

investment in 389–90
  longevity and 327–9, 364
  somatic 327–9
Demes 145
Design
  of muscular and skeletal systems
    11–12
  organismal 1–3, 273
    physiological trade-offs 279
Development
  of *D. melanogaster* 82–3, 105
  effects of hormones on 217–18
  of grasshopper 105
  of vertebrate limb **97**
Developmental conversion 209
Diapause 201–2, 355
Dietary restriction 300
  effects of 366, **372**
  and longevity 368–71
  paradigm 365–73
Differential splicing 74
Differentiation 44
  anterioposterior, in *D. melanogaster*
    86–91
  dorsoventral, in *D. melanogaster*
    85–6
Dimorphism, sexual 118
Diploidy, evolution of 186–7
Disease 286, 384
  see also Pathogens
Disposable soma hypothesis 314, 316,
    321, 326, 333
Disturbance
  environments characterized by
    393–7
  paradigm 393–7
Diversification
  evolutionary 111
  ontogenetic 199, 201–2
  potential 194–6
DNA
  non-coding 28–9, 61
  repair ability and longevity 325
Domain model of eukaryotic gene
    regulation **69**, 70
Drosophila heteroneura 94, **94**
Drosophila melanogaster
  abnormal abdomen mutation 314
  anterioposterior differentiation
    86–91

behavioural activity and longevity 358–9
control of development 82–3, 105
dorsoventral differentiation 85–6
effects of ethidium bromide on 350
gap genes 86–8
gene clusters
  Antennapedia complex 90, 104
  Bithorax complex 90, 92, 104
genes
  abdominal A 90–1
  abdominal B 90–1
  antennapedia 90–1
  bicoid 87–8
  bithorax 225
  buttonhead 88
  cactus 86
  caudal 87, 90
  cornichon 85
  daughterless 84
  deformed 90
  dorsal 86
  doublesex 85
  easter 86
  empty spiracles 88
  engrailed 90, 105
  eve 89–90
  even–skipped 89–90, 105
  exuperantia staufen 87
  fushi tarazu 89–90, 104–5
  giant 88
  gurken 85
  hairy 89–90
  hunchback 87–8
  huckebein 87–8
  intersex 85
  K10 85
  knirps 87–8
  Krüppel 87–8
  labial 90
  nanos 87
  nudel 86
  orthodenticle 88
  oskar 87
  paired 89
  pelle 86
  pipe 86
  pumilio 87
  runt 89–90
  scute-a 84
  sex combs reduced 90
  sex-lethal 84–5
  sisterless 84–5, 90, 118
  snake 86
  spätzle 86
  swallow 87
  tailless 87–8
  Toll 86
  torpedo 85
  torso 87–8
  transformer 85
  transformer 2 85
  tube 86
  Tudor 84
  ultrabithorax 90–1
  vasa 84
  vestigial 127
  windbeutel 86
  wingless 90
genetic correlation in 248
growth in 291
homeotic selector genes 87, 90–1
longevity of 330–2, **353**
  and reproduction 362–3
maternal genes 87–8
pair–rule genes 86, 89–90
segment polarity genes 87, 90
segregation distorter system 140
sex determination 83–5
Drosophila sylvestris 94, **94**
Dynamics
  ecological **268**
  genetic **268**
  metapopulation 267–71

Ecology, non-equilibrium 40
Ecosystem, regulatory 76
Embryogenesis, methylation changes during 70
Endocrine control, *see* Hormones
Endotherms 352, 354
Enhancement 36
Enhancers 67, 73
Enslavement 41
Entification 25, 27
Environmental cycles, plasticity and 201
Epigenetics 61, 132
  importance of 80–1
  insect 95–6
Epistasis 160, 210

Error catastrophe theory of aging 318, 376
Ethidium bromide, effects on *D. melanogaster* 350
Euchromatin 64
Eukaryotes
  cell–cell interactions 75–6
  chromosome structure 64, 67
  conserved biochemistry of 92
  gene **65**
  gene conversion 79
  gene duplication 76–9
  gene regulation in 64–81
    domain model **69**, 70
    enhancers 73
    genomic imprinting 70
    initiation complex 71
    promoters 71
    regulatory recognition sites 71–2
    silencers 73
    transcription factors 71–2
  genome, evolution of **77**
  hormonal effects on regulation 76
  transcriptional regulation **68**
    processing of mRNA 73–4
    rate of release of mRNA to cytoplasm 75
  transposable elements 79
Eusociality 46–7, 263
  evolution of 46–7, 116, 209
  see also Insects, eusocial
Evofacilitation 134
Evolution
  of basal metabolic rate 296
  biochemical 42
  of brain size 12
  of cichlid fish 93–4, 227–8
  concerted 79
  of diploidy 186–7
  of eukaryotic genome **77**
  of eusociality 46–7, 116, 209
  of fish tails 10–11, **11**
  gene segregation, role in 156
  of human societies 48
  of immune system 190–2
  of microcomputer technology 44
  of morphogenesis 44
  of multicellularity 43–4
  of norms of reaction **206**, 231–3, 237–8, **240**

  organizational 32–7, 41–2
  outcrossing, role in 157
  of pathogens 189–94
  of plastic organizations 229–45
  of polymorphic switches 231
  primordial 42
  recombination and 156–7, 172
  regulatory genes and 92–6, 100–1, 103–4
  of senescence 310–35
    antagonistic pleiotropy hypothesis 314, 317, 321
    disposable soma hypothesis 314, 316, 321, 326, 333
    mutation accumulation hypothesis 312–14, 317–18, 380–1
  of sexual reproduction 45–7, 144, 153–7
    genic framework 157–62
    hypotheses of 162–83
    see also Hypothesis
  of social cooperative behaviour 116
  subspecific lineage 112
  of tree snails 94–5
  of tRNA 75
Evolvability hypothesis 154–6, 163
Exaptations 135
Exons 73–4
  shuffling 73–4, 134
Exotherms 352

Fish
  cichlid, evolution of 93–4, 227–8
  tails, evolution of 10–11, **11**
Fitness 32
  geometric mean 32–3
  inclusive 32
    benefits 32
  key features of 194
Foraging costs 299–300
Founder events 259–61, 265–6
Free radicals, *see* Oxygen

Gaia 56
Gastropods, spiral body plan 102
Gene
  clusters, *see Drosophila melanogaster*
  conversion 79
  duplication 76–9, 108
  families 78

regulation
  eukaryotic, *see* Eukaryotes
  prokaryotic 63
  segregation, role in evolution 156
Generalist strategies 276
Genes
  cryptic 126–7, 129
  eukaryotic **65**
  as fundamental units of selection
    114–15
  gap 86–8
  hierarchical organization of 101
  homeobox 104
  homeotic selector 87, 90–1
  inactive 125
  as interactors 109
  maternal 87–8
  pair–rule 86, 89–90
  regulatory 45, 100–1
    complexes 103
    conservation of 100
    and evolution 92–6, 100–1,
      103–4
    linkage of 104
  as replicators 109
  segment polarity 87, 90
  see also Drosophila melanogaster
Genetic
  assimilation 223–9
  code 12–13
  correlations 107, **244**, 246–51
    evolution of 248
  inertia 125–9, 271
    cost 129–30
  retention 125–6, 128
  revolutions
    coadaptation and 258–67
    and metapopulation dynamics
      267–71
Genomes
  coadapted 252–71
    developmental constraints 254
    ecological constraints 254–5
    genetic constraints 253
    genetic polymorphism and 257
    historical constraints 253
    metabolic constraints 253–4
  evolutionary potential of 134
  hierarchically organized, as units of
    selection 115, 119–24

open and closed components
    265–6
  structure 133
Genotypes, extended purpose
    390–2
Germ line, immortality of 383–4
'Goldschmidt toad' 105, **106**
Grasshopper, development 105
Growth
  of *D. melanogaster* 291, **353**
  and longevity 363–4, 370
  population *312*
  of Supermouse 304–5, **372**

Haemoglobin 18, 22
  fetal 22
Helix-loop-helix
  domains 72
  motifs 72
Heredity, basis of 59–62
Heterochromatin 64
Heterochromatinization 67, 127
Heterochrony 106–7
Heterozygosity, maintenance of
    122–3
Hibernation 355
Hierarchies
  biological 8
  control, *see* Hierarchies,
    organizational
  ecological 23
  epigenetic 91–109
  genealogical 23
  levels of life **7**
  lineage structure and 144–52,
    182
  organizational 8, 98–9, 144
    of epigenetics in *Drosophila* **99**,
      99–100
  regulatory 96–109, 112, 132, 182
    selection and 115
  taxonomic 8
  theory 7–8
Histones 67
Holon 7
  organismal 18
Hopscotch, hierarchical 182–3
Hormones
  effects on development 217–18
  in eukaryotic regulation 76

Hybridization 133, 152, 180–1, 262, 391, 396
  advantage of 193
Hypothesis
  antagonistic pleiotropy 314, 317, 321
  balance 253
  disposable soma 314, 316, 321, 326, 333
  of evolution of sex
    centre court advantage 170–4, 258
    mesoevolutionary advantage 155, 162–9, 175
    lineage level 162
  evolvability 154–6, 163
  meiotic drive 158
  mesoevolutionary 155, 162–9, 175
  mutation accumulation 312–14, 317–18, 380–1
  phenotypic balance 297–303, 324, 327
  Red Queen 155–6, 188–94
  safe tuning 294–7
  Tangled Bank 156, 194–6
  see also Theory

Immortality 377, 381–4
  cancer cells 383
  germ line 383–4
Immune system 189–90
  evolution of 190–2
Imprinting, genomic 67, 70, 208–9
Inbreeding 117, 119–20, 265–6, 302
  importance of 147
Inertia, genetic 125–9, 271
  cost of 129–30
Information, transgenerational 88
Initiation complex 71
Insects
  costs of migration 360
  eusocial 33, 80
    ants 37, 242
    caste differentiation 242
    honey–bees 36, 242
    termites 36, 47
  longevity in 329–32, 334–5, 360
  see also Beetles, *Drosophila*, Grasshopper, Milkweed bug
Instinct 245
Interaction effects 14–16
Interactions, predatory 41–2
Interactors 108
Interbreeding 262

Interspecific crosses 208
Introns 73
Inversion polymorphisms 132, 245, 262–3
Isolation 119

Juvenile stages, evolutionary modification of 102

Kairomones 55
Kin
  selection 32, 142, 148
  theory 38–9, 116, 120–1, 148

Lansing effect 385
Leucine zippers 72
Life, levels of **7**
Life–history
  schedules 343–4
  theory 279
Life–span, limitations of 320
Life–time
  energy 339
  limitations 375–86
  scopes 339–40
  and longevity 340
  metabolic 336–42
Liguus fasciatus, see Tree snails
Limbs, loss of in vertebrates 101–2
Lineage
  crosstalk 192
  hierarchy 182
  identity 117
  selection 143, 149, 151
    and sexual reproduction 153–7
  structure, hierarchies and 144–52
Lineages
  asexual **139**, 146, 162, **163**
  parthenogenetic 139–40
  as selective units 116–18, 131
  sexual **139**, 141–2, 144–6, 162, **163**, 182
  subspecific 144
    crosstalk among 174–82, 192
    evolution 112
    reality of in sexual species 131
Linkage 121–2
  disequilibrium 159–61, 164, 247
Linked hitchhiker framework 158

Longevity 376
  antioxidant factors and 323–4, 326, 329, 376
  assurance 310–35
  of bats 355–8
  behavioural activity and 358–61
  body size and 338–9, 369–73
  of clonal organisms 382
  costs of 321–35, 341
  defence and 364
  diapause and 355
  dietary restriction and 368–71
  DNA repair ability and 325
  of Drosophila 330–2, **353**
    and reproduction 362–3
  exercise and 359–61
  growth and 363–4, 370
  hibernation and 355
  in insects 329–32, 334–5
  and life–time scope 340
  maturation time and 343–4
  metabolic rates and 325–7, 329–30, 341, 343, 349–64, **350**
  migration and 360
  of modular organisms 381
  of naked mole-rat 356
  reproduction and 361–3
    in *Drosophila* 362–3
  selection for advanced 320–1
  sleep and 361
  somatic defence and 327–9
  stress and 365–73
  of Supermouse 304, 326–8
  superoxide dismutase activity and 323
  temperature and 350–5, **353**

Macromutations 105–8
Maturation time, and longevity 343–4
Meiotic drive 140–1, 158
  hypothesis 158
Mesoevolution, of plastic organizations 229–45
Mesoevolutionary hypothesis 155, 162–9, 175
Metabolic scope 288, **293**
  limitations on 288
Metabolism, rates of
  and longevity 325–7, 329–30, 341, 349–64, **350**

and sustainable scope 291–3, **293**
Metamorphosis 102–3
Metapopulation dynamics, genetic revolutions and 267
Metapopulations 151–2
Methylation
  changes during embryogenesis 70
  of cytosine 70
Microevolution 112
Migration, costs of in insects 360
Milkweed bug 248
Mitochondria 27–8, 41, 43
Model
  domain, of eukaryotic gene regulation **69**, 70
  ecological triumvirate **388**
  operon 63
  regulatory 110
Modifiers 107–8
  linked, selection of 157–9
Molecular drive 79
Morphogenesis, evolution of 44
Morphology 276–9
Mouse
  genetic correlations in 248
  longevity 307
  marsupial
    programmed death 385
    sustainable scope 291
  spinner 226, 306–7
  t-haplotypes 140
  see also Supermouse
mRNA
  processing of 73–4
  rate of release to cytoplasm 75
Muller's ratchet 44, 154–5, 183–8, 286, 376, 381–4
Muscular and skeletal systems, design of 11–12
Mutation accumulation hypothesis 312–14, 317–18, 380–1
Mutualism, stability of 53–4

Naked mole–rat 263
  longevity 356
Natural selection 17, 19
Neo-Darwinism 60–1
Neurospora crassa, AGE-1 complex 376
Niche
  specialization 194–5, 200–1

variation, diversification of phenotypes in response to 200–1
Norms of reaction 213–18, **214**, 239–5, **241**, **244**
  evolution of 231–3, 237–8
Nucleosome 67, **68**

Oncopeltus fasciatus, see Milkweed bug
Operons 63
Organisms
  clonal, longevity 382
  modular, longevity 381
Organization
  biological 23–6, 26, 28, 42
    evolution of 23–57
  filtering properties of 34–5, **35**
  genomic 62–3, **98**, 98–100
  metalineage 138–44
  organismal 29
  phenotypic, trade–offs and synergisms 272–309
  plastic, evolution of 229–45
  social 33
Outcrossing 176–7, 388
role in evolution 157
Oxygen, free radicals, role in senescence 322

Paley, William 1
Panmixia 144
Paradigm
  dietary restriction 365–73
  disturbance 393–7
  regulatory, genomic units 113
  resource allocation 321–35
  selfish gene 6, 28–9, 62–3
  subspecific lineage 110–37, 145
Parasexuality 29
Parasitism 41, 192–3
Parthenogenesis, failure of in mammals 139–40
Pathogens
  coevolution with hosts 191–2
  evolution of 189–94
Phenocopies 131
Phenotype
  animal, holistic model **274**
  cryptic 202–3
  maternal influence on 205–9
  morphological structure of 276–9

paternal influence on 205–9
Phenotypic adaptive suites 252–71
  developmental constraints 254
  ecological constraints 254–5
  genetic constraints 253
  historical constraints 253
  metabolic constraints 253–4
  proof of the existence of 263–4
Phenotypic balance hypothesis 297–303, 324, 327
Plastic developmental trajectories 215–18
Plasticity, phenotypic 61, 168–9, 198–205, 209–51, 270, 386, 388, 393–7
  and environmental cycles 201
  evolution of 229–245
Plants
  coevolution with herbivores 227
  longevity 381–2
Plasmids 174
Pleiotropy 210, 247
  antagonistic 281, 314, 317, 321
Polymorphism 122–3, 198, **214**, 394
  for chromosome inversions 132, 245, 262–3
  and the coadapted genome 257
  genetic 231
  sexual 199, 204–5, **206**
Polyploidy 78, 388–9, 391–2, 396
Population
  density 284–5
  genetics 59–60
  growth *312*
  structure 145, 147
  subspecific 111
POW domains 72
Predator–prey systems 51
Prokaryotes
  gene regulation in 63
  regulation of metabolism 91–2
Promoters 71
Protein initiation factors 71
Proteins
  B–ribbon 72
  heat shock 354
Pseudogenes 125–6, 129
  maintenance of inactive genes 125
Punctuated equilibria, theory of 258–9

Rate of living theory 321, 326, 332, 336–42, 355
Recognition sites, regulatory 71–2
Recombination 160–1
  role in evolution 156–7, 172
Recombinational load 170
Red Queen
  environments **388**, 392–3
  hypothesis 155–6, 188–94
Reductionism 115
  genic 4, 6, 8
  hierarchical 1–4, 7–10, 22
  molecular 4, 6, 20
Redundancy 27, 33
  morphological 95–6
Regulation
  of chromatin structure 64–71
  epigenetic 76
  eukaryotic gene 64–81
    see also Eukaryotes
  prokaryotic gene 63
  transcriptional **68**, 73–5
  translational 75
Relatedness
  degree of 115–18
  genetic 115–25
  genomic measure of 124
Reproduction
  longevity and 361–3
  sexual
    evolution of 45–7, 144, 153–7
    group selection and 154
    lineage selection and 153–7
    species selection and 154
    metalineage selection and 138–97
Resource allocation 277–8, 329
  matrix *280*
  paradigm, factoring longevity into 321–35
Reverse transcription 78
Risk aversion 33

Safe tuning hypothesis 294–7
Schistocerca americana, see Grasshopper
Selection
  artificial 262
  community–level 48–9
  for enhanced longevity 320–1
  genes as fundamental units of 114–15

group 142–3, 147–51, 153
  sexual reproduction and 154
hierarchically organized genomes as units of 115, 119–24
interdemic 145
kin 32, 142, 148
lineage, and sexual reproduction 153–7
lineages as units of 116–18, 131
metalineage, and sexual reproduction 138–97
and regulatory hierarchy 115
stabilizing 95
subspecific lineage 111, 119, 124, 142
temporal unit of 112–13
transgenerational units of 120
units of 112, 114, 116
Selfish gene paradigm 6, 28–9, 62–3
Senescence 286–7, 386
  avoidable 381
  dysorganizational 319
  evolution of 310–35
    antagonistic pleiotropy hypothesis 314, 317, 321
    disposable soma hypothesis 314, 316, 321, 326, 333
    mutation accumulation hypothesis 312–14, 317–18, 380–1
  free oxygen radicals, role in 322
  genetically programmed 376
  in natural populations 315–16
Sex
  advantages of 138–9, 146, 153, 156, 171–2
  'covert' 187
  determination of
    in *D. melanogaster* 83–5
    male 80
  disadvantages of 153
  evolution of 45–7, 144, 153–7
    genic framework 157–62
    hypotheses of 162–83
    see also Hypothesis
Shields' theory 179
Shifting balance theory 146–8, 175, 178
Sickle cell anaemia 22
Silencers 73
Sleep, importance of 361
Slippage 247
Social systems 48

Society, human, evolution of 48
Specialization 275–6
    niche 194–5, 200–1
Speciation 134, 228–9, 267
    allopatric 394
Species
    diversity 51–2
    interbreeding 94, 152
    selection, sexual reproduction and
        154
Spliceosomes 73
Splicing
    alternative 74
    differential 74
Stability, community complexity and
        52–4
Strategies, generalist 276
Stress
    competition and 390–1
    environments 389–92
        asexuality and 391–2
    and longevity 365–73, **372**
Sublineages, communication among
        174–82, 192
Subspecific lineage paradigm 110–37,
        145
Supergenes 134, 245–6
Supermouse 226, 303–9
    behavioural activity and longevity
        359
    body temperature 305
    fertility 305–6
    growth efficiency 304–5
    longevity 304, 326–8
    specific feeding rate 304
    time budgets 305
Superorganism 46, 54, 148
Superoxide dismutase (SOD) activity,
        longevity and 323
Sustainable scope 288–93, 296
    constraints on 289
    and metabolic rates 291–3, **293**
    and size 291–3
Switches
    allelic 107, **211**
    phenotypically plastic **232**
    polymorphic 209–13
        evolution of 231
        genetic structure of **211**
Symbiosis 39–40

    stability of 53
Symbiotic organisms 286, 391
    coevolution of 46
Symmorphosis 296–7
Synergism 112, 283
Syngameons 152, 181

Tandem duplications 78
'Tangled Bank' hypothesis 156,
        194–6
TATA box 71
Teleology 1
Temperature
    adjustments with **372**
    of body size 352–3
    and longevity 350–5, **353**
Templets
    dynamic genetic 125, 137,
        267–71
    genetic 125, 302
    habitat 273, 279
Termites 36, 47
Theory
    chaos 318
    error catastrophe theory of aging
        318, 376
    hierarchy 7–8
    kin 38–9, 116, 120–1, 148
    life–history 279
    of punctuated equilibria 258–9
    rate of living 321, 326, 332, 336–42,
        355
    shifting balance 146–8, 174, 178
    see also Hypothesis
Thermoregulation
    in bees 36
    in Pieridae 21, 263
Toad, *see* 'Goldschmidt toad'
Torpor 355
Trade–offs 272–309
    physiological 279–87
Transcription
    complexes **66**
    factors 71–2
Transgenerational effects 205–9
Transposable elements 79–80
    meiotic drive and 140–1
    regulation by host genome 136
Transposons, evolutionary significance
        of 135–7

Tree snails, evolution of 94–5
Triad, hierarchical **9**
tRNA, evolution of 75

Variation
  environmental **241**
  genetic 239, **240**, **241**
    and phenotypic 283–4
  genomic 130–7
    reorganizations 133–4

intraspecific
  chromosome number 132
  chromosome structure 132
  niche 200–1
  in resources and storage 284–6
Vitamin E 323–4

Warren truss 4, **5**, 277

Zinc fingers 72, 88